AF664393

The Editor

Dr. Shahid Ahamad, Ph.D. born on July 1,1972 in a village Gaharauli,Teh. Maudaha, Distt. Hamirpur, U.P. Dr.Ahamad completed his B.Sc (Ag.) from N.D.U.A.T. Kumarganj,Faizabad in 1994; M.Sc (Ag.) Plant Pathology from C.S.Azad Unversity of Agri. and Technology, Kanpur, U.P. in 1996 and Ph.D. (Botany) from C.S.J. M. Univ. Kanpur, U.P. in 2003. He worked as a Research Associate with Prof. A.K.Sarbhoy for three years (1997-2000) in the Division of Plant Pathology, I.A.R.I., Pusa, New Delhi. He was appointed as Jr. Scientist (Plant Pathology) in 2001 in the SKUAST-Jammu at Regional Agril. Research Station, Rajouri, for seven years and then appointed as Programme Coordinator Krishi Vigyan Kendra,SKUAST-J., Rajouri-185131 (J&K) in 2008. During short tenure of twelve years, he has published 100 publications in his credit. He has published 4 books, 30 research papers in the journals of national and international repute, 20 popular articles in different magazines, 25 book chapters, 01 review/conceptual articles, 13 bulletins and 18 research abstracts and 10 Reports. He has been conferred with SPPS Fellow-2005 by the Society of Plant Protection Sciences, Division of Nematology, IARI, New Delhi. Recently he has been awarded Young Achiever Award-2011 from SADHNA, H.P. and Distinguished Service Award-2012 from Bioved Research Society, Allahabad, U.P. He has been also choosen as Councillor (2012-15) of Society of Plant Protection Sciences, New Delhi.

Hill Agriculture

Prospects, Constraints and Mitigations

Editor
Dr. Shahid Ahamad
Programme Coordinator
Krishi Vigyan Kendra, Poonch – 185 101
Sher-e-Kashmir University of Agricultural Sciences and Technology-Jammu,
J&K, India

2013
Daya Publishing House®
A Division of
Astral International Pvt. Ltd.
New Delhi – 110 002

© 2013 EDITOR

Publisher's note:

Every possible effort has been made to ensure that the information contained in this book is accurate at the time of going to press, and the publisher and author cannot accept responsibility for any errors or omissions, however caused. No responsibility for loss or damage occasioned to any person acting, or refraining from action, as a result of the material in this publication can be accepted by the editor, the publisher or the author. The Publisher is not associated with any product or vendor mentioned in the book. The contents of this work are intended to further general scientific research, understanding and discussion only. Readers should consult with a specialist where appropriate.

Every effort has been made to trace the owners of copyright material used in this book, if any. The author and the publisher will be grateful for any omission brought to their notice for acknowledgement in the future editions of the book.

All Rights reserved under International Copyright Conventions. No part of this publication may be reproduced, stored in a retrieval system, or transmitted in any form or by any means, electronic, mechanical, photocopying, recording or otherwise without the prior written consent of the publisher and the copyright owner.

Cataloging in Publication Data--DK
Courtesy: D.K. Agencies (P) Ltd. <docinfo@dkagencies.com>

***Hill agriculture :** prospects, constraints and mitigations / editor, Shahid Ahamad.*
p. cm.
Contributed articles.
Includes bibliographical references and index.
ISBN : 978-81-7035-859-6

1. Hill farming--India. 2. Agricultural pests--Integrated control--India. I. Ahamad, Shahid, 1972-

DDC 631.580954 23

Published by : **Daya Publishing House®**
A Division of
Astral International Pvt. Ltd.
81, Darya Ganj, Near Hindi Park,
Delhi Medical Association Road,
New Delhi - 110 002
Phone: 011-4354 9197, 2327 8134
Fax: +91-11-2324 3060
E-mail: info@astralint.com
Website: www.astralint.com

Laser Typesetting : **Classic Computer Services,** Delhi - 110 035

Printed at : **Salasar Imaging Systems,** Delhi - 110 035

PRINTED IN INDIA

Preface

A recent survey by the Economist revealed that the world population has increased by 90 per cent in the past 40 years while food production has increased only by 25 per cent per head. With an additional 1.5 billion mouth to feed by 2020, farmers worldwide have to produce 39 per cent more.

Over 800 million people in the world go hungry each year. The World Food Summit held in June 2002 announced an international alliance against hunger to renew commitments to halve this number by 2015. Poverty alleviation and sustained agricultural production are key strategies. Since the vast majority of the hungry and those living in poverty are in rural areas and dependent on agricultural or natural resources for their livelihoods, attention has focused on how to improve subsistence farming systems, especially in South Asia. By 2015 the region would be food deficient by 16 million tonnes and by 2030 the magnitude would be of the order of 26 million tones.The quality of food available in the region is also a great concern as 40 per cent of world's malnourished, almost 330 million people, are in South Asia. This challenge becomes much more pressing given that 28 per cent of the GDP in the region comes from agriculture and more than 70 per cent of the population is dependent on this sector. Prospects of raising agricultural production through area expansion and application of existing technologies appear to be severely constrained.

Hill agro-ecosystem supports almost half of the country's population either directly or indirectly. These days, agriculture essentially in Asian countries including India where rice and wheat is grown, is posing a number of problems. Hill agriculture is comparatively more vulnerable to insect pest infestation due to the occurrence of varied climatic conditions. Locational specificity in hills is determined by factors like altitudinal differences, steepness of slopes, directions of mountain and valley etc. Micro-climatic specificity of some hilly pockets results in development of hot spot for certain insect pests. Because of rich organic matter, light textured soil and major area (about 90 per cent) under rainfed condition, cut worms, white grubs, termites etc. have become key pest of the region. In addition, the intensive agricultural technologies in recent past have added to pest related problems. Intensive vegetable farming especially in valley areas of hills has witnessed increased pest build up during recent past and the management of pests has been mostly relied on the chemical means, which in turn poses several ill effects such as killing of natural enemies, environmental pollution, health hazard etc.

and therefore, GOI is pressing hard for organic farming. The major challenge in the event of this farming would be the management of insect pests and diseases. Hence, for higher productivity and production of the crops, it is vital to adopt the Integrated Pest Management (IPM) strategy with a minimum use of pesticides, which are considered as one of the major factors for the burgeoning pest problem.

There has been a long felt need for a book exclusively on the "*Hill Agriculture: Prospects, Constraints and Mitigations*". An effort has been made to collect information on various aspects of diseases and their management. Chapters in this book have been contributed by noted Plant Pathologists of India. The Scientists, teachers, students, scholars, extension workers, administrators and policy makers dealing with Agriculture in particular and Hill Agriculture in general will be very useful and informative.

Dr. Shahid Ahamad

Contents

Hill Agriculture: Prospects, Constraints and Mitigations (2013) *Pages* **1–9**
Editor: **Dr. Shahid Ahamad**
Published by: **DAYA PUBLISHING HOUSE, NEW DELHI**

Chapter 1

Agricultural Production Problems and their Amelioration in Submountane Area of Siwalik Hills of Jammu Region, J&K State

***R.D. Gupta*[1] *and A.K. Sharma*[2]**

[1]*Former Associate Dean-cum-Chief Scientist, KVK (SKUAST-J), Jammu*
[2]*Assistant Professor, KVK (SKUAST-J), Jammu*

ABSTRACT

Submountane tract of Sivalik Hills of Jammu region, Jammu and Kashmir State, faces a number of problems in agricultural production. Some of the problems having bearing in agricultural production have been detailed. The study reveals that dip slope and escarpment topography, scarcity of water, soil erosion, poor soil fertility, soil salinity, calcareousness of soils, unbalanced use of fertilizers are the burning issues and infirmities posed by the farmers in the crop production. These problems are becoming more and more complicated owing to backwardness and lack of knowledge regarding real concept of environment due to change in climate. The solution of these problems has also been suggested.

Introduction

The Submountane tract of Siwalik Hills locally known as *kandi*, of Jammu region, Jammu and Kashmir State (located between 74 21–75 45 E Longitude and 32 22 – 32 55 N latitude) is surrounded by the river Ravi on the east and Manawar Tawi on the west within an altitude of 750 m, covering area of Kathua, Reasi, Rajouri, Poonch, Udhampur Samba and Jammu districts. Out of the total geographical area of 4,89.266 ha, 1,16,046 ha (about 24 per cent) is under cultivation with net area sown of 82,880 ha. Gupta and Banerjee (1991) have reported that area lying between 300 to 750m as lower kandi and that between 750 to 1050m as upper kandi. Based on this criterion, Nowshehra and Sunderbani areas of Rajouri district and that of Ramban of Doda district and parts of Poonch can to be included in the studied area.

Agriculture is a main stay of the people of kandi belt, as about 70 per cent of its population is dependent upon land for its living. Eighty to ninety per cent of the farmers are marginal and small having size of land holding less then 1 ha and between 1 – 2 ha respectively. These holdings are scattered and uneconomical for cultivation.

The main crops grown in the area are maize, rice and pearl millet during Kharif and wheat during rabi. Rice is mostly grown as rain fed direct seeded crop. Pulses and oil seed crops are also grown. Hower, the productivity of wheat, maize and other crops is very low. It is 8 q/ha for wheat and 10q/ha for maize as compared to corresponding figures of 16 and 18 q/ha for Jammu province as a whole. A number of factors are responsible to account for this low yield, some of them relating to Shivalik Hills of Jammu having direct bearing on agriculture production, are presented below:

Problems in Agriculture Production

1. Dip Slope and Escarpment Topography

Because of rugged, dip slope and and escarpment topography, vis-à-vis lithosol type of soils (Gupta *et al.*, 1985), the natural vegetation and cultivated crops are largely affected by swift run-off water, there is decrease in moisture content of the soil which eventually become dry un stable, and may even be eroded completely. It leads to loss of soil and plant nutrients. Under such environment the plants cannot establish easily.

2. Dryland Agriculture

Because of over exploitation of fuel wood and timber for local use as well as raw material from the forests for industrial use and fodder from the pastures for cattle use, the area has become denuded. Undulating and rugged topography has further aggravated this situation, resulting in to paucity of available water for plant growth due to excessive run-off, high evaporative demand of atmosphere and unpredictable rain fall vis –a –vis absence of any irrigation. Rainfed or dryland agriculture is, thus, the main practice of the farmers, which is risky owing to erratic uneven and uncertain rains both during summer (monsoon) and winter.

Since the area experiences subtropical climate with very hot summers and cold and dry winters, there remains scarcity of drinking water especially during the months of May to June and October to November but to talk of irrigating the agriculture. The growing of crop by the farmers of kandi area is not only risky but also pays very less and some times even the investments of inputs can not be met with. This calls for an alternative land use management systems. Adoption of Agro-forestry by the farmers of the area, is one such system. This can fulfill the basic needs of the farmers like fodder, fuel wood and small timber besides food and fibre which are the major constraints of the area. Moreover, plantation of trees and grasses in Agro-forestry system will ameliorate the environment besides controlling soil erosion and sustaining ecological balance.

Various agro-forestry systems having great perspectives for the Siwalik hills of the Jammu have been appeared in the literature (Gupta and Raina, 1994). These are: (i) Forestry-Agronomy (ii) Agronomy- Horticulture (iii) Agronomy-Forestry-Animal Husbandry (iv) Forestry-Agronomy-Fishery and (v) Forestry-Agronomy-Animal Husbandry-Fishery. Briefly these have been described in Table 1.1.

Table 1.1: Various Kinds of Agro-forestry Systems and their Perspectives in Siwalik Hills of Jammu

Agro-forestry System	*Combination and Perspectives*
Forestry – Agronomy	Trees either with space of 2x2 m are planted before or after planting. *Leucaena leucocephala* grown 3 m apart in wheat and pulse(moong) crops yielded 12 kg wood after 3 years plantation. The trees when cut at 3 m height, gave 10 kg fodder per tree. The lopped trees again grow and can be cut after a year. The practice could last for 4 – 5 years till the canopy becomes fully closed. Species like *Acacia nilotica, A. modesta, Albizzia lebbek* and *Grewia optiva* are also recommended in the system which can provide fuel wood and timber besides fodder.
Agronomy – Horticulture	Most suitable horticulture plants for this purpose are citrus, mango, guava, and ber planted at 3x3 m spacing. This system is more congenial as horticulture plants require less water, ensure better soil protection and bring more economic returns/ha. Helps to generate more employment to the farmers during the off season when crops are not cultivated. Provides scope for ancillary industries like food processing (canning, preservation and dehydration), transport and refrigeration. Does not require frequent ploughings of land which causes soil erosion.
Forestry – Agronomy – Animal husbandry	In this system, the trees are planted at wider spacing ranging from 5x5m to 9x9m to supply leafy fodder for animal husbandry activities besides fuel wood and small timber. *Leucaena leucocephala, Sesbania* and *Albizzia* species are recommended. Sometimes domestic animals or poultry birds are also allowed to feed harvested fields either between crops or on fallow lands. This will provide manure for trees and crops because of animal/poultry birds excrements.
	Local grasses will come up naturally but *Cencrus cilliaris, C. setigerus, Chrysopogon fulvus*, can be planted. Shrubs *Adhatoda vasica, Carisa spinarium* or Agave can be planted at 1x1 m spacing in contour hedges to conserve soil and moisture.
Forestry – Agronomy – Fisheries	It comprises raising of forest trees as well as agricultural horticultural crops on the terraced lands while ambient ponds/lakes or streams are good for fish culture and other aquatic crops according to depth of water.
	This agro-forestry system has a lot of prospect because of presence of number of ponds in the Siwalik hills which act as abode of fishery. However, the ponds of some of the area have to be revived. Beside rearing the fish, in shallow ponds lotus, wild rice stem and water chestnut can root the bottom soil, where as floating plants like water- lilly, water lettuce can be cultivated for forages. The big deep ponds can be used for breeding fish.
Forestry – Agronomy – Animal husbandry – Fishery	It integrates all the above said systems of agro-forestry together. It is characterized by along path of exchange of material and energy flow within an ecosystem. For instance, if mulberry trees are planted, their leaves are used for feeding silk worms. The excrements of the silk warms become feed to the fish. Their larvae being rich in nutrition can be utilized to feed poultry. Silk is the end product. At the same time different crops can be under planted for desirable purpose in mulberry stands.

Source: Gupta and Raina, 1994.

3. Soil Erosion

The Siwalik hills of Jammu suffer badly with the all kinds of soil erosion and are summarized as follows:

Sheet erosion wherein top soil is removed in thin layers, can be seen in the kandi entire area *i.e.* 4,89,266 ha which includes both lower (300-750m above mean sea level) and upper (750-1050m) zones of Siwalik hills. A large number of villages located on Mansar and Surinsar roads in Samba and Jammu tehsils face a drastic sheet erosion by the *nallah* in rainy season. Similar is the situation on Parmandal to Uttar Bani and Kaluchack to Parmandal road and Jammu to Kote Balwal/Amb Gharota, while going to Akhnoor (Gupta, 2002; Gupta, 2005; Gupta and Sharma, 2007). The cultural waste lands are increasing with not less than 15 per cent every decade.

The extent of rill erosion can be seen in a number of broad longitudinal valleys known as Duns. In Jammu Siwalik range, Duns are present in Udhampur, Jhajar Kotli and Dansal areas which are locally called as *Dadhals*. Due to soft nature of rocks such as clays, mudstones and other fragile rocks accompanied with heavy deforestation and over grazing of these areas, the soils are reported, to loose annually about 106 t/ha/year even in a slope of 2-3 per cent in clay loam soil. The destruction of natural vegetal growth cover causes the loss of top soil, rich in plant nutrients (Gupta, 2002).

Improper land use and denudation accelerates soil erosion, leading to formation of gullies. Gully erosion not only eates away farm lands but also causes erosion in wood lands. Gully erosion is also very common in Siwalik hills of Jammu. One feels sad and gloomy while moving along the *Sahar Khand* or *Khatera Di Khad* in Kathua where the soil has been eroded and forests have totally vanished.

When top soil is eroded, the plant nutrients both native and added nutrients are lost in runoff making the soil un fit for growth of plants. Even the organic matter or humus does not accumulate in the surface of soil which acts as a store house of macro and micronutrients like N, P, K, Ca, Mg, S and Fe, Mn, Cu, Zn, Mo, B and Cl respectively. This renders the soil not only acidic but also deficient in afore said nutrients. The plants, therefore, cannot nourish themselves under such substratum. This is why steep slopes are not devoid of vegetation consisting mostly of Xerophytic shrubs.

The erosion of soil besides reducing the soil fertility and productivity, causing siltation/ sedimentation in the water bodies/reservoirs and floods, also results in multitude side effects such as threats to means of communications and increased run off with less in filtration and percolation.

In light of the fore going crises created by soil erosion, it becomes imperative to adopt soil and water conservation measures. Generally, there are three kinds of measures which can be used for controlling soil and water conservation. They may involve use of various agronomic measures, biological measures (planting of tree and grass species) and engineering structures, depending upon land use pattern.

Various methods used for controlling soil and water conservation in agricultural non agricultural lands are given in Table 1.2.

4. Lack of Irrigation Facility and Use of Mulches for Conserving Moisture

As already mentioned that proper irrigation facility, which constitutes a major input for agricultural produce, is lacking in Siwalik hills of Jammu. There is an acute shortage of water in these hills, especially during May to June and October to November. This calls for adopting the latest dryland agriculture technology devolped at Rakh Dhiansar under Shere-e-Kashmir University of Agricultural Sciences and Technology of Jammu. Use of mulches of local live material such as *Brinker* or *Basuti* (*Adhatoda vasica*) in the form of leaves have proved useful not only in conserving soil moisture for crops but also increased their yield considerably.

5. Poor Soil Fertility, Status

A. Macronutrients

As the study area is liable to suffer severely from soil erosion, so soils are not only thirsty but hungry too. On an average, the consumption of plant nutrients (NPK) through chemical fertilizers is 10-15kg/ha while the estimated removal of these nutrients by various crops is very high, showing, thereby, a nutrients stress in these soils. The low to medium available N and P contents in the soils of Siwalik hills are the evidences to this effect (Gupta and Gupta, 1999). The organic matter is known to be the bank of plant nutrients which has been very much exhausted from the soils of Siwalik hills. This has been corroborated by obtaining very low to low status of organic carbon in these soils (Gupta, 1994; Gupta *et al.*, 1990). It is owing to its rapid oxidation under the prevailing high temperature environment vis-à-vis loss through soil erosion.

Table 1.2: Soil and Water Conservation Technology

Agricultural Lands

Agronomical measures or soil conservation agronomy

Contour farming, strip cropping *viz.* contour strip cropping, mulching, crop rotation, mixed cropping, organic manuring, shallow ploughing, recycling of run off collection, are the simplest agronomical practices. These practices are conducted as anon- monetary inputs for reducing erosion and conserving soil moisture. These have been found very effective not only in controlling soil erosion but also to maintain the fertility and productivity of soil.

Non Agricultural Lands

Biological or soil conservation forestry

Land capability classes v, vi, vii and viii having a number of limitations like slope, erosion, stoniness, rockiness, shallow soil depth, wetness, are not suitable for cultivation. These can be put to productive use as pastures and forest lands by adoption of suitable soil conservation practices, comprising establishment of vegetative cover and closure of forest/ pasture.

Vegetative cover: While establishing vegetativo cover, plantation should be of mixed type and indigenous species be preffered to grow due to their proven suitability to soil. Plants of *Dilbergia sisso, Acacia catechu, Acacia modesta, Acacia nilotica, Dendrocalamus* spp and *Pinus roxburghi* have proved use full for the Kandi belt of Jammu. Rocky areas like Nagrota, Nandani and Jajjar Kotli, have been rehabilitated through *Leucaene leucocephala* and *Dendrocalamus* spp plantation.To meet the fuel and fibre requirement, the grass species like *Chrysopogon fulvus* (75x75 cm) and species like *Dalbergia sisso* (9x9 m) in class v and vi have not only shown a high yield of dry grass (10.5 t/ha/year) but fuel wood also to the tune of 64 t/ha after 19 years.

Closures : In Siwalik hills of Jammu, it has been found that closure of forests at least for 3 years, ameliorate the vegetation vis-à-vis fertility of soil and its physical properties. The consequential grasses like *Dicanthium annulatum* and *Cencrus* spp. These changes have resulted in the reduction of run off and soil loss vis-à-vis improvement in the quality and quantity of grasses.

Biological methods check soil erosion effectively and do not allow plant nutrients to wash away and, there by assist in maintenance of plant nutrient status in soil.

Soil Conservation Engineering

No doubt soil conservation forestry brings an improvement in the area but the process is very slow especially in dry land region like Siwalik hills of Jammu, hence certain engineering structures are essential for accelerating the process of revegetation.

Leveling, bunding, terracing practices are called as engineering measures. Because these practices involve construction or earth movement. These practices provide immediate relief to land against erosion and maintain soil fertility as well build up productivity of the soil.

Single practice may not be sufficient as soil conservation is a multidisciplinary and integrated approach.

Check dams, gully plugging, basin listing, contour trenches, and spurs are the other measures which belong to soil conservation engineering.

Source: Gupta and Banerjee, 1991; Gupta and Sharma, 2007.

Recently fertility status of eroded and non eroded soils of Jammu district from Siwalik hills has been studied in respect of available N, P, K, Ca, Mg and S contests (Gupta *et al.*, 1994; Gupta *et al.*, 2005).Hundred per cent of the studied soils of eroded barren and eroded cultivated un managed lands were found low in available N whereas in respect of available P this figure was 80 per cent. It is point to mention that 100 per cent of the soils of eroded barren lands were low in available K and in case of cultivated unmanaged lands their figure was 75 per cent. The soils eroded barren and eroded cultivated un managed lands were too found deficient in available S, being 90 per cent and 57 per cent, respectively.

A brief account of different available macro-nutrients content in eroded and no eroded soils of kandi belt of Jammu district has been presented in Table 1.3.

Table 1.3: Available Micro-nutrients Status in Eroded and Non-Eroded Soils of Siwalik hills of Jammu District

Nutrients	*Eroded soils (per cent)*				*Non Eroded soils(per cent)*			
	Barren Lands		*Cultivated Un-managed Lands*		*Cultivated Well-managed Lands*		*Forest Lands*	
	Low	*Medium*	*Low*	*Medium*	*Low*	*Medium*	*Low*	*Medium*
Nitrogen	100.0	0.0	100.0	0.0	67.9	12.0	35.7	64.3
Phosphorus	80.7	19.3	71.4	28.6	7.1	71.5	0.0	75.0
Potassium	100.0	0.0	75.0	25.0	28.6	64.3	14.3	85.7
Calcium	-	100.0	0.0	100.0	0.0	100.0	0.0	100.0
Magnesium	-	100.0	0.0	100.0	0.0	100.0	0.0	100.0
Sulphur	90.3	9.7	57.2	42.8	3.6	96.4	0.0	100.0

B. Micro-nutrients

The deficiency of micro-nutrients has also been observed in many areas of Jammu Siwalik hills. Deficiency of zinc is further expected to increase with change of cropping system and intensification of Agriculture. The results of many field trials conducted under Experiments on cultivators Field, ICAR-Project in Siwalik belt of Jammu as well as on dry land agriculture sub station, have clearly showed the response of added $ZnSO_4$ (20 kg/ha) to maize and wheat crops when applied in combination with recommended NPK (60 : 40 : 20 kg/ha) doses. The per cent increase over control was 124.8 with recommended NPK in combination with $ZnSO_4$ against recommended NPK alone where the increase was 122.8 per cent.

The survey of the area shows deficiency of Zinc in citrus plants too. Mottle leaf nutritional disorder in citrus plants in nud area (Dohra village and others of Samba tehsil) is quite prevalent. Similar is the situation in Raya area as well as in Suchani. These plants respond to the application of micro-nutrients, particularly Zn. Cracking of citrus fruits at maturity is another nutritional disorder of copper deficiency. The deficiency of Mn is also being observed.

The fertility of these soils can be improved by taking into account the following measures :

Which ever organic manures like farm yard manure, compost, poultry manure are available must be added to supply the plant nutrient. Wherever, it is possible green manuring with dhaincha (*Sesbania aculeate*) or sanhemp (*Crotolaria juncea*) can be done to improve the organic matter status of the solis.Green manuring crop should be sown immediately after the first showers of monsoon and buried in to soil before flowering.

Introduction of good rotations including legumes will help to remove soil exhaustion and bring about ameliorations in soil fertility.

6. Salinity and Calcareousness

Salinity is due to presence of total soluble salt content. When the total soluble saltexeeds to 0.2 per cent, it affects the plant growth. Some soils of Siwalik hills of Jammu region contain salts as low as 0.15 per cent and as high as 0.45 per cent with Cl, NO_3 and HCO_3 anions (Gupta and Banerjee, 1991) and pH range of 7.4 to 8.1 or some times up to 8.4.

The occurrence of excess free calcium carbonate in the segregated or diffused form causes lime induced chlorosis in number of plants, resulting in their low yield. Lime induced chlorosis is because of iron deficiency.

In soils having excess free calcium carborate, the availability of phosphorus and zinc to plants decrease owing to formation of calcium phosphate and calcium zincate ions. Manny of the soils ofSiwalik hills contain lot of free calcium carbonate, varying from 3.1 to 9.2 per cent (Gupta *et al.*, 1990).

To over come these problems use of organic matter of any kind should be made. Wherever possible, the leaching of salts can be done by frequent irrigations. Salt resistant crops like barley and oats should be preferred.

Trees like *Acacia nilotica*, *Acacia catechu* and *Dalbergia sisso* should be planted. They would make use of alkaline salts and add organic matter which will eventually improve texture as well as calcareousness of the soil.

7. Unbalanced Use of Fertilizers

Most of the farmers of the hilly tracts are poor. Their purchasing power is very low. As such majority of them do not use the fertilizer at all and those who use, they use only nitrogenous fertilizers as top dress. This results in low yield and sometimes there is lodging of crops due to lack of P and K. So, phosphatic and potassic fertilizers should be used in combination with those of nitrogenous.It is interesting to note that in Siwalik hills of Jammu, the soils of some localities under grass cover have soil reaction in acidic range (5.2 – 6.5). The possible reason for the devolpment of acidity in these soils can be ascribed to the continuos use of nitrogenous fertilizer alone in the form of Urea which is known to increase acidity, though less then that of ammonium sulphate. In certain crops, application of phosphatic fertilizers is of more importance being its role in biological nitrogen fixing phenonmenon. Similarly, application of nutrients like sulphur(s) is essential for increasing the yield of oil crops.

8. Landslides and Landslips

Landslides and Landslips are of frequent occurrence in whole of the Siwalik hills. They create threats not only to agricultural lands but also to the hills ecology, recurring loss of crores of rupees. The main causes of land slides are weak and fragile nature of the rocks (Clays, silt, soft sand stones, mud stones and other soft sediments) deforestation, over grazing and unscientific mining. The problem of landslides, land slips and mass wastage accelerate sheet or rill erosion more particularly during monsoon periods.

For aresting the landslides/landslips are to adoption of efficient drainage methods, reduction of steepness of slopes and aforrestation and turfing of bare beds.

9. Effective Means of Communications

Transport is yet another requisite for agriculture development. But in some parts of Siwalik hills of Jammu there is lack of effective means of communication and transportation is a handicap. Even road transport cannot be availed of in an adequate manner at the peak of citrus fruits owing to lack of its frequent service. This results in the deterioration of quality of produce before it reaches export marketing centre.

The most obvious solution for developing these areas is to make accessible by increasing transport facilities and communication according to suitability of local environment. However, the roads and communication ways must be built in well planned manner so that the environment and ecology of the area may not be impaired. In fact, while constructing the new roads blasting of hills/mountains is done vis-à-vis clearing of forests and even the agricultural lands, which as a result cause land slides and land slips of various types and eventually accelerating the severe soil erosion. There is also a creation of imbalance in ecology.

10. Marketing Facilities

The present set of marketing for disposal of agricultural/horticultural produce has not given good dividends to growers. There is no local potential market for horticulture products. The complex pattern of market channel decreases the production returns to the basic growers. So the creation of efficient infrastructure in the Siwalik Hills of Jammu can be fully backed by technical assistance. Quality control measures by the concerned department, is another necessity for the development of agriculture and horticulture.

The state department of agriculture and rural development should go a long way in putting agriculture produce on the modern lines of scientific marketing.

Suggestions and Strategies for the Future

It is concluded from the foregoing study that the Siwalik hills of Jammu region have lot of prospects for overall development despite faced by a number of constraints. A few suggestions are presented hereunder in order to over come the constraints in agriculture development.

1. The soils of the area are liable to suffer from severe erosion, very poor in fertility and lack of irrigation facilities. As such, these soils are more conducive to horticulture, agro-forestry and forestry than arable farming, so priority must be given to these disciplines with their full exploitation. These would be vital for Siwalik hills being more remunerative than production of cereals.
2. In horticulture, drought resistant fruits like ber (*Ziziphus mauritiana*), seedling mango (*Magifera indica*), pomegranate (*Punica granatum*), guava (*Psidium guajava*) and some species of citrus (Kinnow, lime, galgal) should be preferred for plantation.

   Anola (*Phyllanttus emblica*), Jamun(*Syzygtum cumini*), Phalsa (*Grewia asiatica*) and Lasuraha (*Cordia dicotoma*) are some of the local fruits of the area which can be encouraged.
3. For immediate gains to meet out the regular demands of fuel and fodder, the plantation of bushes like *Adhatoda vasica* and *Carissa spinarum*, trees like *Grewia optiva* and *Albizzia* species, and grasses like *Chrysopogon fulvus*, *Cencrus cilliaris* and *Cynodon dactylon* should be grown. The former will stop the erosion of top soil besides providing the fuel wood and the latter unlimited production of grass for stall feeding and hayfor export to the deficient area.
4. There is direct relationship between live stock animals and natural pastures so the devolpment of pasture means, the development of live stock animals. Hence the grasslands should be ameliorated by producing superior grasses like *Cenchrus, Seteria* and *Napier* grass and legumes like *Siratro, Stylosanthes*. Whenever it is possible, plants belonging to leguminous families should be planted. It has been found that *Leucaena* and *Acacia* species fix atmospheric nitrogen to the order of 270 and 500 kg N/ha annually, respectively besides providing fuel and fodder.

5. Much emphasis should be laid on soil and water conservation measures. Agricultural activities on slopes having more than 15 degree should be banned. In order to favour maximum absorption of run off water into the soil during rains, the fields should be levelled, smoothed terraced and graded for minimum earth work. Land leveling plays a significant role in the scientific water management and soil erosion control. Infact field reshapes or modifies the land surface to a planned manner so as to control flow of water to check soil erosion and conserve moisture. Soil erosion can also be checked by mechanical and biological measures. The Government should help the farmers in constructing water harvesting or storage tanks for run off collection, so as to provide life saving irrigation in the years of drought.
6. In order to meet the home consumption of vegetables, the women and un employed youths must be trained for starting therir own kitchen gardens after acquiring training.
7. Village panchyats should be activated to restore common lands/waste lands for growing fodders for animals and trees for fuel.

Refrences

Gupta, G.D., Gupta, R.D., Kher, D. and Sumberia, N.M. (2005). Studies on erodibility indices in relation to soil properties in *kandi* belt of Jammu district. *Journal of Research*, SKUAST-J. 4(2): 209-213.

Gupta, G.D., Gupta, R.D., Kher, D. and Sumberia, N.M. (2005). Studies on the fertility status in relation to soil properties in *kandi* belt of Jammu district. *Journal of Research*, SKUAST-J. 4(1): 67-72.

Gupta, J.P. and Gupta, R.D. (1999). Nutrient management in maize crop. *The Daily Excelsior*, 35(93): 6 (April 4).

Gupta, R.D. (2002). Save the soil, save the nation. *Environment and People*. 8(12): 20-25.

Gupta, R.D. (2005). Soil eorsion: Causes, hazards and control measures. In: *Environmental Degradation of Jammu and Kashmir Himalayas and their Control*. Associated Publishing Company, New Delhi.

Gupta, R.D. and Abrol, V. (2006). Agroforestry: A sound landuse system for Himalayan rainfed ecosystem. *Farmers Forum*, 6(3): 19-22.

Gupta, R.D. and Banerjee, S.K. (1991). *Problems and Management of Soil and Forest Resources of Northwest Himalayas*. Mahajan Book Centre, Gandhi Nagar, Jammu.

Gupta, R.D. and Raina, J.L. (1994). Opportunities for agroforestry in dry land belt of Jammu region. In: *Dryland Farming in India*, (Ed.) J.J. Raina. Pointer Publishers, Jaipur, India.

Gupta, R.D. and Sharma, P.D. (2007). Soil erosion and conservation in Northwest Himalayas. In: *Natural Resource Management for Sustainable Hill Agriculture*, (Ed.) Sanjay *et al.* Soil Conservation Society of India, Jammu Chapter, SKUAST, Chatha, Jammu, J&K, India.

Gupta, R.D., Bali, S.V. and Marwah, B.C. (1985). Soils and Jammu and Kashmir and their management. In: *Soils of India and their Management*. FAI, New Delhi.

Gupta, R.D., Bali, S.V. and Singh, H., Khajuria, M.R. and Koul, V.K. (1990). Ecology of *kandi* belt of the outer Himalayas of Jammu division, Jammu and Kashmir State. In: *Himalayas: Environmental Problems*, (Ed). S.K. Chadha. Ashish Publishing House, New Delhi.

Hill Agriculture: Prospects, Constraints and Mitigations (2013) *Pages* **10–16**
Editor: **Dr. Shahid Ahamad**
Published by: **DAYA PUBLISHING HOUSE, NEW DELHI**

Chapter 2

Organic Farming: Its Modern Concept, Principles, Myths, Economics and World Scenario

***A.K. Sharma*[1], *R.D. Gupta*[2], *Rajeev Bharat*[1] *and Shahid Ahamad*[1]**
[1]*Krishi Vigyan Kendra (SKUAST-J), Jammu*
[2]*Former Associate Dean-cum-Chief Scientist, KVK (SKUAST-J), Jammu*

Introduction

These days, agriculture essentially in Asian countries including India where rice and wheat is grown, is posing a number of problems. Such problems may consist of marginal and small holdings having lack of well-defined credit institutions, shortage of manual labours and markets, high cost of inputs like fertilizers and pesticides, and of sufficient amount of organic manures (Mahajan and Gupta, 2009). Before "Green Revolution" in India which ushered during 1968, the Indian agriculture was mostly organic. In fact, prior to introducing "Green Revolution" the farming practices in the country were traditional and peasants were making use of local seeds and organic manures (FYM, compost, green manure etc). But with an advent of the "Green Revolution" a radical change in inputs use was seen in Indian agriculture. High yielding varieties of various crops especially rice and wheat replaced local seeds, high analysis chemical fertilizers totally replaced the use of organic manures, and chemical pesticides replaced locally prepared pesticides. There is no doubt that use of these new practices brought about a positive response in respect of employment generation, crop yield and income for the farming communities and eventually the country achieved self-sufficiency in food grain production. However, continuous use of high analysis chemical fertilisers and pesticides caused a number of harmful effects not only to physical environment (Land/soil, air and water) but to the biological environment (Plants and animals including human beings) also (Gupta and Sharma, 2006; Gupta and Singh, 2008). The effectiveness of chemical fertilisers and pesticides has been depleting the land resources. The soil is losing its fertility both in available macro- and micro-nutrients. Cost of cultivation has increased, crop yield has fallen and crop profitability has declined (Sahoo, 2009). Not only this, various food grain stuffs were found to be contaminated with pesticides, heavy metals and other chemicals (Gupta, 2005, 2006). Further, contamination of water with fertilizers and toxic pesticide residues has resulted in serious water pollution and heavy hazards, and also caused significant disruption of the normal functioning of the delicacy balanced ecosystem (Gupta *et al.*2010).

In the light of the above, it becomes utmost imperative to make various food stuffs free from all harmful chemicals, safer healthier and tastier. This can be resorted through practising "Organic Farming."

Modern Organic Farming Concept

The modern organic farming concept is not new to India. The idea of organic farming perhaps seems to be emanated from the Indian experience. In true sense, it is a concept or approach to make agriculture near to nature, sustainable, healthy and safe to the society. The concept of organic farming came to fore with an establishment of International Federation of Organic Agriculture Movement (IFOAM) on 6th of November, 1972 in France. The IFOAM, an International umbrella organisation for organic farming, defines organic farming as follows:

"Organic farming or organic agriculture is a production system that sustains the health of soils, ecosystem and people. It relies upon ecological processes, biodiversity and cycles adapted to local conditions rather than the use of inputs having adverse effects. Organic agriculture combines tradition, innovation and science to benefit the shared environment and promote fair relationships and good quality of life for the all involved in the system".

In simple language, organic farming is defined as a production system, which largely excludes the use of synthetically compounded fertilizers, pesticides, growth regulators and livestock feed additives.

- *Principles of Organic Farming*: During recent years, the concept of food quality has been changed totally. In developed countries, there are a number of consumers, who are very particular about the quality of the food and are becoming health conscious these days. To provide clarity on the organic claim, organic agriculture is governed by detailed standards and lists of allowed and prohibited substances vis-a-vis certain principles. Four principles towards achieving organic farming are as follows :
    - *Principle of Health*: Organic farming should sustain and enhance the health of soil, plant, animal and human as one and individual.
    - *Principle of Ecology*: Organic farming should be based on living ecological systems and cycles.
    - *Principle of Fairness*: Organic farming should build on relationships which ensure fairness with regard to the common environment and life opportunities.
    - *Principle of Care*: Organic farming should be managed in a precautionary and responsible manner to protect the health and well being of current generation and posterity.

Myths on Organic Farming

As organic farming came into force because of the growing ill effects of the conventional farming so some analysts supported organic farming while others had less favourable conclusions. But most of them are of the view that more research is required before any definite lesson is drawn. Some of the myths on organic and inorganic farming are as follows (Sahoo, 2009).

Cost of cultivation is lower and crop yield becomes higher after about 3 years of starting of organic farming in contrast to the conventional farming.

- ✰ Organic farming is not economically feasible.
- ✰ Organic farming cannot provide food security to the Nation.
- ✰ Organic farming can withstand at the time of natural disaster.
- ✰ Organic produce contains more vitamins, minerals and enzymes.
- ✰ Organic farming improves the soil quality.
- ✰ Organic farming system has many environmental and ecological benefits.

Authenticity of the Myths of Organic Farming

Many studies addressing most of the aforesaid myths with regard to organic farming have been carried out by various agencies in different countries including India. The findings of these studies have provided a useful guide to authenticate that organic farming may or may not be as profitable as conventional farming.

Economics of Organic Farming

There are lot of variations in costs of cultivation of organic and inorganic or chemicalised farming practices. The main components involving more costs in inorganic farming are seeds of high yielding varieties of various crops, chemical fertilizers and chemical pesticides. Contrary to this, in organic farming, these inputs are replaced by local seeds having low cost, organic manures-FYM, compost, green manure and locally prepared pesticides. Further, the crops grown under organic farming relatively require less irrigation than that in conventional farming. Although labour use in organic farming is higher than that in inorganic farming, the cost of cultivation in organic farming is lower than that in inorganic farming. Thus, inorganic farming is capital intensive, whereas organic farming is labour intensive.

Some agricultural scientists have observed that organic farming is not of much profitable, because whatever it saves in fertilisers and pesticide costs, is not sufficient to compensate additional labour cost required for organic farming. Some others even suspect whether organic farming can achieve the food grain production to meet the demand of the burgeoning human population.

Sustainability and Food Security

Despite an additional labour cost requires in organic farming, a number of studies have demonstrated that the average yields under organic and inorganic farming practices are almost the same (Table 2.1). Moreover, the yields of crops obtained under organic farming do not decline overtime. They are rather relatively stable. Thus, organic farming is more sustainable than that of inorganic farming.

With regard to food security versus organic farming the general belief is that large scale conversion to organic farming would result in drastic reduction in the total food supply. But in reality, this is not the case as many studies have shown that this conversion will prove beneficial as crop yield and production will be stable unlike chemical or inorganic farming where under yields of crops especially of rice and wheat have either become constant or decline (Mahajan and Gupta, 2009). The stability in the food grains production would, thus, contribute food security. Moreover, higher market returns from organic food stuffs can bring forth to local food security by increasing family income.

Table 2.1: Detail of Cotton Yield from Organic and Conventional Farming System

Year	*Status*	*Yield (q/ha)*	*Net Returns (Rs.)*
1	Conventional	10.0	9000.00
2	Conventional	5.0	7500.00
3	Organic	6.2	7000.00
4	Organic	7.5	10500.00
5	Organic	8.7	13500.00
6	Organic	10.0	16500.00

Source: Rajendra *et al.*, 2000.

Health Benefits of Organic Farming

Organic foods or green foods, are in great demand globally today. It is because such foods are healthier, safer, tastier and free from all harmful chemicals (Gupta, 2009). Although Indian farmers have been in practice of producing organic foods since ages yet their faster popularization has become the need of the day. These days, such foods are searched for since the humans have become health conscious owing to availability of choicest food varieties and ill effects of residual toxicity of chemicals.

As most of the fruits and vegetables are taken a fresh, the residues of chemical fertilizers and pesticides in such horticultural produce may lead to various kinds of health hazards (Vishwanath, 2003). Therefore, organic food production offers a better possibility in horticultural crop production also. In addition to this, organic food production is a holistic production management system, which promotes and enhances agro-ecosystem, health including biodiversity, biological cycle and soil biological activities.

As the FAO (Food and Agricultural Organisation) report points out (Burcher, 2007) "Organic foods tend to have higher micro-nutrients content that contributes to better health, lower incidence of non-communicable diseases and boosts plant and animal immunity against disease". The United Kingdom Soil Association carried out a systematic study concerning with trace minerals in organic and non organic food and found that on an average, organic food contains higher levels of Vitamin C (Table 2.2) and minerals like Ca, Mg and Fe. Chromium (Cr) which is found more in amount in organic found, is proved essential for normal utilization of glucose (Ismail, et al., 2009; Jha, 2003). Selenium an antioxidant nutrient that protects against cancer and heart disease is also present in higher content in organic food.

Other Benefits of Organic Farming

The FAO report points out to further benefits such as better animal welfare, wildlife production, avaoidance of pesticides, chemical fertilizers, more jobs and less energy used. Results from the studies conducted by the United States Department of Agriculture support the findings of FAO, showing that organic crops are worth more than conventional crops on the market and on an average farmers could get 50-60 US Dollars more per acre (0.4 ha) by growing organic even with the higher transitional costs (Burcher, 2007). Organic agriculture has the potential to reduce CO_2, NO, and CH_4 green house gases which contribute to global warming. Organic agriculture could double soil carbon sequestration in living stock based system and decrease green house gases by 40-60 per cent.

Table 2.2: Score Cards of Nutrients Parameters

Nutrient	*Per cent Difference in Organic Over Conventional*	*Nutrient*	*Per cent Difference in Organic Cver Conventional*
Vitamin C	+22.7	Iron	+17.2
Calcium	+30.8	Magnesium	+24.4
Potassium	+14.1	Phosphorus	+12.5
Sodium	+19.6	Nitrates	-33.3
β-carotene	-00.3		

Source: Gupta *et al.*, 2010.

On organic farms, increasing soil organic matter and microbial biomass is a fundamental principle to support agro-ecosystems stability. Mandatory crop rotation, the use of traditional seeds and breeds that are adapted to local condition and regeneration of functional biodiversity all play a significant role to maintain ecological balance.

Producing organic food has distinct advantages for farming communities too. Farmer's rights to local seeds and varieties are strengthened, knowledge sharing is promoted, incomes are raised and production increased. Environmental and health protection are improved, natural resources are conserved and outward rural migration is reversed.

Soil Fertility and Biodiversity

Organic farming offers benefits of reduced water use, reduced contamination by pesticides, reduced soil erosion and increased biodiversity. The organic farming builds up soil organic matter better then chemical farming, and as such, it will have long term yield benefits. The soil under organic agriculture has lower bulk density, higher water holding capacity, higher cation exchange capacity, higher microbial biomass and higher soil respiration activities as compared to the traditional agriculture. As organic crops use little or no pesticides and fertilizers so it benefits biodiversity. Owing to the above reasons, the external costs imposed upon the farming communities' by organic farming relatively become lower than that by the in organic or chemical farming.

World Scenario of Organic Farming

Since various studies have evinced that under draught situations crops grown in organic farming produce significantly higher yields as grown to crops grown under chemical farming, so a large number of the farmers has now adopted cultivation of crops through organic farming and still many are adopting this kind of farming. Thus, organic farming no longer to be considered a niche market within developed countries but a vibrant commercial agricultural system in 141 countries of the world (Sahoo, 2009). Its share of agricultural land continues to grow. In 2001, 17.4 million hectares (M ha) of agricultural land were managed organically which arouse 31 M ha in 120 countries between 2005 to 2006 and by 2007, it rose to 32.2 M ha. The organic food crossed from US dollars 20 billion in 2001 to 46.1 US dollars in 2007. Out of the total 32.2 M ha of organic farming landing the year 2007, Oceania region consisting of Australia, New Zealand and Islands states like Fiji and Papua New Guinae share the maximum of about 38 per cent followed by European countries like Italy, Spain and Germany occupying nearly 24 per cent. While Latin America and Asian countries share 20 and 9 per cent of total organic farming, respectively in the world.

There is no doubt that the maximum organic farming are concentrated in Australia, Argentina and Brazil. But the maximum numbers of organic producers are confined to Uganda, India and Ethopia. Among the ten top countries regarding the most of the organic farming land, Australia ranks first having area of 8,361 ha per farmer followed by Argentina having 1,760 ha per farmer and India has been found to be the minimum with an area of 5 ha per farmer.

Between the year 2003 and 2007, the maximum rise under organic farming has been noticed in India (2413 per cent) followed by Brazil (540.8 per cent) and China (415.4 per cent). Similarly between these two years, the number of farmers adopting organic farming practice has increased the maximum in India followed by Uruguay. However, in some countries like Brazil, China, Italy and Argentina, the number of farmers opting for organic farming has declined.

Future of Organic Farming

Although the food crisis has been discussed by a large number of farm scientists yet it is in terms of high prices and accentuating food shortage for the poor. There is, however, another crisis, which stands less elaborated. This relates to the many sided deterioration in the quality of food – in terms of reduced nutrition contents well as increasing hazards implicit in the use of many harmful chemicals used for growing, processing and preserving food.

If anyone thinks that this is an exaggeration, then let him or her see the 1986 report of London Food Commission which pointed out that 92 chemical pesticides used in Britain rare linked either with cancer, birth defects in humans or genetic mutation in animals (Dogra, 2008).

Similarly, the report of National Academy of Sciences USA (1987) indicated that pesticides in food of US citizens caused more then one million cases of cancer over their life.

A number of studies carried out in various parts of India, have also detected residues of pesticides in various food stuffs *vis-a-vis* soils and water as reported by Subba Rao (1999).

It is well known that excessive use of fertilizers causes a loss of flavour of food, its nutritive value and even creates some serious health problems. As there is a lot of evidence of micro-nutrients depletion of soils in some areas (Gupta, 2011).

It is, thus, important to emphasize food production and food processing methods within our country which can provide healthy food having good quality. In this context, the future prospects of organic farming are bright. It is because organic farming provides healthy food. Healthy food provided by organic farming can, therefore, make a big contribution to improving health and nutrition of people as well as farm animals, while also contributing to ameliorate soil quality and farm land fertility and wild life.

References

Burchers, S. (2007). FOA promotes organic agriculture. The Kashmir Times. 64(326): 10 (24th November, Saturday).

Dogra, B. (2008). Unsafe food. The Tribune 128(278): 11 (6th October, Monday).

Gupta, P.K., Singh, J. and Lamba, P. (2010). Livelihood scenario of organic farming for rural India. Agrobios Newsletter, 8(8): 55-57.

Gupta, R.D. (2005). Environment Degradation of Jammu and Kashmir Himalaya and their Control. Associated Publishing Company, Karol Bagh, New Delhi, India.

Gupta, R.D. (2006). Environment pollution: Hazards and Control. Concept Publishing, New Delhi, India.

Gupta, R.D. (2009). Organic agriculture: What, why and how. The State Times. 14(314): 6 (12th November, Thursday).

Gupta, R.D. (2011). Environmental degradation. The Daily Excelsior. 469(5): 6 (5th January, Wednesday).

Gupta, R.D. and Sharma, S.P. (2006). Soil pollution and its solution through integrated nutrient management. In: Balanced fertilization for sustainable crop production (edited by Sharma *et al.*) Department of Soil Science, CSK, HPKV. Palampur, ICAR, New Delhi.

Gupta, R.D. and Singh, H. (2008). Indiscriminate use of pesticides in agriculture : Public health issues an their control. Indian Farmers Digest. 41(1): 8-13.

Ismail, S., Kausadikar, H.K and Walkar, S.L. (2009). Quality of the Agricultural Produce in organic farming. Agrobios News Letter. 7(8): 37-39.

Jha, M.K. (2003). Organic farming: Way to food security. Chronicle, November, pp. 33-34.

Mahajan, A. and Gupta, R.D. (2009). Integrated Nutrient Management (INM) in a sustainable Rice-Wheat Croping System. Springer Science and Business Media BV.

Rajendra, J.R., Venugopalan, M.V. and Tarhalkar, P.P. (2000). Organic Cotton Farming Technical Bulletin No. 1, CICR, Nagpur.

Sahoo, B.B. (2009) Organic vs. Inorganic farming. Agriculture Today (The national agricultural magazine). 12(12): 40-42.

Subba Rao, I.V. (1999). Soil and environment pollution: A threat to sustainable agriculture. Journal of Indian Society of Soil Science, 47(4): 611-631.

Vishwanath, C.S. (2003). Go for green food. *Indian Horticulture*. 47(4).

Hill Agriculture: Prospects, Constraints and Mitigations (2013) *Pages* **17–31**
Editor: **Dr. Shahid Ahamad**
Published by: **DAYA PUBLISHING HOUSE, NEW DELHI**

Chapter 3

Farming System Modules for Hill Agriculture in India

Rajeev Bharat[1], *Anil Kumar Sharma*[2], *Rohit Sharma*[3], *Vivak M. Arya*[1] *and P.K. Rai*[2]
Sher-e-Kashmir University of Agricultural Sciences and Technology of Jammu
[1]*KVK-Kathua,* [2]*KVK-Jammu;* [3]*Farming System Research, Chatha*

Farming system is an optimum combination of different components of a farm like cropping system, poultry, piggery, apiary, fishery, duck keeping, rabbitry, sheep rearing, goat keeping, floriculture, horticulture etc. depending upon socio-economic and institutional factors (Gill *et al.*, 2009). Farming system refers to a set of agricultural activities organized while preserving other set up like land productivity, environment quality, maintenance of desirable level of biological diversity and ecological stability thus emphasizing on system productivity and sustainability rather than on gross returns. It is more of a resource management strategy to achieve economic and sustained agricultural production to meet the different requirements of the farm households with stress on preservation of resource base and maintaining environmental quality. It is a holistic approach to the farm and due consideration is given to the between interaction of the components themselves. For the development of suitable farming system modules it is imperative to understand the intricacies of hill agriculture wherein the farmers are resource poor with restricted mobility and small land holdings, lower levels of farm mechanization and incomes besides limited choice of available technologies and knowledge besides other problems such as soil erosion due to steep slopes and topography, lack of irrigation facilities, varied crop growth season, lack of suitable HYV's of different crops and animals *viz.* cow, goats and sheeps etc.

Objectives of the Farming System

The ultimate objective of the farming system is to evolve farming system model which is pragmatic, technically feasible and economically viable by integrating different components like cropping systems, livestock etc. for generating farm income, employment and recycling of the outputs for sustainable crop production and soil health.

The specific objectives of farming system model are as follows:

1. To find out the existing components of farming system and assess their relative viability in specific area.
2. To formulate suitable farming system models considering various available enterprises under various situations.
3. To have effective recycling of farm residues and resources for optimum utilization of available resources.
4. To maintain the evolved production system and its sustainability without having a negative impact on the resource base.
5. To enhance the profitability and productivity of the farm house hold by judicious mix of different complementary enterprises.

Hills are the lungs of the country. The socio-economic development of the hilly and mountainous areas and protection of their ecology is on priority among the agendas of national concern owing to their huge importance in climate and as repository of rich biodiversity.

One of the major characteristics of the hills and mountainous areas differentiating them from plain areas is topography, elevation, physiographic features, diversity of habitats for flora and fauna ethnic diversity, land use systems and socio-economic conditions.

There is a tremendous scope for cultivation of diverse mix of crops *i.e.* cereals, pulses oilseeds, vegetables, flowers and fruits which form the core of the farming system model for hilly region and contribute to a significant extent in increasing the socio-economic status of hill farming owing to the specificity of these crops, vegetables and fruits to these regions *i.e.* intermediate (800-1500 m m.s.l.) and fruits *viz.* Apple, French Beans, Apricot, Walnut etc. However, increasing population pressure have necessitated a holistic approach level as farming system for increased nutritional and livelihood security in hilly areas.

Traditional agriculture is practiced in hilly areas which combine crop production with livestock, horticulture farm forestry or agro-forestry in integrated fashion which is adopted to the limitations of soil and climatic conditions. The agricultural development in hills is constrained by the lack of appropriate technology, inadequate infrastructure, poor input-output ratio owing to lower resource use efficiency, insufficient extension education and network, decrease in area for forest dwellers and diversity in farming system or land use type. Crop production, livestock, horticulture and agro-forestry are the recognized four major components of hill agro-ecosystem. To meet the demands of increasing population we have to conserve the natural resource base (land, air, water, energy and biogenetic diversity) and ensure its optimal utilization for attaining sustainability.

The basic characteristic feature of hill agriculture is cultivation on terraced fields.

Geographical Distribution

Hilly and mountainous areas in India are vastly distributed all over the country with a major area located in the mighty Himalayas extending to 2500 km in length and 250 to 400 km in breadth. Areas under nomenclature of hill and mountain zones located in various states of the country *viz.* Andhra Pradesh, Assam, Bihar, Gujarat, Haryana, Himachal Pradesh, Jammu and Kashmir, Karnataka, Kerala, Madhya Pradesh, Maharashtra, Manipur, Meghalaya, Mizoram, Nagaland, Odisha, Punjab, Rajasthan, Sikkim, Tamil Nadu, Tripura, Uttar Pradesh, Uttarakhand, and West Bengal over 35 per cent of the total geographic area, the hilly areas have a slope about 15 per cent. However, Himalayas alone under the purview of this chapter is further classified into three major categories namely:

1. Western Himalayas including Jammu and Kashmir and Himachal Pradesh.
2. Central Himalayas including 8 hill districts of Uttarakhand.
3. Northern Himalayas including Sikkim, Manipur, Meghalaya, Nagaland, Tripura, Arunachal Pradesh, Mizoram and hilly areas of Assam, Darjeeling and districts of West Bengal.

Longitudinally, Himalayas are also classified as Shivaliks:

1. Flat sammuts (600-1200 m mean sea level altitude),
2. Middle Himalayas (3000 m mean sea level altitude),
3. Great Himalayas (average altitude 5200 m mean sea level) and
4. Trans Himalayas (4500 m mean sea level altitude)

Prospects of Farming System in Indian Agriculture

Farming system approach has a great scope in hilly areas of Himalayas in India owing to availability of abundant sunshine, long growing season, responsive soils, utilization of surface and ground water and progressive farmers through intensive agriculture using various farming components like cropping systems, dairy, apiculture, sericulture, goatary, sheep rearing etc. to meet the needs of farming families on continuous basis without affecting the ecological stability, sustainability of the farm holdings and equitability. Researches under various ecosystem such as irrigated lowlands, uplands and rainfed dry land areas has demonstrated that such an approach can provide economic viability thereby leading to sustained additional farm income and employment for farm families without going through the risk involved in crop production. For example, less intensive cropping on high terraces with crops like wheat, rice, maize, pulses and oats (fodder crops) besides horticultural crops being the main components of the western and central Himalayan region. Integration of other farm enterprises like dairy, poultry, goat keeping and sheep rearing have been found to increase the farm income considerably besides providing employment to the farm families on continuous and sustainable basis. In Eastern Himalayan region age old practice of shifting cultivation (Primitive agriculture) with cultivation of maize, wheat, mustard, pulses and livestock including dairy, poultry and piggery are in practice in areas of Assam, Sikkim and Nagaland. Forestry and pasture cropping with crops like millets, wheat, fodder, barley along with goat keeping, rabbit keeping etc. are found to be feasible and remunerating in high altitude cold deserts of Himalayas as in some parts of Leh-Ladakh region.

Farming System for Hill Agriculture

Farming system as a concept takes in to account the components of farming enumerated as below with the farm family at the centre managing them independently or in complex diverse mixed form giving rise to multiple a sustainable approach.

1. Cropping systems
2. Livestock (livestock integration, Poultry, Aquaculture, Bee keeping and biogas plant)
3. Horticulture (Vegetable cultivation and fruits)
4. Agro-forestry (fuel wood and fodder)
5. Shifting cultivation

6. Sericulture
7. Mushroom cultivation
8. Grassland

The study of the different farming systems prevalent in the mountain and sub-mountain hill ecosystems of India reveals that in the North western and Central Himalayan region, less intensive cropping on high terraces with crops like wheat, rice, maize, millets, pulses and fodder crops is widely prevalent with horticulture crops as the main components. Amongst other enterprises like livestock activities Piggery, Poultry, Sheep rearing, Goat-keeping, Sericulture, and Dairy cattle are in vogue. The different components of the existing farming systems are discussed under separate heads:

Cropping Systems

Cropping system is a manipulation of crops to convert solar energy, nutrients, water and other inputs in to useful production of biomass. Different species and varieties of crops constitute a cropping system and they are managed as a single unit. Each of a crop component is a sub system of cropping system. Accordingly they may be divided in to different sub heads namely:

1. *Diversified cropping*: Diversified agriculture has different componenets, *viz.* crops, dairy, poultry, apiculture, sericulture, fishery etc.
2. *Multiple cropping* (Sequential cropping): In this system 2 or more crops are taken simultaneously on same piece of land during a year.
3. *Monoculture*: In this system only one crop is grown repetitively on a unit area of land during successive seasons.
4. *Intercropping*: Growing 2 or more crops in inter row spaces of main crop/s in definite row arrangement
5. *Mixed cropping*: Mixing the seed of 2 or more crops and then sowing it in the field or crops grown simultaneously on same piece of land with no definite row arrangement.

The details of prevalent cropping systems in the different agro-climatic zones in hills are given in Table 3.1.

Livestock

Crop-livestock is most common of oldest practice for small farmers of 0.5 to 1.0 hectares of land holding which are primarily focused on ensuring their livelihood security, increasing income, providing organic nutrition to his farm field and to support the needs of growing family size besides few animals to generate sufficient draught and power for his farm. One of the common features of hill agriculture is rearing of milch animals for milk. However, taking this as an enterprise *i.e.* dairying is a labour intensive, round the year employment generating enterprise. It provides farmers regular income and fits well in the cropping system. But at present there is dire need for improvement of dairy component through breeding using exotic breeds like jersey, Brown, Swiss, Holstein Friesian etc. However lean fodder availability and unpredictable weather besides lack of veterinary facilities and socio-economic condition of the farmers are preventing hill farmers of taking it as an enterprise though being an essential part of the existing farming systems in the hilly regions. The most common milch animals being reared in hilly regions are cows and to a lesser extent buffaloes.

Table 3.1: Different Cropping Systems Prevalent in the Hilly States of India (Yadav *et al.*, 1998)

Sl.No.	*Agro-Climatic Zone*	*Cropping System*	
		Irrigated	*Rainfed*
1.	Western Himalayan Zone	Ragi-Wheat/Barley/Pea/Lentil/Chickpea	Maize-Wheat
		Rice-Wheat/Pea/Lentil/Mustard/Chickpea	Ragi-Fallow-Rice-Wheat
		Maize-Wheat/Barley/Pea/Chickpea	Ragi-Fallow-Barnyard Millet-Wheat
		Soybean-Wheat/Barley/Pea/Chickpea	Soybean-Wheat/Barley
		Rice-Wheat/Potato/Chickpea/Lentil	Rice-Mustard/Wheat
		Maize-Potato-Wheat	Frenchbean-Potato
		Maize-Rapeseed/Wheat	Tomato-Cabbage
		Maize-Wheat/Potato	Potato-Pea
		Maize-Wheat-Mungbean	
2.	Eastern Himalayan Zone	Rice-Rice-Rice	Rice-Fallow
		Summer Rice-Urdbean/Mungbean	Rice-Potato
		Summer Rice-Maize-Mustard/Wheat	Rice-Oilseeds/Vegetables
		Jute-Winter Rice	Rice/Soybean-Fallow
		Jute-winter Rice-Chickpea/Lentil or Wheat	Fallow-Chillies
		Jute-winter Rice-Summer Rice	Potato-Fallow-Rice/Millet
		Maize-Maize-Urdbean/Mungbean	Potato-Fallow
3.	Eastern Plateau and Hill region	Rice-Potato	Rice-Wheat
		Rice-Rice-Rice	Rice-Potato-Sesame/Onion
		Maize-Groundnut-Vegetable	Jute/Rice-Maize-Cowpea
		Rice-Mustard	Jute-Chickpea-Fallow
		Jute-Rice-Rice-Khesari	Rice-Rice-Cowpea
		Maize-Wheat-Vegetable	Pice-Potato/Onion
		Early Rice-Miger/Urdbean	Mesta/Jute-Maize-Cowpea
		Rice-Potato-Groundnut	Mesta/Jute-Urdbean/Sesame
			Jute-Rice-Fallow/Urdbean
			Jute-Urdbean
4.	Central Plateau and Hill region	Cotton-Wheat/Berseem/Groundnut	Sorghum-Chickpea/Linseed
			Sorghum/Cotton-Groundnut
		Maize-Wheat	Sorghum-Cotton-Groundnut
		Groundnut-Wheat	Maize-Groundnut
		Kharif Pulses-Wheat	Sorghum/Pearlmillet-Wheat
		Sorghum/Maize-Wheat	Soybean-Cotton
		Sesame-Wheat	Sorghum-Chickpea
		Rice-Wheat/Linseed/Berseem	Rice-Fallow
		Pearlmillet-Chickpea/Rapeseed/Mustard	Fallow-Wheat/Chickpea/Linseed
		Wheat	Rice-Linseed
		Sesame-Mustard/Pea	Fodder-Wheat+Mustard

Contd...

Table 3.1–Contd...

Sl.No.	Agro-Climatic Zone	Cropping System	
		Irrigated	*Rainfed*
5.	Western Plateau and Hill region	Rice-Rice	Pearl Millet-Wheat
		Rice-Wheat-Sugarcane	Mungbean-Wheat
		Pearlmillet-Vegetable/Potato-Onion	Soybean-Wheat
		Cotton-Wheat	
		Cotton-Soybean-Wheat	
		Groundnut-Wheat	
		Sorghum-Wheat	
		Sugarcane+Sorghum/Maize	
		Cotton Rice	
		Rice-Sugarcane-Rice	
6.	Southern Plateau and Hill region	Pearlmillet/Ragi-Rice	Sorghum Maize
		Maize-Rice	Mungbean-Rice-Mungbean
		Rice-Wheat	Groundnut-Pigeonpea
		Sorghum-Onion/Chillies	Maize-Sorghum-Pulses
		Sugarcane (Adsali and eksali)	Sorghum-Chickpea
		Rice-Rice-Rice	Pearlmillet+Groundnut
		Rice-Rice-Sorghum/Cotton	Groundnut+Pigeonpea
		Rice-Maize	Ragi/Sorghum-Pulses
		Rice-Rice	Rice-Pulses/Ragi
			Maize-Potato-groundnut
7.	East Coast Plains and Hill region	Rice-Maize-Pearlmillet	Maize-Maize-Pearlmillet
		Maize-Rice	Rice-Maize+Cowpea (fodder)
		Rice-Cotton	Pearl Millet/Ragi-Rice
		Rice-Maize	Rice-Urdbean
		Rice-Rice-Rice	Rice Pulses
		Rice-Rice-Pulse/Cotton/Maize	
		Rice-Rice-Ragi/Groundnut	
		Rice-Rice-Sugarcane (Adsali)	
8.	West Coast Plains and Hills region	Rice-Rice	Rice-Maize
		Rice-Rice-Pulses	Rice-Pulses
		Rice-Rice-Ragi/Groundnut	Pearl Millet-Wheat/Pulses
		Sugarcane+intercrop	Groundnut-Maize
		Rice-Rice	
		Rice-wheat	
		Rice-Sugarcane	
		Pearlmillet-Vegetable/Potato/Onion	
		Maize-Pulses	

Contd...

Table 3.1–Contd...

Sl.No.	Agro-Climatic Zone	Cropping System	
		Irrigated	*Rainfed*
9.	Gujarat Plains and Hill region	Green manure-Rice Rice-Sugarcane-Rice Rice-Wheat Rice-Cotton+Groundnut-Wheat Sugarcane/Maize-Wheat Groundnut-Wheat/Cumin/Isabgol Castor-Wheat-Cotton Pearlmillet-Wheat-Cotton-Vegetables Sorghum-Cotton Pearlmillet-Garlic/Potato-Sugarcane Cotton-Vegetable	Rice-Chickpea Rice-Maize/Pearlmillet/Sorghum Pearlmillet-Wheat-Groundnut

The livestock components in the form of different enterprise which can help in generation of employment, augmentation of family income for socio-economic upliftment, besides multiple benefits such as organic source of nutrients, alternate source for power, sustainability, recycling of bio-waste etc. are as follows:

Poultry

It can be easily combined with integrated farming and can be adopted by resource poor farmers of the hilly regions as one of the enterprise for income generation *viz.* Backyard poultry and commercial poultry as the demand for these are enormous both for meat as well as eggs. The most prevalent system followed by farmers is battery system and deep litter system. Constructing a perforated platform over a deeper part, the battery system is set up. This is supported by split bamboos and a wooden stand. Over this structure the wire net cage is attached. The poultry breeds Rhodes island and leghorn are suitable for such structure. The droppings of birds falls down through the perforation directly to the fish stock maintained in the pod below. Some of the breeds suitable for backyard poultry in hilly areas are Van raja etc.

Aquaculture

Pond is a multipurpose entity used for providing domestic water for farm family, irrigation to crops and as medium for fishery largely due to availability of springs and waterfalls abundantly available in hilly areas of the country. Traditionally farmer gets lower production of fish to the tune of 300-450 kg tonnes per year which include both wild and cultured fish. However, using modern day knowledge through trainings and demonstrations we can easily overcome the bottlenecks for increased production of the fish. The fish pond of convenient size needs to be worked out depending on the market demand, available resources (financial and water supply) etc. Afterwards the wild and predatory fishes are eliminated by applying Mahua cake @ 250 kg/ha a month before the release of fish stock to the pond. Pond soil and water are analyzed and based on this organic manure and chemical fertilizers are applied to the pond. About 7000-7500 fingerlings consisting of species like rohu (*Labeo rohita*), castlas (*Catla catla*), mrigal (*Cirrhinus mrigala*), silver carp, Common carp and grass carp are released in the pond and supplemented feed for fingerlings @ 2 per cent of composition of fish feed as mustard oil 50 per cent, rice bran 40 per cent and fish meal10 per cent is supplied. Fish culture is very much possible in hill agriculture areas especially in temperate and semi-temperate areas where there is availability of spring water or of natural waterfalls throughout the year like in Nalthi, Sarthangal and Bani areas of Jammu province.

Bee Keeping

It is a dependable and cheap low investment enterprise. The Indian bee species *Apis cereana indica* may give an approximate yield of 1.5 - 1.8 kg of honey per box per collection with an average of 4 collections per year, whereas Italian species *Apis mellifere* hae been found to yield 10 times to that of Indian species. Also, the box can be repetitively used for 5-6 times with approx. income of Rs. 1300/- to 1400/- per box per year from Indian species and much more than by Italian species. However, 6 boxes can be maintained easily in one hectare of land.

Biogas Plant

The cattle shed excreta, along with dung and urine in the form of slurry are fed to the separately located bio gas plant. The biogas produce is utilized for cooking purpose at home, lighting lamp, lighting of fish ponds besides also being used as input to the fishes in the pond. The excess slurry is utilized for composting in a separate pit. The left over digested material is directly used as manure for the field crops. An approximate benefit of Rs. 10-15 per day may be realized from a 2 cubic metre biogas plant.

Horticulture

Vegetable Cultivation

The fruits and vegetables forms a major component of the hill farming systems. The prominent vegetables grown in the intermediate and temperate hill ecosystems are Cabbage, Cauliflower, Tomato, French bean, Turnip, Knol khol, Chilli, Tomato, Potato, Brinjal, Pumpkin, Cucumber, Roundgourd, Bittergourd and Dolichus beans. Many thermosensitive and photosensitive varieties of these vegetables are available and can be grown m successfully in hilly regions thus forming an important componenet of hill ecosystems. Protected cultivation of exotic vegetables like coloured capscicum, button tomato and chilli hold promise for higher income to the hill farmers and also many incentives are given by the central and state governments for promotion of these in hilly areas. Major vegetables and varieties grown and recommended in hills and mid hills of Jammu and Kashmir (Anonymous, 2007) is given in Table 3.2.

Fruit Cultivation

Similarly fruits are one of the major components of the farming systems prevalent in the hilly ecosystems are the prominent fruits grown in the intermediate and temperate areas of hilly regions are Wild Pomegranate, Apricot, Plum, Peach, Pear, Apple, Almonds, Walnut, Cherry and other stone fruits which are a major source of income for the hill farmers. There are many varieties for different regions of hill ecosystems and have higher yield potential to that of old varieties of the region and can be suitable adopted. Some of the varieties are given in Table 3.3.

Table 3.2: Major Vegetables and Varieties Grown and Recommended in Hills and Mid Hills of Jammu and Kashmir (Anonymous 2006)

Climate	*Major Crops*	*Crop Season*	*Varieties Recommended*
Intermediate lower	Potato	September	Kufri Badshah, Kufri Sindhuri, Kufri chipsona 1
	Brinjal	April-July	Pusa purple long, Pusa purple cluster, pusa purple round
	Chillies	March-May	NP 46 A, Pusa Jawala
	Cauliflower	May-June and Aug. - October	Pusa deepali, Pusa snowball, K-1, Snowball-16, Hybrids: Pusa synthetic, Giant snowball, Punam, Pusa hybrid 2, NS 60
	Cabbage	July-September	Golden acre, Pride of India, Pusa mukta, Sri Ganesh gol
	Ginger	Mid April-Mid May	Local
	knolkhol	September/October	White Vienna, Purple Vienna, King of Market
	Onion	August-October	Pusa Red, Punjab selection
	Radish	Mid August-November	Pusa Reshmi, Japanese QWhite, Minowase
	Garlic	Last week of August-October	Local (large segmented, Agrifound parvati)
	Carrot	August-September and Sept.-October	Nantes, Pusa Kesar, P. Yamgadini
	Turnip	Sept.-October	PTWG
	Spinach	June-September	All green, prickly seeded
	Methi	August-September	Pusa early branching/Kasuri methi
	Okra	March-May	Pusa Sawani, Varsha Uphaar
	Bottlegourd	March-May	Punjab summer prolific long, Punjab Long
	Bittergourd	March-May	Pusa doumousami, Coimbatore long, Pusa vishesh, Pusa Hybrid 1
	Cucumber	March-May	Japanese long green, Straight eight
	Pumpkin	March-May	Arka suryamukhi, Arka chandan
	Spongegourd	March-May	Pusa chikni, Pusa Nasadar
	Peas	Mid Sept.-October	Bonneville, AP 1, AP 2 asnd AP 3
Intermediate higher	Chillies	April-June	NP 46 A, Pusa Jawala
	Brinjal	March-May	Pusa Purple long, Pusa purple cluster, Pusa purple Round, Pusa kranti
	Peas	October-November and May	Bonneville
	Radish	March-June and August-September	Pusa Reshmi, Japanese QWhite, Minowase
	Carrot	April-August	Nantes
	Turnip	April-August	PTWG
	Cauliflower	April-June	Pusa snowball, Snowball 16, Pusa Snowball K-1
	Cabbage	May-July	Golden acre, Pride of India, Pusa mukta, Sri Ganesh gol
	Spinach	March-June	Prickly seeded, All green

Contd...

Table 3.2–Contd...

Climate	*Major Crops*	*Crop Season*	*Varieties Recommended*
	Methi	March-June	Pusa Early, Kasuri Methi
	Bottle gourd	March-May	Punjab Summer Prolific Long, Punjab Long
	Sponge gourd	May	Pusa Chikni, Pusa Nasadar
	Bitter gourd	April-May	Pusa Doumousami, Coimbatore Long, Pusa Vishesh, Pusa Hybrid 1
	Cucumber	April-May	Japanese Long Green, Straight Eight
	Pumpkin	May	Arka Suryamukhi, Arka Chandan
	Tomato	February-May	Pusa Ruby, S-1, Marglobe
	Knol khol	March-May	White Vienna
	Potato	March/April	Kufri Jyoti
	French Bean	March-April	Contender, Arka Komal, Pusa Parvati
	Okra	April-May	Pusa Sawani, Varsha Uphaar
Temperate	Tomato	March-May	Pusa Ruby, S-1, Marglobe
	Cabbage	May-July	Golden Acre, Pride of India
	Cauliflower	April-June	Pusa Snowball, Snowball 16, Pusa Snowball K-1
	Knol khol	March-May	White Vienna, Purple Vienna, King of Market
	Potato	March/April	Kufri Jyoti, Kufri Sindhuri
	French Bean	March-April and May-June	Contender, Arka Komal, Pusa Parvati
	Methi	March	PEB, Kasuri Methi
	Brinjal	March-May	Pusa Purple Long, Pusa Purple Cluster, Pusa Purple Round, Pusa Kranti

Table 3.3: Important Varieties of Fruits of Hill Regions of North Himalayas

Sl.No.	*Fruit*	*Varieties*
1.	Apple	Red Delicious, Golden Delicious, Hazratbali, Ambri, Red Gold, Irish Peach
2.	Walnut	Selections-SKU -1, 2, 3, 4, 5
3.	Apricot	Rakcha Carpa, Halman, New Castle, Shakarbara
4.	Wild Apricot	Local Chulai
5.	Wild Pomegranate	Daru, Darueen (Local)
6.	Pomegranate	Kandhari, Arakata, Ganesh, Kabul
7.	Peach	Alberta, JH Hale, Helberta
8.	Plum	Santa rosa, Pakistan frontier
9.	Cherry	Napolean, Vageria, Black heart, River side

Agro-forestry

Forest is composed of distinct biological units called forest communities evolved out of combined action and reaction of cohesive actions of a variety of organisms with complex factor of the habitat that they have evolved both in time and space. Agro-forestry is a suitable land management system increasing the yield of land combining the production of crops including the true crop and forest plants/animals sequentially on the same piece of land and applying management practices compatible with the cultural practices of the local people. In agro-forestry multistoried production of different species of economic plants with appropriate crop geometry are practiced. Agro-forestry provides food for humans, livestock, fibre for clothing fuel for village homes, timber, furniture and income for resource poor farmers of the hilly regions of the country. Agro-forestry aims at conservation and improvement of the site and enhancement of total production and productivity of the system.

Types of Agro-forestry System

There are several classifications for agro-forestry system proposed which howevere can be classified in to 3 major groups (Nair, 1993).

- ✩ *Agri-silviculture*: It consists of growing agricultural crops and forest crops in conjunction *i.e.* between rows of forest crops agricultural crops are grown *i.e.* crops+trees.
- ✩ *Silvi-pastoral system*: This system consists of growing forest crops and pastures simultaneously on the same piece of land *i.e.* Trees+Pasture and/or animals.
- ✩ *Agri-horti-silvi cultre*; it consists of cultuivating agricultural crops in between rows of horticultural fruits and forest trees *i.e.* Trees+Crops+Pasture/animals.
- ✩ *Multipurpose forest tree production*: Here forest trees are regenerated and manged to produce not only wood, but leaves and fruits which are suitable for food or fibre purpose. The common multipurpose forest trees *vis-à-vis* various agroclimates and uses are given in Table 3.4.

Characteristics of Agro-forestry System

The trees to be grown in conjunction with agricultural crops should have inter alia the following characteristics (King 1979):

1. Should be amenable easily to wide spacement
2. Should tolerate relatively high incidence of pruning *i.e.* their photosynthetic efficiency should not decrease with heavy pruning.
3. Should have low crown diameter to bole diameter ratio *i.e.* the width of the crown should be small relative to the bole diameter.
4. Should permit the penetration of light on the ground
5. Should be tolerant of side shading
6. Their phenology, particularly with reference to leaf flushing and leaf fall, should be advantageous to the growth of the annual crop in conjunction which they are being raised
7. The rate of litter fall and litter decomposition should have positive effect on the soil
8. The above ground changes over time in structure and morphology should be such that they refineor improve characteristics, which reduce competition for solar energy, nutrients and water.
9. Their root systems and root growth characteristics should be such that they can explore of soil layers that are different to those being trapped by agricultural crops.
10. Should be efficient root pumps.

Table 3.4: Important Multipurpose Tree Species Grown in Different Climatic Zones of India (Adapted from Gill, 1992)

Sl.No.	Common Name	Botanical Name	Climatic Zones	Important Uses
1	Babul	*Acacia nilotica*	Dry to moist tropical	Timber, fuel and fodder
2	Neem	*Azadirachta indica*	Very dry to wet tropical	Fodder, soil conservation, timber medicinal oil and fuel
3	Aonla	*Emblica officinalis*	Very dry to wet tropical	Fruit and fuel
4	Safeda	*Eucalyptus camaldulensis*	Dry to moist tropical	Industrial timber, oil, gum, tannin, ornamental and soil conservation
5	Pahadi safeda	*Eucalyptus globulus*	Moist tropical and sub tropical	Timber pulpwood, fuel and oil
6	Phalsa	*Grewia* spp.	Sub tropical to temperate	Soil conservation, fodder, fuel and timber
7	Subabul	*Leucaena leucocephala*	Dry to moist tropical	Fodder and soil conservation
8	Mahua	*Madhuca latifolia*	Dry to moist tropical	Oilseed, fruit, timber and fuel
9	Mango	*Mangifera indica*	Moist tropical to sub tropical	Fruit, fodder, timber, packing cases, plywood and fuel
10	Poplar	*Populus* spp.	Sub tropical to temperate	Soil conservation, fodder, packing cases, pulpwood and fuel
11	Apricot	*Prunus armaniaca*	Temperate	Timber poles, fuel, fruits, oil, gum, tannin, soil conservation, ornamental
12	Peach	*Prunus persica*	Temperate	Timber poles, fuel, fruits, oil, gum, tannin, soil conservation, ornamental
13	Oak	*Quercus leucotrichophora*	Temperate	Timber poles, fuel, fruits, oil, gum, tannin, soil conservation, ornamental
14	Karsu willow	*Quercus semicarprifolia* *Salix* spp.	Temperate	Fodder, fuel, fruits, match wood, sports goods, soil conservation, ornamental and medicine

Soil erosin is the major constraint for sustainable agricultural production in steeplands of hilly region. The agro-forestry systems have potential in controlling soil erosin in these lands (Gill and Lal, 1995). The run off barrier function of woody perennials helps in conserving the soil. The trees grown successfully in degraded hill slopes are *Acacia millisina, Acacia catechu, Acacia auriculiformis, Pyrus pasia, Emblica officinalis, Terminalia belerica, celtis australis, alnus nepalensis, Alnus nitida, Morus alba, Acer oblongum, Grewia optiva, Punica granatum, Albizia stipulate, Albizzia lebbeck, Indigofera pluchella, Robinia pseudoacacia, Tonna ciliata* etc.

Shifting Cultivation

Shifting cultivation is a primitive and oldest practice which makes transition from food gathering to food production. It is normally dominanat in the hilly regions of tropics and sub tropics of the world more prone to soil erosin due to steep slopes and are having shallow soils and higher rainfall. It takes into account about 10-12 per cent of population and 30 per cent of the gross area of the earth. In India, it is prevalent in several parts and pockets of North Eastern hilly regions of Arunachal Pradesh, Meghalaya, Mizoram, Manipur, Nagaland, Tripura, parts of Assam, Odisha, Madhya Pradesh, Andhra Pradesh and Bihar. It is popularly known as Jhuming in North East India, Khallur or Karwa in Bihar, Dahiya or Podu in Odisha. In Sweden it is known as Slash and burn system of cultivation and is being practiced on an approx. area of 4.36 million hectares. In this method the prime forest bed is cleared and cleaved to start cultivation and are abandoned as soon as soil fertility is exhausted. This method is composed of three steps namely cleaning and clearing of land, cultivation and restoration. However, this method is not advisable in modern times due to environmental issues and hence cannot be advocated. The ill effects are many amongst which few are enumerated as below:

1. Burning converts the organic forms of nutrients in to inorganic forms
2. Esential nutrients like C, N and S are lost through volatilization in form of CO_2, NO_2, N_2O and SO_2
3. Many nutrients like K, Ca and Mg etc. are lost as a result of pre-monsoon winds in form of ash.
4. Burning results in large scale depletion and danger and denudation of eaces of endangered species
5. Damage to large no. of edible fruit trees, medicinal and aromatic plants thus posing threat to sustainable agriculture.
6. Deteriorate soil properties such as organic carbon percentage and soil microbial pupulation etc.

The important tree species recommended for reclamation of areas used for shifting cultivation are, *Terminalia myriocarpa, Albizia* spp., *Leucaena diversifolia, Toona ciliate, Dendrocalamus hamilton, Duabanga* spp., *Alnus nepalensis, Erythrina indica* and *Pinus kesiya.*

Sericulture

The climate in the subtropical, intermediate and temperate areas of Jammu and Kashmir are suitable to produce mulberry crop and rearing silk through out the year in the areas like Dungara of Billawar tehsil of Kathua district in Jammu province of the state etc. throughout the year. In Kashmir, the climatic condition is suitable for rearing of silk worms during the spring season *i.e.* May to October *i.e.* before the autumn season begins. It requires an optimum temperature range of 21 to 30° C for maximal production and productivity. There are four types of silkworms which can be reared for silk production profitably. These are:

1. Mulberry silkworm (*Bombax mori*)
2. Eri silkworm (*Philosamia ricini*)
3. Tassar silkworm (*Antheraea mylitta*)
4. Muga silkworm (*Antheraea assami*)

Silkworm can be reared in all the three agroclimates *i.e.* Sub tropical, Intermediate and temperate regions with China leading the production tables. However, the different species are grown in different climates, areas and the food trees for these silkworms are different and the details are given in Table 3.5.

Mushroom Cultivation

It has a vast potential for hill farmers to earn their livelihood security and income there by offering a lot of scope for improving their socio-economic condition. Mushrrom is a fungus which is widely being grown for edible purpose. The cultivation of mushrooms can be taken up thrice in a year in hilly

Table 3.5: Details of the Different Silkworm Species Regarding Area of Cultivation, Food Trees and Suitable Climate

Sl.No.	Name of Silk Worm Species	Area of Cultivation	Food Trees	Type of Climate
1	Mulberry silk worm (*Bombax mori*)	Karnataka, Andhra Pradesh, Tamil Nadu, West Bengal, Jammu and Kashmir (Rajouri, Udhampur and Kathua, Pampore)	Mulberry tree (*Morus indica, Morus alba, Morus laevigata, Morus nigra, Morus rubra*)	Sub tropical, Intermediate and Temperate
2	Eri silk worm (*Philosamia ricini*)	Assam, Meghalaya, Tripura, Nagaland and other regions of North East India	Castor (*Ricinus communis*), Kessuru (*Hetropanax fragrans*) for tropical areas whereas *Quercos serrata* and *Q. semicarpifolia*) for temperate areas)	Intermediate and temperate areas
3	Tasser silk worm (*Antheraea mylitta)*	Jharkhand, Chjattisgarh, Madhya Pradesh, Bihar, Kashmir, Himachal Pradesh	Arjuna (*Perminalia arjuna*) and Asian tree (*Perminalia tonentosa*)	Tropical and Temperate
4	Muga silk worm (*Antheraea assamensis)*	Assam, Meghalaya, Tripura, Nagaland and other regions of North East India	Som (*Persea bombycina*) and Soalu trees (*Listea polyantha*)	Intermediate and temperate areas

regions with intermediate and temperate climates. Usully three types of mushroom are cultivated for edible purpose namely:

1. Button Mushroom (*Agaricus* spp.)
2. Dhingri mushroom (*Pleurotus* spp.)
3. Milky mushroom (*Calocybe indica*)

The cultivation of these three mushroom offer wide scope for year round cultivation of mushrooms thus providing round the year employment and regular farm incomes to the farmers of the hill. The low cost mushroom racks made of bamboo sticks offer a wide scope for cultivation of mushrooms in hilly areas owing to abundant supply of bamboo trees in hills. In this bamboo sticks are used to make 1.0-1.5 m wide racks along the walls of existing sheds (Pacca or Kacha) and beds of well decomposed paddy straw enriched with different nutrients and supplements *viz.* urea and *chokker* decomposed through various processes or protocols available with different agricultural universities of different hill states (J&K, H.P. and Uttarakhand) and institutes like National Research Centre for Mushrooms, Solan *i.e.* either by long duration decomposition or by short method of decomposition are made and mushrooms are cultivated accordingly through out the year. Its culture has a lot of promise in diversification of agriculture and generating income for resource poor farmers of hills. It is very low cost requiring only small indoor space, open air and thatched hut thus putting little pressure on the farm. Generally mushroom cultivation gives cost:benefit ratio ranging from Rs. 3 to 4 per rupee invested in the enterprise.

Grasslands

The main objective of growing and developing grasslands (grass and other forages) is to provide cheapest sources of food energy and protein for farm livestock. Grasslands prevalent in the hilly regions of Jammu and Kashmir in particular are uncultivated grasslands. It consists of rough mountain nad hill grazing. Plants of this grasslands are not of great economic value. These consists of fescues, bents, nardus (matgrass), molinia (purple moor grass), cotton grass, heather, bracken and gorse. Regular burning at an interval of 10-15 years can have beneficial effect both on grazing stock and wildlife. Sheep, beef cattle and ponies are the dominantly reared livestock in the areas.

References

Anonymous (2006). Package and Practices of Vegetable crops, SKUAST-Jammu.

Dadhwal, K.S. (1994). Stabilizing of degraded limestone mined lands with special reference to agro-forestry. In *Agro-forestry systems for degraded lands*. Vol. 1 (Eds P. Singh, P.S.Pathak and M.M. Roy). pp. 63-69. New Delhi: Oxford and IBH Pub. Co. Pvt. Ltd.

Dhyani, S.K., Singh, B.P., Chauhan, D.S. and Prasad, R.N. (1994). Evaluation of MPTs for agro-forestry systems to ameliorate fertility of degraded acid alfisols on sloppy lands. In *Agro-forestry systems for degraded lands*. Vol. 1 (Eds P. Singh, P.S.Pathak and M.M. Roy), pp. 241-247. New Delhi: Oxford and IBH Pub. Co. Pvt. Ltd.

Gill, A.S. (1992). Fertiliser management in agro-forestry systems. In *Non traditional selectors for fertilizer use* (Ed. H.L.S. Tandon), pp.129-142. New Delhi: Fertiliser Development and Consultation Organization.

Gill, A.S. and Burman, D. (2002). Production management of field crops in agro-forestry systems. In *Recent Advances in Agronomy*. Vol. 1 (Eds. Guriqbal Singh, Jaspinder S. Kolar and H.S. Sekhon), pp. 523-542. Indian Society of Agronomy, N. Delhi.

Gill M.S., Sarkar, A., Ahlawat, I.P.S., and Gangaiah, B. (2009). Plant Water Nutrient relationship in Crop Production, *Indian Society of Agronomy*, Edn. 1: 154-172

Joshi, N.L., Singh, D.V., R.S. Singh and Anurag Saxena (1998). Climate and Crop Production. In *Fifty years of Agronomic Research in India* (Eds. Yadav, R.L., Punjab Singh, R. Prasad, Ahlawat, I.P.S.) pp. 9-12, Indian Society of Agronomy, N. Delhi

King, K.F.S. (1979). Key note address-Some principles of agro-forestry. In *Proceedings of Agro-forestry Seminar*, pp. 17-26. New Delhi: ICAR

Nair, P.K.R. (1993). *An Introduction to Agro-forestry*, Dordrecht: Kluwer Academic Publishers.

Tewari, D.N. (1994). Agro-forestry for sustaianable development. In *Agro-forestry systems for degraded lands*. Vol. 2 (Eds P. Singh, P.S.Pathak and M.M. Roy), pp. 835-848. New Delhi: Oxford and IBH Pub. Co. Pvt. Ltd.

Hill Agriculture: Prospects, Constraints and Mitigations (2013) *Pages* **32–49**
Editor: **Dr. Shahid Ahamad**
Published by: **DAYA PUBLISHING HOUSE, NEW DELHI**

Chapter 4

Integrated Farming for Food Security and Environmental Quality in North Eastern Hill Region of India

A.K. Singh, L.J. Bordoloi and B.U. Choudhury
School of Agricultural Sciences and Rural Development,
Nagaland University, Medziphema – 797 106, Nagaland

Introduction

North Eastern Hill Region of India represents a unique mountainous, rainfed, complex, diverse and risk prone ethnic region with typical land tenure system. The region is endowed with plenty of natural resources including land, water and vegetation is characterized by climatic, geographical, biological, sociological and cultural diversity. The total forest cover in the NEH region has been estimated to be 1, 41, 652 sq. km, which accounts for 77.1 per cent of the geographic area as against the national average of 19.39 per cent (Anonymous, 2001). Unfortunately, the region is also characterized by lowest utilization efficiencies of natural resources including land. The land use efficiency, which denotes the effective use of land resources for generating livelihood, is a meagre 10.38 per cent in the north east against the national figure of 23.63 per cent. Traditional agricultural practices, irreplaceable being integral parts of their socio-cultural lifestyle, needs to be modified scientifically to turn them into competent systems for ensuring food sufficiency on one hand while maintaining environmental quality on the other.

Natural Resource Base of the Region

Soil and Land Resources

North eastern region is having a geographical area of 26.2 million hectare representing 6 agro-climatic zones *viz.* (i) Alpine zone (ii) Temperate and sub-alpine zone (iii) Sub tropical hill zone (iv) Mild tropical hill zone (v) Mild tropical plain zone and (vi) sub- tropical plain zone. Around 28.3

per cent of the total geographical area has an elevation more than 1200 m, 17.9 per cent between 600 and 1200 m and 10.8 per cent between 300 and 600 m above mean sea level. Beyond the transitional pre-glacial region with increasing altitude, there appears the greater Himalayan region devoid of significant vegetative cover. Rocky surface, alpine vegetation and snow-capped high peaks dominate the physical landscape of the area. The plains mainly comprises of Brahmaputra and Barak valleys. The adjoining areas of Barak river are active flood-plains with marshy lands subjected to extreme annual inundation. Application of USDA proposed land capability classification reveals that there is no class I and class II land in the region; very little class III and a few class IV land. Most of the areas under cultivation fell under class VI and VII (non – arable classes). Areas with considerable slope (hence erosion prone) are already being used for growing subsistence food crops. Sensing the socio-economic complicacies to change the situation, a modified land capability classification have been developed and used in the region (Table 4.1).

Table 4.1: Modified Land Capability Classification for North Eastern Region

Class	*Slope (Per cent)*	*Recommendation*
A-1	0–5	Suitable for cultivation without special soil conservation measures.
A-2	6–50	Suitable for cultivation with special soil conservation measures.
B-1	51–100	With shallow depth (<1.75m), suitable for pastures and fodder crops.
B-2	51–100	Having soil depth (> 1.75m), suitable for orchards, cash crops, plantation crops etc.
C	> 100	Suitable for forest

Source: Borthakur (1992).

Soils of the region broadly fall under the orders inceptisols, entisols, alfisols, ultisols and mollisols. More than 80 per cent of the soils are acidic in nature. Soils are rich in organic matter with medium to high nitrogen, low phosphorus and medium availability of potassium. Widespread deficiencies of calcium, magnesium, zinc, boron and toxicities of iron, aluminium and manganese are the common soil related constraints of the region.

NEH: A Biodiversity Hot Spot

NEH region of India is a part of humid tropics, which is characterized by a dense rainforest ecosystem with high species diversity but with extreme fragility, thereby, subjected to rapid conversion either due to timber extraction, fuelwood extraction and shifting agriculture in particular (Ramakrishnan, 1992). According to a survey carried out by Botanical Survey of India, about 10,000 species represents only 50 per cent of total flora of forest vegetation in the region. Furthermore, the region has been considered as primary or secondary origin to numerous crop plants- their cultivated and wild species; medicinal and aromatic plants; tree fodders, fruit and food producing trees, oil seeds and dye producing trees; spices and ornamentals and many more.

Water Resources

The region receives nearly 16.63 per cent of the national annual precipitation and boasts of 0.68 m ha of inland water bodies (10.23 per cent of the country) and 0.02 m km of riverine area (10.93 per cent of national average). But ironically, the realized irrigation system in the region amounts to only 1.52 per cent of the national average which indicates towards poor water use efficiency of the region. The sloppy undulating terrains of the region without proper soil and water conservation practices naturally has to bear the vagaries of intensive soil erosion, flood and river basin siltation. Roughly, 50–60 per cent of rain water is lost as runoff, 10–13 per cent in percolation and 5–7 per cent is lost through evaporation. As result, only 23–27 per cent of rain water remains available for the agricultural production system (Table 4.2).

Table 4.2: Precipitation (h m), Water Bodies and Other Water Sources of the Region

Source	*Years*	*NEH Region*	*India*	*Per cent of India*
Rainfall	1999	8,668,615	460,216,820	12.75
	2000	38,591,708	331,027,384	11.66
	10 years average	68,886,327	414,195,138	16.63
Rivers and canals (km)		20,875	191,024	10.93
Reservoirs (h)		150,000	2,031,000	7.39
Tanks, lakes, ponds (h)		3,71,000	2,381,000	15.58
Beels, oxbows, lakes and derelict water (h)		159,000	798,000	19.92
Total water bodies (h)		680,000	6,647,000	10.23
Irrigation (h)		838,000	55,143,000	1.52

Source: Dhiman (2005).

Considering the fact that such inefficient utilization of the natural resources has led to poor economic output and various maladies, question arises about what are the strategies to be adopted for agricultural development on a sustainable basis in the region? How to incorporate the traditional wisdom of the farmers in the modern day interventions so as to develop a viable tool for development? To explore answers to these questions, first one has to have parallel knowledge about the reasons behind extensive resource degradations and traditional practices of the farmers and rural folks prevalent in the region.

Natural Resource Degradation

Three major causes have been cited for large scale degradation of natural resources in the region (Bhatt *et al.*, 2005). They are: (i) Shifting cultivation (ii) Population pressure (iii) High biomass requirement and (iv) change in forest cover.

Shifting Cultivation

Whenever arise the issues of natural resource management, the term shifting cultivation (locally *jhum*) seems inseparable to NEH region. As per the state of forest report (1999), cumulative area affected by shifting cultivation is 1.60 million ha. The extent of area under shifting cultivation is maximum (0.39 million ha) in Nagaland, followed by Mizoram (0.38million ha) and Manipur (0.36 million ha). These states account for nearly 65 per cent of the total area under shifting cultivation (Anonymous, 1999). On an average, 3,869sq. km area is put under shifting cultivation every year.

With reduction in *jhum* cycle from 20–30 years to 2–3 years, the land under shifting cultivation loses its nutrients and the top soil. With reduction in crop yield, the families start moving to other virgin areas. Frequent shifting from one land to the other has affected the ecology of these regions. Decline in area under natural forest, the fragmentation of habitat, local disappearance of native species and invasion by exotic weeds and other plants are some of the other ecological consequences of shifting agriculture. The area having shorter *jhum* cycle is more vulnerable to weed invasion compared to the areas under longer *jhum* cycle. Similarly, studies on *jhum* ecology also revealed that water and nutrient losses in shifting-cultivation areas were far greater than in the virgin areas, and areas left for 50 years after *jhuming*.

Population Pressure

As per the reports of last three consecutive census periods (1971, 1981 and 1991), north eastern region has recorded highest decadal growth rates of population in the country. Such burgeoning population has put additional pressure on availability of land, disrupted social norms and resource management systems. All these have led to further deforestation and population growth again (Dasgupta *et al.*, 1986). Over exploitation of natural resources by the exploding population has ultimately brought their degradation.

High Biomass Requirement

Tribal communities of NEH region rely heavily on forest resources for their subsistence and 90 per cent of the population use biomass as an important source of energy (Bhatt *et al.*, 2001). Although there is dearth of basic information regarding supply and consumption pattern of rural energy in the NEH region, firewood consumption has been reported to be quite high (3.90- 5.81kg/capita/day) in comparison to the other hilly terrains of the country (Bhatt and Sachan, 2004).

Change in Forest Cover

Although NEH region can boast of greater per cent forest coverage of geographical area compared to national average, a drastic reduction in forest cover has been reported in most of the states (Table 4.3). Among different states of the region, Arunachal Pradesh has the maximum dense forest (canopy density >40 per cent) cover. Shifting cultivation has been considered as the main culprit for such rapid reduction in forest covers as every year dense forests are being converted to *jhum* fields. In between 1999 and 2001, there was a net loss of 1577 km^2 of forest, which was 5.67 fold higher than the forest loss between 1997 and 1999 (Anonymous, 1999, 2001).

Table 4.3: Trends of Forest Coverage in NEH Region

States	*1991 Assessment*			*2001 Assessment*			*Net Difference*
	Dense Forest Cover	*Open Forest Cover*	*Total Forest Cover*	*Dense Forest Cover*	*Open Forest Cover*	*Total Forest Cover*	
Arunachal Pradesh	57,756	11,091	68,847	53,932	14,113	68,045	(-) 802
Manipur	5,936	11,448	17,384	5,710	11,216	16,926	(-) 458
Meghalaya	5,925	9,708	15,633	5,681	9,903	15,584	(-) 49
Mizoram	3,786	14,552	18,338	8,936	8,558	17,494	(-) 844
Nagaland	5,137	9,027	14,164	5,393	7,952	13,345	(-) 819
Sikkim	2,363	755	3,118	2,391	802	3,193	75
Tripura	2,228	3,517	5,745	3,463	3,602	7,065	1,320
Total	**83,131**	**60,098**	**1,43,229**	**85,506**	**56,146**	**1,41,652**	**(-) 1577**

Source: Adopted from Anonymous (2001)

Traditional Agricultural Production Systems of NEH Region

Whatever may be the kinds and magnitudes of negativity associated therewith, shifting cultivation still is and will continue to be a vital system of agricultural production as it is not only a source of livelihood but also has immense socio-cultural importance among the people of the north east. The shifting cultivation is generally practised in the following sequence:

- ✰ A forest patch is either selected by the household or allotted to them by the village council is used and clear felled the vegetation normally in December- January.
- ✰ The herbs, shrubs, twigs and branches (slashed vegetation) are burnt in February- March. Small, cut-trunk portion and roots are normally not removed.
- ✰ Seeds of cereals, vegetables and oil seeds are sown by dibbling in April–May.
- ✰ Wild edibles grown naturally in the cultivation site are also collected by the household for consumption as well as marketing.
- ✰ Cultivation continued in a similar fashion for a few years.
- ✰ After a few years the cultivated site is abandoned and shifted to other forest sites.
- ✰ After some years, household returns to the former site and starts cultivation again.

The maladies like natural resource degradation, poor and un-sustained productivity has reduced the significance of this traditional practice. The operations of shifting cultivation are manpower intensive; hence the normal system developed is community oriented with pooled manpower whereas different communities have devised different methods of natural and human resource utilization for agricultural production.

Zabo system is an indigenous system practised in the Phek district of Nagaland, which is a combination of forest, agriculture, livestock and fisheries with well developed soil and water conservation base. The system envisages rain water utilization, soil health maintenance through organic recycling and incorporation of livestock as the subsidiary source of income.

The *Apatani plateau system* is a unique system of paddy cultivation prevalent in the Subansiri district of Arunachal Pradesh extending to a stretch of 26 sq. km. The Apatani tribe of the state practises wet rice cultivation following principles based on excellent knowledge of soil and water conservation. Besides other measures, paddy-cum-fish culture and growing millets on terrace risers for their stabilization and overall yield augmentation are some of the unique features of the system. Continuation of the system for centuries without marked reduction in yield itself shows its worth. Rather, this system has played a critical role in making the Apatani plateau, the rice bowl of Arunachal Pradesh.

Bamboo drip irrigation is an efficient system of water use for agricultural production. The system has been developed and used for centuries by the farmers of Jaintia hills in Meghalaya. In this system, water is conveyed from the source through split bamboos to the intended crop fields. Using their wisdom, farmers have also selected crops of low water requirement to be irrigated efficiently by this system.

The terrace rice cultivation or *Panikheti* has been developed and practised by Angami and Chakhesang tribes of Nagaland. In this system, water from the source is conveyed through local channels from the top most to the lower terraces. Water is channelized in a zig-zag fashion so as to break the speed of movement and to maintain desired water level in all the terraces. Almost similar system of terrace rice cultivation (*Dhan kheti*) is prevalent in Sikkim too.

Alder (*Alnus nepalensis*) is an important non- leguminous nitrogen fixing tree of the region, which has been genuinely utilized by the people of the region as an integral component in different agro-forestry systems. For instance, *alder based farming system* (agri- silviculture and horti- silviculture) in Nagaland and *alder- large cardamom system* (horti- silviculture) in Sikkim is of utmost importance in uniqueness and value (Dhiman, 2005).

Bari system (Homestead gardening) is also common in different parts of Assam, Meghalaya and Tripura where households try to integrate diversified components into a system. Farmers try to accommodate different crops, multipurpose tree species, livestock including fisheries with water harvesting and residue recycling measures as per their preference, requirement and suitability. The homestead gardening in itself is a system of component integration near and around houses like agriculture, horticulture including fruits and vegetables, livestock in the form of cattle, goats, pigs, tree species for fuel, fodder and timber with water harvesting for domestic use and fishery- cum- duckery production.

With the present population growth rate, the population of the region is expected to cross 47 millions by 2011 with concomitant widening of the gap between food grain requirement and availability (Table 4.4). Thorough perusal of the above described systems reveal that the non commercial agricultural production systems of the north eastern region are traditionally a blend of all components including agriculture, horticulture, silviculture, livestock and fisheries with suitable soil and water conservation measures. These systems, over the years, have been proved to be viable, profitable, sustainable and eco-friendly. Also keeping the natural resources like land, water, biodiversity and socio- cultural heritage of the region in the fore front, it is also pertinent to go for technologies, which could conserve these resources and also meet the food and nutritional security needs of the households.

Table 4.4: Population, Food Grain Production, Requirement (2001) and Deficit or Surplus in the North Eastern Region

States	*Population (2001)*	*Food Grain Production (2001)*	*Food Grain Requirement (2001)*	*Food Grain Deficit/Surplus (2001)*	*Expected Deficit/ Surplus (2001)*
Arunachal Pradesh	10,91,117	191.30	203.78	- 12.48	- 21.63
Assam	2,66,38,407	3,763.19	4,458.98	- 695.79	+ 567.98
Manipur	23,88,634	369.09	399.31	- 30.22	- 129.76
Meghalaya	23,06,069	191.78	386.01	- 194.23	- 436.09
Mizoram	8,91,058	132.99	149.15	- 16.16	- 48.43
Nagaland	19,88,636	244.48	332.88	- 88.40	- 135.88
Sikkim	5,40,493	86.45	90.35	- 3.90	- 8.60
Tripura	31,91,168	540.43	667.50	- 127.07	- 160.45
Total	**3,90,35,582**	**5,519.71**	**6,687.96**	**- 1,168.25**	**- 372.82**

Source: Dhiman (2005).

Requirements

An integrated approach of conservation and sustainable development for the satisfaction of basic human needs, equity with justice and the maintenance of social, cultural and biological diversity together with ecological integrity of the system could be the most suitable technology for the region.

The land use plans must be based on soil characteristics, land capability, climatic parameters and socio-economic needs of the masses. Land and water resources have to be characterized and grouped according to their production potential and quantum of hazards and have to be managed accordingly. Development programmes must aim at bridging the gap between present level of income and future needs and aspirations of the people on a sustainable basis.

Integrated Farming System (IFS)

Holistic approach of the land use is much more viable for long term sustainable production on one hand and livelihood security on the other, wherein all the components of agriculture, horticulture, forestry, livestock and fishery can be integrated in a complimentary way besides soil conservation measures, vermicomposting, mushroom production and liquid manure preparation etc. (Bhatt *et al.*, 2005). Integrated farming is based on the concept that "there is no waste" and "waste is only a misplaced resource which can become a valuable material for another product" (Edwards *et al.*, 1986). Integrated farming system has the potential to provide the environmentally sustainable and economically viable technology that encompasses rational utilization of available resources of the region. The emphasis in IFS is on optimizing resource utilization rather than maximization of individual elements in the system. Integration of farming basically aims at encashing upon the interdependencies of the systems. It is generally considered relevant to benefit the rural poor. Integration of farming is traditional in Asia, especially in China and is now also applied in Europe and, on small scale, in Africa and some Latin American countries. The suitable tree-crop combinations occupy a key position in any viable integrated farming system. A more advanced and refined form of farming system integration is the Intensive integrated farming system (IIFS). IIFS can be developed by injecting elements of (i) agricultural intensification (ii) diversification and (iii) value addition. The goal is to foster food security at the level of each individual in a household.

Purposes Served by IFS

- To provide some viable alternative to shifting cultivation in the hilly terrains of the north eastern region.
- To realize and utilize the synergistic role of livestock in farming systems.
- Biodiversity enhancement as the 'alternative feeding system' uses mainly local plant resources. It provides comparative advantages to indigenous animal ecotypes and opportunities for greater use of indigenous knowledge.
- To explore the complimentarity of crop- animal- fish- birds- multipurpose trees- horticulture.
- To select and promote crops and farming systems, which optimize use of natural resources rather than trying to maximize it eventually depleting their potential.
- To synchronize the production system intricacies with resources available at hand.
- To ensure maintenance and even build up of soil fertility through various avenues *viz.*, nitrogen fixing hedgerows, vermicompost and liquid manures (prepared on farm with locally available resources), fish pond bed soil, crop rotation with legumes etc.
- To ensure maximum return per unit area per unit time in terms of produce and money.
- To adopt effective soil and water conservation measures as integral components of farming activity.
- To handle farming system components in a better way ensuring proper recycling of their by products thereby curbing the hazards of environmental pollution.

- ☆ To empower the farming community with reduced cost of production and also to benefit the society as a whole by supplying essential commodities throughout the year.
- ☆ To generate year round employment to the farming family.
- ☆ To provide a protective cover/insurance against the failure of individual crop/enterprise etc.

Promising IFS Models for the North Eastern Region

Although the traditional farming practices of the north east have in them elements of IFS, scale of precision and complexities have been standardized scientifically to develop certain well defined models of integrated farming systems in the region. A number of such models developed and tested with success by ICAR Research Complex for NEH Region, Umiam (Meghalaya) are detailed below (Table 4.5).

Table 4.5: Integrated Farming Systems Models for the North Eastern Region

Sl.No.	*IFS Model*	*Integrating/ Subsidiary Components*	*Details of Crops and Livestock Component*
1.	Agri-horti-silvi system	Broiler chicken unit, integrated fish-cum-duck farming	In upland areas ragi, maize and rice bean were grown, whereas, in lowland patches lowland paddy and mustard were cultivated. During *rabi* season, potato, tomato, cabbage, knoll-khol and radish were grown. Nitrogen fixing shrubs were grown on the contour bunds as hedgerow. Fodder grasses (Congo and guinea) were grown on the pond dykes. Duck (breed Assam Local/Indian runner) were reared maintaining a population density of 700 birds per hectare of the pond area.
2.	Agri-silvi-pastoral system	Integrated fish-cum-poultry farming	In upland patches of the farm upland paddy and rice bean were grown during *kharif* and buck wheat during *rabi* season. In low lying areas, paddy, potato and french bean were grown. Fodder grasses and fruit trees were planted on pond dykes. Layer poultry birds (breed White leg horn) were reared maintaining a population density of 700 birds per hectare of pond area.
3.	Agri- horti- silvi- pastoral system	Integrated fish- cum- goat farming	Upland area was used to grow paddy, ginger and turmeric during *kharif* and mustard, tomato and radish during *rabi* season. Fodder grasses and fruit trees were planted on pond dykes. Goats (breed Black Bengal) were reared maintaining a population density of 55 goats per hectare of pond area.
4.	Agri- horti- silvi- pastoral system	Integrated fish- cum- pig farming	In upland areas, paddy, colocasia and maize were grown during kharif and radish, potato and buckwheat during rabi season. Pigs (breed Large Black) were reared maintaining a population density of 34 pigs per hectare of pond area.
5.	Agri- horti- pastoral system	Integrated fish- cum- cattle farming	Upland paddy, broom grass and job's tear were grown in the upland areas. Fruit trees and fodder grasses were grown on the pond dyke. Cattle were reared maintaining a population density of 05 cows per hectare of pond area.
6.	Agri- horti system	Fish farming	Upland paddy and maize were grown during *kharif* season and potato, buck wheat and french bean were grown during *rabi* season. Fruit trees were grown on pond dykes.

N.B.: Stocking density of fish fingerlings was 6000 per hectare of pond area with a combination of: *Catla catla* (Catla)- 20 per cent; *Hypepthalmichthys molitrix* (silver carp)- 20 per cent; *Ctenopharyngodon idella* (Grass carp)- 20 per cent; *Cirrhinus mrigala* (Mrigal)- 20 per cent, *Labeo rohita* (Rohu)- 10 per cent and *Labeo gonius* (Gonius)- 10 per cent.

Source: Adopted from Bhatt *et al.* (2005).

Viability in terms of profitability of all the above described models/systems were ultimately assessed. All the systems were found promising in providing a steady flow of output either in terms of produce or money. Agri- horti- pastoral system with integration of fish-cum- cattle farming was the most promising with an output/input ratio of 2.48. To achieve such success in integrated farming system of course needs extremely careful handling of the various components being integrated. Being able to strike the perfect balance between the components holds the key as any fluctuation in performance of one component might drastically reduce profitability and even jeopardise the system as a whole.

Livestock Component in Integrated Farming System

In the eastern Himalayan region, agriculture is a highly complex production system where crop and animal husbandry constitute an interlinked system. Swine husbandry particularly is integral part of village ecosystem because it is detritus based and depends on the waste generated from agriculture (Maikhuri, 1992). North eastern region, though has a sizable livestock strength (Table 4.6), has not yet been able to make a desired headway in its production status.

Table 4.6: State-wise Livestock Population (2002) in North Eastern Region

State	*Livestock Population (in thousands)*								
	Cattle	*Buffalo*	*Sheep*	*Goat*	*Pig*	*Horse*	*Yak*	*Mithun*	*Total Poultry*
Arunachal Pradesh	453	11	28	183	275	6	9	105	1267
Assam	8000	728	84	2677	1082	12	0	0	18210
Manipur	508	95	8	33	388	2	0	21	3055
Meghalaya	738	18	17	280	351	2	0	0	2152
Mizoram	33	5	1	16	163	2	0	0	1307
Nagaland	383	36	2	161	571	1	0	26	2444
Tripura	1228	18	6	639	237	2	0	0	3595
Total NEH	11343	911	146	3989	3067	27	9	152	32030
Total all India	197708	88832	56766	120774	13581	827	54	152	346693
Per cent contribution by NEH	5.74	1.03	0.26	3.30	22.58	3.26	16.67	100	9.23

Source: Adopted from Anonymous (2002)

In any integrated farming system model, the role of livestock has been identified as synergistic rather than that of a primary producer. This, however, no way reduces the essentiality of livestock in a viable integrated system of farming. The sustainable use of natural renewable resources will be facilitated when the feed is grown, the animals are fed and the excreta is recycled on the farm in a way that minimizes the use of important inputs including energy (Preston and Murgueitio, 1994). Thus, livestock component critically determines the magnitude and direction of resource flow in an integrated farming system (Figure 4.1).

Excellent performance of livestock components under integrated farming systems was recorded. Through series of experiments conducted under IFS in Meghalaya, remarkable improvement in growth performance of livestock/birds over traditional rearing practices were recorded (Table 4.7).

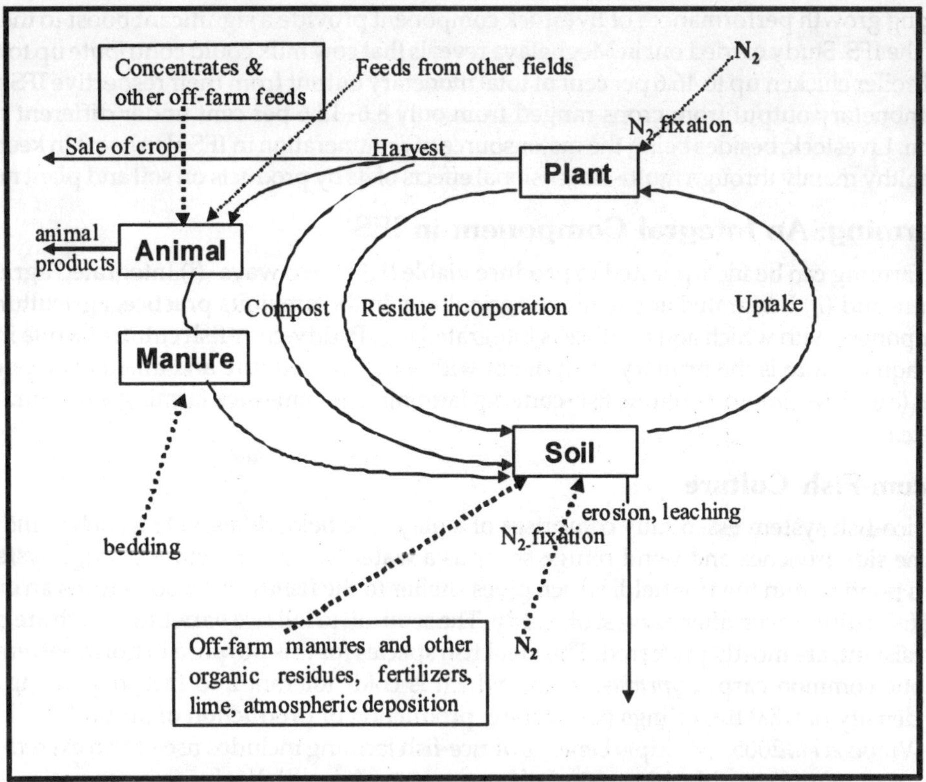

Figure 4.1: Resource Flow in a Livestock Based Integrated Farming System

Table 4.7: Growth Performance of Livestock/Birds (kg/animal) in IFS and Traditional Animal Husbandry Practices

Livestock/Bird	*Growth under IFS*	*Growth in Traditional System*
Goat (2 years)	25.20 (0.035)	22.46 (0.0312)
Pig (1 year)	123.12 (0.342)	89.28 (0.248)
Duck (2 years)		
Indian runner	2.79 (0.00387)	2.29 (0.00318)
Khaki Campbell	2.89 (0.00402)	2.65 (0.00368)
Layer birds (1 year)	2.25 (0.00625)	-
Broiler chicken (6 weeks)	1.69 (0.0404)	1.41 (0.0306)

Note: Data in parenthesis represents daily weight gain (kg) by the animals/birds.

Source: Bujarbaruah and Bhatt (2005).

The weight mentioned in the above table represents weight gained at the time of slaughtering. The slaughtering period was lowest (6–7 weeks) in broiler chicken and highest (more than 2 years) in case of goat and duck. Pig could be slaughtered after one year of rearing in IFS. Such types of encouraging growth performances of livestock component provide a significant boost to the overall output of the IFS. Study carried out in Meghalaya reveals that cow milk could contribute up to 68.7 per cent and broiler chicken up to 46.6 per cent of total monetary output from their respective IFS models, whereas monetary output from crops ranged from only 8.6- 15.6 per cent under different types of integration. Livestock, besides being the major source of remuneration in IFS also helps in keeping the system healthy mainly through multi-dimensional effects of its by products on soil and plant nutrition.

Fish Farming: An Integral Component in IFS

Fish farming can be incorporated to produce viable IFS in two ways- (i) Integrated agriculture-aquaculture and (ii) Integrated aquaculture-agriculture. In the previous practice, agriculture is the main component with which aquaculture is integrated (*e.g*. Paddy-cum-fish culture) while in case of the latter aquaculture is the primary component with which agriculture and allied components are integrated (*e.g*., fish-cum-horticulture, fish-cum-pig farming, fish-cum-duck farming, fish- cum- poultry farming etc.).

Paddy-cum-Fish Culture

The rice-fish system essentially comprises of a main rice field, dikes, side trenches and a pond refuge. The side trenches and pond refuge serve as a water harvesting- cum- storage system. The refuge is a pond within the rice field, which gives shelter to the fishes and also permits an extended period of fish culture even after harvest of paddy. The semi tall to tall rice varieties, which are pest and disease resistant, are mostly preferred. The ideal fish species for this purpose in north eastern region is the exotic common carp (*Cyprinus carpio*), which is cold- tolerant and fast in growing. With a stocking density of 6000 fingerlings per hectare, production of production of around 300 kg fish is possible (Vinod *et al*., 2005). Multiple benefits of rice-fish farming includes use of fish excreta and left over food as manure for paddy, bio-control of insects pests of rice, higher income generation, creating a reliable source of protein etc.

Integrated Fish-cum-Pig Farming

In this system, pig housing is constructed either on the pond or on pond dyke or in the near vicinity of the pond. Pig shed washings are channelled in a controlled manner to the pond to act as manure. Additional pig dung is diverted to compost pit. Supplementary feed is not required for the fishes as pig excreta serves as feed for some fishes and also enriches the pond water resulting in production of plankton (food for fish). Grass carp should, however, be fed with aquatic or terrestrial vegetation. Regular liming of pond is also essential for stabilization of organic matter. Generally 3 pigs have been found to be sufficient to fertilize a pond area of 0.1hectare. With 6000 no. of fish fingerlings per hectare, 1500–2000 kg fish could be produced per hectare per year without any supplementary feeding.

Integrated Fish-cum-Duck Farming

Fish ponds provide the ducks an excellent environment to move around during the day time. Duck shed can be constructed on the pond or on its dyke. Ducks fertilize the pond with their droppings requiring no supplementary feeding for the fishes. Ducks keep the pond clean by feeding on aquatic plants, insects and their larvae, earthworms etc. Hence ducks also need very little feed, which can be supplied through rice bran and kitchen waste. Ducks aerate the pond by swimming and facilitate release of nutrients from pond bed soil by dabbling on it. About 30 ducks are enough to fertilize a pond area of 0.1 hectare. A stocking density of 6000 fingerlings can yield 1500–2000 kg of fish per hectare per year.

Integrated Fish-cum-Poultry Farming

Poultry birds are raised over the fish pond or the pond embankment and their excreta are recycled to fertilize the fish ponds. No supplementary feed is needed for the fishes as they get regular supply of poultry excreta. The feed spilled by chicken also enters the pond. About 50 birds can sufficiently fertilize a fish pond of 0.1 hectare area. Broiler birds reach marketable size of 1.5- 1.8 kg in about 8 weeks and hence it is possible to raise at least 4 batches in a year. Layer birds start laying eggs after 22 weeks and nearly 100 eggs per bird per year can be expected. After 18 months of age, layers need to be replaced.

Maintaining Soil Health in IFS

A basic prerequisite for keeping any IFS viable is to keep its various components/resources healthy. Maintaining soil health in a sound state is a tricky issue the effective handling of which ensures productivity of the system on a sustainable manner.

Livestock *vis-a-vis* Soil Health in IFS

Various animal/bird excreta either raw or processed into certain products are critical for maintaining soil health in IFS. Excreta from poultry, duck etc. and dung of cattle, pig, goat etc. are valuable sources of nutrients for maintaining soil fertility. The excreta of poultry birds, ducks etc. are directly washed down to the pond to serve 2 purposes- first to act as fish feed and second to enrich the pond bed soil with nutrients, which in turn augments plankton production. The dung of the animals are either applied directly to soil as nutrient sources or used as substrates for preparing compost, vermicompost, liquid manure etc. Experiments have shown that these animal manures are not only rich in major nutrients but also contains several micro-nutrients like Mn, Zn, Cu. The demand of livestock manure is very high for agricultural as well as horticultural crops (Gupta, 2006; Dalzell *et al.*, 2007).

The dung production varies with different categories of livestock. The amount of dung produced per animal per day depends upon the amount of forage fed to the animals. In general, the total output of dung (fresh weight) from cattle and buffalo per day is to be about 10 kg and 12 kg respectively giving a total of 3.6 tonnes and 4.3 tonnes of dung per animal per annum respectively. Assuming the animals are stall fed, the annual output of N, P and K for cattle would be 12.6 kg N, 5.4 kg P and 7.2kg K, and from buffalo 15 kg N, 6.5 kg P and 8.6 kg K(per animal) (Kumaresan *et al.*, 2006).

The average dung production per head per year on dry matter basis for domesticated animals is in given Table 4.8.

Table 4.8: Dung Production by Different Categories of Livestock

Livestock	*Dung Production per Animal per Year (Dry Weight Basis) in Tones/Year*	
	Range	*Average*
Cattle	0.4-1.8	1.10
Buffalo	0.8-1.9	1.35
Pigs	0.2-0.3	0.25
Sheep	0.1-0.2	0.15
Goats	0.1-0.2	0.15
Poultry (100 birds)	0.14	0.14

Source: Kumaresan *et al.* (2006)

The nutrient content of animal manure is variable; depending on the types and age of animals, feed sources, housing type, temperature and moisture content. Approximately, 2.36 cattle, 18.25 pigs, or 196 birds/rabbit can provide 10-tone good quality manure every year that contains sufficient quantities of nutrient. Generally, the nutrient value of the manure, in increasing order is cattle, sheep, and goat, followed by pig, chicken, and ducks (Table 4.9).

Table 4.9: Average N, P, and K Level in Dung of Different Livestock

Species	*N (per cent)*	*P (per cent)*	*K (per cent)*
Cattle	0.15	0.01	0.05
Pig	0.60	0.50	0.20
Sheep and Goat manure	0.65	0.50	0.03
Poultry manure (fresh)	0.76	0.63	0.22
Duck manure	0.91	0.38	0.36

Source: Kumaresan *et al.* (2006)

Besides N, P, K content, animal manure also contains significant amount of other nutrient elements like Ca, Mg, S, Mn, Zn, and Cu. Among the different types of animal manure, poultry manure is having the highest nutrient value with respect to both macro (secondary) and micro-nutrients (Table 4.10).

Table 4.10: Secondary and Micr-onutrient Content in Different Sources of Animal Manure (kg/t)

Source of Manure	*Ca*	*Mg*	*S*	*Mn*	*Zn*	*Cu*
Broiler chicken (All type)	18.59	3.63	5.44	0.23	0.23	0.45
Layer (All type)	38.56	2.27	2.72	0.18	0.14	<0.04
Swine (All type)	3.63	0.91	0.91	<0.04	0.04	<0.04
Dairy (All type)*	10.43	3.17	1.36	0.04	<0.04	<0.04

* kg/1000 gallon

Animal urine is also very rich in nitrogen and potassium. The value in cattle urine has been reported to be 0.9 – 1.2 per cent N, a trace level of P and 0.5 – 1.0 per cent K. Urine output is in the range 17–45 ml/kg body weight. The average body weight of cattle, if taken as 200–300 kg the average urine production per day per animal would be about six to nine litres. To use the animal manure efficiently, the animals are to be stall-fed and housing should be designed in such a way that entire urine is collected (Kumaresan *et al.*, 2006).

Vermicompost for Soil Health Improvement in IFS

Vermicompost has been identified as one of the important source of nutrients for maintaining soil fertility/health in any integrated farming system. As the concept of IFS does not permit much use of off farm inputs (including nutrients), recycling of crop residues, weed biomass and spent spawn of mushroom etc. within the farm is of utmost importance. Well decomposed vermicompost could be prepared within 50- 80 days depending on the substrate type, earthworm species used and season. Use of voracious feeder African worm *viz., Eisenia foetida* and *Eudrillus euginae* has been found to compress the composting time remarkably. These earthworm species are easily available now-a-days at a nominal price. Releasing 2000 worms per square meter of biomass has effectively converted the substrate into matured vermicompost within a period of maximum 80 days (Table 4.11). Increasing population density of worm beyond this could not bring about remarkably faster decomposition and hence does not seem economical. Fortifying the biomass with cattle dung etc. improves compost quality besides augmenting a faster rate of decomposition.

Table 4.11: Conversion Efficiency of Different Biomass into Vermicompost in IFS

Biomass/Substrate	Quantity on Dry Weight Basis (kg)*	No. of Worms Released	No. of Worms Retrieved at the End	Effective Decomposition Time	Quantity of Vermicompost Produced (kg)
Mixed weed biomass	250	2000	4250	60	120
Ambrossia artemissifolia	100	2000	3600	50	42
Maize	150	2000	3100	75	55
Paddy straw	200	2000	3200	70	70
Mixed biomass of MPTs**	250	2000	2750	80	105
Spent spawn of mushroom	100	2000	4610	55	45

* Quantity includes that of cow dung used to fortify the substrate.

** MPTs include *Alnus nepalensis, Gmelina arborea, Michaelia champaka, Chukrasia tabularis* and *Erythrina indica.*

Liquid Manure as Nutrient Source in IFS

An avenue for utilizing the nutrient richness of the animal urines is to convert them to liquid manures. Cow urine is collected and stored in small cemented pits. Fresh cowdung is added to the urine and stirred periodically. Within 35- 40 days, cow dung is completely dissolved in the urine producing liquid manure. Liquid manure so produced can be applied to different crops after proper dilution. Liquid manure has been particularly found beneficial against phosphorus deficiency in soil and crop. Besides being a rich source of plant nutrients, liquid manure also bears some bio- pesticidal properties.

Utility of Pond Bed Soil

As we have seen, fish farming is an indispensable component of IFS. In IFS models, fisheries are integrated with other livestock components. At least a portion of the livestock excreta are channelized to the fish ponds as fish feed and for nutrient enrichment of the fisheries to promote plankton production therein. These practices on one hand enrich the pond bed soil with nutrients and on the other gradually reduce the pond depth. After certain years, completely draining the pond water, excavating a layer of pond bed soil and liming the pond etc. becomes imperative. The excavated pond bed soil is nutritionally richer owing to years of deposition of nutrients through animal/bird excreta. Hence, the excavated pond bed soil can be used as a soil conditioner and nutrient source for various crops grown in the farm.

Hedgerows for Soil Fertility in IFS

Hedgerows represent different species of leguminous nitrogen fixing shrubs incorporated in agro-forestry systems for soil fertility build up, soil and water conservation and augmentation of biomass for mulch, fodder and fuel. These potential nitrogen fixing shrubs can be cultivated as a hedgerow on terrace risers, contour bunds and contour trenches (Kadirvel, 2003). Several workers have reported that the hedgerow intercropping is a potentially useful system for humid and sub-humid tropics (Kang *et al.*, 1984; Lal, 1989). For effective use, hedge species are planted closely spaced in parallel rows (4–10 m apart) and crops are grown in the alleys between them. Hedges are to be regularly pruned before onset of rain. The pruned biomass, besides using as fodder or fuel are also used as mulch or incorporated as green manure for soil fertility improvement. The hedgerow also provide mechanical and biological barrier that helps to minimize soil erosion by reducing run off (Singh *et al.*, 2000). Kiepe (1995) noticed that around 98 per cent of the soil could be conserved with combination of hedgerow and mulch compared to control. Hedgerows alone reduced soil loss by 94 per cent and run off by 78 per cent and mulching alone conserved 83 and 42 per cent of soil loss and runoff, respectively. Studies carried out at Changki, Nagaland showed that soil loss was reduced by 22 per cent with the incorporation of hedgerow species like *Flemingia, Tephrosia, Cajanus, Crotolaria* and *Desmodium* in the *jhum* fields compared to that in traditional *jhum* site (38.14t/ha/year). Water run-off and nutrient loss was also considerably reduced with hedgerow planting (Bhatt and Verma, 2002). Screening of hedgerow species needs to be done based on their biomass production and nutrient supplying potential. Biomass produced from some promising hedgerow species in IFS in north eastern region are listed below (Table 4.12).

Table 4.12: Pruned Biomass Produced by Different Hedgerow Species in IFS

Hedgerow Species	*Biomass from Different Parts of the Shrub*(q/ha)*			*Total Biomass (q/ha)*
	Leaves	*Branches*	*Main axis*	
Crotolaria tetragona	62.16	68.54	76.82	207.52
Flemingia macrophylla	20.24	10.22	35.85	66.31
Indigofera tinctoria	55.80	41.62	64.25	161.67
Tephrosia candida	40.75	34.10	48.30	123.15
Desmodium rensonii	13.10	18.45	45.50	77.05
Cajanus cajan	25.16	45.62	58.90	129.68

* Biomass production recorded on fresh weight basis

The nutrient accumulating potential of different hedgerow species is of prime importance for their suitability in any IFS (Table 4.13).

Table 4.13: Nutrient Accumulation by Hedgerow Species through Leaf Biomass in IFS

Hedgerow Species	*Leaf Biomass Added (q/ha)*	*Nutrient Accumulation (kg/ha)*		
		N	*P*	*K*
Crotolaria tetragona	25	87.52	13.33	38.15
Flemingia macrophylla	7	21.12	3.20	8.55
Indigofera tinctoria	16	65.50	14.20	26.48
Tephrosia candida	16	55.10	5.32	26.78
Desmodium rensonii	6	20.15	3.20	8.16
Cajanus cajan	8	26.35	5.10	12.50

Compared to annual cropping, hedgerow systems generally have favourable effects on soil organic matter and they considerably increase nitrogen inputs through high rates of recycling. To maintain crop yield at acceptable levels, however, hedgerow systems must include either fallow periods or limited input of fertilizers (Young, 1997). Prominent characteristics of hedgerow species governing their suitability in IFS include:

- ✰ High leaf biomass production potential.
- ✰ Vigorously growing dense root system.
- ✰ Greater nitrogen fixing ability.
- ✰ Sturdy, tolerance to moisture stress, disease, pest incidence, low soil fertility etc.
- ✰ Having high quality litter (higher in N and lower in lignin and polyphenols content).
- ✰ Greater nutrient content especially in leaf biomass.
- ✰ Absence of toxic substances in litter and root residue.

Epilogue

Thorough perusal of the traditional agricultural practices of north eastern region reveals that for centuries these areas are operating on 'system' basis, intentionally or otherwise, incorporating enterprises like agriculture, horticulture, forestry and livestock including fisheries. However, to achieve a quantum leap in overall production also keeping an eye on environmental quality in these rainfed areas, scientific interventions are required to be made through integrated farming system approach. Gradually intensifying the IFS approach by carefully injecting elements of diversification and value addition and also giving due priority to soil and water conservation shall lead us to a situation where a sound environment with sufficient food grains will be the hallmark of the north eastern region.

References

Anonymous (1999). *State of Forest Report*. Published by Forest Survey of India. Ministry of Environment and Forests, Govt. of India, pp. 113.

Anonymous (2001). *State of Forest Report*. Published by Forest Survey of India. Ministry of Environment and Forests, Govt. of India, pp. 130.

Anonymous (2002). *Wastelands Atlas of India*, published by Department of Land Resources, Ministry of Rural Development, Govt. of India, pp 81.

Bhatt, B.P. and Sachan, M.S. (2004). Firewood consumption pattern of different tribal communities of Meghalaya in Northeast India. *Energy Policy*, **32**:1- 6.

Bhatt, B.P. and Verma, N.D. (2002). Some Multipurpose Tree Species for Agro-forestry Systems, published by ICAR Research Complex for NEH Region, Umiam, Meghalaya, pp. 146.

Bhatt, B.P., Bujarbaruah, K.M and Sharma, Y.P. (2005). Integrated farming system: A sustainable alternative for the benefit of small scale farmers and the environment. In: B.P. Bhatt and K.M. Bujarbaruah (eds.), *Agro-forestry in North East India: Opportunities and Challenges*, pp. 537- 555, ICAR Research Complex for NEH Region, Umiam, Meghalaya.

Bhatt, B.P., Singh, R., Mishra, L.K, Tomar, J.M.S., Singh, Matvar, Chauhan, D.S., Dhyani, S.K., Singh, K.A., Dhiman, K.R. and Datta, M. (2001). Agro-forestry research and practices: An overview. In: Verma, N.D. and Bhatt, B.P. (eds.), *Steps towards Modernization of Agriculture in NEH Region*, pp. 365- 392, ICAR publication, Meghalaya, India.

Bujarbaruah, K.M. and Bhatt, B.P. (2005). Complementarity of livestock and agro-forestry for sustainable agriculture in north east. In: B.P. Bhatt and K.M. Bujarbaruah (eds.), *Agro-forestry in North East India: Opportunities and Challenges*, pp. 581- 591, ICAR Research Complex for NEH Region, Umiam, Meghalaya.

Dalzell, H.W., Biddlestone, A.J, Gray, K.R. and Thurairajan, K. (2007). Soil management: Compost production and use in tropical and subtropical environments. *FAO Soils Bulletin no.* 56. Food and Agricultural Organization of United Nations.

Dasgupta, M., Gangopadhyay, A.K., Bhattacharya, T. and Chaktaborti, M. (1986). *Forest Development in North East India*, Omsons Publication, New Delhi, pp. 274.

Dhiman, K.R. (2005). Integrated farming system for indigenous resource management and sustainability. In: B.P. Bhatt and K.M. Bujarbaruah (eds.), *Agro-forestry in North East India: Opportunities and Challenges*, pp. 525- 535, ICAR Research Complex for NEH Region, Umiam, Meghalaya.

Edwards, P., Kaewpaitoon, K., McCoy, E.W. and Chantachaeng, C. (1986). Pilot small-scale crop/ livestock/fish integrated farm, AIT Research Report, 184, Bangkok, Thailand, pp. 131.

Gupta, J.J. (2006). Livestock waste utilization for organic food production. In: K. M. Bujarbaruah., B. K. Sharma., Narendra Prakash and Rajesh Kumar(eds.), *Organic food production in North East India: Prospects and aspects*, pp. 27- 30, ICAR Research Complex for NEH Region, Umiam, Meghalaya.

Kadirvel, G., Bhatt, B.P. and Das, Anubrata (2003). Agro-forestry- a promising farming system for livestock production in NEH region In: Bhatt, B.P., Bujarbaruah, K.M., Sharma, Y.P. and Patiram (eds.), *Approaches for Increasing Agricultural Productivity in Hill and Mountain Ecosystem*, pp. 139-147, ICAR Research Complex for NEH Region, Umiam, Meghalaya.

Kang, B.T., Wilson, G.F. and Lamson, T.L. (1984). *Alley Cropping: A stable Alternative to Shifting Cultivation*. International Institute of Tropical Agriculture, Ibadon, Nigeria.

Kiepe, P. (1995). No run-off, no soil loss: Soil and water conservation in hedgerow barrier systems. Tropical Resource Management Papers 10, Agricultural University, Wageningen, Netherlands.

Kumaresan, A., Pathak, K.A., Bujarbaruah, K.M., Das, A. and Vinod, K. (2006). Integrated Livestock-Fishculture in Mizoram. Technical Bulletin. ICAR Research Complex for NEH Region, Umiam, Meghalaya.

Lal, R. (1989). Agro-forestry systems and soil surface management of a tropical alfisol. III. Changes in soil chemical properties. *Agro-forestry Systems*. **8**: 113- 132.

Maikhuri, R.K. (1992). Eco- energetic analysis of animal husbandry in traditional societies of India. *Energy*, **17** (10): 959- 967.

Preston, T.R. and Murgueitio, E. (1994). *Strategy for Sustainable Livestock Production in the Tropics*. IInd edition, CONDRIT Ltd. Cali, Colombia, pp. 89.

Ramakrishnan, P.S. (1992). Ecology of shifting cultivation and ecosystem restoration. In: Wali, M.K. (ed.), *Ecosystem Rehabilitation*, Vol. 2: *Ecosystem Analysis and Synthesis*, pp. 19- 35, SPB Academic Publishing Bv, The Netherlands.

Singh, K.A., Satapathy, K.K., Ramchandra, Dutta, K.K. and Singh, J.L. (2000). Bio- terracing in the North- Eastern Hills. *ICAR News Letter*, April- June, 2000. pp. 6- 7.

Vinod, K., Mahapatra, B.K. and Mandal, B.K. (2005). Agro-aquaculture- suitable option for integrated rural development. In: B.P. Bhatt and K.M. Bujarbaruah (eds.), *Agro-forestry in North East India: Opportunities and Challenges*, pp. 571- 580, ICAR Research Complex for NEH Region, Umiam, Meghalaya.

Young, A. (1997). Nutrient Cycling and Nutrient Use Efficiency. In: *Agro-forestry for Soil Management*, CAB International and ICRAF, Nairobi, Kenya, pp. 111- 136.

Hill Agriculture: Prospects, Constraints and Mitigations (2013) *Pages* **50–67**
Editor: **Dr. Shahid Ahamad**
Published by: **DAYA PUBLISHING HOUSE, NEW DELHI**

Chapter 5
Constraints and Interventions in Hill Agriculture of India

Manpreet Kour, J.N. Srivastava and A.S. Charak
Regional Horticulture Research–Sub Station, Bhaderwah, SKUAST-J, J&K

Introduction

Hill agriculture is defined as that livelihood system which includes all activities on which farmers are dependent to make a living, such as, cropping, horticulture, livestock, rangelands, pastures and forests etc. The hill/mountain perspective frame work developed by experts at the International Centre for Integrated Mountain Development (Jodha, 1990), identifies these six mountain/hill specificities *viz.*; inaccessibility, fragility, marginality, diversity, niche, and adaptation mechanisms and it is argued that each of these specificities have three dimensions, physical, biological and socio-economic, which determine the suitability or unsuitability of any intervention. For sustaining livelihoods of hill farmers on agriculture it is necessary that hill agriculture is understood in right perspective (Partap, 1999).

Himalayan region with diverse agro-climatic condition from subtropical plain to semi-temperate to alpine and sub-alpine conditions, inhabit 39 million people and large percentage of hill farming communities (mountains included) (Table 5.1) (NCOF Vol.II. Report, 2005).

Agriculture is the main occupation of the hill farming communities and they practice both settled as well as age-old traditional practice the shifting cultivation (Jhum Cultivation) (Anonymous, 2010). Agriculture is the primary sector of the economy, contributing about 45 per cent to the total regional income of the inhabitants. About 76 per cent of the gross cropped area of entire Himalayan region is under staple food grain crops (Table 5.2) comprising wheat, rice, maize, pulses, millets, barley and buckwheat, fruits comprising apples, citrus, walnuts, plums, peaches, bananas, mangoes and pineapples; vegetables such as tomatoes, radish, potatoes, cabbage, cauliflower, other cash crops like ginger, chillies, cardamom and saffron; and flowers such as orchids, gladioli, marigolds and

Table 5.1: The Himalayan Region; Demographic and Agriculture Indicators

State	Area (sq. km)	Population (No.)	Rural (Numbers)	Population Density (per sq.km)	Area Under Forests ('000 ha)	Net Cropped Area ('000 ha)	Net Cropped Area as per cent of Total Area	Cropping Intensity (per cent)	Av Size of Holdings per Family
North Western Hill	-	-	-	-	-	-	-	-	—
Himachal Pradesh	55673	6077900	5482319	109	1094	551	12.16	174	1.16
Jammu and Kashmir	222236	10143700	7627062	46	2747	733	16.27	147	0.76
Uttarakhand	53483	8489349	6310275	159	3342	788	14.91	164	1.01
Total	331392	24,710,949	19419656	75	7183	2072	14.47	160	0.97
North Eastern Hill	-	-	-	-	-	-	-	-	-
Arunachal Pradesh	83743	1097968	870087	13	5154	166	3.02	159	3.31
Assam	78438	26655528	23216288	339	1930	2701	34.41	152	1.17
Manipur	22327	2388634	1818224	107	602	140	6.33	142	1.22
Meghalaya	22429	2318822	1864711	103	938	240	10.71	111	1.33
Mizoram	21081	888573	447567	42	1599	91	4.31	100	1.29
Nagaland	16579	1988636	1635815	119	875	261	16.73	113	4.82
Sikkim	7096	540851	480981	76	257	95	13.38	127	1.65
Tripura	10486	3199203	2653453	305	606	277	26.41	152	0.6
Total	262179	39,078,215	32987126	149	11961	3971	17.09	145	1.92
India	3,287,240	1028,830,774	742706609	312.98	69024	141231	46.15	134	1.41

Source: NCOF Vol.II. Report, 2005.

chrysanthemums (Pratap, 1998). Apiculture, animal/sheep husbandry, sericulture, pisciculture, weaving are the other economic activities practiced by these farming communities.

Table 5.2: Production Trends in Food Grains, Vegetables and Fruits in the Himalayan States

States	*Food Grains*		*Fruits*		*Vegetables*	
	1990-91	*2003-04*	*1990-91*	*2003-04*	*1990-91*	*2003-04*
N. Western Hill region	–	–	–	–	–	–
Himachal Pradesh	80.27	75.81	14.42	20.29	3.55	3.15
Jammu and Kashmir	83.37	79.00	11.17	12.75	3.88	4.56
Uttarakhand	79.82	77.39	11.93	14.84	4.52	7.05
Regional Average	81.07	77.40	12.49	15.87	4.01	5.06
N. Eastern Hill region	–	–	–	–	–	–
Arunachal Pradesh	76.19	68.71	8.18	15.82	6.92	7.91
Assam	71.30	65.61	1.90	2.73	5.83	5.84
Manipur	81.00	55.74	9.40	8.10	5.90	34.75
Meghalaya	55.46	49.70	10.08	9.06	10.79	13.42
Mizoram	78.35	75.27	11.77	17.27	7.59	6.18
Nagaland	83.14	66.97	2.48	7.96	3.90	8.38
Sikkim	64.54	60.32	5.07	9.76	5.00	11.27
Tripura	65.54	62.24	10.11	5.44	6.82	7.31
Regional Average	71.08	64.41	3.76	4.78	6.12	8.14
India	**68.82**	**65.94**	**1.55**	**2.14**	**3.01**	**3.29**

Source: NCOF Vol.II. Report, 2005.

Himalayan areas also have specific advantages particularly due to the wide diversity and the presence of niches particularly suited to certain crops *e.g.* the apples in Himachal, saffron in Sopore valley of Kashmir, pashmina goats and yak in the highlands of Ladakh or Mithun in Arunachal Pradesh. It offers hope to develop these comparative advantages, promote investment in such niche areas as part of the efforts to improve farm economy in sustainable ways (Partap, 1995; Partap, 1998).

Besides all the diversities hill communities sustain largely on subsistence farming due to the following constraints.

Constraints of Hill Agriculture

Inherent Constraints

Some inherent constraints like inaccessibility, remoteness, fragility, marginality, poor soil conditions and short growing seasons are common in hill agriculture (Partap and Partap, 2010). Inherent constraints can be judged by the unsustainability indicators of hill agriculture which showed increased soil erosion rates from sloping farmlands, abandonment of agricultural land due to decline in soil fertility, appearance of stones and rocks on the cultivated lands, decline in forest and grazing areas and emphasis on mono-cropping that widened the cycle of inadequate food production causing food insecurity and increasing unemployment and frustration (Table 5.3) (Shrestha, 1992). Further

Table 5.3: Understanding Unsustainability of Hill Agriculture through Indicators

Indicators Reflecting Problems Relating to Resource Base/ Production Flow and Resource Management	*Range of Change*
Soil Erosion Rates on Sloping Lands	+20 to 30 per cent
Abandonment of Agricultural Land due to decline in soil fertility	+3 to 11 per cent
Appearance of Stones/Rocks on Cultivated Land	+130 to 100 per cent
Size of Livestock Holding per Family (LSU)	-20 to 55 per cent
Area of Farmland per Household	-30 to 10 per cent
Forest A 8 Forest Area	-15 to 85 per cent
Pasture/Grazing Area	-25 to 90 per cent
Good Vegetative Cover on Common Property Land	-25 to 30 per cent
Fragmentation of Household Farmland (in number of parcels)	+20 to 30 per cent
Size of Land Parcels of Families	-20 to 30 per cent
Distance between Farmland Parcel and Home	+25 to 60 per cent
Food grain Production and Self- Sufficiency	-30 to 60 per cent
Permanent Out migration of Families	None to 5 per cent
Seasonal Migration	High to High
Conversion of Irrigated Land into dry land farming due to water scarcity	+7 to 15 per cent
Average Crop Yields on Sloping Lands	
a. Maize and Wheat	-9 to 15 per cent
b. Millets	-10 to 72 per cent
New Land Under Cultivation	+5 to 15 per cent
Human Population	+60 to 65 per cent
Application of Compost (organic manure)	-25 to 35 per cent
Labour Demand for Falling Productivity	+35 to 40 per cent
Forestry Farming Linkages	Weak to Weak
Food grain Purchases from Shops	+30 to 50 per cent
External Inputs' needs for Crop Production	High to Medium
Fuel wood Fodder Scarcity in terms of time spent in collection	+45 to 200 per cent
Fodder Supply from	
a. Common Land	-60 to 85 per cent
b. Private Land	+130 to 150 per cent
Emphasis on Monocropping	High to High
Steep Slope Cultivation (above 30 per cent)	+10 to 15 per cent
Weed and Crop Herbaceous Products' used as Fuel wood	+200 to 230 per cent
Conversion of Marginal Land into Cultivation	+15 to 40 per cent
Fallow Periods	From 6 to 3 months

Note: A positive sign (+) means increase and negative sign (-) means decline/decreaseTime Frame of changes : 1954-1991 = 37 Years.

Source: Shrestha, 1992.

due to rapid and unplanned development without recognition of these unsustainability issues is destructing the natural resources of the hill areas in the form of:

Soil Loss

It is mostly due to wrong farming practices adapted in the hilly areas without considering the land capability classifications. In many cases, gullies are formed. These gullies are slatternly found in the hill areas. In some cases, the impact was so pronounced that the sub-surface bed-rock was exposed (Islam, 1984).

Nutrient Loss

As continuous process of nutrient loss is going on all over the hill tract areas under high rate of precipitation. The process is accelerated by open cultivation system on steep to very steep land (Farid and Hussain, 1988).

Water Scarcity

About 90 per cent area of the hilly areas are rainfed and further soils of hills are generally undulating with steep slopes, shallow to medium deep, coarse to medium –textured and have relatively low water retentive capacity (Miah, 1993). Spatial and temporal variability in rainfall amount, frequency, intensity, distribution, and also in the onset and recede timings per se is very high thus further increasing the difficulties of hill communities (Anonymous, 2010).

Depletion in Soil Fertility

It is also observed in the hills as soils of these regions are in general, quite shallow, gravelly and not very fertile due to mixing of the soil with fragments of parent rocks occurring within a few centimeters, except in valleys or depressions where they go up to about two metres. Further, large part of relatively fertile soil is under forests and hence cannot be used for agricultural purposes so agriculture takes place mostly in the valleys or scattered pieces of land on the hills that have the requisite fertility.

Vagaries of Nature

Unseasonal or heavy rains, hailstorms sometimes completely destroy the vegetables and significant part of the output from fruit orchards and vegetable farms regularly, before they are ready for harvesting.

Socio-economic Issues

Small and Fragmented Land Holding

Small and fragmented land holding Agriculture is the main occupation of the farming communities residing in the Himalayan regions. For large number of small and marginal farmers of the Himalayan region, shrinking cropland holdings is a key concern for managing food and livelihoods (Partap, 1998). The great majority of the farming households in the Himalayan states are less than 0.5 ha or small landholders with farms of 0.5 to 1.0 ha. While average land holding in Jammu& Kashmir is about 0.76 ha (Table 5.1). Fragmented land holdings of less than 1 ha which is further divided into a range of flat and sloping land types is too little to sustain their livelihoods. it has been estimated that 37 per cent of the cropland is sloping land of various degrees and the Himalayan farmers are even cropping sloping lands beyond 25 and 30 degrees (Partap, 1999). Also, down in the valleys, competition among new human settlements, urbanization, industrialization and government infrastructure development activities, are converting the valley crop land into non farm use.

Low Population Density

Low population density with villages and agricultural areas scattered and situated at great distances from roads and markets prevents the development of market-based institutions. Thus the scattered and small size settlements in hill areas is producing an alarming challenge to development programmes and market formation and this problem is further compounded due to lack of roads and means of transportation (Nakagawa, 1998; Sugaya, 1998).

Declining of Labour

Declining of labour due to lack of employment opportunities in the hills, huge numbers of young people are migrating to the cities and thus there is lack of labour during the peak agricultural demanding season (Sugaya, 1998; Gim, 1998), further, customs of sharing family labour in the labour intensive agricultural operations like transplanting and harvesting is also found to be declining in the rural areas. Hence, in hills the agricultural operations are found to be completely dependent upon the old and female members of the family.

Capital Constraints

Capital Constraints due to lack of capital, technology and market in hills, agriculture is dominated by subsistence level of farming communities (Sugaya, 1998). Even though, there is extension of credit institutions, the interest rates are found to be more in rural sector than in urban sector (*viz.* housing loan, car loans etc.).

Choice of the Farmers

Choice of the farmers Local strains of many agricultural commodities having better quality and taste which sells at premium in the market, many farmers are not ready to adopt improved varieties thus local strains need to be improved.

New Generation of Farmers

New generation of farmers Most of the educated unemployed youth across the Himalayan states belong to the farming families, no doubt, they have acquired traditional knowledge of farming from their families, but they no longer find it remunerative to get engaged in it.(Sugaya, 1998)

Technological Issues

In Adequate Availability of Inputs and New Techniques

In adequate availability of inputs and new techniques from the perspective of the farmers, the major technological constraints are difficulty in availability of new varieties which have wider adaptability and resistant to drought situations, as about 90 per cent area of the district is rainfed. There is also an inadequate availability of the quality seeds, propagated materials of improved varieties of food grains, vegetables and fruit crops and scientific know how of production of these crops (Anonymous, 2010).

Lack of Agricultural Mechanization for Small Holders

Due to the lack of physical facilities (*viz.* road networks and electricity), narrow terrace cultivation and fragmented land holdings in hilly areas, hill agriculture is mainly depended upon human and animal power. The traditional wooden tools and implements have continued to remain in use in the hills and mountains (Pariyar, 2001) and this problem is further aggravated during peak periods when labour services get scared.

Institutional and Policy Constraints

With few exceptions, constraints to improving horticulture crops in the Himalayas include poor orchard management practices, non-availability of the farm inputs (seeds, fertilizers and pesticides) at right time and less remunerative prices due to little access to extension services, proper marketing, post harvest processing and transport facilities (Anonymous,2010). One of the biggest constraints faced by the small farmers is the lack of an effective marketing infrastructure, the predominant practice is to sell or to rent out the produce to middlemen much before the fruits are ready for harvest, as a result, the farmers do not get remunerative prices from the middleman. Because of lack of regular markets and reliable marketing, hills farmers in many areas, are finding it too risky to diversify into more lucrative high value crops.

Veterinary Problems

Shortage of proper feed, fodder and concentrate to milch animals is rampant in the hills. As an estimate, there may be up to 70 per cent shortage of fodder faced by the Himalayan farmers due to infestation of non palatable invasive species, such as lantana, eupatorium and congress grass in the fodder and grazing areas. In many hill regions the problem is of overgrazing due to uncontrolled animal population and poor management.

Improper market prices for egg and broilers. Poor density of dairy co-operative societies. Poor veterinary/artificial insemination facilities. Non-availability of day old chicks as no commercial hatchery exists in these areas. High altitude (Low possible oxygen pressure) cause severe stress during transportation of chicks. Lack of awareness bio-security/insufficient veterinary aids is some of the other Institutional constraints faced by the hill communities.

Sustaining Livelihoods of Hill Farmers through Proven Interventions

Due to various physical, geographical and environmental problems, the scope for' agricultural policies based on modern input-intensive agriculture is severely constrained in the hilly areas resulting in the majority of the hill communities to either survive on subsistence agriculture or migrating to other parts of the country for employment. In other words, policies that might give successful results for any other state in India situated in the plains may not prove to be fruitful in this hilly state. Since the livelihood of the weaker sections in the hill areas is completely dependent on these natural resources, their destruction will make the process of inclusive growth unsustainable in the long run. For sustaining livelihoods of hill farmers on agriculture it is necessary that hill agriculture is understood in right perception and it should include all the possible interventions on those activities in which farmers are dependent to make a living. The diagrammatic presentation of the hill specificities, along with constraints and interventions are depicted in Figure 5.1.

Technological Interventions

Conservation of Natural Resources

Soil and Water Conservation

Soil conservation measures should always be a major consideration for sustainable hill farming as erosion causes rapid fertility depletion, damages crops by sedimentation and raising stream beds/ river channels by siltation which causes flash floods and limits discharge capacity, irrigation and navigation. The common practices in the hill Tracts is to plant crops up and down the slopes which facilitate soil erosion with the onset of heavy rains. However, plantation should be made along the contours. Experiments show that contour farming alone can reduce soil erosion by as much as

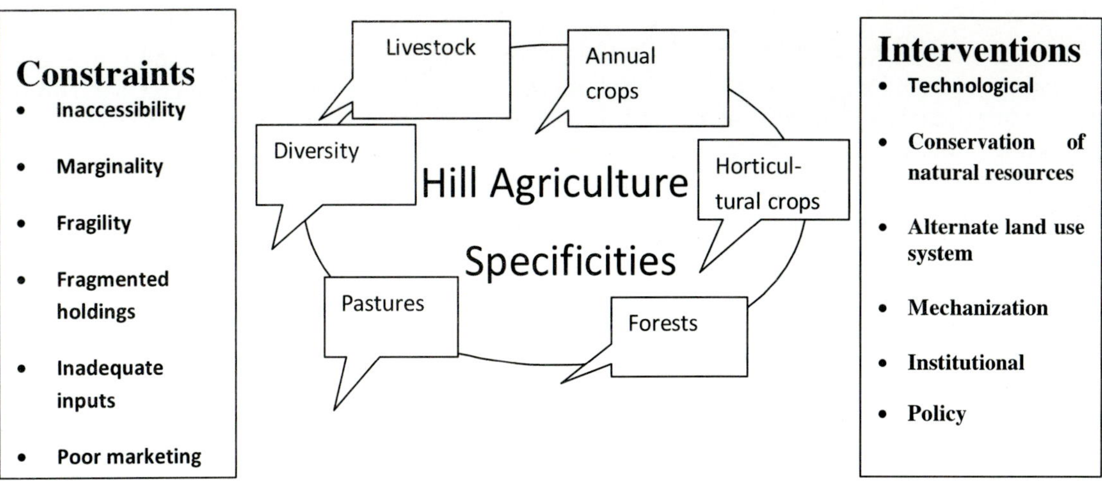

Figure 5.1: Hill Specificities, along with Constraints and Interventions

Figure 5.2: Contour Cultivation

50 per cent on moderate slopes. However, for slopes steeper than 10 per cent, other measures should be combined with contour farming to enhance its effectiveness (Gupta and Gupta, 2010).

Simple conservation measures like mulching strip cropping, cover crops, adopting suitable cropping system, tillage practices, suitable crop rotation, a forestation, controlled grazing, wind breaks and adopting mechanical measure can reduce soil removal considerably. Assessment of degraded and marginal land and their specific use is important in hills (Ahmed, 1984). A rainfed sloping farmland that is marginal for a crop because of continuous irrigation and moisture requirement could be highly productive for perennial fruit crops, may support a productive and sustainable livestock production system and herbal medicines farming (Gim, 1998).

Rainwater Management

The hill farmers generally cultivate their crops under rainfed condition. The availability of surface water resources are very limited due to poor water retention capacity of the sloppy terrains of the region. Thus the the only major source of good quality water in mountaneous areas is rainfall. Therefore, its proper management by rainwater harvesting, inter –row and inter-plot harvesting, water harvesting at surface through ponds, tanks etc. and moisture conservation by mulching, crop covers, use of anti-transpirants, weed control etc. need special consideration for realizing suitable crop production in the region (http://www.rain water harvesting.org/). Further the average yield is low in most cases because of their inadequate knowledge about appropriate sowing time under rainfed cropping system releasing of advanced package of practices can help farmers a lot to know the sowing dates.

Watershed Management Technology

As hills are highly susceptible to water erosion and land degradation, the nature of the soil and the inclined slopes make it difficult for ground water to collect in these areas. Thus watershed management approach becomes the most crucial factor in agricultural performance (Anonymous, 2010). It is nothing but land management and stream management in a better way to improve hydrological regime. It affects the people in every sphere of life. The sustained productivity of food, fuel, forage, fruits and water by the management of vital resources of water, soil, vegetation and phenomenon like floods and drought are determined by nature of watershed functioning. The main aim of the watershed approach is to ensure a holistic view of water and land resources and to prevent further degradation of these ecologically fragile areas (Gupta and Mahajan, 2010).

Alternate Land Use System

The Himalayan region is bestowed with varied landscape features that provide variety of habitats to diverse life forms, thus hills are usually less suited for uni-dimensional land use" but more suited to multiple strategies eensuring a balanced relationship between people and land resources.

Integrated Farming System

Farm economic efficiency of a specific area is an important factor of productivity and growth, where resources are scare and opportunities for developing and adopting better technologies have lately declining. No doubt, conventional farming is risky and farmers are reluctant to invert heavily in crop production. Modest increments in productivity are no longer sufficient to justify the investment of scarce resources as farming sector is small scale and resource poor, but integrated farming system assumes greater importance for sound management of farm resources to enhance the farm productivity, reduce the environmental degradation, improve the quality of life of resource poor farmers and to maintained sustainability (Singh, *et al.*, 2007). Farming system refers to the farm wherein two or more enterprises are integrated with the farm resources for achieving the fuller utilization, realize maximum profit and stabilize returns. It is a resource management strategy to achieve economic and sustained agricultural production to meet diverse requirements of the farm household while preserving the resource base and maintaining high environmental quality. The objectives of farming system in general are converging as to the development of sustainable location specific farm technology to raise and sustain the total farm productivity in terms of food, feed, fodder and fuel (4F) to meet the felt needs of the farmers within the sphere of their agro-socio-political favourites and constraints. Farming systems aim for increased productivity, profitability, sustainability, balanced food, pollution free environment, recycling of unutilized resources, generation of income round the year, adoption of new technology, solving energy, fuel and fodder crisis, avoiding deforestation, increased employment generation, literacy rates, input- output efficiency, enhanced opportunity for agriculture oriented industries and standard of living of the farmers. Generally farming systems practiced by farmers vary according geographical locations, climatologically conditions, infrastructure facilities available, socially acceptable, socio-economic conditions of the farmers (Gill *et al.*, 2007).

Farming Systems in Hilly and Mountainous Areas

Integrated and holistic development of hills and mountainous areas, need to be promoted by resource conservation techniques on watershed basis for improving productivity, profitability and thereby removing hunger and poverty. Due to the fragmentation of holdings, the per capita availability of land is declining day by day, no single farm enterprise is able to meet the growing demands of food and other necessities of the small and marginal farmers (Singh *et al.*, 2006). Integrated farming systems sustained crop productivity by reducing the risk of soil degradation, preserving soil's productive potential and decreasing the level of inputs required. Judicious mix of cropping systems with associated enterprises like fruits, vegetables, flowers, dairy, poultry, duckery, piggery, goatary, fishery, sericulture

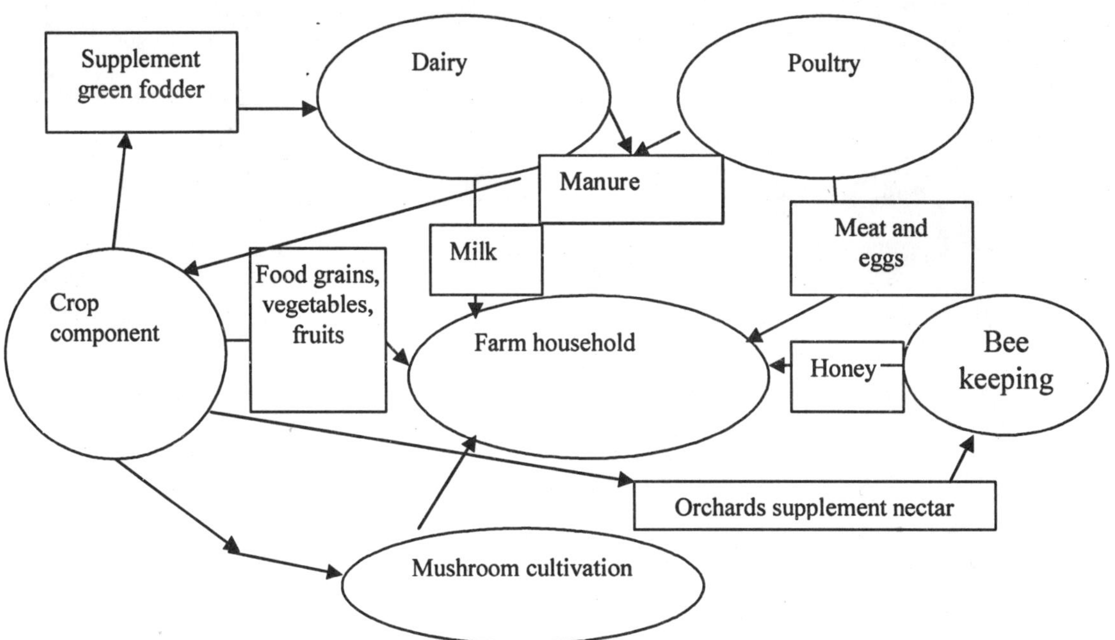

Figure 5.3: Resource Flow Model of Integrated Farming System-Hilly Conditions (0.50ha)

etc. suited to the given agro-climatic conditions and socio-economic status of the farmers shall be able to generate additional employment and income for the small and marginal farmers (Behera *et al.*, 2004). Farming Systems" represent the integration of farm enterprises such as cropping systems, horticulture, animal husbandry, fishery, agro-forestry, apiary etc. for optimal utilization of farm resources bringing prosperity to the farmers.

Crop Diversification on Sloping Land

The movement of soil, water and nutrients is the major problem on sloping land. Cultivation of row crop in sloppy lands permits soil erosion. In this field, the crops particularly cereals, fodder crop etc. should be broadcasted and the plants remain haphazardly in field. As a result, the movement of water gets obstacle and more water is absorbed in the soil, thus reducing soil erosion. Mixed and intercropping (Cowpea-with cotton, maize with rajmash etc.) practice checks the soil erosion and avoids the risks of the crop failure. Cropping systems should be designed in such a way that the soil is almost permanently covered with plant canopy. Leguminous plants such as *Sesbania cannabina, Crotalaria strata, Cassia tora, Vigna indica, Tephrosia candida, Leucaena glauca* and *Medicago hispida*, are used on sloping land either for soil loss reduction or green manure. In arable crops, careful timing of sowing and planting can help to avoid uncovered soil being washed away during the rainy season. After the main crop is harvested, a green manure crop may be sown. On slopes, crops should be grown in contour lines across the slopes (along the contour lines) rather than vertically. Diversified cropping strategies such as mixed/intercropping, strip cropping, alley cropping and agri-horticultural systems are developed to retain maximum amount of rainfall *in situ* and ensure higher production and protection against erosion (Gupta and Gupta, 2010).

Figure 5.4: Intercropping in Maize Crop with Rajmash

Horticulture

The hill areas have a comparative advantage in the production of horticultural products including fruits, flowers and vegetables due to varied climate and thus making it an ideal location for growing temperate, subtropical and tropical fruits and off-season vegetables. Also, the production of fruits like apples requires less moisture than vegetables and hence can be more successfully grown in areas that lack good irrigation facilities and may be more suitable in the relatively higher mountain ranges whereas, vegetables can be grown more successful in the foothills and valleys that provide a Sub-tropical climate, and also have irrigational facilities. The problems of inadequate information about best crops and best practices that are faced by farmers are important for the development of horticulture. The planning for horticultural development has to include supply of appropriate inputs for these best practices to the farmers, a scientific market analysis and the scientific methods of production and processing of horticulture products. (Azad, *et al.*, 1988).

Figure 5.5: Apple Farming

Agro-forestry

Agro-forestry is an intentional integration of agricultural and forestry based land –use-systems to provide tree and other crop products, and at the same time protect, conserve, diversify, and sustain vital economic, environmental, human and natural resources. In agro-forestry systems, there is an ecological and economical interaction among different components. As the hilly areas are encountered with problem of ever increasing demands of fuel, fodder, timber and non-tree products thus agro-forestry is important for meeting fodder, fuel wood and small timber of farmers, conserving soil and water, maintenance of soil fertility, controlling salinity and water logging, positive environment impact and alternate land use for marginal and degraded lands. Selection of proper land use systems conserve biophysical resources of non-arable land besides providing day-to-day needs of farmer and livestock within the farming system. Some medicinal and aromatic plants *viz. Ocium sanctum, Murraya koenogii, Azadiracta indica, Moringa oleifera*, Mints, *Catharanthus roseus* and *Coriandrum sativum* can be commonly grown in Indian households. *Jatropha curcas* –an important source of bio-diesel, *Pinus roxburghii* is an important remunerative aromatic tree species, *Acacia catechu* having high medicinal value in addition to its value for fodder and timber can be grown in hills.the tree species like *Azadiracta indica, Terminalia* spp. *Aegler marmelos* etc. can be intercropped with annual in early years until the tree canopy covered the ground (Gupta, 2010).

Figure 5.6: Olive in Hills

Animal Husbandry

In spite of the opportunities offered by favourable climate, the economic potential offered for dairying, sheep and other animal husbandry in the hill areas has not been fully tapped. In many hill regions the problem is of overgrazing due to uncontrolled animal population and poor management. The scientific management of these lands can increase the yield of fodder and support effectively a large animal population. The animal husbandry programme will need a strong preventive and curative animal health programme, together with processing and marketing of the produce. The scientific management of these lands can increase the yield of fodder and support effectively a large animal population. The animal husbandry programme will need a strong preventive and curative animal health programme, together with processing and marketing of the produce (Anonymous, 2010).

Mechanization

As labour is getting scarce during peak periods, farm holding size and socio-economic background of hills is diverse and is mainly dominated by small farmers and poor farmers, the mechanization need to be focused on appropriate mechanization technologies. Mechanization program should not be limited to promote increased use of tractor but also include improved manual tools; animal drawn implements and appropriate mechanical machinery. As the majority of the farm operation is performed by women, the tools and machinery are to be promoted by catering the needs of women farmers. The research and development on farmer friendly appropriate tools and machinery is also needed to be reoriented (Manandhar *et al.*, 2006).

Institutional Interventions

Institutional interventions are needed to protect watersheds, forbid annual cropping on lands beyond 15 per cent slope and covering the lands having slopes 18-30 per cent with forests under strict regulations, encouraging farmers to convert from inorganic farming to organic farming, Partap and Vaidya (2009) found that there are several factors which are encouraging farmers to convert from inorganic farming to organic farming, namely; improvement in soil quality and fertility, reduced dependence on external sources, productivity and high input costs, health hazards of pesticides and premium prices. The cost benefit analysis indicated favourable economics of organic farming to small farmers of India. It is evident from the land size distribution and economy of the hill agriculture that each household cannot purchase full set of farm machinery for its own use. These custom-hiring enterprises could play an important role on introduction of improved farm machinery and spread the benefit to the farmers in the community. Hence, government should recognize this custom hiring enterprise and provide them support through training and credit (Manandhar *et al.*, 2006).

Infrastructure and Marketing

Hilly and mountaneous areas are characterized by severe infrastructure bottlenecks, especially with regard to roads, transport network, electricity, irrigation, marketing infrastructure for agricultural produce, health, education and financial institutions (Bajracharya, 1992),. There is an urgent need in these areas to address problems with regard to connectivity with *road networks*, particularly of remote and inaccessible areas located in the mid and outer Himalayan ranges. This is particularly important for the marketing of primary produce, including horticultural crops, which requires a well-developed network of roads connecting villages and agricultural areas to the urban areas and mandis. Hills are gifted with a perennial source of water supply throughout the year. This means that there is tremendous scope for the development of small-scale hydro systems for *electrification* of the state, particularly in the far-flung areas of the Central Himalayan region marked by the absence of alternative sources of power, as ninety per cent of the net sown area in hills is rainfed (Anonymous,2010). The hill areas have a comparative advantage in the production of horticultural products including fruits and vegetables. But due to lack of marketing facility, a large part of the total horticultural crops produced by hill communities is wasted each year thus there is a crucial need to develop an efficient marketing infrastructure in order to make it remunerative for the small mountain farmers to grow cash crops including fruits and vegetables (Azad, 1988; Bajracharya, 1992: Jodha, *et al.*, 1992). *Efficient marketing system* would consist of all post-harvest activities including the collection of farm products from the field, processing, packaging of the product, storing and warehousing of the product so as to address the problem of farmers having inadequate storage capacity by setting up of food processing units and cold chains and thus strengthening value addition. Also, there is a need to set up more bank branches at the district and local levels (Anonymous, 2010). Above all there is the problem of quality of education in hilly terrains. There is the need to increase public expenditure on *education* in order to improve the quality by promoting vocational and job-related education such as computer education. Access to health care in the rural parts of mountain districts continues to be poor. Moreover, there are issues related to *women's health* and nutrition, particularly in the remote hilly areas where access to basic health facilities is denied that need to be addressed. There is need to develop institutions that can offer cost effective solutions to problems of access and availability of health facilities for the rural mountainous regions (Anonymous,2010). Hill women are the most important food producers and act as the invisible work force, thus less acknowledged backbone of the family economy. Even though, women fulfill a great number of essential tasks but they have limited access to and control over income, credit, land, education, training and information. It is only recently that *participation of women* in development programs in hilly areas is being considered necessary. The extension approaches and tools may still be gender biased and therefore much needs to be done to encourage co-operation and partnership of women in hill development (Bajracharya, 1992; Rijk, 1989).

Tourism

The main attraction for tourists in the hills and mountainous terrains is the experience of watching the snow capped peaks, singing rivulets, clear blue sky, unparalleled majestic mountains, vast expanses of alpine vegetations, diversified cultures, religious, adventurous wild tourism and much more, so, the tourism infrastructure must make sure that this experience can be provided to the tourists without compromising on comfort, and in new and innovative ways the development of tourism should be a planned approach maintaining the volume and quality of tourism, while trying to maximize the returns to the local economy – in terms of income and employment to the local people. In order to promote high-value tourism in the state, the sector has to provide a high quality tourism experience by developing the infrastructure of hill stations, development of nature and adventure tourism with activities like skiing, paragliding, rafting, kayaking, etc., deploying high quality bus services encouraging car rental facilities, involving aerial tours of the hilly terrains using helicopters and availability of quality hotels. Overall, the tour operators and tourists' agencies must be encouraged to corporatize and become more organized and professional (Driml, and Common, 1995; Hartley and Hooper 1992; Jenkins, and Henry, 1982).

Policy Interventions

To develop and flourish the horticultural sector in the mountaneous states, all the problems listed above have to be addressed by means of appropriate policies. It must also be clearly understood that the overall policy package must try to solve all the problems simultaneously, as any one set of problems, if not addressed adequately, can significantly reduce the development of the sector. The *incentives* that the state can give to the small farmers to shift to horticultural production can also encourage the development of this sector. Thus the state needs to provide these farmers with subsidized inputs like seeds, fertilizers, insecticides, etc. Also, incentives should be there to take care of the uncertainties due to market failure and natural calamities. The state should provide *minimum support prices* in order to make sure that the poor farmers get a remunerative price. Finally, the provision has to be made to minimize farmer's losses due to natural calamites by means of *crop insurance*. For calamities that are relatively moderate in impact, small and marginal farmers must be supplied with poly-houses, poly-tunnels, hail nets, etc. However, for calamities that are severe, the only protection for farmers can be through crop insurance. The state should subsidize the crop insurance for small and marginal farmers The policy should be to provide the farmers to sell their products with alternative options, so that the portion of the profit going to the middlemen is minimized and the farmer gets a better price for his produce.Also farmers producing organic produce should be provided with *organic certification* so that produce can be marketed at a higher price. The small and marginal farmers need both scientific and financial help for their fight against natural calamities (Gim, 1998). Favorable policy is the critical for the promotion of agricultural mechanization in the mountain ranges and hills of the country addressing the needs of different category of farmers. All the different policy *viz.* co-operative farming and land consolidation for agricultural activity, contract hiring of agricultural machinery, support for agricultural machinery manufacturer, standardization and safety of agricultural machinery, promotion for energy efficient machinery?, are needed to be streamlined for promotion of agricultural mechanization. There is some issues in *tourism* sector that need policy intervention for example discounts on standard prices' such as reduced airfares for particular groups, *viz.*, students, senior citizens and so on. The tourism policy must also break the seasonal pattern by developing winter tourism. It is equally necessary to make sure that the potential tourist has adequate information about the facilities available in the state. This will involve innovative campaigns through the media and the Internet that focuses on the factors that attract various types of tourism (Jenkins, and Henry, 1982).

Conclusion

The sustainability prospects for hill communities remain miserable due to constraints found in the mountain agriculture besides all the diversities. The development based on hill view point should be formulated based on farmer response, giving due consideration to the nature of marginality, fragility, diversity and other constraints of each area. For developing right approaches to hill areas development all possible interventions through technology, policy, institutional etc should be framed. About 90 per cent area of the hilly areas are rainfed, so watershed management technology should be introduced to secure the land and water conservation. New approaches introduced should meet the basic needs of hill people *i.e.*, water, food, fodder, feed, fuel and employment. Major constraint in hills are marginal farmers with sloping and fragmented lands and to capture the advantages of available natural resources, integrated farming system with diversified agriculture provides sustainable productivity and profitability through resource recycling in all the micro-farming. It will not only improve the marginalization of hill communities but also help in achieving social equity by building on the comparative advantages of key land resources. Due to isolation of hill communities, marginal farmers having limited access to agricultural technologies and inputs and thus breakdown of isolation by building infrastructure and opening up of hills to the wider market economy, tourism have great impacts on the livelihoods of small hill farmers.

References

Ahmed *et al.* (1984). Soil Conservation needs in the hilly areas of Bangladesh. Task-Force report, resource loss and economic loss in the hill tract areas of Bangladesh BARI, report, Joydebpur, Bangladesh.

Anonymous (2010). Report of the evaluation study on hill area development programme in Assam and West Bengal organized by Planning Commission, Government of India.

Azad, K. C. R Swarup and B. K. Sikka (1988). Horticultural Development in hill Areas, Mittal Publications. Delhi.

Bajracharya, B. B. (1992). Infrastructural Development Imperatives for Sustainable Mountain Agriculture, In N.S. Jodha, M. Banskota and T. Partap (eds), Sustainable Mountain Agriculture, Vol. 1, Oxford and IBH Publishing Co., pp. 235-252.

Behera, U. K., Jha, K. P. and Mahapatra, I. C. (2004). Integrated management of available resources of the small and marginal farmers for generation of income and employment in eastern India. *Crop Research* **27**(1): 83-89

Driml, Sally and Mick Common (1995). Economic and Financial Benefits of Tourism in Major Protected Areas". *Australian Journal of Environmental Management*, 2:19-29.

Farid, A. T. M. and Hossain S. M. S. (1988). Diagnosis of farming practices and their impact on soil resource loss and economic loss in the hill tract areas of Bangladesh BARI, report, Joydebpur, Bangladesh.

Gill, M. S., Toor, M. S and Ramasundaram, P. (2007). Integrated Farming System –a success story. *Journal of Farming System Research and Development* 13(1), 10-16.

Gim, U. H. (1998). Republic of Korea- A Country Report. In: Proceeding of the Workshop on Perspectives on Sustainable Farming Systems in Upland Areas. Tokyo, Asian Productivity Organization.

Gupta, A. and Gupta, N. K. (2010). Agronomic practices in a water shed management project.Training manual of fourteen days training of trainers on 'scaling of water productivity in agriculture for livelihood through teaching cum demonstration organized by Water Management Research Centre, Directorate of Research, SKUAST-J.

Gupta, N. K. and Mahajan, P. (2010). Concept of watershed and watershed management. Training manual of fourteen days training of trainers on 'scaling of water productivity in agriculture for livelihood through teaching cum demonstration organized by Water Management Research Centre, Directorate of Research, SKUAST-J.

Gupta, S. K. (2010). Agro-forestry for sustainable farming in the hills. Training manual of fourteen days training of trainers on 'scaling of water productivity in agriculture for livelihood through teaching cum demonstration organized by Water Management Research Centre, Directorate of Research, SKUAST-J.

Hartley Keith and Nicholas Hooper (1992). Tourism Policy: Market Failure and Public Choice, In Peter Johnson and Barry Thomas (eds), Perspectives on Tourism Policy, London: Mansell Publishing Ltd., 15-28.

Islam, M.A. (1984). Soil erosion and conservation in the hilly areas of Bangladesh BARI Project report.

Jenkins, C.L. and B. M. Henry (1982). Government involvement in Tourism in developing countries, *Annals of Tourism Research*, 9 (4), 499-521.

Jodha (1990). Mountain Perspective Framework. Discussion Paper, MFS 1/1990. www.icimod.org

Jodha, N.S., M Banskota and T. Partap (1992). Strategies for Sustainable Development of Mountain Agriculture: An Overview, In N.S. Jodha, M. Banskota and T. Partap (eds), Sustainable Mountain Agriculture, Vol. 1, Oxford and IBH Publishing Co., New Delhi,3-40.

Manandhar, GB, KB Shrestha and SK Adhikary (2006). Mechanization Strategy to Commercialized Agriculture. Paper Presented at a Workshop Organized by FNCCI during Nepal Agro Mechanization and Technology Development Expo-2006. March 2-8, 2006. Biratnagar, Nepal.

Miah, M. M. U. *et al.* (1993). Fertilizer and Soil Management Guide for the Hilly Regions of Bangladeshan unpublished report of BARC, Bangladesh.

Nakagawa, S. (1998). The Current Situation and Future Challenges in Slope Land Agriculture in Japan and Asia. In Proceedings of the Workshop on Perspectives on Sustainable Farming Systems in Upland Areas. Tokyo, Asian Productivity Organization.

National Commission on Farmers (2005). Serving Farmers and Saving Farmers: Crisis to Confidence. Hill Agro-ecosystem, Second Report pp.142-228. Ministry of Agriculture, Govt of India.

Pariyar (2001). Baseline Study on Agricultural Mechanization needs in Nepal. Rice wheat consortium for Indo-Gangetic plains, New Delhi, India.

Partap, T. (1995). High Value Cash Crops in Mountain Farming: Mountain Development Processes and Opportunities. MFS Discussion Paper, Series, No. 95/1. ICIMOD, Kathmandu: International Centre for Integrated Mountain Development.

Partap, T. (1998). Sloping Land Agriculture and Resource Management in Semi-arid and Humid Asia: Perspectives and Issues. In: *Perspectives on Sustainable Farming Systems in Upland areas.* Asian Productivity Organization, Tokyo. pp. 38-84. www.apo.org.

Partap, T. (1998). Marginal Lands and Marginalized Farmers of the HK-Himalayan Region: Sustainability Challenges and Opportunities. Paper Presented in the DSE-ZEL-ICIMOD, International Conference on Mountains 2000 and Beyond: Sustainable Development of the HK-Himalayan Region. Wildbad Kreuth, Feldafing. June 21-23, 1998.

Partap, T. (1999). Sustainable Land Management in Marginal Mountain Areas of the Himalayan Region. *Mountain Research and Development*, Vol. 19, No.3, pp 251-260.

Partap, T. and Partap, B. (2010). Mountain farmers Adaptive Strategies in Response to Impact of Climate Change on their Livelihood Options. In Proceedings of the International Symposium on Benefitting from Earth Observation : Bridging the Data Gap for Adaptation to Climate Change in the HKH region, held during Oct 4-6,2010. Pub. ICIMOD, Kathmandu, Nepal. www.icimod.org.

Partap, T. and Vaidya, C.S. (2009). Organic farmers Speak on Economics and Beyond: A nationwide Survey of Organic Farmers Experiences in India. Book. Published by Westville Publishing House, New Delhi. pp161.

Rijk, A.G. (1989). Agricultural Mechanization Policy and Strategy. Asian Productivity Organization, Tokyo, Japan.

Shrestha (1992). A study on indicators of unsustainability of hill farming in Nepal.

Singh, Kalyan, Bohra, J.S., Singh, Y. and Singh, J.P. (2006). Development of farming system models for the north-eastern plain zone of Uttar Pradesh. *Indian Farming* **56** (2): 5-11.

Singh, K., Bohra, J.S. and Singh, Y. (2007). Development of sustainable farming system model for irrigated condition of North eastern plain zone of Uttar Pradesh. *Journal of Farming System Research and Development* 13(1), 10-16.

Sugaya, H. (1998). Japan: A Country Report. In: Proceedings of the Workshop on Perspectives on Sustainable Farming Systems in Upland Areas. Tokyo, Asian Productivity Organization.

Hill Agriculture: Prospects, Constraints and Mitigations (2013) *Pages* **68–84**
Editor: **Dr. Shahid Ahamad**
Published by: **DAYA PUBLISHING HOUSE, NEW DELHI**

Chapter 6

Integrated Nutrients Management for Sustaining Soil and Crop Productivity

Tahir Ali[1], M.H. Chesti[2], M.A. Bhat[2] and Uzma Bashir[1]
[1]*Division of Soil Science, SKUAST-K, Shalimar, J&K*
[2]*Regional Agricultural Research Station, SKUAST-J, Rajouri, J&K*

Limited availability of additional land for crop production, along with declining yields for major food crops, have increased the concerns about agriculture's ability to feed a world population expected to exceed 7.5 billion by the year 2020 (Ghuhn *et al.*, 2000). Decreasing soil fertility has also raised concerns about the sustainability of agricultural production at current levels. Integrated management of the nutrients needed for proper plant growth, together with effective crop, water, soil, and land management, will be critical for sustaining agriculture over the long term. Integrated nutrients management is an approach that seeks to both increase agricultural production and safe-guard the environment for future generations.

The basic concept of integrated nutrients management (INM) is the maintenance and increasing soil fertility for sustaining crop productivity through optimizing all possible sources of plant nutrients, *viz.* inorganic and organics required for crop growth and quality in an integrated manner appropriate to each cropping system and farming situation in its ecological possibilities. Since the concept also deals with the supply of nutrients to crops, it is also considered as Integrated Nutrients Supply and Management (Palaniappan, 2002). The latter aims at sustainable productivity with minimum deletions effects of chemical fertilizers on the health of soil and quality of environment. It relies upon a sound crop husbandry and nutrient saving management options. Another important aspect is that it takes into account the cropping system as a whole against individual crops involved in the system. Thus, the INM conceptually integrates all sources of plant nutrients and also improves crop production technologies into a productive agricultural system. This chapter is discussed under the following headlines:

1. Components of INM
2. INM strategies for different cropping systems and
3. INM in relation to soil quality and crop sustainability

Components of INM

Following are the main components of integrated nutrients management system:

- ☆ Nutrients reserves in the soil
- ☆ Organic resources
- ☆ Use of biofertilizers
- ☆ Chemical fertilizers
- ☆ Inclusion of legumes in cropping systems
- ☆ Crop residues
- ☆ Green manures and mulches
- ☆ Sewage and sludge.

Nutrients Reserves in the Soil (Native)

Soil is a store bin of several plant nutrients. These are present in available and reserve forms and a dynamic equilibrium always exists between these forms of nutrients in soil. If the available form of a given nutrient element is depleted by crop uptake or by any other means, there will be immediate replenishment from the reserve source. In this context, it is appropriate to say that soil is not eternal source of plant nutrients. In a bid to produce more, soils are exploited very badly and they are gradually becoming deficient in several nutrients in recent years. There are several soil related constraints coming in the way of achieving sustainability in crop production. These are:

Physical Constraints

Excessively coarse texture, deterioration in soil structure, surface crusting, excess soil compaction, poor permeability, sub-soil pans, poor infiltration, water logging, soil erosion and shifting cultivation.

Chemical Constraints

Soil acidification, salinization in irrigated agriculture and dry-lands, soil fertility interactions, elemental toxicities arising from build up of heavy metals etc.

Biological Constraints

Decline in microbial biomass, soil borne diseases and decline in quality and quantity of organic matter.

To overcome the above mentioned constraints and to enhance the nutrient supply to crops, adoption of soil management and conservation practices, amelioration of problems soils, improving soil conditions for achieving higher fertilizer use efficiency, choosing crop species to exploit the reserve forms of nutrients, choosing appropriate cultural practices and cropping systems for maximizing utilization of available nutrients, growing legumes and tree crops in the cropping systems, inoculation with P solubilizing bacteria to mobilize unavailable P for crop production need greater attention.

Organic Resources

Organic resources are largely biological in origin and they have several nutrients in their composition, which on decomposition are released into soil. It has been well established that the applied organic resources not only increase soil fertility but also improve soil physical condition which is helpful for proper growth of plants. Increasing water holding capacity, aeration, permeability, soil aggregation and nutrient holding capacity and decreasing bulk density and soil crusting are attributed to the continuous use of organic manures. In this way, the productivity of soil is improved to serve over a long period and sustain the crop production. The health care and environmental safety are also ensured by application of organic wastes to soil. Organic manures are of different origin and hence exhibit wide variability in their nature and chemical composition. Apart from higher quantities of N, P and K, they also contain secondary and micronutrients. All these nutrients are released into soil system when organic manures are applied to field. Several organic wastes are recycled through rural and urban composting. Vermicomposting is gaining popularity during recent years because of its environmental friendly and economic reasons.

Future agriculture is poised again to make efficient and exploitive use of organic manures for crop production as it has been established that they cut down the expenditure on chemical fertilizers at least by 50 per cent. Organic farming is gaining momentum for reasons of improving the quality of environment. However, it has to be borne in mind that organic manures are slow releasers of plant nutrients and will not on their own be in a position to meet the requirements of crops, particularly when higher yields are expected. Certain social problems in this connection, storage and usage would also come in the way of their use in agriculture.

Use of Biofertilizer

Biofertilizer form the other important component of INM. They contain actively living cells of bacteria, fungi and serve as microbial inoculants which would participate in biological nitrogen fixation either by symbiotic or non symbiotic means and also solubilize the phosphate in soil from insoluble compounds besides mobilizing other nutrients like carbon and sulphur.

Among the biofertilizers available *Rhizobium* inoculants are noteworthy. Leguminous crops fix atmospheric nitrogen through symbiotic association with *Rhizobium* which harbours in either root or stem nodules (Table 6.1). The quantities of nitrogen fixed by legumes due to inoculation with *Rhizobium* varies with the host species. *Rhizobium* inoculants are manufactured on large scale and are made available to farmers. Since they are host specific, maximum benefit can be derived only where the respective cross inoculation groups are considered. *Azospirillum and Azotobacter* are useful for improving production of crops resulting in nitrogen economy of 25 to 50 per cent and increase in the crop field 1-45 per cent (Handoo *et al.*, 2004). There are biofertilizers which can also fix N by non-symbiotic means. In rice fallows, use of blue green algae (BGA) is proving beneficial. *Azolla,* an aquatic fern, which harbours *Anabaena* appears to be promising in submerged fields as well as for crops grown under ideal conditions, when applied as a green manure. The quantities of nitrogen made available by these microbial inoculants, on an average, range from 20 to 30 kg N $ha^{-1}year^{-1}$ (Sreenivasa, 2002). Other biofertilizers such as phosphobacteria and mycorrhizae (VAM) are useful for mobilizing P in soils for use by crops. Phosphorus solubilizers can save in general 40 per cent phosphatic fertilizers and can enhance the crop yield from 10-50 per cent (Handoo *et al.*, 2004).

The response range of crops to different biofertilizers is furnished in Table 6.2.

Table 6.1: Nitrogen Fixation by Legumes

Legume		*Quantity Fixed (kg N $ha^{-1}year^{-1}$)*
Forage legumes	Clover - temperate	23-620
	Lucerne- temperate	164-300
Grain legumes	Peas - temperate	46
	Lupins - temperate	128
	Pegionpea- temperate	41-150
	Cowpea- temperate	73-354
	Soybean- temperate	17-206
	Cluster bean- temperate	37-196
	Groundnut- temperate	33-111
	Chickpea- temperate	41-270

Source: Sreenivasa, 2002.

Table 6.2: Response (Per cent yield increase) of some Crops to Different Biofertilizers

Biofertilizer	*Crop*	*Response Range*
Rhizobium	Legume crops	4-139
Azotobacter	Several crops	0-31
Azospirillum	Pearlmillet, Sorghum	0-160
BGA	Rice	1-21
Mycorrhizae	Several crops	10-55
Azolla	Rice	0-24

Source: Sreenivasa, 2002.

Chemical Fertilizers

There has been a big gap between annual drain of nutrients from the soil due to crop removal and soil erosion and the nutrient inputs from existing sources. Presently the gap between removal and their addition to the crop is established to be 0.5 million tones and is likely to go beyond 2 million tones by the year 2025 (Manna and Ganguly, 1998). The country will need to produce at least 300 million tones of food grains by the year 2020, which will necessitate the use of 30 to 35 million tones of NPK of various sources (Rathore and Rathore, 2006). In addition, high value crops having export potential, which may even claim fertilizer use on priority basis, will need 14 to 15 million tones of NPK (Table 6.3).

Table 6.3: Projected Demand of Major Nutrients (NPK) by the Year 2025

Crop Group	*Nutrient Demand (million tones)*
Food grains	30.0
Oilseeds	1.5
Vegetables	2.0
Potato	1.0
Horticultural crops	3.0
Cotton	3.1
Sugar cane	0.9
Plantation crop	3.2
Total	44.7

Present consumption is 14.31 million tones.

Source: Manna and Ganguly, 1998.

Widespread deficiency of micronutrients such as Zn, Fe, Mn, Cu and B indicates that the crop demand for these elements by 2025 will be very high (Table 8.4). Fertilizers are of great help to the crops and soil-plant system if they are properly used. They have certain positive impacts from which both plant and soil can be benefited.

Table 8.4: Projected Micro-nutrient Demand for Agricultural Production by 2025

Micro-nutrient	*Requirements ('000 tones)*	*Present Production ('000 tones)*
Zn ($ZnSO_4$)	559.32	58.44
Fe ($FeSO_4$)	169.50	35.53
Mn ($MnSO_4$)	54.34	4.10
Cu ($CuSO_4$)	121.11	12.47
B (Borax)	126.34	21.95

Source: Manna and Ganguly, 1998.

The use efficiency of nutrients varies with crop and soil. Use efficiency of fertilizer nutrients is very low in India (Table 8.5). The low efficiency indicate that the added fertilizer nutrient is not fully utilized by the crop and a major portion of it is lost either through different ways and made unavailable for crop use.

Table 8.5: Nutrient Use Efficiency in India

Nutrient	*Efficiency (per cent)*
N	30-50
P	15-20
K	70-80
Zn	2-5
Fe	1-2
Cu	1-2

Source: Sankaran *et al.*, 2001.

Increasing the use efficiency of fertilizer nutrients has become obligatory and ways and means to increase the use efficiency are being studied. In this connection, all the factors *viz.*, soil, crop, fertilizers, etc. which affect nutrient use efficiency need consideration for achieving sustainability through INM. The nutrient use efficiency can be achieved by scheduling fertilizer recommendations, choosing appropriate nutrient source depending on soil and crop, selecting appropriate time and method of application, correction of deficiencies, improving water use efficiency through appropriate water management practices and controlling weeds, pests and diseases.

Nutrients management practices for efficient utilization of fertilizers can be made in following ways.

1. Nutrient supply to be matched to the crop requirement and uptake. (top dressing, optimizing soil and plant nutrient factors like soil moisture, soil structure, soil test recommendations etc.).
2. Improvement in nutrient application methods (broadcasting, band placement, split application etc.).
3. Improvement in physical characteristics of fertilizers and use of inhibitors to reduce losses (urea super granules, nitrification inhibitors, mineralization of organic matter and efficient use of other plant nutrient sources like FYM, phosphocompost etc.).
4. Soil conditions, crop and water management (puddling, cropping system, residue management, break/catch crop, soil moisture regime and irrigation schedule etc.).

Inclusion of Legumes in the Cropping Systems

It is well experienced that if the same crop is grown season after season in the piece of land, the soil fertility and crop yield declines. This also increases weed, pest and disease incidence. A suitable cropping system practiced in a particular area according to soil and environment could be very helpful to restore the soil fertility and quality, besides it increases yield, quality, marketability as well as water and nutrient use efficiency in an eco-friendly way and reduces the load of chemical fertilizer and pesticides. Legumes used in crop rotation play a dual role in agro-ecosystems by protecting the soil from erosion and enriching organic matter and N through biological nitrogen fixation. The inclusion of leguminous crop is desirable in the agriculture system, they are somehow more environmental friendly than mineral fertilizers. Different crops trap different soil layers for meeting their nutrient and water requirement depending upon their root system. For sustainability of intensive cropping system it is desirable not to grow a particular crop or a group of crops on the same soil for a long period. Productivity of soil can be prolonged if the crops are changed over seasons or years (Perumal *et al., 1991*). Raising a grain legume or green manure in rotation with cereals has been shown to increase the system productivity, improves soil fertility and reduce fertilizer N requirement. Following principles are to be taken into consideration while making systematic schedule for effective crop rotation.

1. Grow a heavy nutrient feeder crop followed by a less nutrient feeder crop.
2. Grow a high water requirement crop followed by a low water requirement crop.
3. A deep- rooted crop is followed by shallow rooted one.
4. A crop requiring more tillage operations is followed by a crop requiring less tillage.
5. A legume crop is followed by non-legume one.

Following crop rotations can be adopted for increasing production and productivity of crop at minimum cost and risk.

- ☆ Potato – Tomato – Palak – Radish
- ☆ Potato – Beans – Turnip – Palak
- ☆ Okra – Radish – Cauliflower – Squash – Cowpea
- ☆ Potato – Onion – Green manuring crop
- ☆ Brinjal – Pea – Bittergourd
- ☆ Tomato – Garlic + Pea
- ☆ Radish – Pea – Okra
- ☆ Radish – Pea – Bittergourd
- ☆ Brinjal – Pea – Okra
- ☆ Okra – Pea – Tomato
- ☆ Tomato – Green manuring crop – Brinjal
- ☆ Potato – Beans – Kale - Radish
- ☆ Capsicum – Pea
- ☆ Capsicum/Tomato – Cabbage/Cauliflower/Kale/Knol Khol
- ☆ Three year crop rotation like 1st year Solaneceous crop – 2nd year cucurbits – 3rd year cole crop

Legume crop like beans, peas, cowpea etc are known to improve soil fertility by fixing atmospheric nitrogen resulting 30-35 per cent increase in yield (Narayan *et al.*, 2004). Growing of crops like beans, peas, cowpea in rotation with other vegetables can enhance their yield and improve their quality. Cowpea fixes 80-85 kg, fenugreek 44 kg, pea 52-57 kg, pegion pea 168-200 kg, black gram 50-55 kg, cluster bean 37-196 kg and chickpea 85-100 kg N ha^{-1} (Papaniappan and Annadurai, 1999).

Crop Residues

Large numbers of crop are grown annually in India. After making use of their economic parts, the remaining portion is mostly wasted. The adoption of mechanized farming in many advanced regions in the country has resulted in leaving a sizeable amount of crop straw/stalk in field after harvesting. The potential of crop residues of major cereals, oilseeds and commercial crop for recycling of valuable plant nutrients for sustained crop production is enormous. About 100 million tones of crop residues are available for recycling in agriculture annually in India (Rathore and Rathore, 2006). Crop residues are a large reservoir of plant nutrients supplementing approximately 0.5, 0.6 and 1.5 million tones of N, P_2O_5 and K_2O respectively). Tables 6.6 and 6.7 embodies the data showing potential of crop residues in India. Most often these residues are disposed off by burning *in situ*.

Table 6.6: State-wise Production of Crop Residues and their Availability for Incorporation

State	*Total Crop Residue Production (million tones)*	*Crop Residues Available for Incorporation (million tones)*
Andhra Pradesh	24.600	12.400
Assam	5.000	1.900
Bihar	17.700	6.500
Gujarat	11.100	6.000
Haryana	15.450	6.310
Himachal Pradesh	2.100	0.710
Jammu and Kashmir	2.040	0.750
Karnataka	12.800	7.100
Kerala	1.240	0.470
Madhya Pradesh	29.400	13.000
Maharashtra	27.800	13.300
Odisha	12.280	5.500
Punjab	29.900	10.500
Rajasthan	19.800	9.800
Tamil Nadu	19.700	8.100
Uttar Pradesh	60.000	26.000
West Bengal	18.440	6.870
Manipur	0.396	0.136
Meghalaya	0.226	0.087
Mizoram	0.199	0.048
Nagaland	0.305	0.133
Sikkim	0.776	0.283

Contd...

Table 6.6–*Contd...*

State	Total Crop Residue Production (million tones)	Crop Residues Available for Incorporation (million tones)
Tripura	0.730	0.260
Andaman and Nicobar	0.051	0.022
Arunachal Pradesh	0.286	0.059
Delhi	0.205	0.109
Dader and Nagar Haveli	0.035	0.011
Daman and Diu	0.003	0.001
Goa	0.236	0.080
Total	312.718	136.439

Source: Singh and Mcdgal, 1977.

Table 6.7: Crop Residues from Different Crops, their Nutrient Content, Nutrient Available for Utilization and their Fertilizer Replacement Value (mt = million tones) Based on Crop Yield

Crop	Total Crop Residue Production (mt)	Crop Residues Available for Utilization (mt)	Nutrient Content (per cent)			Total	Nutrient potential (mt) Available for Utilization	Fertilizer Replace-ment Value
			N	P_2O_5	K_2O			
Rice	119.25	39.75	0.61	0.18	1.38	2.590	0.860	0.430
Wheat	93.90	31.30	0.48	0.16	1.18	1.710	0.570	0.280
Sorghum	19.80	6.60	0.52	0.23	1.34	0.410	0.140	0.070
Pearl millet	10.80	3.60	0.45	0.16	1.14	0.190	0.060	0.030
Maize	14.40	4.80	0.52	0.18	1.35	0.295	0.098	0.050
Pulses	17.95	6.00	1.29	0.36	1.64	0.590	0.190	0.100
Groundnut	11.48	11.48	1.60	0.23	1.37	0.367	0.367	0.184
Others	29.94	29.94	0.80	0.21	0.93	0.581	0.581	0.290
Sugarcane	28.05	28.05	0.40	0.18	1.28	0.522	0.522	0.261
Potato	7.60	7.60	0.52	0.21	1.06	0.136	0.136	0.068
Barley	2.50	0.83	0.52	0.18	1.30	0.050	0.017	0.008
Total		**169.95**	–			**7.441**	**3.541**	**1.721**

Source: Bucha Reddy, 2002.

Crop residues when incorporated into the soil also improves the physical, chemical and biological properties of soil and protect the soil from wind and water erosion. Positive effects of crop residue incorporation on soil properties like organic matter, cation exchange capacity and water holding capacity are recorded (Planiappan and Siddeswaran, 1995). When crop residues are incorporated in the field, the grain yield is higher (Sharma, 2002). Application of one-third N dose along with residues hastens the decomposition process which eliminates the immobilization of fertilizer N during the peak demand of the crop. Fertilizer N application along with residues reduces the C: N ratio which favours fast mineralization of the residue N and thus increases the availability of N to the crop. Apparent recovery of the fertilizer N with residue incorporation is high due to low N losses and synchronization of N supply and the crop need. Also the agronomic efficiency is higher when residue are incorporated than removal or burning.

Compost can be made out of the crop residues for suppression of various soil borne pathogens. Inoculation of these wastes with efficient celluloyric micro-organisms produces enough simple sugars for growth and multiplication of beneficial microbes like free living N fixers and phosphate solubilizers in soil and increase crop yields. Residue incorporation is an environmental friendly practice and need to be promoted.

Green Manures and Mulches

Both green manures and mulches have a great potential to be used for the betterment of Indian Agriculture. Raising a green manure crop even for short period of 45-50 days would add a substantial quantity of nutrients upon incorporation into the soil. Green leaf manuring from the avenue and bund grown trees like *Pongamia*, Neem, Glyceridia and *Leucaena* should form a part of the crop management system.

Green manuring crop like sun-hemp, daincha, black gram, green gram, cluster beans and cowpea can add huge quantities of green matter. Upon decomposition besides releasing nutrients, they also result in the production of enzymes, vitamins, hormones and antibiotics. Green manuring is known to increase the yield to an extent of 15-20 per cent, besides improving the protein and vitamins content (Yadvinder Singh *et al.*, 1991). The economy of fertilizer N through green manuring ranges from 30-100 kg N/ha but in most of the instances it is around 50-60 kg N ha^{-1}.

The beneficial effects of green manures are directly related to the amount of fresh biomass production and their nutrient supplying capacity when they are incorporated into the soil. This biomass can be either the green manure grown *in situ* and incorporated or green leaf manures grown elsewhere and brought in for incorporation in the field. Nitrogen accumulation ranges from 25 to 199 kg ha^{-1} for tropical legumes, 56 to 330 kg ha^{-1} for winter legumes, 134 to 295 kg ha^{-1} for tree legumes, 40 to 240 kg ha^{-1} for green legumes and 98 to 532 kg ha^{-1} for stem nodulating legumes. In Table 6.8 the relative biomass production and N accumulation is presented.

Table 6.8: Biomass Production and N Accumulation in some Common Green Manures

Local Name	*Botanical Name*	*Season*	*Output in 45-60 Days*	*N Accumulation (kg ha^{-1})*
			Green Matter (tones ha^{-1})	
Sunhemp	*Crotolaria juncea*	Wet	21.2	90
Daincha	*Sesbania aculeate*	Wet	20.2	86
Pillipesara	*Phaseolus trilobus*	Wet	18.3	201
Green gram	*Vigna radiate*	Wet	8.0	42
Cowpea	*Vigna unguiculata*	Wet	15.0	74
Guar	*Cyampsis tetragonoloba*	Wet	20.0	68
Senji	*Meliotus alba*	Dry	28.6	163
Khesari	*Lathyrus sativus*	Dry	12.3	66
Berseem	*Trifolium alexandrium*	Dry	15.5	67

Source: Ramasamy *et al.*, 2001.

The N concentration in different cultivated annual legumes ranges from 2.50 to 3.93 per cent. Food legumes like cowpea and mungbean have comparatively low N content. Nitrogen content in the foliage of some tree legumes is usually high. However, different parts of green manure plants vary in N content as well as other constituents such as C : N ratio, soluble carbohydrates, cellulose and lignin. Similarly, the concentration of other nutrients also vary, with the type of green manures. The P content of green manures ranges from 0.20 to 0.34 per cent; the green leaf manures have comparatively low P content (0.20 to 0.24 per cent) than the leguminous crops. The concentration of K, Ca, Mg and S is quite variable among different legume species. The content of micronutrients in both leguminous and non-leguminous crops is quite similar. Table 6.9 presents chemical composition of certain green manure crops.

Table 6.9: Nutrient Content and C : N Ratios in the Green Leaf and Green Manures

Crop	*Total N (per cent)*	*C : N*	*Total P (per cent)*	*Total K (per cent)*
		Conventional green manures		
I Annuals				
Sesbania aculeate	3.97	21.1	6.37*	4.80
Crotolaria juncea	1.90	44.1	0.34	3.60
II Perennials (Leaves and Twigs)				
Gliricidia sepium	2.76	31.1	0.28	4.60
Pongamia glabra	2.78	33.1	0.27	0.19
		Non conventional green manures		
I Trees (Leaves and Twigs)				
Azadiracta indica	2.83	70: 1	0.28	0.35
Delonix elata	3.51	27 : 1	0.31	0.13
Delonix regia	2.76	32 : 1	0.46	0.50
Pheltophorum ferrugenum	2.63	34 : 1	0.37	0.50
II Weeds				
Adathoda vasica	1.32	60 : 1	0.38	0.15
Parthenium hysterophorus	2.68	30 : 1	0.68	1.45
Echhornia grassipes	3.01	29 : 1	0.90	0.15
Trianuthema portulacastrum	2.64	32 : 1	0.43	1.30
Ipomoea portulacastrum	2.64	32 : 1	0.43	1.30
Calotrophis gigantean	2.06	64 : 1	0.54	0.31
Cassia pistula	1.60	120 : 1	0.24	1.20

Source: Yadvinder Singh *et al.*, 1992.

Decaying mulches release polysaccharides and related substances which promote aggregation. Addition of organic matter through tree litter biomass is also one of the major sources of N in the soil. They also release secondary compounds which may volatilize from leaves as lechates, decay of plant litter, decay from sloughed tissues, from the roots and exudates. Among various tree species such as *Mangifera*, *Leucaena* and *Syzygium*. *Leucaena* is more beneficial due to presence of photochemicals like mimosin guercetin, galic protocatechulic, p-OH benzene, p-OH phenylacetic, vanillic ferulic, caffeine and p-coumaric acids.

Sewage and Sludge

Municipal sewage, which is rich in nutrients, is a potential source of nutrient and water. This minimizes the water pollution problems resulting due to the direct discharge of sewage in water bodies. In general sewage contains more than 90 per cent water. The solid portion contains 40-50 per cent organics, 30-40 per cent inert materials, 10-15 per cent bio-resistant organics and 5-8 per cent miscellaneous substances on oven dry basis. It contains a considerable amount of N, P, K and micro-nutrients like Fe, Zn, Cu and Mn. Paper mill sludge (C : N ratio of 28-42 : 1, 29 per cent of the total N in inorganic form) is an effective source of N. It has a considerable effect on the yield of succeeding crop also.

INM Strategies for Different Cropping Systems

The basic concept underlying the principle of integrated nutrient management is to maintain or adjust plant nutrient supply to achieve a given level of crop production by optimizing the benefits from all possible sources of plant nutrients. The basic objectives of INM are to reduce the inorganic fertilizer requirement, to restore organic matter in soil, to enhance nutrient use efficiency and to maintain soil quality in terms of physical, chemical and biological properties. The INM strategies developed for different cropping systems are presented in the Table 6.10.

Table 6.10: INM Strategies Developed for Different Cropping Systems

Cropping System	*INM Strategy*
Rice – wheat	Green manuring of rice with sunnhemp equivalent to 90 kg fertilizer N along with 40 kg N ha^{-1} produces yield equivalent to 120 kg N ha^{-1}.In an acid Alfisol soil, incorporation of *Lantana camera* 10-15 days before transplanting of rice helps to increase the N use efficiency.Apply 75 per cent NPK + 25 per cent NPK through green manure or FYM at 6 t ha^{-1} to rice and 75 per cent NPK to wheat. Inoculation of BGA @ 10kg ha^{-1} provides about 20-30 kg N ha^{-1}.
Rice – rice	Use of organic sources, such as FYM, compost, green manure, *Azolla* etc. meet 25-50 per cent of N needs in *kharif* rice and can help curtailing NPK fertilizers by 25-50 per cent. Apply 75 per cent NPK + 25 per cent NPK through green manure or FYM at 6 tones ha^{-1} to *kharif* rice and 75 per cent NPK to *rabi* rice. A successful inoculation of blue green algae (BGA) @ 10 kg ha^{-1} provides about 20-30 kg N ha^{-1}.
Rice-potato-groundnut	Use 75 per cent NPK with 10 tones FYM ha^{-1} in rice and potato.
Sugarcane based cropping systems	Combined use of 10 tones FYM ha^{-1} and recommended NPK increases the cane productivity by 8-12 tones ha^{-1} over chemical fertilizer alone.
Maize based cropping systems	Apply 50 per cent recommended NPK as fertilizer and 50 per cent of N as FYM in maize and 100 per cent of recommended NPK as fertilizer in wheat.
Soybean – wheat	To get 2 tones soybean and 3.5 tones wheat, apply 8 tones FYM ha^{-1} to soybean and 60 kg N + 11 kg P ha^{-1} to wheat or apply 4 tones FYM + 10 kg N+ 11 kg P ha^{-1} to soybean and 90 kg N+ 22 kg P ha^{-1} to wheat.
Pulses	Integrated use of FYM at 2.5 tones ha^{-1} and 50 per cent recommended NPK fertilizers plus *Rhizobium* inoculation helps in saving of 50 per cent chemical fertilizers.
Sorghum based cropping system	Substitute 60 kg N through FYM or green *Leuceana leaucocephala* loppings to get higher yields and Fertilizer use efficiency.
Cotton	50 per cent of recommended NPK can be replaced by 5 tones FYM ha^{-1}.
Oil seeds (Mustard, Sunflower etc.)	Substitute 25-50 per cent of chemical fertilizer through 10 tones FYM ha^{-1} to get higher yield and Fertilizer use efficiency.

Source: Subba *Rao et al.*, 1995; Acharya *et al.*, 2003.

INM in Relation to Soil Quality and Crop Sustainability

Soil quality affects and is affected by food, feed and fiber production practices. It is also directly linked to environmental quality. Unfortunately, moderate to severe soil degradation through erosion, compaction, leaching, continuous and indiscriminate use of inorganic fertilizers (without the addition of organic manures), loss of biodiversity, structure and tilth continues in India and around world due to previously unrecognized consequences of traditional soil and crop management practices. Traditional agricultural practices have had enormous direct and indirect consequences on productivity, profitability and environmental quality. Maintaining good soil quality is a prerequisite for sustaining higher productivity under intensive agricultural production systems. Soil quality is a complex subject, encompassing the many valuable services humans derive from soil and the many ways soils impact terrestrial ecosystems. Different definitions of soil quality have been proposed, each reflecting a different perspective on the use and value of soils quality (Doran and Parkin, 1994; Carter *et al.*, 1997; Johnson *et al.*, 1997).

- ☆ The potential utility of soils in landscapes resulting from the natural combination of soil chemical, physical, and biological attributes.
- ☆ The capability of soil to produce safe and nutritious crops in a sustained manner over the long-term, and to enhance human and animal health, without impairing the natural resource base or harming the environment.
- ☆ The capacity of a soil to function within ecosystem boundaries to sustain biological productivity, maintain environmental quality, and promote plant and animal health and
- ☆ The capacity of soil to function.

Soil quality can be determined from a variety of soil properties known as indicators, the selection of which may be partially dependent upon land use. Indicators of soil quality will reflect important soil functions:

- ☆ Producing vigorous and healthy plants.
- ☆ Cycling and retaining globally important nutrients, *e.g.* (a) storing nitrogen in soil and releasing it to roots for efficient crop production and (b) storing carbon in soil and releasing it to the atmosphere in a dynamic balance that stabilizes atmospheric concentration of CO_2.
- ☆ Supplying plants with water, nutrients, and plant-growth promoting compounds.
- ☆ Protecting water quality (both ground water and surface water) from nutrient and pathogenic contamination.
- ☆ Providing physical stability and support for vegetation, buildings, and roads.
- ☆ Enabling animal habitat and serving as a reservoir for biodiversity (microscopic and visible).
- ☆ Buffering against toxic accumulation and transport of natural and synthetic compounds.
- ☆ Filtering elements to protect animals, plants, and the environment from undesirable exposure.

Soil quality assessments often use a small group of indicators (*i.e.*, a minimum data set) to economically and efficiently characterize selected key soil functions. Scientists do not have unlimited time and resources to study all of the potential functions served by soil in a region, nor can they predict future needs or demands on soil resources. A minimum data set for soil quality assessment is designed to establish a reliable estimate of the capacity of soil to perform a defined set of functions and to assess changes in soil quality over time. This approach has become popular and is being used by scientists worldwide to assess soil quality. Under this approach the indicators are grouped into three distinct categories: physical, chemical and biological indicators. Soils productivity and its environmental moderation capacity depend on interactive effects of these groups of indicators. The minimum data set of physical, chemical and biological indicators of soil quality is given in Table 6.11.

Table 6.11: Proposed Minimum Data Set of Physical, Chemical and Biological Indicators for Screening the Condition, Quality and Soil Health

Indicators of Soil Condition	*Relationship to Soil Condition and Function (Rationale as a priority measurement)*	*Ecologically Relevant Values/Units (Comparisons for evaluation)*
Physical		
Soil texture	Retention and transport of water, nutrients and chemicals, susceptibility to erosion	Per cent Sand, silt, and clay; less eroded sites or landscape positions
Depth of soil, top soil and rooting	Water and nutrient availability, rooting volume for crop production	cm or m; non-cultivated sites or varying landscape positions
Infiltration and soil bulk density	Runoff and leaching potential, volume of pore space, drainage and compaction	Per cent ($cm^3 cm^{-3}$, $g cm^{-3}$); precipitation intensity
Water holding capacity (Water retention characteristic)	Availability of air and water, retention and transport of water and chemicals	Per cent ($cm^3 cm^{-3}$); precipitation intensity
Chemical		
Soil organic matter (OM) (total organic C and N)	Defines soil fertility, stability and erosion extent, use in process models and for site normalization, organic matter reserves, nutrient cycling and soil structure	kg C or N ha^{-1}/30 cm; non-cultivated or native control
pH	Acidity or alkalinity of soil, nutrient availability	Compared with upper and lower limits for plant and microbial activity
Electrical conductivity	Presence and quantity of soluble salts	dSm^{-1}; compared with upper and lower limits for plant and microbial activit
Extractable N, P&K	Plant available nutrients and potential for N loss, productivity and environmental quality indicators	kg ha^{-1}/30 cm; seasonal sufficiency levels for crop growth
Biological		
Microbial biomass C and N	Microbial catalytic potential and repository for C and N, modeling: Early warning of management, effect on organic matter	kg N or C ha^{-1}/30 cm; relative to total C and CO_2 produced
Potentially mineralizable N (anaerobic ncubation)	Soil productivity and N supplying potential, process modeling, (surrogate indicator of biomass)	kg N ha^{-1}/30 cm/day; relative to total C or total N contents
Soil respiration, water content, and temperature	Microbial activity measure (in some cases plants), process modeling, estimate of biomass activity	kg C ha^{-1}/day; relative microbial biomass activity, C loss verses inputs and total C pool

Source: Larsen and Pierce, 1994, Doran and Parkin, 1996.

Many scientists agree that organic matter content is the best single indicator of soil quality and the main constituent that differentiates top soil form subsoil. Soil organic matter is unique in that it is a part of the physical, chemical and biological properties of soil. Loss of organic matter results in loss of soil tilth and productivity. Organic matter serves as the major granulating agent in soil and provides structural stability and optimum air and water relations. The physical factors of soil are influenced, to a large extent, by the structure and aggregation of the primary particles. The characteristics of soil structure that is most important to plants is the porosity, which results from aggregation and aggregate stability. Porosity influences gas diffusion and nutrient, water and movements in soil. The total porosity is made up of pores of different sizes. Macropores will be defined as pores larger than 30 μm in diameter, mesopores as 0.2 to 30 μm, and micropores as < 0.2 μm diameter. The difference between total porosity and the volumetric water content at -0.01 MPa is considered as the macro-porosity, the difference in water contents between -0.01 and -1.5 MPa as the meso-porosity, and the water content at –1.5 MPa as the micro-porosity.

As a generalization, a medium-textured soil with good aggregation contains about 50 per cent pore space and 50 per cent solid particles by volume. Porosity values can range from 25 per cent in compacted soils to 60 per cent in organic, well-aggregated soil. Porosity greatly influences water relations in soils. Soils with high clay content usually have a greater total porosity than sandy soils. However, a high percentage of the total pores are micropores that do not permit rapid water movement. Therefore water infiltrates slowly into and out of these high-clay soils. During a rain event, water may run off the surface of these soils more rapidly than it moves downward into them. Once they become wet, they dry out slowly. Sandy soils, with a higher percentage of macro-pores, have a high water infiltration rate. Water moves rapidly through the profile, and the soils generally dry out rapidly after rain.

The porosity in soils is strongly influenced by soil characteristics such as texture and organic carbon content and by management. The total porosity generally increases with increasing organic carbon content. Macro-pores can represent a third of the total porosity of soils. Pores in this size class include biopores, shrinkage cracks and other interaggregate pores, and the largest pores within aggregates or peds. This porosity is also referred to as structural porosity or aeration capacity. These pores influence soil mechanical characteristics, water and solute flow and root development. Macropores are the most dynamic of any size class of pores. There exist strong correlations between macroporosity and organic carbon (Kay, 1998). Mesopores are less strongly influenced by management (particularly tillage and traffic) and exhibit less temporal variation than macro-pores. These pores are also referred to as textural pores and the water retained by these pores is plant available water. An increase in the organic carbon content results in an increase in the available water content. Increasing organic carbon content results in increasing water contents at field capacity (-0.01 MPa) as well as at the permanent wilting point (-1.5 MPa) with the largest increase occurring at field capacity. Micropores are least influenced by management of any of the pore classes. Although these pores remain water-filled for a much larger proportion of time than the other pore classes, little of this water is available to plants and the rate of water flow through them is very slow. These pores remain largely biologically inactive because they are not penetrated by roots or by most micro-organisms and are, therefore, only accessible to molecular sized byproducts of biological activity through diffusion. The relative inaccessibility of these pores to microorganisms may also give rise to one of their most important functions, *i.e.*, the physical protection of organic carbon. The microporosity of soils, as for the other pore classes, tends to increase with increasing organic carbon.

Soil organic matter is considered to be the key attribute of soil quality. Soil chemical properties are associated with the organic matter content of soil. The organic matter content of cultivated soils of the sub-tropics is comparatively low because of high temperature and intense microbial activity. Therefore, soil organic matter has to be replenished through periodic addition of organic manure for maintaining soil productivity. Apart from nutritional effects application of organic manures occasionally affects plant growth physiologically provides growth-regulating substances and modifies soil physical behaviour. Organic materials such as crop residues, farmyard manure, composts, poultry manure, etc. are available in abundance but are not fully exploited for various reasons. Again microbial biomass is one of the essential living components of all terrestrial ecosystems. It regulates many critical ecosystem processes, including decomposition of organic material, nutrient transformations and cycling and biophysical integration of organic matter with soil solid. Use of organic manures is generally seen as a key issue for soil health and sustainability, both in terms of maintaining the amount of soil organic matter and in terms of supplying important micro-nutrients.

The long-term sustainability of the productivity of crops and cropping systems is directly related to maintenance of an adequate level of soil organic matter (Ingram and Fernandes, 2001). The maintenance of soil organic matter in agricultural soils is generally governed by annual temperature, precipitation and many interacting factors such as soil types, tillage, application of fertilizers, quality and quantity of organics returned to soil, and the method of residue management. Continuous use of the monocropping, imbalanced and inappropriate use of chemical fertilizers, and minimum or no use of organics year after year are major constraints to organic carbon accumulation in soil. As agriculture became more settled, with a permanent land-base, and as cultivation practices become intensified, soil fertility became severely depleted. Long-term fertilizer experiments proved that balanced use of chemical fertilizers either maintained or slightly enhanced the soil organic carbon content over the initial values and application of farmyard manure improved soil organic carbon, which was associated with increased crop productivity. Assessing of soil organic carbon accretions/sequestration under intensive cropping with different management system plays an important role in long-term maintenance of soil health. Sustainable agriculture is an important issue worldwide. The International Research Institutes all over the world are deeply concerned with sustainability of the intensive agricultural system prevalent in several parts of the world. The rapidly increasing population, shrinking good quality land resources for crop production and increasing concern for declining soil fertility and environmental degradation create awareness that resources are limited and highlight the urgency of continuously enhancing and sustaining productivity of land. An agriculture system to be sustainable must meet the changing food, fibre, fodder, and fuel needs of a nation and should not be detrimental to its natural base. Integrated nutrient management is an alternative to maintain and to enhance soil organic carbon (Singh, 2005). Though it is an age-old practice, its importance was not very much realized in the pre-green revolution era due to the low nutrient demand of the existing subsistence agriculture. This approach of nutrient management aims at efficient and judicious use of all the major source of plant nutrients in an integrated manner, so as to maintain and improve soil organic carbon for sustained crop productivity without any deleterious effect on physico-chemical and biological properties of the soil on a long-term basis.

References

Acharya, C. L., Subba Rao, A., Biswas, A. K., Sammi Reddy, K., Yadav, R. L., Dwivedi, B. S., Shukla, A.K., Singh, V. K. and Sharma, S. K. (2003). Methodologies and Package of Practices on Improved Fertilizer Use Efficiency under Various Agro-Climatic Regions for Different Crops/Cropping Systems and Soil Conditions. Indian Institute of Soil Science, Bhopal.

Bucha, R.B. (2002). Crop residue management for sustainable crop production. *Compendium*: Approach to plant nutrition for sustainable crop production (Sept 18-Oct 8, 2002). Directorate of Oil Seeds Research, Rajendranagar, Hyderabad, pp. 87-95.

Carter, M.R., Gregorich, E.G., Anderson, D.W., Doran, J.W., Janzen, H.H. and Pierce, F.J. (1997). Concept of soil quality and their significance. In : *Soil Quality for Crop Production and Ecosystem Health* (Ed. E.G. Gregorich and M.R. Carter), pp 1019, Elsevier, Amsterdam.

Doran, J.W. and Parkin, T.B. (1994). Defining and assessing soil quality : **In** : *Defining Soil Quality for Sustainable Environment* (Ed. J.W. Doran, D.C. Coleman, D.F. Bezdicek and B.A. Steward), pp 3-21. Soil Science Society of America Special Publication No. 35, Madison, Wisconsin.

Doran, J.W. and Parkin, T.B. (1996). Quantitative indicators of soil quality : A minimum data set. **In** : *Methods for Assessing Soil Quality, SSA Special Publication No.* 49 (Ed. J.W. Doran and A.J. Jones), pp 25-37. Soil Science Society of America Special Publication No. 35, Madison, Wisconsin.

Ghuhn, P., Goletti, F. and Yedelman, M. (2000). Integrated Nutrient Management, Soil Fertility and Sustainable Agriculture: Current Issues and Future Challenges. *IFPRI Home>>2020 vision Publications and Resources Catalog >> Briefs.http://WWW.ifpri.org/2020/briefs/number 67.htm.*

Handoo, G.M., Ghulam -ud-Din. And Chattoo, M.A. (2004). INM system for sustainable vegetable production. Compendium: Organic farming in vegetables. (March 25-29, 2004) Division of Olericulture, SKUAST-K Srinagar pp 38-43

Ingram, J.S.I. and Fernandes, E.C.M. (2001). Managing carbon sequestration in soils : concepts and terminology. *Agriculture, Ecosystems and Environment* **87** : 111-117.

Johnson, D.L., Ambrose, S.H., Basset, T.J., Bowen, M.L., Crummey, D.E., Issacson, J.S., Johnson, D.N., Lamb, P., Saul, M. and Winter, N.A.E. (1997). Meaning of environmental terms. *Journal of Environmental Quality* **26** : 581-589.

Kay, B.D. (1998). Soil Structure and Organic Carbon- A Review. In: Lal, R, Kimble, J.M., Follet, R.F. Steward, B.A. Soil Processes and the Carbon Cycle, Advances in Soil Science, pp.169-192.

Larsen, W.E. and Pierce, F.J. (1994). The dynamics of soil quality as measure of sustainable management. **In** : *Defining Soil Quality for Sustainable Environment Special Publication No. 35* (Ed. J.W. Doran). America Special Publication No. 35, Madison, Wisconsin.

Manna, M.C. and Ganguly, T.K (1998). Influence of inorganic and *in situ* phosphocompost on soil microbial activities in typic Haplustest under rice wheat cropping system.

Narayan, S.l, Ahamad, N., Narayan, R. and. Faheema, S. (2004). Crop rotation, mulching and inter cropping as components of Organic Farming in Vegetables. (March 25-29, 2004) Division of Olericulture, SKUAST–K, Srinagar, pp. 17-21.

Palaniappan, S. P. (2002). Concepts of system approach to plant nutrition. *Compendium*: Approach to plant nutrition for sustainable crop production. (sept 18- oct 8, 2002). Directorate of Oil Seeds Research, Rajendranagar, Hyderabad, pp. 1-11.

Perumal, R., Francis, H., Kandaswamy, P. and Planiappan, S.P. (2001). Integrated nutrient management in Tamil Nadu, Department of Soil Science and Agricultural Chemistry, TNAU, Coimbatore, pp. 50.

Planiappan, S.P. and Siddeswaran, K. (1995). Nutrient recycling and soil management in diversified lowland rice based cropping system. Proceedings. International Rice Research conference on fragile ecosystem. 13-17 Feb., 1995. IRRI, Manila, Philippines, pp. 325-334.

Ramasamy, S., Magudapathy, K. and Suseendran, K. (2001). Green Manures and soil fertility. *Advances in Integrated Nutrient Management for Sustaining Crop Productivity and Soil Fertility*, pp. 119-206.

Rathore, Kapila and Rathore, S.S. (2006). Comprehensive Nutrient Management for optimum soil health and sustained crop productivity. *Indian Journal of Fertilizers* **2**(9): 141- 150.

Sankaran, N., Somasundaram, E. and Poonguznalan, R. (2001). Soil Fertility Management under irrigated cropping systems. Advances in Integrated Nutrient Management for Sustaining Crop Productivity and Soil fertility. pp 15-23.

Sharma, S.N. (2002). Nutrient management in relation to wheat residue management in rice. Indian Journal of Agricultural Science **72**: 449-452.

Singh, A.K. (2005). Assessment and economic analysis of integrated plant nutrient system and its effect on soil quality changes. *Indian Journal of Fertilizers* **1**: 53-60.

Singh, M.P.and Modgal, S.C. (1997). Preliminary studies on rice-wheat criopping system in mid hills of north western Himalayas. In: Proc. Symposium on Cropping Systems Research and Development. For Asian Rice Farmers IRRI. Los Banos, Laguna, Phillippines, pp. 220-226.

Sreenivasa, R.A. (2002). INM concept and need for sustainable crop production. *Compendium* "Approach to plant nutrition for sustainable crop production" Sep 18–Oct 8, 2002. Directorate of Oilseeds Research, Rajendranagar, Hyderabad, pp. 12-17.

Subba Rao, A., Sammi Reddy, K. and Takkar, P. N. (1995). Phosphorus management- A key to boost productivity of soyabean- wheat cropping system on swell-shrink soils. Fert. News, **40**: 87-95.

Yadvinder, S. Khind, G.S. and Bijay, S. (1991). Efficient management of leguminous green manures in wetland rice. *Advances in Agronomy* **45**:135-189.

Yadvinder, S., Bijay, S. and Khind, G.S. (1992). Nutrient transformation in soils amended with green manures. *Advances in Soil Science* **20**: 237- 309.

Zahid, P.B. (1999). Preparation of organic fertilizer from seaweed and its effect on the growth of some vegetable plants. *Hamdard Medicus*. **42**: 93-108.

Hill Agriculture: Prospects, Constraints and Mitigations (2013) *Pages* **85–104**
Editor: **Dr. Shahid Ahamad**
Published by: **DAYA PUBLISHING HOUSE, NEW DELHI**

Chapter 7

Agro-forestry in North Eastern Hill Region of India

***A.K. Singh*[1] *and Mandhata Singh*[2]**

[1]*Department of Agiculture Chemistry and Soil Science, SASRD, Nagaland University, Medziphema – 797 106*

[2]*Department of Agronomy, IAS, Banaras Hindu University, Varanasi – 221 005*

Agro-forestry is a tool for sustainable agriculture production and supply of essential commodities such as food, fodder firewood, timber, oilseeds and industrial material etc. In India, the Agro-forestry assumes great significance in view of the fact that about 50 per cent of its area suffer from one or the other type of land degradation and is not suited for intensive agricultural system. Under such type of situation, to achieve the target of 33 per cent of land area under forest and grasses cover and to achieve the sustainable yield, the only alternative available before researcher and planners is to encourage farmers to grow more and more Multi-purpose tree species (MPTs) under farm forestry and other forestry. Multi-purpose tree species provide five F *i.e.*, fuel, fodder, fiber, fruit and fertilizer are recognized as a valuable tool in sustainable development and in enhancing livelihood.

Agro-forestry is an ancient practice of growing trees and agricultural crops in intimate combination with one another for the welfare of farm family. Agro-forestry has the capability to increase the productivity and at the same time maintain the nutrient balance as well as protect the nature. Due to its technical and economic potential, it can sustain agricultural production. It has two major roles to play, the productive and the service role. Appropriate definition of agro-forestry can be that "it has a collective name for land use system and technologies in which woody perennials, trees, shrubs, bamboo etc. are deliberately combined on the same land management unit with herbaceous crops and animal in same form of spatial arrangement of temporal sequence".

Tree has the dominant role to play in all agro-forestry systems for sustainable agriculture and environmental protection. Tree planting reduces salinity, improve soil fertility, control and prevent erosion, controls water logging, reduces catchment eutrophication and possible checks acidification, increases local bio-diversity and reduces the green house effect.

India has been at the forefront since organized research in Agro-forestry started worldwide in 1979 at Imphal, Manipur, 1st seminar on Agro-forestry was organized by the Indian Council of Agricultural Research, New Delhi to accumulate and compile the data on the research and development of Agro-forestry in India.

The All India Coordinated Research Project (AICRP) on Agro-forestry with 20 centres all over country was launched in 1983 to implement the major recommendations of the aforesaid seminar. Further, National Research Centre for Agro-forestry was established on 8th May, 1988 at Jhansi to accelerate the basic strategic and applied research in Agro-forestry. At present there are 36 centres under All India Coordinated Research Project on Agro-forestry with project coordination unit at National Research Centre for Agro-forestry, Jhansi. These centres represent almost all the agro-climates of the country.

Why Agro-forestry

Agro-forestry systems make maximum use of the land. Every part of the land is considered suitable for plants that are useful. Emphasis is placed on perennial, multiple purpose crops that are planted once but yield benefits over a long period of time. Furthermore, systems of agro-forestry are designed for beneficial interactions of the crop plants, and to reduce unfavourable interactions. They are designed to reduce the risks associated with agriculture, small scale or large, and to increase the sustainability of agriculture. Agro-forestry practices normally help conserve, and even improve the soil.

Agro-forestry includes recognition of the interactions of crops, both favourable and unfavourable. The most common interaction is competition, which may be for light, water, or soil nutrients. Competition invariably reduces the growth and yield of any crop. Yet competition occurs in monoculture as well and this need not be more deleterious in agro-forestry systems. Interactions may be complementary, as in the case of trees, pasture, and foraging animals, where trees provide shade and/or forage, and animals provide manure.

Agro-forestry systems are designed to produce a range of benefits including food, feed, fuels, often fibers, and usually renewed soil fertility.

Agro-forestry systems take advantage of trees for many uses, to hold the soil, to increase fertility through nitrogen fixation, or through bringing minerals from deep in the soil and depositing them by leaf-fall, to provide shade, construction materials, foods and fuel.

Agro-forestry systems may be thought of as principle parts of the farm system itself, which contains many other sub-systems that together define a way of life.

Need and Scope of Agro-forestry

Agro-forestry, in true sense, has been realized as a need of the day. It does not confine to the regional, geographical or agro-climatic boundaries. Agro-forestry concept is a universal application.

Agro-forestry for Food, Fuel and Fodder Needs

Owing to increase in population of human and cattle, there is increasing demand of food as well as fodder, particularly in developing countries like India. Each year the farmers of the world must now attempt to feed 81 million more people irrespective of weather conditions. It is important to note that there will be 19 per cent decline in crop land per head by the end of this century due to population explosion.

In general, there is acute shortage of food and fodder in developing countries. However, the situation in India is more alarming than in other countries of the third world. Asia posses the smallest area of the "potentially arable land" (672 million ha compared with 681 million ha in South America and 734 million ha in Africa) and has the highest population. Out of this, 590 million ha (83 per cent) is already under cultivation. Therefore, there is slight scope to increase food production by increasing the area under cultivation. Hence, food production is to be increased from the land already under cultivation or from land not conventionally considered arable. A management system therefore needs to be devised that is capable of producing food from marginal agricultural land and is also capable of maintaining and improving the quality of producing environment.

In India too a similar situation with regards to food, fodder, fuel and fibre exists. There is a global crisis of energy and man is striving hard to find out same alternative source of energy. We may have sufficient food but not the fuel to cook. Fuel wood is one of the established sources to meet energy requirement.

At present we are utilizing 60-80 million tonnes of dry cow dung, equivalent to 300- 400 million tonnes of freshly collected manure.

The agro-forestry has both productive and protective potential, and it can play an important role in enhancing the productivity of our lands to meet the demand of ever- growing human and livestock population. The different aspects in which agro-forestry can help are given below:

Food

1. Enhanced sustainability of cropping systems through soil and water conservation by arrangements of trees to control run-off and erosion.
2. Enhanced food and feed production from crops associated with trees through nitrogen fixation, better access to soil nutrients brought to the surface from deep tree roots, improved availability of nutrients due to high cation exchange capacity of the soil and its organic matter and mycorrhizal associations.
3. Food for man from trees in the form of fruits, nuts, cereal substitutes etc.
4. Feed for livestock from trees.
5. Micro-climate improvement due to trees, particularly shelter-belts and wind-breaks.

Water

1. Improvement of soil-moisture retention in rainfed croplands and pastures through improved soil structure and micro-climate effect of trees.
2. Regulation of stream flow, reducing flood hazards and a more even supply of water through reduction of run-off and improvement of interception and storage in infiltration galleries.
3. Improvement in drainage from waterlogged or saline soils by trees with high water requirements.

Energy

1. Fuel-wood for direct combustion.
2. Pyrolytic conversion products such as charcoal, oil and gas.
3. Ethanol produced from fermentation of high-carbohydrate fruits.
4. Oils, latex and other combustible saps and resins.

Shelter

1. Building materials for shelter construction.
2. Shade trees for people, livestock and shade-loving crops.
3. Wind-breaks and shelter-belts for protection of settlements, crop lands, pastures and roadways.
4. Fencing: live fences and fence posts.

Raw Material for Industries

1. Raw material for pulp and paper industry.
2. Tannins, essential oils and medicinal ingredients.
3. Wood for agricultural implements and various crafts.
4. Fibre for weaving.

Cash

1. Direct cash benefits from sale of tree products.
2. Indirect cash benefits from increased productivity.

The North Eastern region of the India comprises of eight states *viz.* Arunachal Pradesh, Assam, Manipur, Meghalaya, Mizoram, Nagaland, Sikkim and Tripura and lies between 20°57, and 29.5°N latitude and 89.4° and 97.5°E longitude. The North Eastern region of India is bounded by China and Tibet in North, Myanmar in the East and Bangladesh in the South occupies 5.6 per cent of the total geographical area of India. Nearly 90 per cent of the population of this region is rural and agriculture is the most dominant vocation of the people. Among workers of the region 60.08 per cent are farmers, 9.28 per cent agricultural labourer while 7.28 per cent workers are connected with livestock, forestry, fishery and such other allied activities. In north-eastern region in India, mountainous terrain coupled with high precipitation form the unique natural resources and about 70 per cent of the total geographical area of the regain is comprised of hilly areas. Shifting cultivation and deforestation have been identified as a major constraints in developing sustainable food production system in the regain, due to their adverse effect on soil and water resources. As a result of this 19.8 per cent of the total geographical area of this region has become degraded or wasteland. Physiographical, the region encompasses a wide variation in altitude, climate soil and rainfall. Demographically, more than 120 diverse ethnic group, who are maintaining their own heritage. Inhabit this region. The cropping system, cropping pattern, cultivation and storage practice differ from State to State, over all; about 70 per cent of land area is cultivation at one time or the other each year in the region. So, any technology helping to support Jobs are encouraged in this region shifting cultivation is prevalent in the region on a scale and annual area under this practice is 386.5 thousand ha. (Borthakur, 1992). Efforts are being made to wean away the farmers from shifting cultivation. But it is not easy to eliminate, besides this there are some potential indigenous farming system in the region developed by the tribal farmers using their ingenuity and skill, which have sustainable Agriculture bare and are practiced since centuries'. The traditional people in hilly terrains of north-eastern India have been practicing Agro-forestry in a variety of forms. The region harbors at least 138 herbaceous and 59 tree species of medicinal and aromatic importance, which are used to cure various diseases at indigenous levels (Arunanchalam and Khan, 2003). A number of unique and important plant species occur in this region. A few of them are *Gossipium arborium, Rynchostylis retusa, Schumannianthus dichotimus, Nepenthes khasiana, Citrus indica, Ensele superb,*

Mangifera sylvatica, Parkia roxburghii, Heliconia dasyntha, Lilium muckiniae, Musa velutina, Psidium guinensis and *Antherium* species.

The traditional people in hilly terrains of North Eastern India have lived closely with the natural resources; their utility requirements also depend most on them. Agro-forestry in Arunachal Pradesh, where it has been practiced traditionally, is a mixed and often apparently random poly culture of indigenous trees and crops that form a multi-strata system fairly like a natural forest. In Arunachal Pradesh the thinly populated and biodiversity rich in the Indian Eastern Himalayas, shifting cultivation has been projected as an environmental problem contributing significantly to the loss of lush green forest in the state. The practice of food production has evolved in close conjunction with traditional and socio-cultural value that the people perceived towards livelihood substance in the mountains. This practice has, however been refined through experience and requirement. Although this age old practice has been projected harmful, no good alternative was found suiting to the local conditions.

With the growing human population, greater ethnic diversity, loss of forest cover decreasing shifting cultivation, changing rural lifestyle, expanding township, no industries, ban on timber logging and increasing dependence on rice and vegetables, the state of north eastern region would run in the dark in terms of food security as most existing production systems are less productive, but the shifting cultivation harboring greater crop/genetic diversity. Thus, challenge lies in providing a minimum of traditional practice of *Jhum* and modern Agro-forestry. Dhyani and Chauhan (1993) has already worked an alternate land use system with introduction of MPTs for amelioration on soil fertility of degraded soils under identical climate conditions.

Efforts to deal with the people of shifting cultivation of the past three decades or so hence met setback mainly because of the magnitude of the problem and some other constraints. A judicious application of all these can ease out the age-old practice to be replaced by the new scientific farming systems.

Table 731: Wasteland in North-Eastern Hill Region (000' ha)

State	*Total Geographical Area*	*Barren and Uncultivated Land*	*Cultivable Wasteland*	*Fallow Other than Current Fallow*	*Total*	*Per cent of the Total Area*
Arunachal Pradesh	8358	37	149	118	304	3.6
Assam	7852	1533	130	122	1785	22.7
Manipur	2236	1419	24	-	1443	64.5
Meghalaya	2249	230	456	266	952	42.3
Mizoram	2109	201	74	259	534	25.3
Nagaland	1653	627	42	426	1095	66.2
Tripura	1048	120	2	2	124	11.8
Total	22505	4167	877	1193	6237	24.4

Source: Directorate of Economics and Statistics, Ministry of Agriculture, GOI (1986).

Area under Agro-forestry

The greatest role of Agro-forestry initiatives in the country has been in creating awareness among the common man for protection and preservation of trees, bushes and grasses. However, there is always confusion on the statistics of the area under Agro-forestry in the country. As per one of the estimate about 25.72 million ha is under various types of tree plantations which include Agro-forestry, social forestry, farm forestry and others (Table 7.3).

Table 7.2: Deterioration of Land in North Eastern Hill region (area in per cent)

States	*I*	*II*	*III*	*IV*	*V*	*VI*	*VII*	*VIII*	*IX*	*X*	*XI*
Arunachal Pradesh	3.97	0.05	3.69	1.69	2.55	0.01	0.37	1.51	0.01	8.04	21.88
Assam	1.08	2.08	10.70	3.97	2.83	0.00	4.80	0.07	0.00	0.00	25.52
Manipur	0.01	1.45	53.81	2.73	0.00	0.00	0.00	0.00	0.00	0.00	58.00
Meghalaya	18.68	0.07	9.30	16.10	0.00	0.00	0.00	0.00	0.00	0.00	44.16
Mizoram	0.00	0.00	17.84	1.43	0.00	0.00	0.00	0.00	0.00	0.00	19.31
Nagaland	9.63	0.00	31.51	9.55	0.00	0.00	0.00	0.00	0.00	0.00	50.69
Tripura	2.74	0.00	3.82	5.61	0.00	0.00	0.00	0.00	0.00	0.00	12.17

Source: Directorate of Economics and Statistics, Ministry of Agriculture, GOI (1986).

I: Upland with or without scrubs; II: Waterlogged and marshy land; III: Shifting cultivation area; IV: Underutilized/degraded notified forest land; V: Degraded pastures/grazing land; VI: Degraded land under plantation crop; VII: Sands-Inland/ coastal; VIII: Barren rocky/stony waste/shut rock area; IX: Hill slide slope; X: Snow covered and or/Glacial area; XI: Area under wastelands.

Table 7.3: Area Under Various Types of Tree Plantation in India

Sl.No.	*Plantation Types*	*Area (million ha)*
1.	Agro-forestry and Social forestry	23.0
2.	Externally added Agro-forestry	
	(a) Farm forestry	1.24
	(b) Village woodlots	0.50
	(c) Strip plantations	0.07
	(d) Rehabilitation of degraded forests	0.24
3.	By NGOs/Vas	0.04
4.	Tree growers co-operatives	0.04
5.	Conservation forestry	0.19
6.	Afforestation (Windbreak/shelterbelt) under irrigation development programme of Indira Gandhi Canal Project	0.40
	Total	25.72

Source: Directorate of Economics and Statistics, Ministry of Agriculture, GOI (1986).

Getting Started with Agro-Forestry Systems

Steps in the decision-making process:

1. Decide whether agro-forestry systems are appropriate
   - ☆ Describe family and community needs.
   - ☆ Find the limiting constraints in agriculture, including markets and marketing.

Table 7.4: Prominent Agro-forestry Systems in Different Agroclimatic Regions

Sl.No.	*Agro-climatic Zone*	*Agro-forestry System*					*Specialized*
		Agri-Silvi-culture	*Agri-Horti-culture*	*Silvi-Horti-culture*	*Agri-Pas-toral*	*Horti-Silvi-pastoral*	
1.	Western Himalayan Region	4	4	5	4	4	-
2.	Eastern Himalayan Region	5	3	5	4	3	Jhum
3.	Lower Gangetic Plains Region	5	-	3	2	2	Energy
4.	Middle Gangetic Plains Region	4	-	2	-	-	Homestead
5.	Upper Gangetic Plains Region	5	-	5	-	-	Agrihorti-silviculture
6.	Trans-Gangetic Plains Region	5	-	5	-	-	Agrihorti-silviculture
7.	Eastern Plateau and Hills Region	5	-	4	3	4	Block plantations
8.	Central Plateau and Hills Region	4	-	5	4	4	Block plantations
9.	Western Plateau and Hills Region	4	-	5	2	4	-
10.	Southern Plateau and Hills Region	5	-	3	2	4	-
11.	East Coast Plains and Hills Region	4	5	4	2	2	Home garden/ Shelter-belt
12.	West Coast Plains and Hills Region	4	5	4	2	2	Home garden
13.	Gujarat Plains and Hills Regions	4	2	4	3	5	Block plantation
14.	Western Dry Region	5	3	4	-	5	Shelter-belt
15.	The Island Regions	4	5	4	1	2	Home garden

Source: Handbook of Agriculture, 2010.

Nos. 1–5: indicate the priority practiced mode; 1: Least and 5: Highest.

- ☆ List the potential benefits of an agro-forestry system in the region in question, and their relative importance.
- ☆ Then decide if it is worth the effort to develop one.
- ☆ Consider whether the people of the region are willing or capable of adopting a system.

2. Design a system
   - ☆ Select the area.
   - ☆ Characterize it (describe it, its strengths, weaknesses) with respect to existing soil, water, and crops.
   - ☆ List the needs that could be met with an agro-forestry system.
   - ☆ Characterize the crops desired by minimum space requirements, water and fertilizer needs, and shade tolerance.
   - ☆ Select the trees, shrubs, or grasses to be used.
3. If the system is temporary
   - ☆ Plan the features of soil erosion control, earthworks, and gully maintenance first.
   - ☆ Plan spacing of fruit trees according to final spacing requirements.

- ☆ Plan a succession of annual or short-lived perennials beginning with the most shade tolerant for the final years of intercropping.

4. If the system is permanent
   - ☆ Plan the proportion of the permanent fruit and lumber trees on the basis of relative importance to the farmer.
   - ☆ Plan the spacing of long-term trees on the basis of final space requirements times 0.5.
   - ☆ Plan succession of annual and perennial understory crops, including crops for soil protection and enrichment.
   - ☆ As large permanent trees grow, adjust planting plan to place shade tolerant crops in most shady areas.
5. In temporary and permanent systems
   - ☆ Keep all ground in crops or protective covers at all times.
   - ☆ Try the system on a small scale first.
   - ☆ Measure the inputs and outputs of the system.
   - ☆ Evaluate whether the benefits expected have been achieved.
   - ☆ Look for the desired plant materials or for suitable substitutes locally (Table 7.5).

Expand or extend any new system cautiously.

Agro-forestry Practices

Alley Cropping

Alley cropping involves growing crops (grains, forages, vegetables, etc.) between trees planted in rows. The spacing between the rows is designed to accommodate the mature size of the trees while leaving room for the planned alley crops. When sun-loving plants like corn or some herbs will be alley cropped, the alleyways need to be wide enough to let in plenty of light even when the trees have matured. Alternatively, the cropping sequence can be planned to change as the trees. Growth decreases the available light. For example, soybeans or corn could be grown when the trees are very small; then, as the tree canopy closes, forages could be harvested for hay; finally, when the trees are fully grown and the ground is more shaded, grazing livestock or shade-tolerant crops like mushrooms or ornamental ferns could occupy the alleyways. Like all integrated systems, alley cropping requires skillful management and careful planning. Both the crop and the trees have requirements that sometimes necessitate trade-offs between them. The design must allow sufficient room for the equipment needed to service each enterprise. If either crop requires chemical herbicides or insecticides, the other must be tolerant of these treatments. In the case of livestock, there may be periods during and after chemical use when animals must be withdrawn from the area. Livestock can cause damage, even when the trees are fully grown; roots injured by livestock hooves are susceptible to disease. Soil compaction is a danger in wet weather. These examples indicate how crucial planning is to the ultimate success of an agro-forestry system.

In most alley cropping systems, trees are planted in straight rows, sometimes with no regard for slope or contour. There are, however, advantages to planting the trees in curves or on the contour. These include the slowing of surface water movement and the reduction of soil erosion.

Figure 7.1: Alley Cropping

The trees can be planted in single rows or in blocks of multiple rows between alleys. The first row in a block is planted on the contour line; subsequent rows are planted below the original line according to the slope of the land. The final row of trees in one block is planted parallel to the contour line on which the next block of trees will begin. The width of the tree blocks varies, but the cropping alleyways between them have parallel edges. This design avoids creating point rows within the alleys, thus simplifying crop equipment maneuvers. The width of the alleys is determined by the size of this equipment. If planting on the contour is impractical, another option is to plant trees in curved zig-zags so that water running downhill is captured or at least slowed. Islands of trees can offer some of the same advantages if they don't interfere with cropping operations.

In large plantings, fast-growing hardwoods or pines are interplanted as trainers to ensure that the crop trees develop upright, unbranched trunks. Alternatively, the crop trees can be planted close together in the rows, to be thinned and pruned several times as they grow. Although these early-harvested trees may have little market value, their presence during the first years of growth has increased the main crop's value. The goal is to produce long, straight sawlogs with few lower branches, for maximum profit at final harvest. Whatever the planting design, trees on the outside edge of a group will grow more side branches, or even a lopsided trunk, resulting in lower-value saw logs.

Silvopasture

Tree and pasture combinations are called *silvopastoral agro-forestry*. Hardwoods (sometimes nut trees) and/or pines are planted in single or multiple rows, and livestock graze between them.

Although both the trees and the livestock must be managed for production, some systems emphasize one over the other. Usually, in the early years of establishment, crops or hay are harvested from the planting. Grazing generally begins after two or three years, when the trees are large enough that the livestock cannot damage them. In other instances, tree tubes and electric fencing protect the young trees, and grazing begins immediately.

Figure 7.2: Silvopasture

Grazing livestock on silvopasture eliminates some of the costs of tree maintenance. With good grazing management, for example, herbicides and mowing may become unnecessary. Grazing also enhances nutrient cycling and reduces commercial fertilizer costs; the animals remove few nutrients, and their waste is a valuable input for the trees. Well-managed grazing will increase organic matter and improve soil conditions.

However, controlling the number of animals per acre, limiting the number of days those animals remain on each site, and avoiding compaction are critical for a successful silvopasture system. Competition for water between the pasture and the trees may be a concern. In a silvopasture with nut trees, for example, seasonal water shortages during late summer can negatively affect nutfill and the production of fruit buds for next year's harvest. Irrigation is justified in such a situation if the trees are being managed for nut production.

Water competition may not be as critical for *timber* silvopastures. Further information about silvopastoral systems is available from the National Agro-forestry Center and other resources listed at the end of this publication.

Windbreaks or Shelterbelts

Extensive research on windbreaks, also called *shelterbelts,* has been carried out in the U.S. Trees are planted in single or multiple rows along the edge of a field to reduce wind effects on crops or livestock. Windbreaks have been shown to reduce wind impact over a horizontal distance equalling at least ten times the height of the trees. Wind and water erosion are reduced, creating a moist, more favourable microclimate for the crop.

In the winter the windbreak traps snow, and any winter crops or livestock are protected from chilling winds. Beneficial insects find permanent habitat in windbreaks, enhancing crop protection. Although the trees compete for available water along the edges between the windbreak and the crop rows, potentially reducing crop yield near the windbreak, the net effect on productivity is positive. In fact, even on land that's well suited for high-value crops, a windbreak can increase the crop yield of

Figure 7.3: Windbreaks or Shelterbelts

the entire downwind field by as much as 20 per cent, even when the windbreak area is included in the acreage total. Windbreaks can be designed specifically for sheltering livestock. Studies have shown the economic advantages of providing protection from wind chill, a major stress on animals that live outside in the winter. Reduced feed bills, increases in milk production, and improved calving success have resulted from the use of windbreaks.

Besides providing protection to crops and livestock, windbreaks offer other advantages. They benefit wildlife, especially by serving as continuous corridors along which animals can safely move. Farmers can even develop windbreaks into additional profit centers for the farm. Hunting leases, selective timber harvests, firewood sales, and special forest products are some of the possibilities (these marketing options are discussed below). Any tree species can be used in a windbreak. However, deciduous species, even in multiple rows, will lose effectiveness when they lose their leaves. For year-round use, some of the species selected should be evergreen. Fast-growing trees should be included; it's best to plant deep-rooted, non-competitive species along the edges. Regular deep chisel-plowing along the edges will keep roots from spreading into the crop rows. If some of the trees are harvested periodically, replacements can be planted, establishing a long-term rotation within the windbreak.

Riparian Buffer Strips

Trees, grasses, and/or shrubs planted in areas along streams or rivers are called *riparian buffers* or *filter strips*. These plantings are designed to catch soil, excess nutrients, and chemical pesticides moving over the land's surface before they enter waterways. Such plantings also physically stabilize

Figure 7.4: Riparian Buffer Strips

stream banks. On cropland that is tiled to improve drainage, polluted water can flow directly into streams; constructed wetlands installed in the buffers can capture and clean this drainage water before it enters the stream.

Forested areas along streams fulfill other needs of the community at large by storing water and by helping to prevent stream bank erosion, which in turn decreases sedimentation downstream. These areas protect and enhance the aquatic environment as well. Shading the water keeps it cooler, an essential condition for many desirable aquatic species. Buffer strips also provide wildlife habitat and can be managed for special forest products. Crop and livestock farmers, as well as local communities, have become aware of the threat that agricultural practices can pose to pure drinking water. Consequently, there are federal, state, and local government programs to assist in the design and planting of riparian buffer strips. The federal Continuous Conservation Reserve Program can be used for this purpose. The local Farm Services Administration office can advise on this program and other options. Conservation organizations are another potential resource. Some offer conservation easements or trusts when land is permanently withdrawn from agricultural production.

Forest Farming and Special Forest Products

When a natural forested area is managed for both wood products and an additional enterprise, it becomes an agro-forestry system. Besides producing saw timber and pulpwood, woodlands can generate income from many other products. Established forests offer many non-timbers. Special forest products. that contribute to cash flow without requiring the one-time harvest of old trees. For example, landowners can manage established woods to encourage naturally occurring patches of berries or bittersweet/or they might plant understory crops adapted to the forest type and climate. Growing mushrooms on logs is another, more labour-intensive, possibility; a canopy of either hardwoods or pine will provide the shade needed to maintain moisture for fruiting.

Berries and vines for crafts or basketry are examples of products that can be harvested and marketed without any costs of establishment; on the production end, they may require only that the canopy be managed for optimal light conditions.

Basic Approach to Agro-forestry Intervention

The economic output under a production system can be increased as long as, we fully understood current productivity in term of sustainability and ecological viability. Synthesis of balanced tree/ grass/arable crops production systems control run-off, water supply and support soil fertility to maintain the systems sustainability. Considering the domains of agro-forestry systems in the eastern hills of India.

Three areas have become important:

1. Development of alternatives to shifting cultivation.
2. Improvement in traditional homestead agro-forestry.
3. Rehabilitation of degraded hill slopes.

Indigenous farming systems particularly on the hill slopes suggest that a diverse product based complex agro-forestry have been quickly occurring in jhuming area of different states.

Until recently, potential of these agro-forestry practices to solve land use problems and enhance productivity was not recognized by research and extension agencies with the passage of time wise spread interest has mounted in implementing agro-forestry development programme. Agro-forestry has been popularized among decision makers as a conservation farming solution to sustain productivity of hill agro-ecosystems (Garrity, 1995; Singh, 2002). Consequently there is enormous demand for sound agro-forestry technology.

Tribal Wisdom and Agro-forestry

Over the years, land users in this region have accumulated knowledge of their environment comprising land, water, trees, livestock etc. This knowledge helped them to evolve many appropriate agro-forestry practices tested over long periods and proved beneficial in the management of resources to meet a range of needs from food to shelter (Singh, 1999). What urgently needed is a more quantitative/ understanding of their extrapolability and agronomic improvements to enhance crop productivity.

Large Cardamom Plantation on Age Old Agro-forestry System

The indigenous tribes *i.e.* lepcha and limbu used to collect large cardamom from natural forest. As these forests processed to village ownership, a large cardamom growing in the natural forests were domesticated it becomes a plantation crop (Singh *et al.*, 1989). Farmers evolved a classical tree cutting schedule in potential area to get continuous supply of fuel wood and fodder without affecting the shade requirement of large cardamom plants. When trees attain 16 cm basal diameter, they are cut to release young trees growing out of natural regeneration and provide shade to the large cardamom capsule in November in such a way that the lopped twigs are not allowed to fall directly on the cardamom plantations comprising of trees and perennial herbs resemble to a natural forest ecosystem. It is a economically viable cash crop with export potential.

Alder Agriculure Productive Conservation of Trees

The practice of using alder trees for raising food crops is a time tested practice of Nagaland (Gokhale *et al.*, 1995). The alder is used for raising the primary crops of maize, job's tear, millet, potato, barley, wheat and secondary vegetable crops chillies, pumpkin and colocaasia. In *Khonoma* village, people have resorted to terracing the alder field to further reduce soil erosion and retain soil fertility. Spreading nature of root systems of alder also help to retain and keep intact the soil on hill slopes. If a village with 100 families could set aside about 120 hectares of land to grow alder trees, all families would be able to get sufficient fuel wood every year and at the same times raise crops under the alder in about 30 hectares area every year.

Agro-forestry for Tree Fodder Production

Fodder trees, mostly native to the place; provide green fodder during the dry winter months from October to March. Trees leaves constitute 25 to 60 per cent to total green fodder feed to livestock in the state like Sikkim (Singh, 1996). Many fodder trees are planted on terrace riser. There is further need to generate information on different tree-crop interactions and to develop techniques of lopping/planting of trees to minimize tree effects on the agricultural productivity.

Different tree species produce fodder at different times of the year. The sustained supply of fodder was turned to the advantage of creating multispecies plantation. It was also suggestive to know number of livestock kept and the way feeding systems were formulated. Lopping commences in October or November and completed in March or April. Generally lopping of evergreen trees is extended up to April and the lopping of deciduous or semi-deciduous trees are completed before the leaf shed. Some of the fodder trees are lopped during May to September, a peak period for agricultural operation in the hills. This is done to save time for agricultural activities. Productivity of trees are being governed by month and intensity of lopping (Singh, 1995). The effect of light, moderate and heavy lopping treatments on a three year old alder (*Alnus glutinosa*) showed that within 24 hours of lopping, transpiration rates decreased and water potential increased and the effects remained for 12 days resulting into enhanced water use efficiency during early period of regrowth (Singh and Thompson, 1995). Therefore, fodder tree may not tolerate more than 30-40 per cent of leaf area removal to produce fodder on sustainable basis. Among the introduced tree species, *Leucaena leucocephala* variety S 11 dose well at low mid hills (Singh, 1996). It is completely free from leaf spot disease caused by *Camptomeris leucacena*, most common under humid climate of eastern India (Singh, 1995).

Mixed Homstead Garden

Around the house, a number of trees species *viz.* arecanut, tree tomato (*Cryptomandra betacea*), citrus, guava, jackfruit, drumstick (*Moringo spp*) and some fodder tree are invariably grow with under story crops such as vegetables beans, cucurbits, ginger, turmeric, black papper and betel having largest variation in their combinations from house to house depending upon area available, site conditions (Slope and soil depth) and choice of family (Singh and Pradhan, 1993). The characteristics of these agro-forestry systems can be defined by a range of plant species producing diverse products.

Agri-Horti System

In low to mid hills, mandarin is a commercial fruit crop of this region. The fruit orchard is intensively intercropped with annual food crops grown on uplands (Singh and Pradhan, 1993).

Tea Plantation

Tea plantation also exists in small pockets up to 1500m MSL. They are grown without shade trees due to low light intensity and duration of bright sunshine available at the site but wild cherry (*Prunus ceresoides*) is planted as live boarder (Singh and Pradhan, 1993). At foot hills and low hill up to 500 m MSL there is scope to select shade trees with desirable characteristics for tea production. Shade trees like *Albizzia chinensis* add 2500-5000 kg organic matter per hectare to the soil annually, contained 63-126 kg N 18-36 kg P_2O_5, 22-44 kg K_2O, 32-64 kg CaO and 16-32 kg MgO per hectare. Under shade trees, fodder root of tea develop at a higher rate than without shade and they absorb large amount of harmful infrared radiation from solar spectrum and diverts larger assimilates towards plucking shoot and the leaf.

Agro-forestry for Soil Conservation and Amelioration

Agro-forestry plays a key role in keeping the soil resource productive which is one of the major sustainable issues. Closely spaced trees on slopes reduce soil erosion by water through two main processes; first as a physical barrier of stems, low branches, superficial roots and leaf litter against running water and secondly as sites where water infiltrates faster because of generally better soil structure under trees than an adjacent land.

Agro-forestry played a major role in the recent past in rehabilitation of wasteland such as desert and land that have been degraded by salinization and ravines, gullies and other farms of water and wind erosion hazard. The example of rehabilitation of more than 3 million ha. (out of 7.1 million ha) of salt affected soils, particularly the village level community lands, area along road side, canals and railways tracts through Agro-forestry intervention. Silvipostoral system is the most appropriate land use system for these degraded lands. A silvipostoral system comprising *Leucaena leucocephala and Cenchrus ciliaris* and *Stylosanthes hamata* has shown the way to rehabilitate some of the degraded and marginal land. Similarly *Albizia lebbeck, Cenchrus stylo* and *Albizia procera, Cenchrus stylo* based Agro-forestry systems have recorded significant improvement in soil organic matter, available N, P and K.

Singh *et al.*, 1988 reported higher annual turnover of organic matter, total nitrogen and organic sulphur under silvipostoral management in semi arid region. Research efforts made at NRCAF, Jhansi for over a decade has resulted in development of appropriate technology for the rehabilitation of degraded lands.

Agro-forestry for Carbon Sequestration

The earth's terrestrial vegetation plays a pivotal role in the global carbon cycle and tremendous amounts are actively exchanged between vegetation and atmosphere. In any land use practices that increase vegetation cover or reduce its removal, could have an influence on the global carbon budget by increasing the terrestrial carbon sink. In the context of carbon cycle, Agro-forestry plays an important role in carbon dioxide sequestration due to following reasons.

Agro-forestry has importance as a carbon sequestration strategy because of carbon storage potential in its multiple plant species and soil as well as its capability in agricultural lands and in reforestation. Carbon storage depends on several factors including climate, edaphic and socio-economic conditions. Perennial systems like home gardens and Agro-forestry can store and conserve considerable amount of carbon in living biomass and also in wood products. Agri-horti-silviculture and silvipostoral systems of Agro-forestry have been observed as the potential sinks for sequestering surplus carbon from the atmosphere. Since Agro-forestry increases the income of farmers, it reduces the incentive for future extraction from natural forest for income generation (Singh *et al.*, 2007).

Agro-forestry for Food Security

The country's food production has increased four folds in the past five decades but recent improvement in food supply have been insufficient to fulfill the nutritional need of the average person in the country. Increasing the production and consumption of protein rich food, pulses in particular, will be necessary if the country is to meet its protein need. Agro-forestry with appropriate tree crop/ legume combination is one option in this regards.

Agro-forestry for Fodder Production

Trees and shrubs often contribute substantial amount of leaf fodder in arid, semi arid and hill regions during lean period through lopping/pruning of trees, particularly known as top feed. The leaf fodder yield depends on species, initial age, lopping intensity and interval as well as agro-climatic conditions. This fodder is usually rich in protein, vitamins and minerals like calcium through in general low in phosphorus and crude fiber (Singh, 1990). Such top feed species play an important role in human food security through their function as animal feed resource. The top feeds are considered very important in vegetation stabilization and sustained productivity of rangelands. They also play an important role as windbreaks and by providing shade for grazing animal.

Agro-forestry for Energy Plantation

The main biomass energy is source in rural areas which are being used in the households; include wood from trees and shrubs. Due to the programme of social forestry and agro-forestry initiatives large amount of woods are now being produced from outside the conventional forestlands. Small landholdings and marginal formers, through short rotation forestry and agro-forestry are now providing the bulk of country's domestically produced timber products and is competing with annual imports values at about US $ 1 billion. This has a great significant as the extraction of timber from the state owned forests has been banned to reduce deforestration and to conserve the environment. *Prosopis juliflora* is the major source of fuel for the boilers of the power generation plants in the A.P. The fuel wood potential of indeginous (*Acacia nilotica, Azadirachta indica, Casuarina equisetifolia, Delbergia sissoo, Prosopis cineraria* and *Ziziphus mauritiana*) and exotic (*Acacia auriculiformis, A tortilis, Eucalyptus camaldulensis* and *E. tereticornis*) tree revealed that calorific value ranges from 18.7 to 20.8 MJ/kg indigenous tree species and 17.3 to 19.3 MG/kg for exotic. Species such as *C. equisetifolia, Prosopis juliflora, Leuceana leucocephala* and *Calliandra calothyrsus* have become important due to their potential for providing wood energy at the highest efficiency, shorter rotation and also their high adoptability to diverse habitats and climates.

Table 7.5: Some Important and Potential Agro-forestry Systems/Practices for the North Eastern Hill Regions and their Area of Application (Dhyani and Sharda, 2005)

Agro-climatic Zone and AFs/P	*Brief Description*	*Areas*
Temperate pure Agri-silviculture		
Pines with field vegetable crops	Khasi pine (*Pinus kesiya*) with pea radish, potato, sweet potato, Cabbage, turnip, Cauliflower, mustard, maize	North Eastern Hills
Hortiagriculture		
Plum with vegetables	Plum with pea, radish, cabbage or cauliflower, trees scattered or on field boundaries	Arunachal Pradesh
Subtropical hill zone		
Horti-silviculture		
Alder with large cardamum	Alder (*Alnus nepalensis*) and Chilloni (*Schima wallichii*) with large cardamom	Sikkim, Nagaland, Manipur
Chilloni+pineapple	Chilloni (*Schima wallichii*) with pineapple	Meghalaya, Manipur
Agri-silviculture		
Chilloni with crops	Chilloni (*Schima wallichii*) with ginger/turmeric	Meghalaya, Manipur
Agriculture with alder	Alder on field margines with maize, job's tears (*Coix lacryma*), millet, potato and wheat and secondary crops *e.g.* Chillies, pumpkin, taro (Colocsia) etc.	Nagaland
Multipurpose trees on crop lands	MPTs such as *Prunus cerasoides, A. nepalensis, Ficus, Schima wallichii, Artocarpus, Bambusa, Gmelina arborea, Morus alba,* Banana etc. Scattered on bunds terrace riser or field boundaries.	Nagaland, Meghalaya, Arunachal Pradesh
Contour hedgerow inter-cropping and sloping agric land technology (SALT)	*Desmodium rensonni, Tephrosia canadida, Indigofera, Leucaena leucocephola, Gliricidia, Robinea, Cassia siamea* and *Crotalaria tetrogona* with crop and fruit plants.	Arunachal Pradesh, Nagaland
Silvipasture		
Chilloni/pine with broom grass	Broom grass is cultivated under the trees of khasi pine and chilloni.	Nagaland, Meghalaya, Arunachal Pradesh
Sub tropical hill		
Apiculture with trees	Tassar silkworm rearing on the leaves of oak (*Ouercus* sp.) and mulberry plants on terrace risers.	Manipur, Meghalaya, Nagaland, Arunachal Pradesh
Horti-agriculture		
Khasi mandarin with pineapple/vegetables	Mandarin with pineapple/beans/radish/ginger/turmeric/ cole crops in Meghalaya and other states	Meghalaya

Table 7.5–*Contd...*

Agro-Climatic Zone and AFs/P	*Brief Description*	*Areas*
Sub-tropical plain zone		
Tree gardens or Homesteads		
Fruit trees on kitchen garden or farm boundaries	Tree tomato (*Cytomaddra betacea*), guava, banana and moringo grown iin kitchen garden or farm boundaries.	Meghalaya, Nagaland, Manipur
Mild tropical hill and plain zone multistoried.		
Arecanut based system	Arecanut+pineapple+betelvine/blackpipper, arecanut and betelvine+pineapple and black pepper. All these system are practiced on the most degraded boundary lands with high slopes and shallow soil depth conditions. Bamboo drip irrigation is used for watering the plants.	Meghalaya, Tripura
Tea with shade trees and betel vine/black pepper	Tea cultivated on hill slopes with shade trees *e.g. Alnus nepalensis, Derris robusta,* Silver oak, Albizia and betelvine or black papper.	Arunachal Pradesh, Manipur, Nagaland, Tripura.
Horti-silvipasture		
Mixed land use	Arecanut+black papper+mandarin; arecanut+ jackfruit+mandarin; bamboo+broom grass+ *Schima wallichii*; broom grass+tapioca+sweet potato. These practices are based on the production of fruits, nuts, betel leaf integrated with woody perennials.	Meghalaya, Tripura
Horti-silviculture		
Fruit trees on degraded lands	Jackfruit+pineapple; mandarin+jackfruit; arecanut+ jackfruit+betelvine	Meghalaya, Tripura
Agri-silviculture		
Plantation crop combination	Multipurpose trees such as *Prunus cerasoides, Alder, banana, Schima wallichii, Artocarpus, Gmelina arboria,* mulberry, banana etc. scattered on bunds, terrace risers or field boundaries.	North Eastern hills
Horti-silviculture		
Coffee and black pepper with hollock and hollony	This is a recent introduction. Coffee and black pepper cultivated with hollock (*Terminalia myriopcarpa*) and holony (*Dipterocarpus mocrocarpus*) in forest area and with *Erythrina indica* on crop land.	Meghalaya, Arunachal Pradesh
Home gardens	Intensive integration of MPTs (Albizia, Gmelma, Jackfruit, Banana, Papaya, Coconut, Arecanut) with annual crops and livestock	Meghalaya, Arunachal Pradesh
Small production systems	Silkworm rearing, livestock honeybee and fish production in and around homesteads	North Eastern hills.

Tree Speciese Based on Specific Utilization Purpose

Fodder-cum-Fuelwood Species

Albizia amara, Albizia procera, Albizia lebbeck, Erythrina indica, Gliricidia sepium, Hardwickia binata, Leaucanea leucocephala, Pithecellobium dulce, Prosopis cineraria, Sesbania grandiflora, Sesbania sesban.

Fuel-wood and Timber Species

Acacia nilitica var. cupressiformis, Acacia nilotica, Albizia lebbeck, Albizia procera, Azadirachta indica, Cassia siamea, Casuarina equisitifolia, Dalbergia sissoo, Dendrocalamus strictus, Pongamia pinnata, Melia azadirach, Parkinsonia aculeate, Thespepsia populnea.

Softwood and Pulpwood Species

Ailanthus excels, Ailanthus tryphysa, Bombax ceiba, paraserianthes falcataria, Populus deltoids, Bamboo species.

Fruit and Vegetable Species

Annona reticulate, Annona squamosa, Artocarpus heterophyllus, Emblica officinalis, Moringa officinalis, Ziziphus mauritiana.

Agro-forestry for Employment Generation in North Eastern Hills

A number of agro-forestry systems as enumerated in Table 5 can be developed for different agro-climatic situations involving agricultural crops, legumes, grasses, fruit plants, flowering and medicinal plants, vegetables livestock etc. with the support of local natural resources. Sericulture, apiculture, fishery, tea, flowering and medicinal plants as component of agro-forestry systems have great potential in the region because of its specific environmental conditions and several micro-situations. Marketing of silk cacoon, tea leaves, fish, mushroom in now being done by co-operatives. These vocations provide rural population on attractive source of income and help to stop their migration thereby preserving their skills and way of life. Besides, permitting gainful utilization of the vast natural wealth and help arrest forest destruction.

References

Arunachalam, A. and Khan M.L.(2003). Traditional agro-forestry as a viable choice to conserve agro-biodiversity in the north-eastern India- Pathak, P.S. and Ram Niwaj (Editors), Agro-forestry: Potential and Opportunity,@2003 Agrobios (Indai) and Indian Society of Agro-forestry.

Borthakur, D.N. (1992). Agriculture of the north-eastern region. Bee Cee Prakashan, Guwahati (Assam), India.

Dhyani, S.K. and Chauhan, D.S. (1993). Micro planning for integrated wasteland development programme in North Eastern India. Training programme, NEHU, Shillong, 1719 June:1-16.

Dhyani, S.K. and Sharda, V.N. (2005). Agro-forestry systems as rural development options for the Indian Himalayas, *J. Tree Sciences* 24 (1): 1-19.

Garrity, D.P. (1994). Developing conservation strategies for resources poor farmers in South east Asia. Paper presented in 8th Int. SoilvConserv. Organization Conf. Held from December 4-8, 1994 at New Delhi, India.

Gokhale, A.M., Zeliang, D.K., Kevichasen and Angami, R. (1995). *The use of Alder Trees*. State Council of Educational Research and Training, Education Department, Govt. of Nagaland.http://attra.ncat.org.

ICAR (2009). Hand book of Agriculture, ICAR, New Delhi.

Singh, G., Kolar, J.S. and Sekhon, H.S. (2002). *Recent Advances in Agronomy*. Indian Society of Agronomy, IARI, New Delhi.

Singh, K.A. (1995). Biomass productivity of multipurpose tree species on the hill slopes of Barapani, Meghalaya. *J. Hill Research.* 8 (2):226-228.

Singh, K.A. (1996). Effect of age, cutting interval, nitrogen and farm yard manure on root growth parameters of forage grasses in an acid inceptisol. *Indian Journal of Agricultural Sciences.* 66 (1):38-44.

Singh, K.A. (1996). Growth, biomass productivity and coppicing behaviour of multipurpose tree species at a mid hill of Sikkim. *Range Mgmt.& Agro-forestry.* 17(1):93-97.

Singh, K.A.(1999). Nutrient content in tree fodder and the leaves of bamboo used during scarcity period in eastern Himalaya. *Indian Journal of Animal Nutrition.* 16 (2):178-182.

Singh, K.A. (2002). Application of bio-terracing for land development in North Eastern Hills of India. *Indian Journal of Soil Conservation.* 30 (3):226-231.

Singh, K.A. and Pradhan, I.P. (1993). An appraisal of some indigenous farming systems with special reference to agro-forestry in Sikkim. *Indian J. Hill Farming.* 6(1):113-122.

Singh, K.A. and Thompsion, F.B. (1995). Effect of looping on water potential, transpiration, regrowth ^{14}C- photosynthesis distribution and biomass prediction in *Alnus glutinosa. Tree Physiology.* 15:197-202.

Singh, K.A., Rai, R.N., Patiram and Bhutia, D.T. (1989). Large cardamom (*Amomum subulatum*) plantation-an age old agro-forestry system in Eastern Himalayas. *Agro-forestry System.* 9:241-257.

Singh, P. (990). Agri-silviculture systems in India. In: shrubs and tree fodder for farm animals (Eds. C.Devendra), IDRC, Ottawa (Canada): 186-195.

Singh, V.P.; Dhyani, S.K., Verchot, Lou and Lasco, R. (2007). Climate change and agro-forestry. National Conference on Climate Change and Indian Agriculture. Organized by ICAR at NASC, New Delhi, Oct.12-13, 2007.

Hill Agriculture: Prospects, Constraints and Mitigations (2013) *Pages* **105–120**
Editor: **Dr. Shahid Ahamad**
Published by: **DAYA PUBLISHING HOUSE, NEW DELHI**

Chapter 8

Role of Backyard Poultry Farming in Hilly Areas in Food Security

***Pranav Kumar*[1], *Devesh Thakur*[2], *Dhirendra Kumar*[1] and *Prem Kumar*[3]**

[1]*Faculty of Veterinary Sciences and Animal Husbandry, R.S.Pura, SKUAST-Jammu*
[2]*Veterinary College, CSK H.P. Agricultural University, Palampur, Himachal Pradesh, India*
[3]*SMS, KVK, R.S. Pura, SKUAST-Jammu*

Introduction

"India has reached its target of cereal security to a great extent, now what is required is nutritional security, which can be achieved by making available animal protein in adequate quantity"

–M.S. Swaminathan

Poultry has influenced man's civilization in many ways. Eggs and meat of birds are being consumed since pre historic times. Compared to eggs, there is no other single food of animal origin which is eaten and relished by so many people in the world over and none is served in such a variety of ways. Indigenous or backyard poultry rearing is a traditional and age-old practice in downtrodden and under-privileged people of hilly areas. Till today, they prefer to keep non-descript, multicolored birds because these birds are eco-friendly and able to provide small but regular source of income over many generations. It has a good potential in our country especially in rural areas of hilly region to improve the socio-economic condition and overcoming protein deficiency among the poor. The per capita consumption of egg and chicken meat in urban areas is ranged between 80-120 eggs and 2.25-3.5 kg, respectively against 5-20 eggs and 750 g in rural areas. Further, these poultry products are expensive (10–40 per cent) in rural/tribal areas due to their non-availability. Rural families in our country consume rice or wheat as staple food, which is rich in energy and low in protein. Due to protein deficiency in people of hilly regions, particularly pregnant women, nursing mothers and growing children are becoming vulnerable to many malnutrition types of diseases. The backyard in rural/hilly areas is rich with "natural food base" (like fallen grains, insects, earthworms, kitchen waste, green grass, etc). These waste food materials can be ploughed back into the human food chain, by converting them in to nutritionally balanced and delicious egg and chicken meat. Adapting the rural poultry farming can also alleviate the protein hunger in the hilly areas. In areas where plenty or natural feed resources are available, a small number of birds (10-20) can be reared for egg purpose under free-range conditions. If the local demand is for meat, they can be reared in large number under intensive/semi intensive conditions by providing all in outs similar to the commercial broilers.

Poultry is one of the fastest growing segments of the agricultural sector in India today. While the production of agricultural crops has been rising at a rate of 1.5–2.0 per cent per annum that of eggs and broilers has been rising at a rate of 8–10 per cent per annum (Mehta *et al.*, 2003).

National annual consumption is 37 billion eggs and one billion broilers. Estimates of income elasticity for meat and eggs strongly suggest that consumption of these products can be expected to continue to grow strongly. Per capita consumption of eggs in rural areas is less than half that in urban areas (Mehta *et al.*, 2003).India has one of the world's largest and fastest growing poultry industries, ranking third in egg production and sixth in poultry meat production. (USDA, Agricultural Marketing Service, 2011). Poultry farming is recognized as a potential tool to fight poverty and malnutrition.

Though poultry development in the country has taken a quantum leap in the last three decades, the growth has been mainly restricted to commercial poultry. Rural backyard poultry including hilly areas, though still contributing nearly 30 per cent to the national egg production, is the most neglected one. This is in spite of the fact that their poultry eggs and meat fetch a much higher price than that from commercial poultry. About 70 per cent of the poultry products and eggs are consumed in urban and semi urban areas and the rural consumption is quite low. The percentage of native birds in the total poultry population has dropped from 50 per cent about 30 years ago to about 10 per cent now *(Rangnekar and Rangnekar, 1999)*. So, the large scale commercial and small scale rural family sectors need not be mutually exclusive.

Trends in the poultry sector provide a striking example of how sector growth does not necessarily go hand in hand with poverty reduction and more so with food security to rural livelihoods. Family poultry (or the 'traditional scavenging' system), which is based almost entirely on native birds, has been by-passed by the poultry revolution, with virtually all the growth occurring in the large-scale 'confined and intensive' (or industrial) sub-sector. By contrast, traditional poultry-keeping appears to be a stagnant low-productivity sub-sector.

Backyard Poultry Farming in India

In India poultry farming under backyard system is as old as its civilization. Randhawa (1946) reported that number of terracotta have been discovered from Mohenzo-daro and Harappa, which indicated that, the people domesticated number of birds and the domestic fowl (*Gallus gallus domesticus)* originated in India and its ancestor's, the red jungle fowl (*Gallus gallus*) is still found in Northern India from Kashmir to Assam and in Madhya Pradesh, West Bengal, Orissa, Visakhapatnam and parts of Godavari district of Andhra Pradesh. He also reported that the Indus valley people kept the fowl only for sports, and that its breeding for flesh occurred later.

It is a traditional system of rearing poultry with an average flock size of 5-15 native breed or non-descript fowls by farm families to meet their dietary needs or small cash needs. Here the birds are mainly reared for food security and even small cash if they get out of rearing will be used for food most of the times. Poultry farming is possible in widely different agro-climatic environment including hilly regions (National Commission on Agriculture, 1976), as the fowl possesses marked physiological adaptability. Requirement of small space, low capital investment, quick return from outlay and well distributed turn over throughout the year make poultry farming remunerative in hilly areas, rural and urban areas. The rearing of poultry provides an excellent opportunity for gainful employment to idle or unemployed members of rural communities. Additionally, chicken meat consumption is a significant protein source which helps to cover the nutritional needs of the rural population.

Need of Backyard Poultry in Hilly Regions

In true terms 'livestock revolution' aims at not merely increasing the quantum of production but to have an holistic approach wherein problems of malnutrition, gender inequality and depraved socially-backward sections of the society are also dealt with.

- ☆ Poultry, due to its least demanding nature in terms of infrastructure in the backyard set up, has been widely accepted by the rural poor. More than 50 per cent of the landless and marginal farmers at the bottom end of the smallholder spectrum eke out their living from poultry and small ruminants. (DRAFT NATIONAL POULTRY POLICY-2005)
- ☆ Taking into consideration the recommendation, the poultry production is to be increased 5 folds for egg production and nearly 11 folds for broiler production. The growth of the poultry industry in such tune seems to be impossible as industrial poultry is in direct competition with human beings and milch animals for grains and cakes (protein) due to which either these feed ingredients are not available to poultry or become costlier rendering poultry keeping economically unviable as happened during 1991-1992, 1998-1999 and 2000-2001 when Indian poultry industry was almost paralyzed and 60-70 per cent small farmers shut their operations (Sharma *et al.*, 2001).
- ☆ It may also be mentioned that groups of small rural producers cater to the needs of consumers who have a specific preference for coloured birds and brown-shelled eggs, both of which are mostly produced in the rural sector/backyard poultry.
- ☆ The high yielding exotic birds are very prone to various emerging diseases. A huge investment is needed for establishing modern diagnostic and feed testing facilities to prevent frequent epidemics.
- ☆ It ensures availability of egg and meat even in remote rural areas. The per capita availability of eggs and poultry meat in rural areas which is much lower than the urban areas can be increased.
- ☆ To provide food security to rural folk, there by making development more holistic.

Poultry thus has been one of the core support system to subsistence farmers, providing supplementary income and is taken care of by the family, especially the women folk.

Importance of Backyard Poultry in Hilly Regions

Backyard poultry a traditional system of poultry keeping is a part of livestock rearing practiced by rural folk since the time immemorial. It is a type of organic farming with no harmful residue in egg and meat. It is eco-friendly approach (Panda and Singh, 2002). Further, Village chickens are very active in pest control, provide manure, are required for special festivals and are essential for many traditional ceremonies (Alders *et al.*, 2003). Family poultry acts as a biopesticide by eating insects and worms while scavenging (Rajini, 2004). A recent study in Mozambique (Harun, 2001) showed how village poultry play a key role in the local economy, and how increased production has the potential to improve food security, assist in poverty alleviation and mitigate the adverse economic impacts of HIV / AIDS for rural populations.

Sources of Income and Employment

Poultry industry is one of the most important industries in India as it contributes 2.5 per cent to the National GDP, 33,000 crores to the GNP, employs 30 lakh people either directly or indirectly and earns 200 crores of foreign exchange. India has witnessed a rapid development of poultry industry and the poultry production has increased from 258.34 million in 1987 to 661.5 million in 2003. At present, annual growth rate is about 10 per cent in meat production and 6 per cent in egg production. Broiler production recorded an annual output of over 1.9 billion tonnes in 2005, contributing around 2.1 per cent to global output and a corresponding average annual growth rate of 12-15 per cent. In India, if proper atmosphere is provided for the growth of this industry provide remunerative jobs to 3.7 million people in rural areas. For *e.g.* it can be adopted as whole time business on large scale, it can fit very well in mixed farming system to provide additional income to farmers engaged in arable farming, it can be taken as part time business on small scale by under-employed persons involved in service profession. Even house -wife and children can look after a small flock of birds in their extra time for small earning.

Sources of Food in Hilly Regions

- ☆ Egg and meat are, perhaps, the cheapest source of protein to fight protein malnutrition in hilly regions of India. Backyard Poultry System can act as solution to this problem of malnutrition. Two eggs provide 160 calories of energy and more than 20 per cent of the daily requirement of proteins, Vitamins A, D and $B_{12,}$ Riboflavin, Folic acid, Pantothenic acid, Phosphorus, Iodine along with fat.

Advantages of Backyard Poultry in Hilly Regions

Backyard poultry has got many advantages compared to other small scale farming. They are:

- ☆ It provides supplementary income in shortest possible time.
- ☆ It is very simple in operation.
- ☆ It needs very minimum or practically no capital investment.
- ☆ It gives highest rate of return on land and investment.
- ☆ As the Indigenous germplasm is used mostly, they got better adaptability in tropical areas and protect themselves from predators.
- ☆ People have more preference for eggs and meat of indigenous poultry even in urban areas.
- ☆ It ensures availability of egg and meat even in remote rural areas.
- ☆ It saves money on eggs, meat, garden fertilizers.
- ☆ It is considered as a tool for rural development, utmost need of present day.

Need of Improved Germplasm for Backyard Poultry Farming

Although indigenous birds are desirable but they are poor performers in terms of meat and egg production. The indigenous birds are mostly non-descript and are of very little value as layer because of their low production potential (70-80/year), which is far less compared to either improved strain of birds or layers. The following desirable major characteristics of scavenging chicken can be considered while selection of breed for backyard poultry farming.

a. *Colour of the bird*: Bird with coloured plumage escape from the predators being camouflaged by their colour as against white or light coloured bird. Better immune competence has been reported for coloured birds as compared to white plumage.
b. *Body conformation*: There is always danger of predators like wild cats, stray dogs, mongoose etc. in free range systems. A bird with low body weight, low shank and strong wings has a greater chance of escaping from predators by fast running and flying till they reach to safer place.
c. *Productivity of the birds*: The heavy dual purpose birds cannot perform well as compared to layer bird both from the survival and economic parameters points of view. Less nutrient availability and long distance to be covered for collecting the feed suggests the choice of light layer bird than heavy bird.
d. *Disease resistance*: The scavenging birds must have immune competence as there is always lack of quality feed and drinking water and they have to roam into dirty surrounding in search of feed. In addition appropriate prophylactic measures are not always possible in remote.
e. *Self propagation*: The capacity of self propagation is an essential requirement for the scavenging bird as it might not be possible always to get the replacement chicks from the hatchery for the remote areas and for weaker section of the society due to financial constraints.

The concept of improved colour feathered bird is found to be fowl of choice for backyard poultry production. Vanaraja, Gramapriya, Giriraja, Girirani, Gramshere, Krishipriya, Krishibro, CARI-Gold,

Since all the improved bird varieties developed for backyard poultry farming are hatched out artificially, they essentially require initial brooding. Electrical heater or kerosene lamp can be used for heating the brooder house. Bukharies with burning of woods charcoal or saw dust are useful in remote areas where availability of electricity, gas or kerosene is limited. Heat source of 2 watts/chick is adequate upto 6 weeks of age.

Feeding of Poultry

Birds under free range can easily pick up its food from the backyards once it learns to scavenge in the fields. The need for additional feed supplementation depends on the free area available in the backyards. Generally, the birds under free-range conditions can meet their protein requirement through scavenging. Therefore, feeding the birds with cereals (bajra, ragi, jowar, korra, broken rice, rice polish, rice bran, etc.) available is always beneficial to sustain the production. The nature of supplements fed depends on the purpose of rearing. For meat purpose, feeding the birds with commercial broiler/layer chick feed is suggested. If the purpose of rearing is for egg production, the birds should largely depend on food available in free-range conditions. Care should be taken to restrict the weight of pullets (female) between 2.2 to 2.5 kg at 6.0-6.5 months of age. Excess body weight may reduce egg production. The broken/shell-less eggs can be minimized by supplementing the calcium sources (lime powder, shell grit, stone grit, etc.) @ 3-4 g/bird/day. As the backyard birds scavenge on grass field to have an access to novel protein and energy rich sources like insects, white ants, green grass, grass seeds, waste grains etc. thereby the requirement of concentrate feed is much less than those reared under intensive poultry farming. Feed supplementation in the form of scratch usually given in the morning/evening to develop habit to reach owner's place for laying eggs and for night shelter. Depending on the

availability of free range area and also the intensity of vegetative growth, the requirement of supplemental feed varies between 25 to 50 g/bird/day. These birds can also perform well on whole grain feeding under scavenging conditions. However, for better shell quality, shell grit or marble stone chips need to be supplemented @ 5–7 g/bird/day during laying period. The poultry feed should be well balanced with respect to energy, protein, minerals, vitamins, fiber and water in proper proportion. Use of feed stored for a long time should be avoided. Two feeding regimen should be followed (morning and evening) and feeders should not be filled more than 2/3rd of their capacity to avoid wastage.

Health Care

The most important disease that affects birds under free-range farming is the Newcastle diseases. Night shelter should have good ventilation, required light and protection from the predators. The material used for night shelter such as wood and bamboo offers a good hiding place for external parasites. Therefore periodic cleaning of night shelter is essential. Since the chicks move in free range, there is a possibility of parasitic infestation. The deworming at 2-3 months interval is required. Under free-range conditions adults Vanaraja birds should be vaccinated against Newcastle disease at 6 months interval, preferably one should be before the onset of summer. Vaccination of native birds along with Vanaraja is recommended. Controlling the incidence of diseases in the free-range and backyard chicken is a challenging task as they are exposed to adverse environmental conditions such as weather changes, poor quality feed, contaminated water and air, predators etc. contacts between flocks of different households, and the common practice of buying and selling chicken are the important sources of disease transmission. Ranikhet disease (RD) and fowl pox are most important diseases. Vaccination against these diseases on community basis can minimize their occurrence and resultant economic losses.

Types of Rural Poultry Production System

Family poultry are kept under a wide range of conditions, which can be classified into one of four broad production systems (Bessei, 1987):

- ✰ Free-range extensive,
- ✰ Backyard extensive or scavenging backyard poultry (*Commonly practiced in India*),
- ✰ Semi-intensive or semi scavenging system, and
- ✰ Intensive (commercial, very rarely practiced in house holds).

Free-Range Extensive

It is commonly practiced in Africa and Latin America. Under free-range conditions, the birds are not confined and can scavenge for food over a wide area. Rudimentary shelters may be provided, and these may or may not be used. The birds may roost outside, usually in trees, and nest in the bush. The flock contains birds of different species and varying ages.

Scavenging Backyard Poultry

Under this system, Poultry are housed at night but allowed free-range during the day. They are usually fed a handful of grain in the morning and evening to supplement scavenging. As such no specific feeding is required. It is not an occupation but a supplementary household activity.

Usually non-descript fowls are used. High yielding varieties fail to give better results. The Central Avian Research Institute (CARI), Izatnagar, premier poultry institute developed 4 types of indigenous germplasm developed for scavenging system. CARI Nirbheek, CARI Shyama, Hit CARI, UPCARI.

Semi Scavenging System

This is a combination of the extensive and intensive systems where birds are confined to a certain area with access to shelter. They are commonly found in urban and peri-urban as well as rural situations. This system is used for small poultry flocks. Here birds are reared under partly controlled management. This is a self sustainable complete package system developed by Q.M.E. Haque (1996) which is successfully practiced in Bangladesh. This system requires following additional care to increase returns.

- ✰ High yielding birds which are capable to scavenge in the field are required in this system.
- ✰ Substantial supplementary feeding and Proper housing is required for better production.
- ✰ Proper vaccination and medication schedule should be followed to reduce mortality due to diseases.

Usually non-descript fowls or indigenous germplasm developed for this system are used. Comparatively high yielding varieties with better adaptability to the tropical conditions are used in this system. Giriraja, Vanaraja, CARI debendra, Grampriya, etc., are the some such indigenous germplasm developed by various institute involved in poultry research across the country.

Food Security in Hilly Regions

Food Security is defined as a situation that exists when all people, at all times, have physical, social and economic access to sufficient, safe and nutritious food that meets their dietary needs and food preferences for an active and healthy life (FAO, 2002, *The State of Food Insecurity in the World*). Food security is said to be existed in household when quality food is accessible to it and at the same time it is affordable to that household

Elements of Food Security

- ✰ Availability of food at national level: through domestic production and/or through imports.
- ✰ Access to food at house hold level: issue of purchasing power.
- ✰ Access at intra- house hold level: more an issue of gender discrimination and sociology at house hold level.
- ✰ Absorption of food: clean drinking water, environmental hygiene and primary health care

Status of Food Security in India

India, being the world's largest democracy given freedom of speech, religion, and the press ever since inception of constitution (January 26, 1950). Within this democracy people still live everyday being food insecure. Though India's overall economy and trade relations are improving, the poor people are still struggling to survive. Poverty easily coexists with food insecurity and is the main cause of hunger and malnutrition. More than 850 million people all over the world live everyday being food insecure. Malnutrition not only denies people their right to health; it also has serious economic implications. Malnourished adults are less able to work effectively – thus undermining productivity and economic growth. On the other hand population is growing at galloping speed. Part of the reason India's population is growing so rapidly is because family is very important to the Indian people. So, the food and nutritional security can be improved through developments in domestic policies targeting individual house holds or families. Then improving backyard poultry can become one important tool to come out of this complicated mess of malnutrition, hunger death, poverty etc in general and food insecurity in particular. Today the very 'big' nutritional deficiencies noticed in India are,

- ☆ Protein energy malnutrition among the school children
- ☆ Vitamin A deficiency among children
- ☆ Iron and Iodine deficiency anaemia among pregnant woman

(Reference-Indian Council of Medical Research)

Links between Backyard Poultry and Food Security in Hilly Regions

Food security tends to be a problem for the poor and much less for the better off households and type of livestock kept by a household may therefore be used as an indirect indicator of food security. In many countries, scavenging poultry is one of the few assets owned by poor households and almost as a rule, it is the women who decide over income from poultry. In one study in Bangladesh, it was found that no less than 80 per cent of the bullock holders were on holdings with more than 1 acre while more than 50 per cent of the total number of goats and chickens were kept by households with less than 0.5 acre of land or no land at all. In an example from Africa it was found that in Swazi rural households backyard poultry keeping is much more widespread than of cattle, goats and pigs and they are food secure. The conclusion is that one link between livestock and food security is that the type of livestock kept in a household may be taken as an indicator of food security in case of poor households (Frands Dolberg, 2004).

The role of family poultry in poverty alleviation, food security and the promotion of gender equality in developing countries is well documented (Guèye, 2000). Family poultry production represents an appropriate system to contribute to feeding the fast growing human populations and to provide income to poor small farmers, especially women. (Gujit, 1994; Alders, 1996; Kitalyi and Mayer, 1998). Targeting small-scale, family-based poultry systems as an effective entry point for poverty alleviation programmes is gaining widespread acceptance (S. MACK *et al.*, 2005).

Advantages of Poultry as a Human Food in Hilly Regions

- ☆ It is found that 75 per cent of non-vegetarian food consumed in India comes from the poultry products, as it is acceptable to all non-vegetarian individuals.
- ☆ Poultry and poultry products constitutes an important component of human diet in most of the developing countries of the world.
- ☆ Poultry products have no religious taboo.
- ☆ Poultry products are the most economical source of animal protein.
- ☆ Egg is considered as complete food with all nutrients required for human growth and has got highest Biological Value.
- ☆ Poultry is the least cost alternative only next to fish and produces more of animal protein from the same amount of feed compared to milch cow, sheep, Goat and Pig.
- ☆ Egg is less perishable compare to other animal foods and it is the only unadulterated food of animal origin.
- ☆ This consumption is also increasing at a rapid rate due to low fat content, easy availability and cost effectiveness.
- ☆ Poultry farming require less area with high return than any other animal husbandry and agriculture activities.
- ☆ Land topography and soil fertility is never a criteria for poultry like agriculture.

☆ In the primary sector, Agriculture provides about 100 to 120 days employment to the rural poor, however, poultry farming can be a viable option for rural poor to overcome the issue.

A very interesting fact in recent study from Kenya has shown that children who get some animal source foods have better cognition and activity levels in school than children who did not receive such food and interestingly, the children getting meat did better than the children getting milk. *(Charlotte Heumann and Diane M. Harris, 1999).*

Problems Faced by Backyard Poultry Farming

The major reason for failure of earlier programmes on backyard farming is higher early chick mortality. Brooding and immunization programmes could not be adopted during earlier programme. Due to lack of facilities for scientific management of the chicks during the initial 4 to 5 weeks in the village conditions is the major hurdle for success of the backyard farming. Though these birds have better general immune competence they need to be protected against Newcastle disease. The earlier reports also indicated that the most of the early chick mortality was due to Newcastle disease (Bell, 1996; Rangnekar and Rangneker, 1996). Protecting the chicks with inactivated vaccine is the effective way to the control the disease in rural poultry production (Bell, 1996). Realizing the importance of initial brooding and vaccination, the Project Directorate on Poultry (ICAR) has taken initiative to supply the grown up chicks (about 4 to 6 weeks of age) to the rural/tribal populations either directly or through any non-govt. or govt. agencies after protecting the chicks with Marek's and Newcastle diseases. This resulted in higher survivability up to 97 to 98 per cent under field conditions. In addition to following the prescribed vaccination schedule the farmers are advised to practice routine deworming of the birds at every 35 to 40 days interval.

The growth of backyard poultry is declining due to various problems. In short they can be listed as follows,

☆ Poor disease control (single most important disease- ND and others like internal and external parasites, gumboro disease, mycoplasma, *Salmonella*).

☆ There are less genetic improvements in finding suitable germaplasm to different agro-climatic regions of India.

☆ Very poor managemental awareness is leading to high mortality. High Mortality in first weeks of life under village and backyard conditions is mainly a nutrition and management problem and not a disease problem (Source: Matthewman in Smith 2001, p. 199).

☆ Non availability of quality and good feed in local market suitable for chicks of backyard farming.

☆ Rural and tribal farmers are very less access to production inputs, infrastructure and capital.

☆ Lack of proper linkage between various govt. departments, research institutes. SAUs, veterinary colleges and NGO's working in poultry sector and rural upliftment.

☆ Transfer of technology for spread of backyard poultry is not adequate.

☆ Poor Marketing and training facilities are also equally contributing to the problem and, foremost, there are no conducive institutions and governmental policies for backyard poultry development. However, addressing any one or several of these constraints without attention to all will do little to improve the situation (Permin *et al.*, 2000).

SWOT Analysis of Indigenous (Backyard) Poultry Rearing in Hilly Areas

Strengths

- Indigenous poultry enterprises do not need high start-up costs to establish a profitable business.
- Good source of income for disadvantaged women in efforts to emancipate and empower marginalize women.
- Emergency source of income.
- Traditional poultry rearing knowledge.
- Market for indigenous (desi) eggs and chicken is strong and largely unexploited

Weaknesses

- Lack of sensitization of the potential of poultry as a income generating activity.
- Low education level of indigenous poultry keeper.
- Lack of good management practices during disease condition.
- Increased work load on women.
- Local breeds have low productivity in terms of egg production and weight gain due to genetic limitations.

Opportunities

- Poultry can be used as a tool to alleviate poverty while also ensuring food security.
- Maximizing of production to take advantage of economies of scale.
- Processing to add value to chicken products like chicken pickle.
- Marketing of live birds and eggs could be organized.
- Self employment and entrepreneurship.

Threats

- Attack by predators.
- Lack of adequate scavenging land.
- An outbreak of bird flu and other epidemics.
- Ill health management practices.
- Discontinue poultry keeping once the government schemes/project are over.

Govt. of India Initiatives

1. The Department of Animal Husbandry has taken up in association with ICAR initiated a targeted program, AICRP on Poultry for up-gradation of desi germplasm. Birds like Vanaraja, Krishibro, Krishilayer, Caribro, Carired, Gramapria, Naked neck broiler, Dwarf broiler etc. are result of this initiative.
2. The Central Poultry Breeding Farms located at Bangalore, Mumbai, Bhubaneswar and Chandigarh has been re-oriented towards the goal of making rural poultry a viable income generation proposition. Accordingly, the basic purpose of the scheme is to increase egg and

meat production especially in the rural areas of the country without putting substantial strain on feed resources and other inputs the cost of which the small farmers are not able to absorb. This will also meet the specific rural consumer preference as well as provide subsidiary occupation to rural poor, women and unemployed youth.

3. Assistance to state poultry farms is a scheme initiated by GOI with states. Under this scheme, one time assistance is provided to State Poultry Farms to suitably strengthen them in terms of hatching, brooding and rearing of the birds with provision for feed mill, their quality monitoring and in-house disease diagnostic facilities. These farms maintain only the parent stock of low-input technology birds duly identified by this Department in consultation with ICAR and State Governments.
4. NATP project on small farm rural poultry production is started at CARI, RAU (Bikaner), CCSHAU (Hisar) and GAU (Anand).

The Vision: Holistic Approach

Keeping the above problems in mind it is now proposed that to have

(i) Rural Production and Social Organization

A more holistic approach with the following new initiatives:

The pattern envisaged now will also cater to intermediate rungs of poultry farmers who will rear the chicks in so-called 'mother units' and also help in providing the backyard poultry rearers with 4-8 week old reared chicks so as to minimize the otherwise high chick mortality. The training will be more intensive and services will be provided at their doorsteps. The formation of self-help groups amongst the poor will be encouraged to build self-confidence and social security in the long run.

(ii) Cluster Approach

An exercise, involving as small as 20 birds in the backyard can also be viable but not always monitorable. Further, we need to ensure that whatever little surplus production is there should effectively be channelized into a well-defined marketing set-up. This will open vistas for expansion and upgradation of activities at the backyard level. For this a cluster approach is necessary.

(iii) Backward and Forward Linkages

The required minimum infrastructure, inputs, feed, health service dissemination etc. as backward linkage and marketing of egg, meat, poultry products' as forward linkage have to be provided by a resource agency or link agency. Preferably, these units should concentrate on inputs and services including procurement, rather than production itself. The failure of most of the poultry corporations/ federations has been generally attributed to their giving more attention towards production and less towards marketing.

(iv) Mother Units and Mobilizing NGOs/SHGs

Suitable NGOs may be involved for operating the mother units, as far as possible. At the backyard rearer level it is necessary that they may form Self Help Groups (SHGs) as it can address the problems of their micro-financing. The financial institutions will also come forward easily to assist these highly motivated and confident homogenous groups. Economic viability of the mother units and their deep commitment are both essential for the success of the programme.

(v) Convergence of Poultry Activities

While reviewing the work done so far, it was seen that there are many development programs for rural farmers, women, tribals and other weaker sections of the society wherein poultry keeping is also one of the components. NABARD and NCDC are also re-financing many poultry projects and, though lately the projects try to incorporate all major inputs and services like feed, health care, training etc., subsequent follow-up and constant interaction are mostly lacking. Further, the programs implemented by different agencies are not complementing each other and have over-lapping objectives at times. Therefore, there is a clear need to coordinate other poultry development projects to work towards a common goal of strengthening all the spheres of backward and forward linkages and Research and Development associated with this sector. The present proposal envisages linkages with other development agencies and programs through networks between agencies/programs and research institutes, State Agriculture Institutions/other ICAR Institutions etc. This will not only help steering the course of the project for rapid and efficient implementation but also offer a platform for trouble-shooting and problem management together.

(vi) Outreach of Services

The skill and training dissemination as well as health care outreach are very limited through the present set up. It is, therefore, necessary to work in close liaison with the District Rural Development Agencies (DRDAs) who may help in increasing the outreach of activities District/Block Extension teams. Poultry Link Workers (PLWs), may be selected through mass contact programs and trained intensively for skill and service dissemination. These link workers will not only increase the outreach but will develop the necessary rapport within and between the villages through their link and help sort out common problems. This will go a long way in ensuring sustainability.

Conclusion

The overall aims of development are to improve food security and rural livelihood in hilly regions. To develop effective strategies for backyard poultry development, traditional but inefficient methods of production must be replaced by more suitable measures with the main objective of improving food supply. Family poultry is very important means of food of animal origin to women and children in hilly areas who have little or no access to other factors of production. India has tremendous potential in organic poultry in these areas because of traditional backyard system practiced here since time immemorial and organic poultry happens to be the need of modern world in coming days. India is a developing country in which majority of people are involved in agricultural practices. Some are merely poor farmers, marginal farmers, and landless labours. In this world where the prices are soaring like anything with people not able to make even 2 meals per day. Here poultry farming specially, the Backyard poultry farming comes to rescue in which not only man but also women, children and even old people can help and add a hand to its own development as well as the nation's development. Backyard Poultry Farming can provide Food – in the form of meat and egg, Protein Source – Cheapest one *i.e.*, egg, Manure – Poultry waste, Income and Employment.

Don't think that a Poultry Bird is so small in size, how it will help me in gaining profit. "Sometimes good things come in small package."

So the Backyard Poultry in hilly regions is poised to transform poultry sector in particular and hill animal-agriculture in general, if regulations, infrastructure facilities, transfer of technology and sectoral cum target oriented development programmes are brought in practice with basic thrust towards food security, poverty alleviation and improving the rural livelihoods"

References

Alders, R., Cambaza, A.B. and Harun, M. (2003). Village chickens, food security and HIV/AIDS mitigation [Online]. Australia. AusAID Southern Africa Newcastle Disease Control Project, Maputo.<http://www.vsap.uq.edu.au/RuralPoultry/content/SANDCP/HIV-AIDS%20paper%20Nov03.pdf> Accessed 2003 March 10.

CARI perspective plan, vision-2025 (2007).

Chakravarty, Sujoy (2006). "Food Insecurity in India: Causes and Dimensions." Indian Institute of Management Ahmedabad, Research and Publication Department.

Dolberg, Frands. Links between livestock, environment and food security. LEAD E-Newsletter 04 Volume 1-4.

Draft National Poultry Policy, Government of India (2005)

Food Insecurity in India. Jamie Kaczor, Student Participant, Anthony Wayne High School, Whitehouse, OH.

Harun, M. and Massango, F.A. (2000). Village Poultry Production in Mozambique: Farming Systems and Ethnoveterinary Knowledge in Angonia and Tsangano Districts, Tete Province [Online]. Canberra [Aust].Australian Centre for International Agricultural Research (ACIAR). <http://www.aciar.gov.au/web.nsf/JFRN- Accessed 2004 March 10.

Heumann, Charlotte and Harris, Diane M. (1999). Contribution of animal source foods in improving diet quality for children in the developing world. University of California. Paper prepared for the World Bank. *Institute on "Modern Techniques and Practices in Poultry Production and Breeding"*. GAU Anand, India.

Jadav and Siddiqui. *Handbook of Poultry Production and Management*, Jaypee publications, New Delhi.

Mack, S. *et al.* (2005). *World's Poultry Science Journal*, Vol. 61, March.

Permin, A., Pederson, G. and Riise, J.C. (2000). Poultry as a Tool for Poverty Alleviation: Opportunities and Problems Related to Poultry Production at Village Level [Online]. Canberra [Aust].Australian Centre for International Agricultural Research (ACIAR). <http://www.aciar.gov.au/web.nsf/doc/JFRN J479R/$file/PR103% 20 Chapter% 2029.pdf> Accessed 2004 March 10.

Rao, C.K. (2002). Appropriate poultry strains for weaker sections – why and how? In souvenir of 2nd National Seminar on Rural Poultry for Adverse Environment, University of Agricultural Sciences, Bangalore, pp 15-21.

Sasidhar, P.V. (Ed). (2006). Poultry Research Priorities to 2020, CARI, Izatanagar.

Singh, D.P. (2003). Problems and prospects of traditional backyard poultry production. *Proceedings Winter*

Singh, D.P. (2003a). Breeding strategies for scavenging chicken. *Proceedings of XXII IPSACON-2003*, 27-28 March, CARI, Izatnagar, Bareilly, India, pp. 40–41.

Singh, D.P. *et al.* Impact of constraints minimization on productivity and popularity of traditional backyard poultry production. *Central Avian Research Institute, Izatnagar, India.*

Sonaiya, E.B. (2007). *World's Poultry Science Journal*, Vol. 63, March.

The Socio-economic Importance of Family Poultry, FAO corporate Document Repository, originated by Agriculture and Consumer Protection.

Hill Agriculture: Prospects, Constraints and Mitigations (2013) *Pages* **121–129**
Editor: **Dr. Shahid Ahamad**
Published by: **DAYA PUBLISHING HOUSE, NEW DELHI**

Chapter 9
Genetic Improvement of Cereal Crops in North-West Hills

R.K. Salgotra

Dryland Research Sub Station,
Sher-e-Kashmir University of Agricultural Sciences and Technology
Dhiansar, Bari Brahmina, Jammu – 181 133, J&K, India

Over the last 50 years, Indian agriculture has witnessed spectacular advances in both production and productivity of food crops. India produced over 240 million tonnes of food grains in 2010-2011 (Anonymous, 2009). However, all these achievements did not receive required appreciation till the impact of dwarf wheats and rices became evident after the mid-sixties. This could be effectively achieved because of the realization to exploit indigenous and exotic germplasm. Introduction of promising genetic resources to widen the genetic base required in crop improvement for diversified national requirements, and trials of straight introductions and acclimatization of exotic genetic stocks have proved rewarding. Hybrid research in maize resulted in the release of several high yielding hybrids. India was also the first to release the hybrids of basmati, cotton and pearlmillet with remarkable yield potential. Production is almost doubled in crops, such as cotton, jute and oilseeds since 1950. There is tremendous increase in productivity of all most of the crops after introduction of green revolution but in hills it is still need to do more. Increased crop production remains an important step toward the goal of increasing the profitability and sustainability of hill agriculture. Efforts are made here to enlist some of the major achievements relating to crop improvement programmes in hills as experienced during the last fifty decades, and it is attempted to cover all major crops.

Maize

Maize (*Zea mays* L. ssp. *mays*) is a diploid 2n=20, in the tribe *Maydeae*, *Tripaceae* or *Androponeae* in the sub-family of *Panicoideae* of the graminae. It is one of the most diverse grain crop found in nature. It is also known as the queen of the cereals, cultivated all over the world and is an economically important crop for human populations. It is grown from 58°L North to 40°S South and extends from sea level upto altitudes of 3800 meters amsl, covering regions with rainfall varying from 250 to 10,000 mm per annum (Hallauer and Miranda, 1988). It has worldwide significance as human food, animal feed, pharmaceutical and a source of large number of industrial manufactures. Besides, it is the most important crop in the hills for food and fodder.

A common classification of maize based on endosperm characteristics distinguishes seven types

Dent Corn (*Zea mays* var. *indentata*)

Because the crown of the kernel consists of floury starch, moisture loss from this area upon kernel maturation causes a slight collapse in volume that produces a characteristic dent. This is the most produced type of maize on a global basis, accounting for 73 per cent of commercial production and is used as livestock feed and for industrial manufactures (starch, syrup, oil, alcohol).

Flint Corn (*Zea mays* var. *indurate*)

The kernels of flint maize are hard and smooth and contain little starch. Flinty maize was probably developed from pop types by selection for grain size (larger than pop and smaller than dent) and greater yield. This type is produced in areas where cold tolerance is required or where storage and germination conditions are poor.

Pop Corn (*Zea mays* var. *everta*)

Consisting of a small spherical grain with floury (soft) starch core and a flinty (hard) endosperm shell. Moisture trapped in the floury starch expands upon heating and bursts through the hard shell, creating the popular confection.

Flour Corn (*Zea mays* var. *amylacea*)

Flour maize remains the preferred form for direct human consumption, as it consists of soft endosperm that is easily ground to produce meal. They usually develop no dent.

Sweet Corn (*Zea mays* var. *saccharata*)

This type contains a sweetish starch and is characterized by a translucent honey appearance when immature and a wrinkled condition when dry. The ears are picked green for table use and canning.

Pod Corn (*Zea mays* var. *tunicata*)

This is probably one of the earliest domesticated types. Each kernel is enclosed with in the lemma and palea, which are will developed. It is not grown commercially.

Waxy Corn (*Zea mays* var. *ceratina*)

The kernel has a waxy appearance. The starch is gummy and has some of the characteristics of tapioca.

Maize Improvement Programme

First Phase

Maize was the first major cereal crop in India to be affected by hybridisation. Initial efforts were made to identify yellow flint type inbreds with disease and pest resistance, which could be combined in hybrids. In 1961, the first four double cross maize hybrids (Ganga 1, Ganga 101, Ranjit and Deccan) were released for commercial cultivation. Emphasis soon shifted towards composite breeding in view of the problems associated with hybrid seed production, such as replacement of seeds. India is the first country in the world to have developed composite varieties in maize, thus helping the farmers to avoid the replacement of the seed every year (as in the case of hybrids). The release of six maize composites (Amber, Jawahar, Kisan, Vikram, Sona and Vijay) in 1967 has been another significant milestone in the history of maize research in India. Most of the hybrids and composites released for cultivation have fairly high levels of resistance to foliar diseases: In the early 1970s, significant improvement in protein quality was also brought about by the use of the opaque-2 gene. This led to the release of opaque-2 composites in 1971, namely, Rattan, Shakti and Protina, which contain 100 per cent and 60 per cent lysine and tryptophan (essential amino-acids), respectively than the normal maize varieties.

Second Phase

In the mid 1970s, research was carried out to develop more productive, disease resistant and nutritionally superior composite varieties for direct use and to serve as the basic materials for generating superior composite varieties for the hybrid programme. At the same time, special emphasis was given to develop and improve early and very early maturing varieties for rainfed cultivation and for regions where succeeding crops were to be grown. In the early 1980s, one of the components of the population improvement programme was oriented towards inbred enrichment and enhancement to develop heterotic pools for hybrid development. In this, important features were introgression of materials having multiple disease resistance, increase in frequency of desirable alleles for stalk borer resistance, and improvement in tolerance to soil-moisture stress. Some of the desired inbreds like CM 111, CM 119, CM 202, CM 208, CM 300, CM 400, etc., have been used in the development of several hybrids. Five pools, *viz.*, AB yellows, AB white, BC yellow, CD yellow and CD white, were formulated, representing a wide array of maturity levels and grain colour (yellow and white) to develop promising varieties. VL 88, an early-maturing composite was developed from the CD yellow pool and gained wide popularity among farmers. Realizing the potential of winter (*rabi*) maize, a separate winter maize-breeding programme was initiated. The development of heterotic pools exclusively for winter maize was initiated in 1980. Systematic work on *rabi* maize was taken up in the south in the mid 1970s and in the north in the early 1980s. With the availability of high yielding maize hybrids, it was demonstrated that maize could be grown during the winter season. Breeding work on fodder maize started in the 1970s, resulting in the release of the popular fodder composite African Tall in 1980.

Third Phase

Work on sweet corn and popcorns, initiated in late sixties, resulted in composites like Madhuri sweet corn (released in 1990), Amber Pop (1971) and VL Amber Pop (1983). Earlier bad effects of opaque 2 versions of maize were modified to hard opaque 2 composite and were released in 1997, in the name of Shakti 1. Some of the QPM hybrids of CIMMYT, Mexico germplasm were tested and the experimental hybrids possessed high yield potential and were superior in protein quality. In 2001, India released two QPM white hybrids Shaktiman-1 (CML142 x CML150) x CML186 and Shaktiman-2 (CML176 x CML186). Directorate of Maize Research released two new yellow QPM hybrids HQ-1 and HQ-2 in 2003 using CML 161 and CML 193. For purposes of baby corn, 27 varieties and 10 hybrids of maize have been recommended by DMR. These cultivars can be grown in *kharif* and *rabi* season across the country. Baby corn is the ear of maize plant harvested young, especially when the silks have either not emerged or just emerged, and no fertilisation has taken place, depending on the cultivar grown. The dehusked young ears of baby corn are eaten as a vegetable.

Table 9.1: Some of the Promising Maize Hybrids and Composites Released in Hills for Commercial Cultivation

Variety	*Maturity Type*	*Grain Type*	*Yield Potential (q/ha)*
Composites			
Mansar	Medium (90-95 days)	Orange-yellow semi-flint	35-40
VL 88	Early (75-80 days)	Yellow semi-flint	25-30
Kiran	Early (80-85 days)	Orange-flint	30-35
Vijay	Full-season (95-100 days)	Orange semi-flint	40-45
Navjot	Medium (85-90 days)	Orange-yellow semi-flint	35-40
Hybrids			
Ganga 5	Full-season (100-110 days)	Orange semi-flint	40-45
Ganga 2	Full-season (100-110 days)	White semi-flint	40-45
Ganga 11	Full-season (100-110 days)	Orange semi-flint	45-50
Sartaj	Full-season (95-100 days)	Yellow-orange semi-flint	45-50
Deccan hybrid Makka I	Early (80-85 days)	Yellow semi-flint	35-40
VL 42	Early (80-85 days)	Orange flint	35-40

Rice

Rice is one of the most important crops worldwide, as it feeds over half of the world's population. Domesticated rice is an important supply of the world's rice. However, these strains are genetically static and cannot adapt to changing growing conditions. Traditional varieties or landraces of rice are genetically evolving and provide a pool of traits that can be tapped to improve crops worldwide. Ironically, the hill population is today no longer able to sustain itself at the level of self-sufficiency that was present over a hundred years ago. The support area is very much smaller and of poorer quality. The human and livestock population is larger, cultivation more widespread; fields are more fragmented and are on marginal lands, overgrazing persists as also over cutting of forests for fuel wood, fodder and timber. Reduction in the density of forest cover and overgrazing has led to soil erosion of unprecedented proportions. The land has not only depleted its productivity but has fragmented into smaller holdings. Land and water resources are becoming even scarce.

Upland rice is grown on three continents, mostly by small or subsistence farmers in the poorest regions of the world. Grain yields are generally low ranges from 0.5 to 1.5 tons per hectare (t/ha) in Asia. As upland rice is mostly grown in rotation with other crops, the actual area under upland rice based systems is much larger. The upland rice area in India is up to 36 per cent of the total rice area (Huke and Huke 1997). High yields under upland conditions are indeed possible. Much of the future expansion of the world's rice land will probably be in upland rice in hilly regions because most of the land suited to irrigated paddy culture is already planted in lowland rice.

All evidence suggests upland rice research is at a stage where it can make important and rising contributions to increasing rice productivity in the uplands. Rice research for the uplands can provide substantial research outputs for understanding the uplands and the role of rice in it, raising rice yields substantially and developing improved germplasm with traits and genes for meeting specific requirements of the uplands. At the same time, the emphasis on favourable environments in research is bypassing large parts of the uplands, that is major poverty areas, particularly those that are remote, not linked to infrastructure and communication systems and thus not integrated into markets and not reached by agricultural service institutions, such as extension, credit and modern input supply. While some remote areas with expanding infrastructure are getting more and more connected to the favourable regions, there remain substantial areas in the mountainous regions, where extreme remoteness still exists and is unlikely to be overcome for considerable time. Moreover, there is substantial evidence that in areas populated by ethnic minorities ethnic barriers, language and lack of education prevent people to migrate (ADB 2001). If rice research is to benefit also these poorest groups, it has to work on rice in these difficult environments.

Breeding Methods for Hill Rice

Because improved hill rice varieties are intended for cultivation under a cultural system that involves direct seeding, selection among hybrid progenies should be carried out under such conditions. In this respect, the techniques for planting and field evaluating improved lines are similar to those for small grains such as wheat. But hill rice breeding can be accelerated by planting F_1 hybrids, selected early-generation lines and promising selections under lowland culture for maximum seed production. In tropical areas where irrigation water is available during the dry season, it is possible to grow and select two generations of hybrid progenies in a year.

They generally lack, however, resistance to bacterial blight, virus diseases, leafhoppers and planthoppers, and stem borers. So improved as well as traditional lowland varieties should be used to provide genes for extreme earliness, tillering ability, resistance to the major diseases and insects, superior grain characteristics and above all, a higher yielding potential because of a nitrogen-responsive and lodging resistant plant type. Some lowland varieties also have tolerance to problem soils. The guiding principle in breeding, therefore, is to accumulate favourable genes such as for drought resistance and yield, from both variety groups, and to eliminate unfavourable genes such as for low tillering and pest susceptibility.

To accumulate genes that transmit favourable attributes for areas of unreliable rainfall and other adverse environmental conditions, the greatest possible number of such favourable genes should be assembled by every conceivable breeding method, whether it calls for composite crosses, backcrosses, or recurrent selections. Initially, three-way or composite crosses may bring together a large number of gene combinations within a relatively short time. Other approaches should be adopted, however, when applicable. As Hurd (1971) pointed out breeding for high yield in semi-arid climates would largely require many genes to be assembled, each having a small effect. Recurrent selection may therefore prove an efficient way of obtaining the desired genotypes. So little has been published about the varietal development of upland rices that the breeder must mobilize his own resources and imagination to organize hybridization and selection programs. Drought-resistant strains of *O. glaberrima* and of the wild relatives of *O. sativa* are also potential parents. The underlying principle is to use the best genetic sources available regardless of variety designation.

Field resistance to drought is so important in most upland rice areas that breeding population and lines should be screened for this trait as early as the F_2 and not later than, the F_3 generation. Before a breeding line becomes reasonably homozygous, tests for drought resistance should continue for several generations. An effective method is to select F_2 plants in the wet season in an up land field, test the F_3 lines in the dry season, further select among the F_4 lines in the wet season, and so on. Since drought resistance and yield are related traits, estimates of yield performance in the wet season, when feasible, will increase the reliability of dry season findings. This approach is, of course, subject to the unpredictable occurrence of drought in the wet season. Planting on different or several dates in the wet season improves the chances of subjecting plants to drought periods.

As hill culture encompasses a wide range of climatic and cultural conditions, breeding populations should be tested and selected in diverse environments to provide arrays of genetic lines adapted to different ecological niches. The important consideration is the choice of the most suitable parents and of the test sites in relation to the environmental conditions of the production areas for which the genetic lines are intended. When the breeders use diverse parents large breeding populations and multi site testing at an early stage, a wide array of genotypes may be developed that can fit into environments ranging from rainfed lowland culture to dry upland culture. The breeder should therefore fully use the genetic potential of the hybrid populations by exploring different environments during selection and testing.

Future Emphasis on Upland Rice

During its early years, the primary emphasis on improving rice production under favourable soil, climatic and cultural conditions. In the first production technology, the availability of water and nutrients was not a seriously limiting factor. This emphasis was appropriate; it was necessary to determine the potential of the rice plant under favourable as well as under limiting conditions. But from the beginning, scientists were investigating ways to overcome several limiting factors including moisture stress. The short-statured, stiff-strawed varieties developed by various national and international programs tillered profusely and responded remarkably to fertilizers and pesticides.

Upland rice accounts for from 15 to 20 per cent of the world's rice area. In many countries, such as Brazil or Nigeria, most of the rice is grown under upland conditions. The problems faced by the upland farmer are many. Drought is a constant threat to upland rice, dictating a need for varieties with much higher drought tolerance than in lowland rice. Certain diseases, such as blast, are more severe under upland than under lowland conditions. Techniques by which large numbers of cultivars can be screened to determine sources of genetic resistance to these diseases are essentials in any effective research program. Insect control may sometimes be easier under upland than under lowland conditions, but the upland farmer is less likely to have the money to purchase and apply the chemicals.

An intensified research can be concentrated on a number of plant character needed for the improvement of upland rice. For example by expanding upland collection of drought-tolerant cultivar from around the world. These materials along with thousands of accessions already in available in the germplasm bank will be screened for resistance, tolerance, and avoidance of drought. New and improved screening techniques are being developed for both greenhouse and field conditions. The relationships of drought resistance and morphological characteristics, such as rooting habits, need to be studied.

Appropriate technology must be developed for upland rice farmers. More effective weed control through chemical, mechanical, and cultural methods will be investigated. There is need to study the relationship among growth habits, plant height, and ability to compete with weeds. Response to various levels and methods of fertilizer applications under different soil and climatic conditions can be determined, along with improved methods of stand establishment. Physiological factors like hormonal control of the mechanisms by which rice plants escape and tolerate drought; factors that affect the opening and closing of leafs, the effects of high temperature, wind, and humidity on plant performance; and the epidemiology of major upland rice diseases, such as blast need to be studied thoroughly like lowland rices. Morphological characteristics, such as rooting habits, leaf angle, tillering potential, and length and size of panicles, need to be studied in relation to actual field performance, under different cultural and soil conditions. In summary, upland rice research is a major component for rice breeders. Also natural resource and landscape management issues need to be taken into account. Cooperatively with rice scientists around the world, we hope in time, to help provide upland rice farmers with varieties and technologies that will better equip them to cope with their harsh environments.

Wheat

Wheat (*Triticum aestivum*) is a staple food crop and contributes a major portion in dietary value of food in India. The systematic wheat research in India started with the establishment of the Indian (Imperial) Agricultural Research Institute in 1904 at Pusa in Bihar. Major emphasis was laid towards breeding disease resistant varieties with good grain quality for chapati making. Wheat varieties representing several series, such as NP (New Pusa), Pb. C (Punjab Cross), K (Kanpur), RS (Rajasthan), Hy (Hybrid of Madhya Pradesh), NI (Niphad), etc. were developed. These wheat varieties were, however, not suited for intensive cultivation because of their tall and weak straw which tended to lodge under high fertility and irrigated conditions. These varieties mostly included pure line selections, such as NP4, NP 12, Type 8A, Type 9D and also the disease resistant varieties developed through hybridization, namely, NP 165, NP 710, NP 718, NP 770, NP 824, NP 783, etc. These varieties did provide at that time a degree of stability against rusts prevailing in different parts of the country. However, bottleneck genes for tallness did not permit any real breakthrough in productivity till the dwarfing genes in Norin 10 were exploited by Dr. Norman E. Borlaug and sent for multilocation testing through International Wheat Rust Screening Nursery in 1961-62.

The yield potentials of these dwarf wheats were immediately realized and Dr. Borlaug, based on his visit to India in 1962, supplied one quintal seed of each of the four most promising strains. These were extensively tested and later huge quantities of two varieties, namely, Lerma Rojo and Sonora 64 were imported in 1965 and 1966. These although had red grain but possessed a number of desirable attributes, such as dwarfness, relatively stiff straw, earliness and good resistance to rusts. Later, Kalyansona and Sonalika, both amber-grained varieties, were developed by wheat breeders in India through selection from advance breeding material supplied by CIMMYT to suit Indian consumer's preference. Both these varieties became most popular in the country and helped India achieve a green revolution. The dwarf wheats thus have provided both, increase in stability in yield as well as production. Researches in other disciplines also generated best possible production technology and the strategy to combat successfully the races of three wheat rusts causing major damage to wheat in the past. Table 9.1 provides details of the most promising varieties released for different agro-climatic zones in India.

Table 9.2: Present Wheat Varieties Recommended in North-West Hills

Variety	*Maturity Type*	*Grain Type*	*Yield Potential (q/ha)*
Northern Hills	Hills of Jammu and Kashmir, HP, UP	**Irrigated**	
		Timely sown	CPAN 1796, HB 208, HS 240
		Rainfed	
		Timely sown	VL 421, CPAN 1796, HB 209
		Early sown	VL 616,
		Late sown	Raj 3765

Barley

Barley (*Hordeum vulgare* L.) is an important crop of hills. Both two-row and six-row barley is preferred by the people. Barley is an important cereal crop of India, which play a major role in the barley producing countries. Ranking of barley is next to the maize, wheat and rice both in acreage and in production of grain. In India the major states growing barley are U.P., Rajasthan, Punjab, Haryana, M.P., H.P., Bihar, Uttarakhand, Jharkhand and Jammu and Kashmir. It is also grown in small pockets in other states like Chhatisgarh, W.B., Nagaland, Maharashtra, Sikkim and Delhi.

The social and economic situations of Sikkim have accorded great importance to the barley crop. A major share of barley grain is used as animal feed either alone or in combination and also consumed as a food by poor classes as pure flour or mixed with wheat or gram flour for making chapattis. Barley chapattis is highly palatable and digestible as compared to the wheat. It can be digested easily even by the person like suffering from stomach ailments because it does not contain gluten. A limited portion of barley is consumed as parched grain to make *sattu*. As a fodder crop, barley has tremendous potential and variation for production of very high amount of digestible dry matter as well as protein yield/ha.

Fertility of the lateral spikelets means number of row of grain and their arrangements in the ear head form the basis of barley classification. The cultivated barley is classified into following groups:-

1. *Hordeum vulgare*: It is 6-row barley which is widely grown and is most important species. In this species all 3 spikelets at each node of the spike are fertile. These are dense and lax forms, depending upon the length of the rachis internodes. It is grown only at higher altitude, where barley is used as a major article of food.
2. *Hordeum distichum*: It is a 2-row type of barley and only the medium spikelet are fertile, although the lateral spikelets are infertile, they passes all the floral organ. It is preferred for malting because of their bold, plump grain, uniform germination, higher malt extract and other desirable traits.
3. *Hordeum irregulare*: This is irregular barley with tough rachis. Their central florets are fertile whereas the lateral florets are reduced to rachilla in some cases and these are distributed irregularly on the spike. The remainders of the lateral florets are either fertile, sterile or sex less.

Breeding Programme

In early days barley was believed to be grown on saline – Alkaline land or land with no irrigation facilities having marginal soil fertility. The grain yield used to be very poor due to following reasons:-

1. The old variety had very poor response to management practices and where grown under slightly enriched condition they cause severe lodging which is a serious problem as they ultimately reduce the grain yield.
2. The variety had high susceptibility to disease like smut, yellow rust, foliar disease, foot and root rot, molya disease and pest like nematodes, termites, aphids etc. These problems cause a high instability in production.
3. It was regarded as second thought crop and was not given due care.
4. Lack of dwarf and input responsive variety.
5. Lack of scientific production technology.

All the varieties recommended for cultivation upto this time possessed restricted adaptation. Systematic work on the identification of varieties with wide adaptation on national level was initiated in 1964 with the organization of 'All India Co-ordinated Barley Improvement Project" in 1967 which tackle various production oriented problems on a multi – disciplinary basis. Rapid progress has since been made in the development and release or pre-release of several varieties which are an improvement upon the existing ones for yield potential, disease and pest resistant.

A brief account relating to crop improvement research in India has been presented. The crop production technology has enabled India achieve self-sufficiency in a number of crops and considering still the realizable productivity gaps, much higher production and productivity is expected to be achieved in the near future. Looking at the research carried out for over the past four decades, one would derive satisfaction that the Crop Science research in India, directed towards breeding high yielding varieties to suit different agro-climates, particularly in hills and agricultural patterns/needs etc. has stood the test of time in achieving self sufficiency in food grains.

References

ADB (2001). Participatory poverty assessment: Lao People's Democratic Republic. Asian Development Bank, Manila.

Anonymous (2009). Bulletin of Statistics. 2: 85, 22.

Hallauer, A.R. and Miranda, J.B. (1988). Quantitative Genetics in Maize Breeding, 2nd ed. Iowa State University Press, Ameserdam. Pp 345-356.

Huke, R.E., and E.H. Huke (1997). Rice Area by Type of Culture: South, Southeast and East Asia. International Rice Research Institute.

Hurd, E.A. (1971). Can we breed for drought resistance? In: Drought injury and resistance in crops (Eds. K. L. Larson and J. P. Eastin). Crop Science Society of America, Wisconsin, pp. 77-88.

Hill Agriculture: Prospects, Constraints and Mitigations (2013) *Pages* **130–144**
Editor: **Dr. Shahid Ahamad**
Published by: **DAYA PUBLISHING HOUSE, NEW DELHI**

Chapter 10

Vegetable Cultivation in Mid Hills of Jammu Region: Problems, Prospects and Future Strategies

Anil Bhushan[1], Susheel Sharma[1], Sanjeev Kumar[2] and B. Singh[3]
[1]Regional Agricultural Research Station, Rajouri, SKUAST-J,
[2]Division of Vegetable Science and Floriculture, Chatha, SKUAST-J
[3]RHRSS, Bhaderwah, SKUAST-J

The state of Jammu and Kashmir is fortunate to have a wide range of agro-climates ranging from low altitude sub-tropical, mid altitude intermediate, temperate (dry and wet) and very high altitude cold arid zone. The diverse nature of agro-climate of our state presents not only a great challenge but also a huge opportunity for agriculture in general and vegetable production in particular. At present more than 50 vegetables are grown in different regions. Vegetable types, their varieties and management practices followed also vary with the region. A total of 41700 hectare, which is 4 per cent of the total agricultural area, comes under vegetable cultivation in the state. J&K ranks 11th among all the states in the country with 14.5 tonnes per hectare vegetable productivity. However, the state does not fall in the major vegetable growing states of the country despite the fact that productivity of some major vegetables *viz.* brinjal, tomato, peas, okra, Cole crops etc. are comparable to or even better than national average.

Jammu region, with 32500 hectare area, accounts for 66 per cent of the total vegetable growing area of the state. It has registered a consistent increase in area since the year 1998-99 when it was only 17600 hectare. The region comprised of subtropical plains (Jammu, Samba, Kathua and some parts of Udhampur and Reasi) and lower to mid altitude mountainous regions (Rajouri, Poonch, Bhaderwah, Kishtwar, Ramban, Reasi and some parts of Kathua and Udhampur). Of all the districts, Rajouri and Poonch districts account for lowest productivity of 9.29 and 8.25 tonnes/hectare respectively. The area under vegetable cultivation in Rajouri and Poonch districts are approximately 822 and 375 hectare with annual production of 7638 and 3096 tonnes respectively.

The major rainfed cropping systems are maize-wheat, rice-wheat, and intercropped pulses especially rajmah and oilseeds in maize and wheat, while rice-wheat and vegetable based crop sequences are dominant under irrigated conditions. Only one cropping season is feasible in the high-hill temperate zone where crops are grown during the summer (April to September), as snow cover during winter does not permit sowing in month of October onwards. Further rise in production is desired in vegetable production to meet the huge gap between present level of 185 g/C/day consumption and recommended level of 300 g/C/day. It can be achieved either through horizontal expansion *i.e.*, to bring more area under vegetables or through vertical expansion *i.e.* to enhance vegetable productivity/unit area. Both the ways seem effective particularly in context of intermediate zone as there is ample scope of increasing production as well as productivity. It could be done through introduction or development of new high yielding varieties/hybrids in combination with efficient technologies dissemination strategies.

Import figures of state are also high with monthly inflow of 10-12 thousand metric tonnes of vegetables from neighbouring states. In case of twin districts of Rajouri and Poonch 70-80 per cent demand is met through supply from Jammu plains. This clearly indicates that present area and levels of production of this zone are unable to sustain the local vegetable consumption.

Vegetable Production Constraints

In spite of suitable agro-climatic conditions and large potential for growing vegetables, it has not been exploited and still it remains in an underdeveloped state and no organized industry exists due to following constraints:

Environmental Factors

Assured irrigation is essential for irrigating vegetable crops at critical growth stages. Most of the agriculture in this area is rainfed (70-80 per cent) and vegetables are no exception. Assured irrigation is available only in some patches alongside local nullah/streams. More area under such belts needs to be brought under vegetable cultivation. Whereas, maize and wheat withstand prolonged dry spells, the vegetables being tender and succulent are unable to survive under water stress conditions. The low and erratic rainfall during summer months is responsible for crop damages or even failure. Moreover, high rainfall during monsoon leading to intense leaching results in soils with poor base status. The mean average temperature during winter months *i.e.*, November to March is very low which is not favourable for proper vegetative growth and development of vegetable crops. The problem is more prominent in leafy vegetables *viz.* spinach and methi while foliage growth in knolkhol, broccoli, cauliflower, turnip, pea is also checked in December and onwards which will have directly affect the size of edible parts. Therefore, proper sowing time and specific variety for such agro-climate needs to be identified as severe climatic conditions during winter months restrict the choice of crops.

Topography

The area is characterized by hilly terrain with undulating topography. The rugged physiography of this zone hampers the vegetable cultivation in this region. The land is inaccessible, and infrastructure, communications and mobility are obstructed by different physical, climate, biological and socio-economic factors. Moreover, the hilly terrain makes it difficult to market the produce at proper time. Steeply sloping landforms encourage heavy runoff resulting in severe erosion hazards leading to decrease in fertility of the soil and the productivity. Therefore, the fields have to be properly levelled to facilitate irrigation and proper drainage.

Soil Factors

It is evident that vegetables grow best in fairly deep soils. However, the soils in hills are clayey in texture besides being shallow and stony, low in organic matter and nutrient content. The tilth or physical condition of the soil is important in relation to plant growth. If the soil is poor in physical condition, it will be hard and crusty when dry, and sticky when wet. Such soils are not favourable for vegetable cultivation. In addition, due to slopes, soils are prone to erosion, which is aggravated by heavy migratory grazing which has also led to soil degradation. Excessive moisture leading to water stagnation in valleys during post-monsoon period limits the choice of crop. Deforestation for shifting cultivation also leads to severe soil degradation problem.

Small and Fragmented Land Holdings

The average land holdings of farmers of hill region are very small. The problem is further compounded by their fragmented and scattered nature in far furlong and inaccessible areas. Thus, small and fragmented land holdings make it impossible to go for vegetable cultivation on commercial scale.

Lack of Improved Production Technology for Mid Hills

The improved technology regarding vegetable cultivation is available for subtropical plains only. The same technology is not suited for intermediate zone. Therefore, there is urgent need for standardization of production technology of vegetable crops for mid hill conditions.

Lack of Awareness among Farmers

Farmers of this zone are practising traditional cereal based maize- wheat cropping system. Vegetable cultivation is confined as kitchen gardening activity for their own consumption at limited scale. The main vegetables are:

Winter vegetables: Cauliflower, Knolkhol, Garden Pea, Carrot, Turnip, Cabbage, Onion, Garlic and Potato etc.

Summer vegetables: Tomato, Brinjal, Chilli, Capsicum, Bottle gourd, Cucumber, Bitter gourd, Sponge gourd, Okra etc.

However, there is lack of awareness and interest among farmers to realize the full potential of vegetable cultivation and its subsequent benefits in improving their socio-economic status.

Unavailability of Quality Seed

Quality seed is most important prerequisite for successful vegetable cultivation. One of the most apparent reasons hindering vegetable production scenario in this zone is unavailability of quality seed of different vegetable varieties/hybrids. Vegetable growers have to depend upon substandard, unauthenticated and poor quality seed and un-reliable vegetable seedlings sold in local market. Farmers need support from government agencies and agricultural universities to ensure the availability of quality seed of recommended varieties of selected vegetables for this zone.

Lack of Trained Personnel

There is a dearth of trained personnel who can guide and disseminate the latest technical know-how of vegetable gardening like protected cultivation, organic farming, marketing, value addition and other related aspects which make vegetable cultivation a profitable venture for them. There is an urgent need for extension network of the State department and State Agricultural Universities to refocus their research and extension priorities keeping themselves abreast with the latest trends and technologies of vegetable cultivation and demonstrate their practical adaptability under local conditions particularly among the rural people living in the remote far-flung areas of the region.

Table 10.1: Varieties/Hybrids of Major Vegetable Crops Grown in Mid Hills of Jammu Region

Crop	*Varieties/Hybrids*
	Summer season vegetables : March-April to August-September
Tomato	Pusa Ruby, Selection-120, Swarna Lalima, Swarna Naveen, Punjab chuhara Hybrids: Naveen, Rupali, Pusa Hybrid-2
Brinjal	PPL, PPR, PPC, Punjab Bahar, Pant Rituraj, Pant Samrat Hybrids: Pusa hybrid-5, Pusa hybrid-6
Chilli (Hot pepper)	Pusa jawala, Pant C-1, DKC-1, DKC-8, Pusa Sadabahar, Punjab Lal Hybrid: CH-1, CH-3, Agni,
Capsicum (Bell pepper)	California wonder, Yolo wonder, Arka Mohini, Arka Gaurav Hybrid: Nishat-1, Bharat, Indira
Okra	Arka Anamika, Varsha Uphaar, HISSAR Unnat, PARBHANI Kranti, Pusa Padmini, Punjab-7, Pusa A-4
Bottle gourd	Pusa Summer Prolific Long, Pusa Summer Prolific Round, Punjab Komal, Punjab Round
Cucumber	Japanese Long Green, Pusa Sanjog, PoinsetteHybrid: Pusa Sanjog, Solan Hybrid-1
Bitter gourd	Coimbatore long, Arka Harit, Priya, Punjab-14
Sponge gourd	Pusa Chikni, Kalyanpur Chikni
	Winter Season vegetables: September-October to March-April
Cauliflower	Early: Early Kunwari, Pusa KatekiMid: Pusa Synthetic, Pusa Shubra Late: Pusa Snowball-1, Snowball-16, Giant Snowball
Cabbage	Golden Acre, Pride of India, Pusa Mukta, Drum head
Knolkhol	White Vienna, Purple Vienna, G-40, King of North
Broccoli	Early Green, Palam Samridhi, Palam Kanchan
Peas	Arkel, P-89, Bonnievillae, AP-1, Palam Priya, Ageta-6
French bean	Contender, Pusa Parvati
Turnip	PTWG, Pusa Chandrima, Pusa Kanchan
Radish	Pusa Chetki, Japanese White long, Pusa Himani, Pusa Reshmi
Carrot	Pusa Kesar, Pusa Meghali, Early Nantes, No-29
Onion	Pusa Red, N-53, Punjab Selection, Agrifound Dark Red, Agrifound Light Red
Garlic	Agrifound White(G-41), Yamuna Safed(G-1), Local
Potato	Kufri Jyoti, Kufri Badshah, Kufri Chandramukhi, Kufri Sindhuri, Chipsona-1, Chipsona-3
Spinach	All Green, C-13, Banerjee Gaint
Methi	Pusa Early Branching, Kasuri Methi, Khushboo

Prospects of Vegetable Cultivation

Seed Production of Winter Vegetables

Hilly regions of Jammu are bestowed with congenial mild climate during summer months with maximum temperature ranging between 30-35°C. There is at least a gap of 10°C when compared with summer peaks in Jammu plains. Moreover, extended winter season allows the winter crops grown for seed production to mature properly and attain full seed size without facing the difficulties of pod shattering and shrivelled seed due to sudden rise in temperature. Therefore, there is a huge advantage in terms of climate for seed production of temperate vegetables especially Cole crops *i.e.*, cauliflower, knolkhol, broccoli and temperate radish and carrot. Valley areas and cool temperate pockets need to be developed for seed production of cabbage, snowball cauliflower and brussel sprout. Besides this there are many vegetables of temperate nature, which require chilling temperature to get transformed into reproductive stage from vegetative stage, seed of such vegetables can be successfully produced in the hilly areas due to natural low temperature prevailing in such areas. Besides climate, there are many other factors which favour quality seed production in hilly regions of Jammu. These are

1. *Soil*: Soils of hills are generally well drained and rich in organic matter. Thus making them most suitable for quality seed production
2. *Isolation distance*: Topography of hills, presence of forest and river catchment areas, act as natural barrier between the valley pockets or seed producing areas, rendering the genetically pure high quality seed.
3. *Natural pollinators* : The pollinating agents, especially bees flora-honey bees, and other types of bees, butterflies are available in plenty naturally in hills, which brings out pollination by carrying pollen grains actively.
4. *Irrigation*: Hilly regions are full of natural water resources like rivers, nallahs, springs etc. Thus enough water is available for quality seed production, especially in those areas where regular rains are not received.
5. *Pest*: Vegetables grown in the hill areas are less attacked by the insect pests, due to the prevailing mild climate conditions, which do not favour the growth and development of the pest and diseases. Thus it saves the expenditure on insecticides and pesticides.
6. *Availability of labour*: Vegetable seed production is a labour intensive work and thus availability of labour and that too at cheaper rates in the hill make it more economic in such areas.

The seed production of winter as well as summer vegetables would widen the scope for expanding vegetable seed industry not for indigenous consumption only but also for export to European and Western countries where seed production is becoming a costly affair day by day with the increase in costs of labour. This profitable venture shall not only lead to cheaper cost of seed production but shall also provide employment opportunities to the hilly people. Identification of potential pockets with suitable climate conditions in these areas will also overcome the problem of many diseases and insect pests which otherwise have become quite predominant due to continuous growing of vegetable seed crops in the same tract/land.

Off Season Vegetable Production

Generally, there is an increased demand for fresh vegetable in plains during summer months as production levels of plains are unable to meet the expected demand and thus straining the supply chain in cities. Rich agro-climatic and topographical conditions are most suited for growing of off season vegetables in this region during mild summer months. Off-season vegetable production and marketing is the most profitable farm business giving very high production and income to the farmers per unit of land than even fruits. Thus hills have definite comparative advantage in producing aristocratic off-season vegetables, European and vegetable seeds which either cannot be produced at all in plains or cannot be produced during summer season because of high temperature thus these are

Table 10.2: Prevalent Diseases of Major Vegetable Crops and their Management Under Mid Hills of Jammu Region

Crop	*Disease*	*Symptom*	*Management*
Tomato	Damping Off (*Pythium, Phytophthora, Fusarium, Rhizoctonia* spp.)	Seedlings are killed at pre and post emergence stages and diseased seedlings topple	☆ Partial sterilization of soil ☆ Drenching nursery beds with formalin solution (1 part formalin and 7 parts water) ☆ Seed treatment with captan or carbenazim @ 2-3 g/kg seed
	Buckeye rot (*Phythophthora nicotianae var. parasitica*)	Brown spots with light and dark concentric zonations appear on green fruits	☆ Adopt field sanitation measures ☆ Spray the crop with Ridomil MZ (0.25 per cent) or Copper oxychloride (0.3 per cent) at 10-15 days interval
	Early Blight (*Alternaria solani*)	Concentric or dark brown spots appear on leaves and later on on fruits; defoliation occurs in advanced stage	☆ Seed treatment with Thiram(3g/kg seed) or Carbendazim (2g/kg seed) ☆ Spray with Mencozeb(0.25 per cent) or Copper oxychloride (0.3 per cent) on appearance of disease
	Bacterial wilt (*Pseudomonas solanacearum*)	Wilting, stunting, yellowing of foliage and finally plant collapse	☆ Follow three year crop rotation in the infected field ☆ Always transplant disease free seedlings
Brinjal,	Phomopsis blight (*Diaporthe vexans*)	Brown lesions developed on leaves and shoots. Infected fruits show brown discolouration.	☆ Hot water treatment of seeds at 50°C for 30 minutes ☆ Crop rotation
	Little leaf (*Phytoplasma*)	Leaves are reduced to very small size resulting in stunted growth and bushy appearance; plants fail to bear any fruit.	☆ Prophylatic spray of Metalyxl + Mencozeb 0.25 per cent just before onset of monsoon followed by spray with 0.3 per cent Copper fungicide (300g/100 litres of water) at 10-15 days interval.
	Fruit rot (*Phytophthora parasitica*)	Fruit start rotting from the apical portion.	☆ Seed treatment with Thiram (3g/kg seed) ☆ Spray the crop with zineb 0.25 per cent or Mencozeb of Copper fungicides (0.3 per cent)
Chilli and Capsicum	Wilt complex (*Fusarium* sp., *Phytophthora* sp., *Pseudomonas* sp.)	The wilt complex includes root rot, collar rot and true wilt symptoms with sudden death symptoms while green. The wilt complex usually appears severe with the onset of monsoon when plants are in bearing state, indicating high sensitivity to high soil moisture.	☆ Disease tolerant/resistant varieties ☆ Seed treatment with Carbendazim + Thiram (1:1) @ 3g/kg seed. ☆ Seedling dip with Carbendazim 0.1 per cent + Streptocycline (100ppm) for 30 minutes before transplanting ☆ Transplant seedlings in well drained field and provide without allowing water stagnation.
	Fruit rot and Dieback (*Colletotrichum capsici*)	Necrosis of twigs from tip to backward. Circular spots on the fruit appear which become elliptical in shape.	☆ Prophylactic spray of Ridomil MZ 0.25 per cent just before the onset of monsoon followed by spray with Copper fungicide @ 0.3 per cent at 10-15 days interval.

Contd...

Table 10.2–*Contd...*

Crop	*Disease*	*Symptom*	*Management*
Cucurbits	Powdery mildew (*Sphaerotheca fuliginea* and *Erysiphe cichoracearum*)	Appearance of small white patches on the leaves which later on spread to cover whole plant with white floury mass on stem, tendril and even fruits.	✰ Spray with 0.05 per cent or 0.1 per cent Carbendazim or Wettable Sulphur@ 0.2 per cent can also be sprayed in morning and evening to avoid sulphur injury.
	Downey mildew (*Pseudoperonospora cubensis*)	Disease is recogonised by presence of yellow spots on upper surface of leaves and purple to brown on the corresponding lower sides with white downey growth.	✰ Spray the crop with Mancozeb or Zineb @ 0.25 per cent or Copper fungicide 0.3 per cent and repeat the spray at weekly interval.
	Anthracnose (*Colletotrichum lagenarium*)	It causes dark brown lesions on leaves, stem and fruit in some cucurbits like bottle gourd and melons.	✰ Avoid contact of fruit with soil-Mix Carbryl dust 15 per cent @ 37.5kg/ha at the time of last ploughing.
Okra	Seed/Seedling rot (*Fusarium* sps. *Rhizoctonia* spp. and *Phytophthora* spp.)	Rooting of seeds result into failure of germination or pre-emergence death. Emerged seedlings exhibit dark brown rotting of roots and collar portion with sudden or protracted death or decline.	✰ Use healthy and plump seeds-Seed treatment with Carbendazim 0.3 per cent before sowing.
	Yellow vein mosaic	Leaf veins turn yellow, with leaf lamina light green and later on the whole leaf turn yellow resulting into stunted growth and reduced setting of fruits which become yellowish white in colour	✰ Use resistant/tolerant varieties (varsha uphaar, A-4 etc.) especially during rainy season. ✰ Spray the crop with systemic insecticide to control insect vector (white fly) and rogue out infected plants in early stages.
Beans	Anthracnosa (*Colletotrichum* spp.)	Sunken black spots with reddish yellow margins appeared on almost all parts of the plant including pods.	✰ Seed treatment with Thiram @ 3g/kg seed before sowing ✰ Spray the crop with Copper fungicides, Mencozeb or Zineb @ 0.25 per cent when disease appears on the foliage.
	Yellow vein mosaic	Very severe disease on beans transmitted by white fly (*Bemisia tabaci*)	✰ The disease can be managed by spraying insecticides like Thiodon (0.1 per cent) or Metasystox (0.1 per cent).
Cole crops	Damping Off (*Pythium, Phytophthora, Fusarium, Rhizoctonia* spp.)	Seedlings are killed at pre and post emergence stages and diseased seedlings topple	✰ Partial sterilization of soil-Drenching nursery beds with formalin solution (1 part formalin and 7 parts water) ✰ Seed treatment with captan or carbenazim @ 2-3 g/kg seed

Contd...

Table 10.2–*Contd...*

Crop	Disease	Symptom	Management
	Downey mildew (*Peronospora parasitica*)	Characteristic water soaked lesions on under surface of leaves and cause curd rot. Rotten tissue become brown and are surrounded with black discolouration at the base of the curd stalk.	☆ Spray the crop with Metalaxyl + Mancozeb 0.25 per cent at 10-15 days interval starting from just appearance of disease in problematic area.
	Stalk rot (*Sclerotinia, Sclerotiorum, Alternaria, Botrytis* spp. and *Erwinia carotovora* and *Bacillus polymixa*)	Leaves droop down, stalk rots from inside becomes hollow and filled with black sclerotia, curds if formed develops white rot.	☆ Follow cauliflower-paddy rotation ☆ Spray Carbendazin (0.05-0.1 per cent) from earthing up stage at 10-15 days interval till bolting disease prone field/area. ☆ For curd protection spray with mixture of Mancozeb (0.25 per cent) and Streptocycline 100ppm (1ml/ litre) if there is likelyhood of frost injury.
Onion and Garlic	Damping off (*Rhizoctonia solani, Fusarium* spp., *Phythium* spp.)	Seedlings are killed at pre and post-emergence stage.	☆ Partial sterilization of soil-Drenching nursery beds with formalin solution (1 part formalin and 7 parts water) ☆ Seed treatment with captan or Carbenazim @ 2-3 g/kg seed
	Downey mildew (*Peronospora destructor*)	Light yellow spot on root and downey growth on leaves and flowering stalks followed by necrotic lesions.	☆ Spray the crop with Ridomil MZ (250g/100litre of water) at the time of disease appearance and repeat sprays at 10-15 days interval with Mancozeb (250g/ 100 litre of water).
	Purple blotch (*Alternaria porri*)	Purple zonated spots appear on leaves and flowering stalks which usually break at the point of attack.	☆ Immerse bulbs, before sowing, in Mancozeb (250g/ 100 litre of water) at the time of disease appearance and repeat sprays at 10-15 days interval.
	Stemphylium blight (*Stemphylium vesicarium*)	Small yellow to orange spots or streaks appear in middle of leaves and on flower stalks on one side	☆ Spray Mancozeb (0.25 per cent) at 15 days interval in bulb or Copper oxychloride + Sanovit (0.1 per cent) at 15 days interval in seed crop.
	Black mould (*Aspergillus niger*)	Black powdery masses of spores are borne on exterior.	☆ Protect bulbs from moisture after harvest and store them properly.

Table 10.3: Brief Production Technology of Important Vegetable Crops Under Mid Hill Conditions of Jammu Region

Crop	Sowing Time	Transplanting Time	Seed Rate	Spacing	Fertilizer Dose	Yield per Hectare
			(A) Summer season vegetables			
Tomato	February- April	March-May	500g	60cm x 45cm	120:60:60	300-400q/ha
Brinjal	March-April	April-May	500-600g	60cm x 45cm 90cm x 60cm	100:60:30	300-400q/ha
Chilli	March-April	April-May	1-1.25kg	45cm x 30cm	100:60:30	100-120q/ha
Bell Pepper	March-April	April-May	1-1.25kg	60cm x 30cm	100:60:30	120-150q/ha
Potato	September	–	20-25q (2.5cm dia.)	60cm x 20cm	120:60:120	200-300q/ha
Muskmelon	February	–	1.0 kg	2-2.5m x 60-70cm	120:60:60	125-150q/ha
Watermelon	Mid February	–	3.25kg	2-2.5m x 60-70cm	60:40:40	80-100q/ha
Round Gourd	Mid February-March	–	5.0kg	3m x 75-90cm	60:30:30	100q/ha
Bottle Gourd	March	–	5.0kg	2-2.5m x 75-90cm	100:50:50	200-250q/ha
Bitter Gourd	March	–	5.0kg	1.5-2.0m x 60-70cm	100:50:50	100-130q/ha
Cucumber	March	–	2.5kg	1.5-2.0m x 60-70cm	40:25:25	125-150q/ha
Pumpkin	March	–	5.6kg	2.5-3.0m x 75-90cm	100:50:25	300-350q/ha
Sponge Gourd	March	–	5.6kg	2.5-3.0m x 75-90cm	40:30:30	100-110q/ha
French Bean	March-April	–	85kg (bush) 40kg (pole)	60cm x 10cm	50:100:50	100-125q/ha
Cowpea	March and June	–		45-60cm x 10-15cm	50:60:50	50-100q/ha
Okra	March-MayJune-July	–	20-25kg 12-15kg	45cm x 30cm 60cm x 30cm	60:30:30	50-60q/ha (Summer) 100-120q/ha (Rainy)
Ginger	Mid April-Mid May	–		20cm x 15cm	75:50:75	200-250q/ha
Turmeric	March-April	–	20-25q	60cm x 20cm	60:30:90	300q
			(B) Winter season vegetables			
Radish	Mid-August-November	–	10-12kg (Kera) 4-5kg (Dibbling)	45cm x 5cm	60:30:50	250-300q (Asiatic) 60-80q (European)
Carrot	August-October	–	5-6kg	30cm x 10cm	60:30:50	200-300q (Asiatic) 100-150q (European)
Turnip	September-October		4-5kg (Kera) 2-3kg (Dibbling)		60:30:50	250-300q
Cauliflower						250-300q/ha
Cabbage	July-August	August-September				250-300q/ha
Knolkhol						150-200q/ha
Onion	August	October	10-12kg	15cm x 7.5cm	100:50:50	200-300q/ha
Garlic	Last week of August to first week of October		550-600kg	15cm x 7.5cm	100:50:50	100-200q
Peas	Mid-September-October		100kg	40cm x 10cm	50:60:50	200-250q/ha

produced in hills fetching high prices. Generally, there is an increased demand for fresh vegetable in plains during summer months as production levels of plains are unable to meet the expected demand and thus straining the supply chain in cities. Thus farmers in hill can obtain as high income as in plains or even more per ha, if they adopt highly intensive vegetable based cropping systems, new technology and complete package of recommended practices. The joint and concerted efforts of government and private sector organisations in collaboration with the hill farmers made steady progress in this direction and it made possible production of off-season vegetable and greater awareness of the farmers in the interiors through dispersal of production technology among the farmers through various developmental agencies and programme, net work of roads, both main and link roads,

Figure 10.1: Broccoli var. Early Green

Figure 10.2: Evaluation of Garden Pea Under Intermediate Conditions of Rajouri

improvement in irrigation potential and making available sufficient literature on vegetables, quality seed of vegetables/varieties and acceptability of the farmers to the latest technology and financial assistance through banks or any other agencies in the form of subsidised inputs, advancement of loans on easy interest rates and organisation of co-operative societies etc.

Cultivation of Exotic/Minor Vegetables

The agro-climates of Jammu region coupled with suitable soil physiographic conditions presents a unique opportunity for introduction and cultivation of many lesser known exotic/minor temperate vegetables. The important ones include broccoli, kale, collards, asparagus, parsley, lettuce, Chinese cabbage, leek, Swiss chard etc. A few of these vegetables have already been introduced in mid hills of Rajouri district and production technology for their cultivation and seed production is being standardized. However, there is a huge potential for introduction, popularization and cultivation of these lesser used vegetables to cater the local as well as nearby plains.

Vegetables as Important Component of Intercropping

The agricultural lands of mid hills of Jammu region are mainly occupied either by horticultural trees or field crops like maize. In both the cases, vegetables can be used as a viable source of ensured additional income when used as intercrop. They are most suited to utilize the left over space between fruit trees in most efficient manner without offering any competition to the fruit plant. The most

Figure 10.3: Knolkhol Seed Production

Figure 10.4: Beet Leaf var. C-13

important crops suited in fruit orchards are cucurbits like bottle gourd, bitter gourd, cucumber, sponge gourd, pumpkin, watermelon and muskmelon. There is already a well established practice of growing rajmash along with maize crop which acts as support to the vine. Other pole type cultivars of French bean and garden pea can be successfully intercropped with maize which enriches soil besides providing additional income to the farmers.

As Source of Nutritional Security

There is still a wide gap between minimum vegetable consumption per head per day (300g) and actual consumption (185g) on national level. The gap seems even more evident in case of hilly regions for various reasons. Dietary habits of inhabitants of hilly regions indicate their preference for cereals with maize being their staple food. However, vegetables are gaining their due importance in the local diet with the increasing awareness among locals and realization of importance of vegetables in providing balanced diet comprising of vitamins and minerals in addition to carbohydrates and proteins which are being provided through cereals and pulses.

Future Strategies

The region has been depending on its neighbouring areas to meet its increasing need for vegetables despite a vast potential for vegetable cultivation in the region. Thus there is a need to adopt the new research strategies to meet the challenges of vegetable productivity and production.

Identification, Collection, Improvement and Characterization of Indigenous/ Local Germplasm

The mountainous areas of Jammu region are rich in genetic diversity of some major vegetables especially cucurbits, beans, alliums, ginger and turmeric, brinjal and chillies. However, there is hardly any comprehensive research initiative been initiated so far to exploit this valuable local germplasm for crop improvement programme. Therefore, there is an urgent need to identify gaps in germplasm collection, to survey and collect land races, ecotypes and traditionally grown cultivars of different vegetables possessing quality traits, yield and yield contributing traits resistance to biotic and abiotic stresses.

Protected Vegetable Cultivation

Agriculture in the hilly regions of the country is at subsistence level and gains of green revolution have not reached the hill farmers. It has been established that the protected cultivation increases the yield manifold along with better quality produce by achieving higher water and nutrient use efficiencies, increasing photosynthetic efficiency and reducing transpiration losses. The need of the protected structure directly depends upon the climate of the region, growing season and the kind of the crop to be grown. Greenhouses have tremendous potential in increasing production and productivity of vegetables like normal tomato, cherry tomato, coloured sweet peppers, green sweet peppers, slicing cucumbers, muskmelon, summer squash including some rare vegetables even under adverse agro-climatic conditions. Our country has varied agro-ecological situations in different regions, so the cost effective low and medium-cost greenhouses, low tunnels and other related facilities have to be developed accordingly. Therefore, adoption of systematic approach based on the involvement of greenhouse engineers, breeders, crop production and protection scientists and post harvest technologists will help in increasing the sustainability and relevance of protected cultivation across the globe. Creating awareness among the greenhouse growers about safe food production and judicious use of safe pesticides can be instrumental in providing quality products without polluting the environment.

Efficient Water Management Strategies

During the peak summer months there is an acute shortage of water on the hills due to longer dry spells. It is, therefore, important to conserve or harvest rain water by storing surface run-off, building check dams, percolation tanks and sub-surface dykes. Plastic storage tanks may also be used where water tanks are not feasible.

Drip Irrigation

Vegetable cultivation requires lots of irrigation water as well as frequent and controlled irrigation compared to the cultivation of cereal crops. Thus, better access to water and especially access to controlled irrigation technologies is sometimes the most critical factor in a farmers' decision for increasing vegetable acreage. Drip irrigation is a low pressure, slow application direct to the root region of the plant and highly efficient system. The method is economical as it reduces water loss due to percolation and evaporation and its operational cost and water requirement is less. Drip irrigation should, therefore, be popularized among the farmers and the equipments be given free or at much subsidized cost to lure the farmers to take up vegetable production enterprise.

Post-harvest Management

The vegetables produced in hills suffered major post harvest losses due to wrong harvesting practices, poor or lack of proper storage, lack of efficient postharvest facilities, poor marketing and transportation etc. There is urgent need to minimise these postharvest losses by evolving efficient post harvest strategies of value addition, cold storage facilities, developing efficient marketing chains, encouraging contract farming and improving transport system.

Training Programmes

Proper Training should be given to the farmers of the region for scientific cultivation of vegetables. The young farmers should be trained in the art of growing and nursing vegetable crops under open as well as protected conditions. In the initial years, free supply of seeds/plants, manures and pesticides/ fungicides may be adopted and they should be made available on his farm and trained staff should visit regularly to know their difficulties and remedial measures.

In addition to these strategies, there is also need to prioritize the vegetable cultivation in existing cropping system, fine tune research on long term fertilizer trials/plant nutrient dynamics particularly nutrient and water use efficiency should be strengthened, developing protocol for organic farming of selected vegetable crops and developing database and initiating socio-economic studies on vegetable based production system. Simultaneously it will be important to educate people, including children especially in the rural interiors of the mid-hill areas regarding the usefulness of vegetables in human diet for healthy nutrition. Establishment of small vegetable gardens in rural schools with participation of school children should also be encouraged. The overall emphasis should not be on improving quantity only but quality of the vegetable produce also in order to attain the objective of achieving the nutritional security of the people of the region.

References

Arya, P. S. (2000). Off season vegetable growing in hills. APH Publishing Corporation, New Delhi. 427p.

Sharma, J. P. (2007). Vegetable Production Technologies Under Rainfed Condition. Model training course on sustainable agriculture development in degraded and rainfed areas, held on 23rd to 30th July, 2007 organized by Directorate of Extension Education, SKUAST-J.

Sharma, J. P. (2006). Fundamental Approaches In Sustainable Agriculture. Kalyani Publishers. 288p.

Anonymous (2009). Research and Development scenario in vegetable production with reference to Jammu region- *Status paper, 2009.* Division of Vegetable Science and Floriculture. SKUAST-J, FOA, Chatha, Jammu.

Kumar Sanjeev, Puja Rattan, Gupta R.K. and Sharma J.P. (2009). Efficient resource management for sustainable vegetable farming under temperate hill conditions. In: ICAR 21 days Summer School on Integrated Farming for High Hills held in Division of Agronomy, SKUAST-J w.e.f. 5th to 25th February, 2010. *pp-181-190*

Anonymous (2006). Package of practices for vegetable crops. Published by Directorate of Extension Education, SKUAST-J. 107p.

www.nhb.gov.in/statistics and production. Indian Horticultural Database (2010). National Horticultural Board, Gurgaon, Harayana.

Hill Agriculture: Prospects, Constraints and Mitigations (2013) *Pages* **145–170**
Editor: **Dr. Shahid Ahamad**
Published by: **DAYA PUBLISHING HOUSE, NEW DELHI**

Chapter 11

Common Bean (*Phaseolus vulgaris*): Challenges and Opportunities in Hill Agricultural System

***Parvaze A. Sofi*[1], *S.M. Razvi*[1], *Shafiq A. Wani*[2], *M.A. Khan*[1], *Kamaluddin*[1] *and M.H. Chesti*[2]**

[1]*Faculty of Agriculture, Wadura, (SKUAST-K), Sopore – 193 201*
[2]*Director Research, SKUAST-K, Shalimar – 191 121*
[3]*Regional Agricultural Research Station, SKUAST-J, Rajouri – 185 131, J&K*

Introduction

Legumes constitute the third largest family of higher plants, with 20,000 species and second in agricultural importance based on area and total production. These are unique crops in having an inbuilt capacity for fixing atmospheric nitrogen. Thus these crops meet their own nitrogen requirement to a great extent. They also leave nitrogen in the soil that enrich it and become available to the succeeding crop. They are also important constituents in the diets of a very large number of people, especially in the developing countries and are a good sources of protein, which help to supplement cereal diets, improving their nutritive value. Protein content of pulses is twice that of cereals (>20 per cent), whereas their by products provide nutritious fodder for livestock. With the growing demand for nutritional security, pulses are becoming ever more important as a plant based source of protein in human diet. Besides, the newly emerging, strong consciousness for health among the human population is creating a genuine need for adopting a nutritionally complete, but a predominantly vegetarian diet, which is healthy, eco-friendly and energetically less expensive. Consequently the vegetarian food profile with pulses as the major protein source is likely to become more widespread leading to a greater demand for plant based proteins. Unfortunately, unlike those of cereals, the existing production levels of pulses cannot meet the emerging demands.

Pulses are the second most important food crops after cereals. In 2008, total global pulse production stood at 56.97 million tones from an area of 72.45 million hectares and average productivity of 780.81 kg/hectare (FAO, 2009). India is the largest producer and consumer of pulses in the world. At present, it accounts for 33 per cent of the world area and 22 per cent of world production of pulses. During 2009-10, pulses were cultivated over an area of 23.60 million hectares with a production of 14.80 million tones and productivity of 6.27 q ha^{-1} (Anonymous, 2009). Global production of pulses increase has been marginally by 0.2 per cent annually. However, this falls short of projected growth in pulses demand of 2.7 per cent annually and is not adequate to ensure per capita availability of 80g which is a minimum level recommended by the World Health Organization (WHO) and the Food and Agriculture Organization (FAO). Analysis of factors responsible for low production and productivity levels indicate a multitude of constraints for the dismal scenario that are biological, geographical, agronomic, institutional and socio-economic in nature. High yielding seeds of pulses have not been made available and many of them have been susceptible to pests, diseases and weather fluctuations. The possibility of improving yield *per se* through genetic manipulations has been showing encouraging results in recent years.

The Genus *Phaseolus*

The legumes called as beans refer to the food legumes of genus *Phaseolus*. The genus *Phaseolus* is of American origin and comprises about 90 species (Table), out of which only five have been domesticated, namely common bean (*Phaseolus vulgaris*), lima bean (*Phaseolus lunatus*), scarlet runner bean (*Phaseolus coccineus*), Tepary bean (*Phaseolus acutifolius*) and yearlong bean (*Phaseolus polyanthus*) and a few additional show signs of domestication (Salinas *et al.*, 2006). However out of these species, only common bean holds the largest acerage and accounts for about 85 per cent of the total area under *Phaseolus* species (Singh, 2001).

Table 11.1: Systematics of *Phaseolus*

Kingdom	*Plantaea*
Division	*Magnoliophyta*
Sub-division	*Magnoliophytina*
Class	*Rosopsida*
Order	*Fabales*
Family	*Faboideae (Fabaceae)*
Sub-family	*Papilionoideae*
Tribe	*Phaseoleae*
Sub-tribe	*Phaseolinae*
Genus	*Phaseolus*

Out of about 90 odd species, only five have been cultivation and a very few are showing the signs of incipient domestication. The domesticated species of the genus *Phaseolus* are briefly discussed below:

Common Bean *(Phaseolus vulgaris)*

Common bean is the most important food legumes of the genus *Phaseolus* and accounts for about 85 per cent of all the *Phaseolus* species grown over the world and is the second most important commercial legume crop after Soybean. It is also known as *Dry bean, Green bean, Snap bean, String bean, Wax bean, Field bean, French bean, Pinto bean, Haricot bean, Kidney bean, Black bean, Turtle bean, Navy bean, Great Northern bean* etc. Bean plants are polymorphic, meaning that they can have a variety of different forms. They may be erect and bushy (up to 60 cm in height) or climbing (with stems up to 3 m long). They have compound leaves, with three smooth edged oval leaflets that taper to a point. Small variegated, white, pink or purplish flowers, about 1 cm long, develop in clusters. Flowering continues for two to three weeks, so pods develop at different times on the same plant. Depending on the variety, the pods can be green, yellow, black, or purple. They range in shape from cylindrical to flat and usually contain

Figure 11.1: A Typical Bush Type and Climbing/Twinning Common Bean Plant (Courtesy: Dr. P.A. Sofi)

Table 11.2: Composition of Seed of Dry Common Bean (per 100 g)

Constituent	*Quantity*	*Constituent*	*Quantity*
Energy	333 Kcal g	Water	11.6 g
Protein	24 g	Fat	1 g
Carbohydrates	60 g	Fibre	15 g
Copper	0.3 mg	Calcium	143 mg
Magnesium	140 mg	Phosphorus	68.1 mg
Sodium	11 mg	Iron	8 mg
Zinc	3 mg	Thiamine	0.7 mg
Riboflavin	0.5 mg	Niacin	5.4 mg
Vit B6	0.2 mg		

Table 11.3: Composition of Seed of Green Common Bean (per 100 g)

Constituent	*Quantity*	*Constituent*	*Quantity*
Energy	129 Kcal g	Water	11.6 g
Protein	1.8 g	Fat	1 g
Carbohydrates	7 g	Fibre	3.4 g
Copper	0.44 mg	Calcium	37 mg
Vit A	35 micro gram	Vit C	16 micro gram

from two to four seeds. The seeds come in a variety of shapes, sizes and colours. Beans need up to four months of warm weather and are not frost tolerant. They do poorly in very wet or humid tropical climates because of susceptibility to bacterial and fungal diseases. They need well-drained soils with a pH between 6.5 and 7.0 and are sensitive to deficiencies or high levels of minerals in the soil.

Lima Bean (*Phaseolus lunatus*)

Lima bean is one of the four other cultivbated species of *Phaseolus* and is named after the capital of Peru. It is also known as *Butter bean, Java bean, Burma bean, Guffin bean, Hibbert bean, Sieva bean, Rangood bean, Madagascar bean, Paiga bean, Civet bean, Sugar bean, Haba bean, Hendersen bean, Chad bean and Pallar bean*. It is of dual origin like common bean. The large seeded Andean types known as Lima types have originated in about 2000 BC and are distributed in North Peru while as small seeded Sieva types have originated about 800 AD and are distributed from Mexico to Argentina. Some authors have divided it into two types: the cultivated lima bean (*Phaseolus lunatus var. lunatus)* and the wild ancestral types (*Phaseolus lunatus. var sylvistris)*. On the basis of seed size and shape, lima beans have been classified into three types namely Sieva (flat moderate seeded), Potato (small round seeded) and Big lima (large flat seeded). In USA, small seeded Sieva types are called as Dixie or Henderson types. Lima beans are both annual and perennial. The former are small bushes, 30-90 cm while as latter are twinning, 2-4 m tall with enlarged rootstocks. The flowers are usually pale green but occasionally violet and are smaller thasn common bean or scarlet bean. Pods are oblong in shape and the seeds are variable in shape usually 1-3 cm long. The root system is tap root with nodules. The stem is erect, < 1m tall and is solid. The leaves are pinnately trifoliate. Bracts and bracteoles are present. There are 9-10 stamens (diadelphous, 9 united and 1 free). The style is spiraly coiled and hairy.

USA is the global leader in cultivation of Lima bean with about 21000 ha followed by Madagascar (19000 ha) and Peru (6000 ha). The crop is grown for its dried seed. The raw seeds contain hydrocyanic acid which gets destroyed upon cooking. The seeds also contain a toxic lectin, Haemagglutinin. It also contains cyanogenic glucosides called as phaseolunatins.

Table 11.4: Composition of Seed of Lima Bean (per 100 g)

Constituent	*Quantity*	*Constituent*	*Quantity*
Energy	290 Kcal	Water	11.6 g
Protein	19.1 g	Fat	1.7 g
Carbohydrates	52.9 g	Fibre	19.4 g
Copper	0.44 mg	Calcium	85 mg
Magnesium	190 mg	Phosphorus	320 mg
Iron	5.9 mg	Zinc	2.8 mg
Thiamine	0.45 mg	Riboflavin	0.13 mg
Niacin	2.5 mg	Vit B6	0.51 mg
Tryptophan	180 mg	Methionine	280 mg
Threonine	800 mg	Leucine	1560 mg
Lysine	14440 mg	Phenylalanine	1160 mg
Valine	980 mg	Isoleucine	950 mg

Scarlet Runner Bean (*Phaseolus cocineus)*

Runner bean or Scarlet runner bean as it is called due to the characteristic red colored flowers and multicolored seeds though many varieties have white flowers and seeds. It is also called as *Oregon Lima bean, Case knife bean and Botil bean*. It differs from common bean in that its cotyledons stay on ground upon germination. Runner bean is said to have originated in Central American mountains of Meexico and Guatamela. It is a twinning perennial crop with attractive flowers and oftenly visited by Humming bird and its flower is called as Humming bird flower. The flowers are borne in clusters and are pea shaped. The leaves are heart shaped. It is a very fast growing plant and can attain a height of 5 meter. Runner bean is susceptible to frost. Both green pods and seeds are edible. However in some cases, only seeds are edible as the pods are highly fibrous.

Figure 11.2: Scarlet Flower and Growth Habit of Scarlet Runner Bean (Courtesy: Dr. P.A. Sofi)

Tepary Bean (*Phaseolus acutifolius)*

Tepary bean is a highly drought tolerant species of *Phaseolus* but is susceptible to frost. It is also called as *Texas bean, Pawi, Yori Mur, Mexican Haricot bean, Escomite, Dinawa* and *Rice haricot bean*. It is native of South America, North America and Mexico. Among *Phaseolus* species, its protein content is high at 24-25 per cent. Plant can be bush type or vining. The former can attain a height of 1-3 feet and spread of 3-4 feet while as latter are 12 feet tall. The leaves are pointed, trifoliate while as the seeds are slightly hairy, 3 cm long with 5-6 seeds per pod. The species is characterized by long pods as well is a valuable source of drought tolerance and disease resistance.

Table 11.5: Composition of Seed of Tepary Bean (per 100 g)

Constituent	*Quantity*	*Constituent*	*Quantity*
Energy	162	Protein	9 g
Carbohydrates	31g	Fibre	9 g
Calcium	144 mg	Phosphorus	405 mg
Potassium	1653 mg	Vit C	1 g

Year Long Bean (*Phaseolus polyanthus)*

Year long bean is of Mesoamerican origin and is also known as *Botil, Piloya, Dzich, Piligua, Petaco, Cacha and Matatropa*. Its germination is epigeal and posseses firous and fasciculate roots. Inflorescens

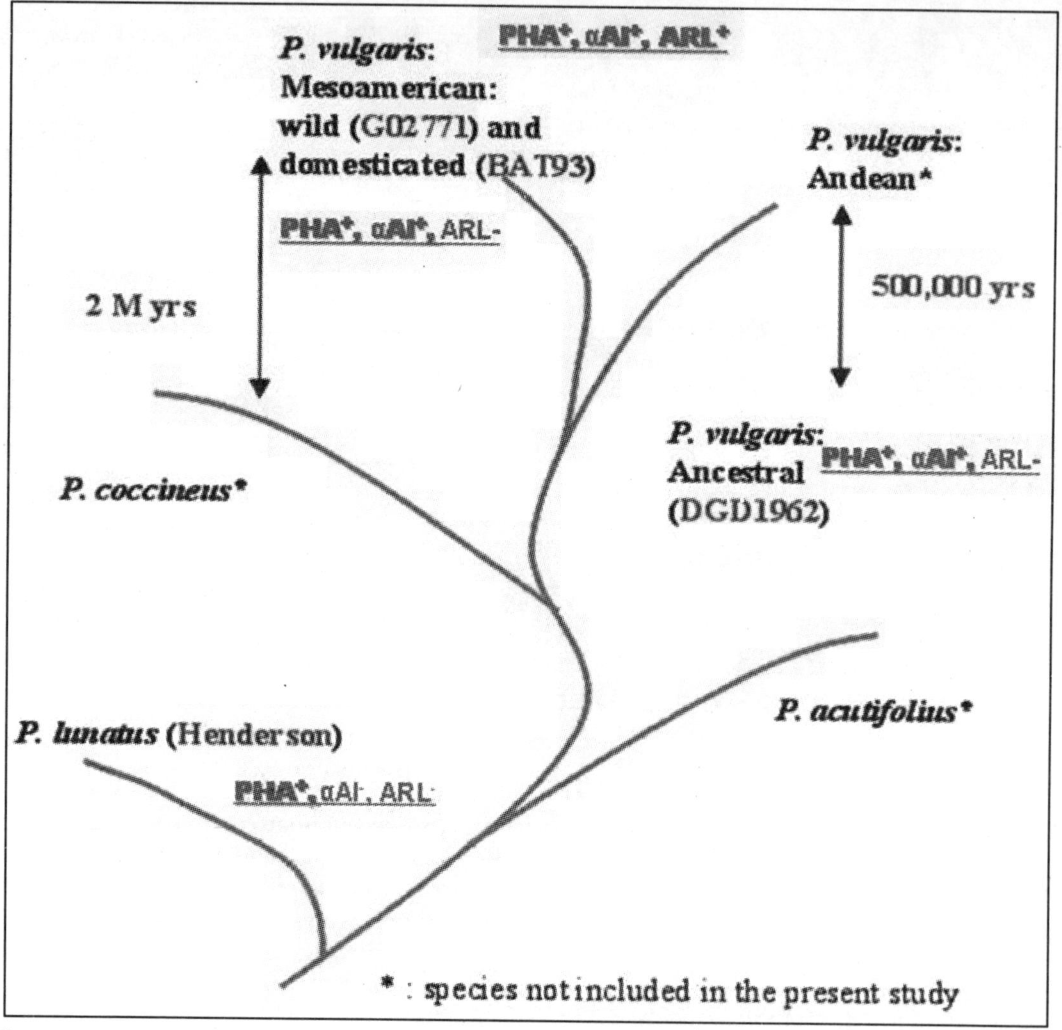

Figure 11.3: Phylogenetic and Genealogical Relationships among Domesticated *Phaseolus* Species (*Source*: Mclean *et al.*, 2004)

contains 6-16 fruit bearing stems having white or lilac flowers. There is great variation in 100-seed weight in cultivated and wild types and ranges from 70-100 and 16-25 grams respectively. This species has a long flowering period ranging from 2-5 months. *Phaseolus polyanthus* is the least evolved species in the genus and it has not been cultivated on a scale that would ensure sizeable evolutionary changes, the greatest limiting factor being its poor digestibility. However the species is a valuable source of genes especially the Aschochyta blight resistance.

Common Bean (*Phaseolus vulgaris*)

Among pulses, common bean (*Phaseolus valgaris* L.) is the most widely distributed and consumed legume species of the genus *Phaseolus*, a genus comprised of some 70 species (Freyt and Debouck, 2002) that has contributed to human welfare with five cultigens domesticated in pre-Columbian times *viz.*, the common bean (*P. vulgaris*), the year bean (*P. dumosus* Macfad.), the runner bean (*P. coccineous* L.), the temporary bean (*P. acutifolius* A. Gray), and the lima bean (*P. lunatus* L.). Among the five domesticated species, the common bean (*P. vulgaris*) accounts for more than 90 per cent of the cultivated crop worldwide and is by far the most widely consumed grain legume in the world (Singh, 2001). It is grown in subtropical or temperate regions throughout the world and during the cool, dry season in tropical areas. Like other legumes, they supply proteins, carbohydrates, vitamins and minerals and complement cereals, roots and tubers that compose of the bulk of diets in most developing countries. The seeds have a protein content of 22 to 24 per cent. Beans are low in sodium and gluten free making them suitable for low salt and gluten free diets. The straw that remains after harvest is a valuable source of organic nitrogen.

Brazil is the world's largest producer of common bean followed by India, China, Myanmar, Mexico and the United States and produces more than 2.2 million tons, which represent 17.3 per cent of the world's production. However Brazilian average yield (8.52 q ha^{-1}) is much lower than that observed in the United States and China, where average productivities reach 19.23 and 15.60 q ha^{-1}, respectively (FAO, 2008).

Table 11.6: Common Names of Common Bean

Sl.No	*Country*	*Common Name*
Sl.No	*Country*	*Common name*
1.	Sudan	Fasulia
2.	Spain	Mongeta
3.	Netherlands	Boon
4.	Germany	Haricot
5.	Greece	Fassolia
6.	Italy	Fagiolini
7.	Rwanda	Ibishimbo
8.	Burundi	Ibiharage
9.	Kenya	Maharagwe
10.	Tanzania	Maharagwe
11.	Zaire	Madesu, buhoti, Tshilusa
12.	Uganda	Ebijanjalo, ebihimba
13.	Iran	Lubiya
14.	Portugal	Feijao
15.	Zimbabwe	Bhimzi
16.	India	Rajmash
17.	Pakistan	Rajmash

Rajmash or common bean is regarded as "*Grain of hope*" as it is an important component of subsistence agriculture and feeds about 300 million people in tropics and 100 million people in Africa alone. Besides it is emerging as an important income generation especially in Central America where beans are No. 1 income generators among field crops. In fact in countries like Myanmar, common bean contributes about 10 per cent to total export earnings. Globally, with 21 million tones, it accounts for about half of the total pulse production (FAO, 2010). In certain developing nations it is the major source of dietary protein along with maize and cassava, while as in many African nations, it not only provides the sizeable portions of dietary protein but is also an important source of family income. In poor nations, where economic accessibility to animal protein is limited, beans offer a cheap source of dietary protein. In countries like Burundi, Rwanda, Uganda and Brazil, the contribution of beans to dietary protein is to the tune of 40 per cent, 31 per cent, 15 per cent and 9 per cent respectively. Beans also offer an alternative for diversification of agricultural system but poor yield renders them poor competitors of cereals, oilseeds and other cash crops.

Table 11.7: Area (Mha), Production (Mt) and Productivity (T/ha) of Dry and Green Beans in World and India (*Source*: FOASTAT, 2010)

Region	*Dry Bean*			*Green Bean*		
	Area	*Production*	*Productivity*	*Area*	*Production*	*Productivity*
World	25.56	20.69	0.81	1.30	19.04	14.18
India	6.00	2.44	0.41	0.20	0.54	2.71

Table 11.8: Trends in Area, Production Yield and Import of Pulses in India (2000-01 to 2009-10)

Year	*Area (M Ha)*	*Production (Mt)*	*Yield (kg/ha)*	*Import (Mt)*
2000-01	**20.4**	**11.1**	**544**	**0.4**
2001-02	**21.7**	**13.4**	**609**	**2.2**
2002-03	**20.5**	**11.1**	**543**	**2.0**
2003-04	**23.7**	**14.9**	**637**	**1.7**
2004-05	**22.8**	**13.1**	**577**	**1.3**
2005-06	**22.4**	**13.1**	**585**	**1.6**
2006-07	**23.2**	**14.2**	**612**	**3.7**
2007-08	**23.6**	**14.8**	**688**	**2.8**
2008-09	**23.0**	**14.2**	**617**	**2.3**
2009-10	**22.8**	**14.6**	**643**	**4.3**

Origin, Evolution and Domestication of Common Bean

There have been many reports about the origin and evolution of common bean. However, the exact origin has not been elucidated (Gentry, 1969). The archeological records of bean cultivation have been found in both Andean and Mesoamerican regions. In Andean region, the records have been found in Peru, Chile, Ecuador and Argentina dating back to 800-9600 years whileas, in Mesoamerican region, the records have been found in Mexico, Central America, South-west America and North-east America dating back to 800-7000 years.It is assumed that a wild leguminous vine growing in Mexico and Central America is the progenitor of cultivated common bean. Vavilov (1935) stated that it might have originated in Mesoamerica. Burkart (1952) proposed *Phaseolus aborigineus* from Argentina as the progenitor of cultivated beans, but it was later reduced as a sub-species of *Phaseolus vulgaris*. Another wild progenitor has been identified as *Phaseolus vulgaris* ssp. *mexicanum*. Although, there have been separate wild progenitors for *Andean* and *Mesoamerican* gene pools, the respective wild progenitors have a common ancestor in Ecuador and North Peru. The earliest record of a wild progenitor being collected is in 1896 when Edward Palmer found a stripling vine in Durango as the wild progenitor of *Phaseolus vulgaris*. The wild progenitor of common bean is a annual, more rarely a short lived perennial vine flowering and fruiting in the first year. The vines are generally 8-10 feet long. The flowers are more often lavender, sometimes pink and very rarely white. Over the last 7000-8000 years common bean has evolved from such wild progenitors into one of the most important legume crops.

The wild common bean is presently distributed in the western mountain ranges from Chihuahua in Mexico to San Luis in Argentina at about 1400-2200 masl. It is relatively abundant in Western Mexico and much less in Central America (Debouck, 1996). In the Andes, it ranges from Western Venezuala and extends upto South-western Columbia and Cundinamarca. The wild common bean is a vine and is usually found in open, sunny pine-oak forests and oak grasslands along with the members of Compositeae and Solonaceae in Mexico and Central America. While as in case of Andes, it is relatively more shady and found in humid montane forests along with *Podocarpus, Alnus, Celtis* and *Schinus*. The variation in the wild progenitors of both gene pools exhibit differences at ecological, morphological, and physiological as well as biochemical and molecular level also. The phaseolin seed storage protein variation (Table 11.9) has been reliably used, but the level of variation is not so much as to detect subtle variations.

Table 11.9: Morphological Differences in Wild and Cultivated Common Bean (Modified from Schoonhoven and Voysest, 1998)

Trait	*Wild*	*Cultivated*
Roots	Fibrous, Fasciculate	Fibrous, different sizes and morphologies as per growth habit
Stem	Climbing (multinoded with 20-30 nodes)	Climbing or upright (multinoded ranging from 3 nodes to 20-30 nodes)
Internodes	Generally long (10-20 cm)	Varies from short (2-5 cm) to long (20-30 cm)
Leaves	Small or intermediate sized	Small to large sized
Plant height	2-5 m	Depending on growth habit (20 cm to 10 m)
Inflorescence	Always lateral	Depending on growth habit (lateral or terminal)
Flower	Reflexed standard	Standard usually erect
	White wings rare	White wings common
	Bracteole shape and size variable	Bracteole size variable
Pod	Strongly dehiscent	Dehiscent in dry beans Indehiscent in snap beans
	6-8 cm long	4-30 cm long
	5-8 ovules	2-9 ovules
Seeds	Small 6-14 g/100 seed	Small to large 20-100 g/100 seed

Domestication has taken place separately in the two progenitors of the already diverged gene pools distributed from Northern Mexico to Columbia (*Mesoamerican gene pool*) and from Southern Peru to Northwest Argentina (*Andean gene pool*). The two separate domestication events have led to the divergence of two separate gene pools in common bean under the natural forces of evolution and those driven by human interventions (Kwap and Gepts, 2009). The process of evolution has been characterized by changes forced by domestication, cultivation practices and other evolutionary forces such as mutation, migration, selection as well as genetic drift (Gepts and debouck, 1998). There have been a significant amount of morphological changes in the wild and cultivated beans during domestication. The climbing habit has been an evolutionary feature for conferring greater fitness in the growth habitat. Since wild common bean was found as a climbing vine in bushes and shrubs, the bush habit would reduce the fitness on account of less competitive ability to harness light. The process of domestication has led to many changes such as shortening of growth cycle, ease of harvesting, compact growing habit, hush types that do not require support, fewer but strong internodes as well as reduced twinning.

Divergence of common bean into two separate gene pools is a result of the separate domestication events in their ancestors as well as the selection pressures under domestication. The events of domestication have caused two important implications: one that there has been partial reproductive isolation between two gene pools evidenced by the partial lethality of F_1 between the two, and second that there has been appearance of ecogeographic races in each of two gene pools, so much so that it is being suggested that *Phaseolus vulgaris* may be in a state of incipient speciation (Gepts and Bliss, 1985). Except for the Chilean landraces, no significant recombination has taken place between the two gene pools. The Chilean race from Andean gene pool might have undergone introgression from Durango race from Mesoamerican gene pool (Gepts, 1998). Another example is from snap beans, which basically having originated from Andean gene pool are actually intermediate between the two gene pools. These introgressions have been established by seed storage phaseolins and isozyme analysis.

In the current scenario, many such events must have taken place largely due to free exchange of germplasm, internationalization of gene banks and the extensive crossing programmes being undertaken in national and international bean breeding programmes. In fact, the existence of separate lineages in common bean offer both opportunities as well as present challenges for plant breeders. It is assumed that new recombinants could be derived from such inter gene pool crosses provided the reproductive isolation could be overcome to a fair extent. In this endeavour, breeders ability to isolate desirable segregants for gene complexes of two gene pools holds great importance. An important consideration also, is that pathogenic divergence has to be taken care of as the seem to follow the Andean-Mesoamerican divergene. Same may be the case with other associated organisms such as soil rhizosphere microbes.

Genetic Diversity in Common Bean

In terms of use, there are two commercial classes of common bean *viz., dry* and *green* bean. Dry beans, that predominate the global common bean production, have large seeds, pods which are flatter with more fibers and are used in form of dry seeds, seldom as pods. In contrast the green beans further classified as *string* or *runner* beans, *stringless* or *French* bean based on presence or absence of fibrous strings in the pods, snap beans that are largely harvested for fresh pods as they possess thick succulent pods with no or little fibres in pod walls, *shelling* beans in which beans are removed from pods before cooking and the *popping* beans that have small round seeds and is classified as a separate subspecies *Phaseolus vulgaris* ssp. *nuna* (Singh, 2001).

Genetic diversity in common bean is organized into two major gene pool classes. The large seeded (100-seed wt= >40g) Andean types and small (<25g) and medium (25-40g) Mesoamerican types. The differentiation of two gene pools is further evidenced by relationship of seed size with the difference in *Dl* gene (*Dl-1* and *Dl-2)*, Phaseolin seed proteins, isozymes, morphological traits as well as genetic markers. The Andean and Mesoamerican gene pools have further been divided into six sub classes. Singh *et al.* (1991) proposed that within each gene pool three races could be distinguished. The races arte differentiated on the basis of differences in seed size and growth habit, environmental adaptation regimes, infertility barriers, resistance to diseases, isozyme diversity, phaseolin types,

Figure 12.4: Market Classes of Common Bean

Figure 11.5: Patterns of Variation in Seed Shape, Size and Colour (*Courtesy*: Dr. P.A. Sofi)

Contd...

Figrue 11.5–*Contd...*

Contd...

Figrue 11.5–*Contd...*

Contd...

Figrue 11.5–*Contd...*

molecular markers such as RFLP, RAPD, AFLP and SSR. (Becerra *et al.*, 2011). Within the Middle American gene pool, flowing races are recognized:

Mesoamerica (M)

It is common to both Mexico and Central America and is characterized by relatively small seed and warm low land adaptation. Most races of M land races have habits of type 2 or 3, although some

Figure 11.6: Patterns of Variation in Flower Colour
(*Courtesy*: Dr. P.A. Sofi)

Figure 11.7: Patterns of Variation in Leaf Shape and Size
(*Courtesy*: Dr. P.A. Sofi)

have type 4 habits. Commercial classes within race M include small black, small Central American red and navy beans.

Durango (D)

It is composed principally of growth habit type 3 genotypes with small leaves, medium seed size and adaptation to dry high lands of Mexico. Commercial race D classes includes pinto, great northern and small red Mexican beans.

Jalisco (J)

It is found in the more humid high land area of Mexico and is composed of mostly climbing type 4 genotypes with medium seed size. Beebe *et al.* (2000) has added one more race to Mesoamerican group comprising climbing bean accessions from Guatamela that did not group with any previously defined races.

The Andean gene pool was like wise subdivided into three races on the basis of morphological and ecological criteria (Singh *et al.*, 1991). In addition to these races, Andean gene pool also includes the popping bean (*Phaseolus vulgaris* ssp. *nuna*) which are thought to be the ancient landraces.

Nueva Granada (N)

It represents the medium to large seeded accessions of bush growth habits, and includes the major of the commercial large seeded genotypes in use today. Race N is the most widely cultivated Andean race, and is grown at both mid-attitudes of the Andes and Africa, in warm low land environments of Brazil, Mexico and the Caribbean, and in temperate climates of North America and Europe.

Figure 11.8: Patterns of Variation in Growth Habits of Common Bean (*Courtesy*: Dr. P.A. Sofi)

Peru (P)

It consists of Andean climbing beans, most of which are adapted to high land environments.

Race Chile (C)

It consists of plants with small leaves with a prostrate type III growth habit, light pink to white flowers, less fibrous pods, medium sized, round to oval seed and usually red colours. The race is somewhat similar to race Durango of Andean gene pool. This race is often found at higher latitudes of Turkey, Iran and China (Beebe *et al.*, 2001).

The divergence of common bean into two gene pools through different domestication processes is a unique situation among crops except in case of rice where two separate gene pools namely Asian rice (*O. sativa*) and African rice (*O. glabberina*) have been established through different domestication events following different lineages from the pre-perrenis ancestor. In fact in rice also, *O. sativa* has undergone further speciation into *indica*, *japonica* and *javanica* types. A large number of studies have been undertaken to elucidate certain key aspects of common bean divergence, such as origin, relationship between two gene pools, qualitative and quantitative differences in genetic diversity, levels of linkage disequilibrium and the extent to which different loci have been subjected and have responded to selection under domestication (Kwak and Gepts, 2009). In 1980's, a wild ancestors of common bean were discovered in Ecuador and Northern Peru (Debouck *et al.*, 1993). Based on the DNA sequence of phaseolin seed proteins (Kami *et al.*, 1995) and chloroplast DNA (Chacon *et al.*, 2007) it was established that the discovered segment of bean germplasm is the most probable progenitor of *Phaseolus vulgaris*. Another important feature of bean domestication has been that there has been greater loss of diversity and divergence in case of loci that were involved in the process of domestication (Papa *et al.*, 2005).

Importance of Common Bean for Hill Agriculture

Pulses have been the mainstay of Indian agriculture in general and hill agricultuire. Rajmash or common bean is usually cultivated at higher elevations in hilly states of North and North east. In Northern India they are largely cultivated in Chamba, Kinnaur, Rajouri, Bhaderwah, Kishtwar, Uri, kupwara, Shopian etc. There is great variation in farmers preference for the growth habit, and seed shape, colour and size. In Ropa Valley of Kinnaur, red capsule types are more preffered, while as small seeded red types are preferred in Salooni, Teesa, Kishtwar and Bhaderwah and upper tracts of Uttarkashi and Chamoli, yellow types are liked in Spiloo and Moorang areas, small white seeded are preferred in Kalpa and Rogi. In Chauhar valley in HP and Chadar and Dharam valley in Uttarakhand, round mottled types are preferred. Common bean like other pulses does have certain unique features that make it a favourable competitor to cereals in hills:

- Ability to sustain a certain level of yields under marginal and poor management system that can ensure nutritional security. Globally it provides 15 per cent calorie and >30 per cent protein requirement of the population.
- It can act as a significant revenue generator given the fact that such examples are already available in developed countries such as USA, Brazil, Canada and developing nations such as Myanmar.
- Cheapest source of protein (22-25 per cent) for poor farmers who cannot afford animal protein due to higher cost. More importantly common bean contains a protein that inhibits the activity of HIV-1 reverse transcriptase.

- ☆ Good source of minerals and essential amino-acids. In fact it has certain minerals like iron and zinc that are reported to delay the progress of AIDS especially in malnutritioned people in poor farming systems.
- ☆ Ability to fix nitrogen, thereby not only minimizing the requirement of input support but also has the ability to improve soil properties in association with the soil microbes.
- ☆ Have deep root system, thereby can extract water from deeper;layers of soil and are fairly drought tolerant.
- ☆ Can use water judiciously and have lower water requirement as compared to cereals, thereby would be ideal to compete with cereals if genetic enhancement is undertaken to increase yield levels of existing varieties.
- ☆ Can be conveniently fitted in various cropping sequences and patterns due to availability of a range of maturity groups and its ability to yield better in intercropping.

Constraints in Common Bean Production

Pulses in India have long been considered as the poor man's only source of protein and have been called as "Poor Man's meat" a stigma that has been attached to this valuable food crop. Pulses are grown on 22-23 million hectares of area with an annual production of 15 million tonnes (mt). India accounts for 33 per cent of the world area and 22 per cent of the world production of pulses. The major pulse crops grown in India are chickpea, pigeonpea, lentil, moongbean, urdbean and fieldpea. About 90 per cent of the global pigeonpea, 65 per cent of chickpea and 37 per cent of lentil area falls in India, corresponding to 93 per cent, 68 per cent and 32 per cent of the global production, respectively (FAOSTAT 2009). Due to stagnant production, the net availability of pulses has come down from 60 g/day/person in 1951 to 31 g/day/person (Indian Council of Medical Research recommends 65 g/ day/capita) in 2008. If we take into account the WHO recommendation of 80 g/person/day, then the pulse demand in India by 2018 will be about 38 million tonners, which when compared to the total global production of 55 million tones, poses a daunting task to scientists, policy makers and farmers.

Among pulses, common bean (*Phaseolus vulgaris* L.) is the most widely distributed and consumed legume species of the genus *Phaseolus*, a genus comprised of some 70 species (Freyt and Debouck, 2002) that has contributed to human welfare with five cultigens domesticated in pre-Columbian times *viz.*, the common bean (*P. vulgaris*), the year bean (*P. dumosus* Macfad.), the runner bean (*P. coccineous* L.), the temporary bean (*P. acutifolius* A. Gray), and the lima bean (*P. lunatus* L.). Among the five domesticated species, the common bean (*P. vulgaris*) accounts for more than 90 per cent of the cultivated crop worldwide and is by far the most widely consumed grain legume in the world (Singh, 2001). It is grown in subtropical or temperate regions throughout the world and during the cool, dry season in tropical areas. Like other legumes, they supply proteins, carbohydrates, vitamins and minerals and complement cereals, roots and tubers that compose of the bulk of diets in most developing countries. The seeds have a protein content of 22 to 24 per cent. Beans are low in sodium and gluten free making them suitable for low salt and gluten free diets. The straw that remains after harvest is a valuable source of organic nitrogen.

Brazil is the world's largest producer of common bean followed by India, China, Myanmar, Mexico and the United States and produces more than 2.2 million tons, which represent 17.3 per cent of the world's production. However Brazilian average yield (8.52 q ha^{-1}) is much lower than that observed in the United States and China, where average productivities reach 19.23 and 15.60 q ha^{-1}, respectively.

In India, common bean is considered as a minor pulse crop which is confined to the hilly tracts of Himachal Pradesh, Uttar Pradesh, Jammu and Kashmir and North Eastern States, as it requires moderate temperature for growth. Lack of diversity and non-availability of determinate types is a major bottleneck in bean improvement programmes in India. However, the genetic improvement of common bean in India has been accomplished primarily by targeting breeding strategies based on consumer preference for seed size, shape and colour. The state of Jammu and Kashmir is a very rich source of common bean diversity for seed, colour, shape, plant type and other important traits. Under Kashmir valley conditions, the crop is domesticated throughout the region and very good quality beans are grown fetching a premium price in the market. However the main drawback for the crop is low production and the main obstacle for higher production is the narrow varietal profile, inefficient plant architecture, higher rate of flower drop, non-synchronous maturity and susceptibility to diseases and pests. Since, only one dry bean variety (Shalimar Rajmash-1) of determinate type is released for its cultivation in Jammu and Kashmir.

In the state of J&K, common bean or Rajmash is the most important and widely grown warm season pulse crop, that makes an important component of subsistence agriculture based in rural areas especially in the low input farming system, where it not only augments the livelihood and ensures nutritional security but also gives sustainable yields. However, the yields over a period of time have been disappointingly low at 5-6 q/ha. Even at the national level, the pulse yields have been hovering around the same level, despite more than 200 varieties being released since 1970, with the high yielding varieties hardly making any dent. The broader bottlenecks that have resulted in the continued dismal performance in pulse sector in India as well as J&K State are:

1. The green revolution has witnessed research, policy and institutional support more biased towards cereals, resulting in the substantial gains in the yields of cereals. During the green revolution era, whereas the productivity gain for cereals was about 280 per cent, it was a meager 32 per cent for pulses.
2. There have been hardly any major technological breakthroughs in case of pulses that could have enhanced the yield levels. This is largely due to less focused R&D support for genetic enhancement of pulses.
3. The procurement price policy of the Government has been more favourable for cereals rather than pulses.
4. Pulses are inherently low yielding due to various factors. Pulses are physiologically more complex than cereals in terms of nitrogen use and its relation to photosynthesis. Moreover the nitrogen fixation process implicates energy demand that limits the yield levels.
5. Pulses have complex market classes based on shape, size, colour of seed that has limited the pace of genetic enhancement.

Constratints in Hill Agriculture

At the regional level, common bean as well as other pulses in J&K continue to enjoy the status of secondary crops with potential to support subsistence and augment livelihood especially in the hilly tracts of Jammu and Kashmir. The cultivation is largely confined to the areas of Shopian, Pulwama, Bandipora, Uri, Rajori, Poonch, Baderwah, Doda, Batote, Kupwara, Gurez, Kihtwar with the common feature that cultivation is under marginal low input system. The major constraints of common bean cultivation in hill farming systems are:

1. *Inherent Bottlenecks of Hill systems*: hill agricultural systems are characterized by certain unique features that limit the production potential of farming systems and as such are the major constraints. These include:
   - ☆ *Marginality*: Small land holdings with minimum support and maximum constraint. In HP, 63.8 per cent of about 80000 farm families have less than 1 ha of land and only 5 per cent have more than 4 ha of land.
   - ☆ *Fragility*: since the cultivation is invariably dependent on rainfall and inherent resources, the yields are seldom sustainable. The undulating topography cases erosion losses as well as wastage of rainfall water, a situation that has been accelerated by migratory grazing in the hills. Moreover the soils are shallow and invariably vulnerable to water stress more often at critical growth stages.
   - ☆ *Inaccessibility*: hill areas are largely inaccessible and the infrastructure, communication, implenments and mobility are obstructed by various physical, climatic, biological and socioeconomic factors. In fact the improved technologies have seldom reached hill areas
   - ☆ *Vulnerability*: hill areas are vulnerable to natural disasters such as rainstorms, hailstorms, floods, disease epidemics, and above all erratic rainfall patterns
2. *Lack of high yielding varictics*: The current varieties being grown in the Rajmash growing areas largely consist of low yielding landraces which have adapted to lower yield levels. In fact the varieties that have been released by SKUAST-K like Shalimar Rajmash-1, Selection-3, French Yellow and the varieties recommended such as Canadian Red and Contender have yield levels of 1-1.2 T/ha under optimal management which are hardly realized under farmer's conditions.
3. *Lack of adequate quality seed*: in case of pulses, the Seed Replacement Rate has been dismally low as compared to cereals and other cash crops. This is largely due to the greater efforts at producing quality seed in cereals and cash crops. The situation in hills has been all the more worse due to the inaccessibility of technology delivery systems.
4. *Lack of genetic variability in the usable gene pool*: This is a problem in all the bean improvement programmes, where only few parents have predominated in cultuivar development. The narrow genetic base makes the cultivars less resilient to climatic uncertainities and therefore yield levels that are not sustainability.
5. *Relegation to harsh environments*: In fact this is one of the major constraints in enhancing the yield levels of pulse crops. About 88 per cent of pulse area is rainfed and hardly receives input support as is enjoyed by cereals. Thus invariably, this crop suffers from drought and moisture stress at critical crop growth stages. The relegation of pulses to low input marginal environments further relegates them to harsher environments, causing further lowering of the yield, thereby significantly reducing the competitive ability of this valuable group of crops to replace cereals.
6. *Inherent susceptibility to diseases and pests*: Common bean is attacked by a large number of diseases such as common blight, angular leaf spot, anthracnose, alternaria leaf spot, rust, aschochyta blight, powdery mildew, white mould, BCMV, golden mosaic virus, yellow mosaic virus, curly top as well as certain insects especially pod borer, that greatly reduce the yield levels.
7. *Inefficient input and technological support and technology delivery system*: As already discussed, There have been hardly any major technological breakthroughs in case of pulses as is

evidenced by the fact that varietal development has not been speeded up to develop area specific crops, nor have been the production technologies been optimally standardized to enhance the yield levels with current varietal profile. More disappointingly, the existing technologies have not been efficiently delivered due to lack of focus *vis-a-vis* pulses in general. Ironically, while as there are sizeable germplasm repositories of crops like rice, maize, oilseeds etc, there have been hardly any efforts to collect and characterize the local germplasm as well as procure exotic germplasm.

Opportunities for Enhancing Common Bean Production in Hills

While the yields are disappointingly low in farmer's field, there are considerable prospects for substantial improvement in the productivity. This is evidenced from the large gap in the farmers yields and those attained in experimental trials under minimum constraint system. The continuing major challenge, therefore is to cost effectively enhance present yield levels so that pulses in general and common bean in particular may compete with cereals. This is important as the climatic factors and the cultivation conditions cannot be substantially modified at farmer's fields.

Kashmir valley as well as other hill areas of H.P and Uttarakhand has the ideal conditions for cultivation of warm season pulses such as common bean. Therefore appropriate efforts are needed at the genetic enhancement to develop varieties that are attractive to farmers in terms of their yield advantage, resilience to biotic and abiotic factors as well as have acceptable quality standards.

Enormous genetic variability exists in the state of J&K, especially in the potential areas of variability for growth habit, use category, seed size and color, quality parameters. Similarly the genetic resources available in the national and international gene banks such as NBPGR, CIAT, AVRDC, USDA-ARS, IPK etc contain desirable traits that could be used to broaden the spectrum of varietal profile.

The ideal approach will be to utilize the local and exotic genetic resources, to develop varieties for specific farming systems prevalent in bean growing areas and develop appropriate plant type for intercropping and mono-cropping systems, following a two pronged strategy: one to increase the production levels in harsh environments by developing varieties that can yield better under minimum management and maximum constraint situations to enhance the livelihood opportunities to farmers, and the other to develop varieties for favourable environments with optimum input support and minimum constraint situation so that common bean can be a favourable competitor to *kharif* cereals.

Strategies for Enhancing the Yield in Hills

Collection and Conservation of Germplasm

As already stated that genetic base of the existing varieties of common bean is narrow, making them less efficient in responding favorably to environmental fluctuations and management levels. There is an urgent need to broaden the genetic base of common bean varieties by using the genetic resources that are available locally or elsewhere. Thus the four other domesticated species of *Phaseolus* as well as the wild species can be used to introgress valuable genes into cultivated varieties. Since many traits are not available in the local germplasm and are only found in exotic accessions, pre-breeding is an important component of germplam management and utilization.

Breeding for Higher Yield

Unlike cereals, where yield increases have been significant to make sizeable impact on productivity enhancement, the same has been modest in pulses and even in some cases has been disappointing.

Neither the improved partitioning nor responsive cultivars have happened to legumes (Kelly *et al.*, 1998). Especially in case of common bean varieties for hills, emphasis has toi be given to appropriate plant type as the crop is grown seldom as a monocrop and largely as an intercrop along with cereals such as maize and other legumes such as cowpea, faba bean, moong etc. crop varieties can perform differentially under mixed cropping system largely prevalent in hills. For such systems climbing and twinning types are more preffered as they have better competitive ability to harness solar radiation and bush types are at a disadvantage. However for the areas where crop is grown as a mono crop, appropriate plant type should be designed to perform optimally. Breeding for higher yield in common bean has not met with the rate of success as has been realized in legumes as there are many inherent constraints such as growth habit, maturity, quality, market classes and then diseases and pests.

Table 11.10: Varieties of Rajmash Released for Northern Region

Variety	*Institution*	*Year of Release*	*Area of Adaptation*
VL Bauni bean 1	VPKAS Almora	1973	NW hills
VL Bauni bean 12	VPKAS Almora	1973	Uttarakhand hills
VL Bauni bean 17	VPKAS Almora	1973	Uttarakhand hills
VL Rajma 63	VPKAS Almora	1978	Uttarakhand hills
VL Rajma 125	VPKAS Almora	2008	Uttarakhand hills
Shalimar Rajmash 1	SKUAST-K	2005	Hills of J&K
Shalimar French bean 1	SKUAST-K	2005	Hills of J&K
Arun	IIPR Kanpur	-	NEPZ
Uday	IIPR Kanpur	-	NEPZ
Utkarsh	IIPR Kanpur	-	NEPZ
Amber	IIPR Kanpur	-	NEPZ

Standardisation of Production Technologies

The management and technological needs of hill farming are diverse and many more and the isolated research efforts at the experimental stations may not be able to take care of farmer's diverse needs and complexities. Therefore the technologies need to be developed and validated under the hill systems preferably under a participatory mode where farmers have a major say in the technology development, refinement, selection parameters as well as follow up studies. Such an approach will enhance the rate of adoption of improved varieties and technologies and have a sizeable impact. One of the important technological interventions would be to develop formulations of effective microbes that synergistically interact with pulses and enhance yields. There is wealth of evidence about the efficience of such approach as pulses are invariably responsive to bio-agents.

Disease and Pest Management

About 50 diseases and pests have been reported in common bean that cause severe decline in the common bean yields. Currently about 80 per cent of management strategies are based on cultural practices and use of chemicals both of which are unsustainable in hills due to obvious limitations. Therefore resistance to diseases has been one of the focal areas in various national and international breeding programmes with little focus on yield per se. Since the control measures are seldom put in

A. Alternaria Leaf Spot

B. Angular Leaf Spot

C. Common Blight

D. Curly Top Virus

E. Bean Yellow Mosaic Virus

F. Common Rust

Figure 11.9: Common Diseases of Rajmash in Kashmir Valley (*Courtesy*: Dr. P.A. Sofi)

place due to inaccessibility, and other features of hill farming, development of resistant varieties is the most economical and sustainable option.

Abiotic Stress Mitigation

Common bean cultivation in hills is invariably challenged by various abiotic stresses at various critical stages of growth. These stresses include drought, salinity, nutritional deficiencies, flooding, hailstorms, soil problems etc, and drastically limit the production limits that can be achieved under hill system. These stresses can be managed by genetic enhancement as well as cultural practice management, water harvesting, contour farming etc. CIAT has very recently developed a highly drought tolerant variety INTA SEQUIA in collaboration with National Agricultural Research organization of Tanzania. Similarly CIAT and EMBRAPA have jointly developed a drought tolerant line A774 that has exhibited 30q yield advantage in national trials under severe drought.

Mechanisation

Non-availability of labour is a major constraint in hills therefore mechanization is an important mitigating strategy for enhancing returns from farming as well as livelihood. However, the tools and equipments have to be either specifically designed or suitably modified to be effective in the undulating contours of hills.

Organic Production

Common bean has been grown in northern states of India as well as north eastern state since decades under low input system which can be synonymous to organic culture. However due to absence of proper certification standards, the product cannot fetch premium price in domestic or export market. There are certain areas in the hills where common bean has been growing without fertilizers, pesticides and other inputs and are purely organic. These areas need to be promoted for establishment of organic farms for which adequate institutional and policy support is needed from the research institutions as well as the government.

Policy Support

There should be a strong policy support from the government to promote the cultivation of pulses in the hilly areas. This needs provision of quality seed, credit facilities, implements and tools as well as marketing support. There is also a need to strengthen and promote the informal seed production system prevalent in the hills. This could in turn enhance the employment and livelihood opportunities for the poor hill farmers.

Acknowledgement

Part of work discussed has been funded by the J&K State Council for Science and Technology, DST, J&K Govt. vide grant #25-ST/2011.

References

Anonymous (2008). Digest of Statistics, Directorate of Economics and Statistics, Planning and Development Department, J&K, pp. 107-112.

Anonymous (2009). Economic Survey of India, Ministry of Finance, Govt. of India, New Delhi, pp. 16-18.

Beebe, S, Skroch, J., Tohme, J, Duque, M, Pedraza, F and Nienhuis, J. (2000). Structure of genetic diversity among common bean landraces of Mesoamerica based on correspondence analysis of RAPD. Crop Sci. 40: 264-73.

Beebe, S., Rengifo, J. Gaitan, M, Duque, M and Tohme, J. (2001). Diversity and origin of Andean landraces of common bean. Crop Sci. 41: 854-62.

Burkart, A. and Brücher, H. (1953). Phaseolusa bo rigineusB rtkart, die mutma3liche andine Stammform der Kulturbohne. Zichter 23 (3): 65 -7 2.

Chacon, S, Pikersgill, B and Debouck, D. (2007). Phylogeographic analysis of chloroplast DNA variation in wild common bean in Americas. Plant Sys. Evol. 110: 432-44.

Debouck, D., Toro, O, Paredes, O, Johnson, W and Gepts, P. (1993). Genetic diversity and ecological distribution of *Phaseolus vulgaris* in NW South America. *Economic Botany*. 47: 408-23.

FAO (2008). Food and Agricultural Organization of the United Nations. Available at :<http:// faostat.fao.org/site/339/default/aspx.

FAOSTAT (2009). FAO Rome.

FOASTAT (2010). FAO Rome.

Gentry, H.S. (1969). Origin of the common bean, *Phaseolus vulgaris*. *Economic Botany* **23** : 55-69.

Gepts, P. (1998). Orifin and evolution of common bean: past events and recent trends. Hort Sci. 33: 1124-1130.

Gepts, P and Bliss, F. (1985). F1 hybrid weakness in common bean: differential geographic origin suggests two gene pools between *Phaseolus vulgaris* and its fungal pathogens. J. Heredity. 76: 447-450.

Kami, J, Beccera, V., Debouck, D. and Gepts, P. (1995). Identification of presumed ancestral DNA sequence of phaseolin in *Phaseolus vulgaris. PNAS.* 92: 1101-1104.

Kwak, M and Gepts, P. (2009). Structure of genetic diversity in two major gene pools of comonbean. TAG. 118: 979-992.

Kelly, J., Kolkman, J. and Schinider, K. (1998). Breeding for yield in dry bean. *Euphytica*. 102: 343-356.

Mclean, P., Kami, J. and Gepts, P. (2004). Genomics and genetic diversity in common bean.

Papa, R., Acosta, J., Salinas, A. and Gepts, P. (2005). A genome wide analysis of differentiation between wild and domesticated *Phaseolus vulgaris* in Mesoamerica. TAG. 111: 1147-1158.

Salinas, A., Bibler, R. and Lavin, M. (2006). Phylogeny of genus *Phaseolus vulgaris*: a recent diversification in an ancient landscape. Sys. Bot. 31: 779-92.

Singh, S.P., Gepts, P. and debouck, D. (1991). Races of common bean (*Phaseolus vulgaris*). Econ. Bot. 45: 379-96.

Singh, S.P. (2001). Broadening the genetic basis of common bean cultivars : A Review. *Crop Science* **41**: 1659-1675.

van Schoonhoven and Voysest, O. (1998). *Common Beans : Research for Crop Improvement* (Eds.). CAB International, Wallinford, UK and CIAT, California, Colombia.

Vavilov, N.I. (1939). The important agricultural crop of pre-Columbian America and their mutual relationship. Izd. Gos. Geogr. O-va (Public. Nat. Dept. Geogr. USSR) 71(10): 1-25.

Hill Agriculture: Prospects, Constraints and Mitigations (2013) *Pages* **171–178**
Editor: **Dr. Shahid Ahamad**
Published by: **DAYA PUBLISHING HOUSE, NEW DELHI**

Chapter 12

Field Pea Production and Protection in Hills

***M.C. Dwivedi*[1] *and V.K. Singh*[2]**

[1]*Division of Agronomy, S.K. University of Agricultural Sciences and Technology, Seed Production Farm, Chakroi-Jammu (J&K)*

[2]*Division of Plant Pathology, S.K.University of Agricultural Sciences and Technology, DLRSS, Dhiansar – 181 133, Jammu (J&K)*

Introduction

Pea (*Pisum sativum* L.) also called as pois proteagineux in French, guisante in Spanish, pisello in Italian, futtererbse in German and matar in Hindi, is an important cool season crop largely confined to cooler temperate zones between the tropic of cancer. There are two types of cultivated pea, *i.e.* garden pea and field pea. Garden peas are harvested in an immature condition and cooked as fresh or canned for subsequent uses. Field pea is generally grown for dry seeds which are used for a variety of snack preparations and dal. It is one of the important pulse crop mostly all states of India, grown in area of 0.68 million hectares producing 0.62 million tonnes of dry grain. The average Indian productivity of field pea is 911kg/ha. The major field pea-growing states are U.P, M.P, Bihar and Maharashtra. Besides these states, it is also cultivated in Delhi, West Bengal, Punjab, Haryana and H.P.

It is consumed both as green immature seeds as well as dry seeds. The dry grain is generally consumed as chhole after soaking, cooking and mixing with spices or as snack after roasting or frying. It is used along with cereals such as rice, wheat, maize or millets so as to balance the level of essential amino acids, because of the fact that pulses are deficient in methionine and rich in lysine, where as cereals are rich in methionine but deficient in lysine (Srivastava and Ali, 2004). Field pea is a rich source of protein, carbohydrate and certain minerals. The seed coat contains most of the non digestible carbohydrates (dietary fibre) and is relatively rich in calcium and phosphorus. The genotypic variations exits in protein and other nutritional parameters of seed of field pea (Adsule *et al.*, 1989), but no consolidated information is available on protein, crude fibre and fatty acid composition of different field pea genotypes, which are responsible for lowering cholesterol level in human beings.

Soil

Although pea is a crop of winter season, it is susceptible to frost. It is grown on wide variety of soil types, ranging from sandy to heavy clay soils but sandy loam soils are the best. The crop is tolerant to alkali soils but does not do well under acidic soil conditions, unless liming is done. A well drained soil is essential for successful production of peas. Pea is highly sensitive to water logging conditions; hence a well drained loam soil is considered best for pea cultivation. They tolerate a moderate soil pH range 6.0-7.5. The optimum pH is 6.5. It requires a well pulverised seed bed.

Land Preparation

Field preparation is very similar to any rabi crop but the soil should not be very pulverised and fine, however, it must be free from all established weeds and stubbles of the kharif crop grown earlier. Under rainfed conditions it must be treated with 20-25kg of Chloropyriphos dust so that there is no attack of termites. The bed should be thoroughly levelled, especially in case of irrigated conditions for proper distribution of irrigation water. The field should be ploughed with disc or mouldboard plough. Where tractor is available one deep ploughing followed by three or four harrowing and planking should be given to prepare a well pulverised seedbed. Where bullocks are the source of power, deep ploughing with local plough should be done. Avoid powdery seedbed. Field should be well levelled.

A good seedbed should have friable but compact soil with adequate moisture, and should be free from weeds and plant debris. Tillage requirements depend on soil type, moisture content, preceding crops, planting methods, weed infestation and the target crops. Generally one deep ploughing followed by cross harrowing and planking ensure good seedbed. In irrigated conditions the land preparation is generally done with pre-plant irrigation if moisture content in the top soil layer after harvest of proceedings crop is not adequate for germination of winter crop.

Seed and Seed Treatment

Quality seed is the key to plant establishment and crop productivity. The seed should be genetically and physically pure with 90-95 per cent viability free from disease and physical damage. Prolonged and faulty storage decreases seed viability. High moisture and temperature are conducive for poor viability of seeds. Pulse seeds should be treated with different fungicides *viz.* Carbendazim, Thiram, Captan, Carbendazime + Thiram (1:1), Carbendazime, + Captan (1:1), and *Trichoderma harzianum* to control seed borne diseases. Mostly fungicide 2–3 g per kg is used while in case of bioagent 5–6 g/kg of seed. The seed treatment of legumes is also done with rhizobium culture to capitalize on their intrinsic capability and potentiality to trap atmospheric nitrogen in the root nodules through BNF. It is estimated that annual BNF is about 175 million tonnes of which close to 79 per cent is accounted for by terrestrial fixation.

The free living rhizobia infect the host legumes through infection thread or at the point of emergences of lateral roots and are transferred into bacterioids which are the sites for nitrogen fixation. The nitrogen fixation is governed by the host, micro symbiont, physical and chemical properties of soil, soil moisture and nutrient status, soil microflora and environmental factors. Legume species not only differ in nodulation but cultivars within a species also differ significantly, suggesting thereby that the host factors are important determinants of nodulation. Moisture stress as well as excess moisture (water logging) adversely affects rhizobia population and BNF. Poor BNF under waterlogged conditions is due to anoxia (lack of oxygen). Deficiency of essential plant nutrients, specially phosphorus and molybdenum appreciably decreases BNF. Molybdenum is an integral component of nitrate reductase enzyme and therefore its deficiency directly influences the BNF. Increasing soluble salt and extremes of soil pH reduce BNF. However, rhizobium has high degree of tolerance to soil salinity and alkalinity than the host plant.

Sowing Time

Planting time is the most important non-monetary input having profound effect on crop growth, phenological development, occurrence of pests and crop productivity. The plant environment such as temperature photoperiod, moisture availability (in dry land area) etc. is significantly changed with the time of planting. It also influences availability of plant nutrients, BNF and weed infestation. The optimum time of planting of filed pea which is generally sown with pre plant irrigation and fallows a regular kharif crops lies between mid Octobers to first week of November.

In central zone where grow the period of winter pulses is restricted due to high temperature and low profits moisture, early planting is generally advantageous (Singh *et al.*, 1996) working at Ambikapur (M.P.) found that planting of field pea in the first week of october produced 2225 kg/ha as compared to 1292 kg/ha from mid november planting. Planting time is generally dictated by environmental temperature. However, in rainfed areas where soil moisture is the major constraint and for seed germination and plant establishment, the availability of moisture in surface layer also plays key role besides temperature. The optimum time of planting differs from location to location and also within same location depending on the cropping systems in which pulses are grown. Early as well as delayed planting decreases yield of winter pulses. Early planting induces excessive vegetative growth an account of high temperature and humidity and poor pod-set. Seedling mortality due to high temperature is also commonly observed in central India. The delayed planting restricts vegetative growth and pod bearing branches, decrease BNF may also lead to forced maturity as well as increased incidences of pests.

Field Pea Varieties

Several field pea varieties are recommended for Indian conditions such as T-163, Rachna, Pant pea-5, P.G-2, HUP-2, Swarn rekha, Hans and P-48.

Plant Population, Seed Rate and Planting Method

Adequate plant stand is the basic pre-requisite for higher productivity. Plant population is generally determined by growth pattern and canopy structure of a cultivar, planting time, growing conditions and the cropping systems. Seed rate is the key determinant of plant population. Seed rate is decided on the basis of seed size, seed purity and germination percentage. Planting method has also significant influence on seed rate. In broadcast method of sowing, seed rate is kept higher than row planting on account of uneven distribution and poor germination of seeds in the former case. In broadcast method, many seeds do not get properly placed in the moist soil layer and are also liable to bird damage. Yields are generally higher in row planting than broadcast due to better plant establishment and minimum inter and intra plant competition.

In the field pea 50-60 kg/ha seed rate for small seeded genotypes (13–16g/100 seed wt.) and 80-90 kg/ha for bold seeded (20g/100 seed wt.) has been found optimal. Planting method is an important management factor which has direct effect on seed requirement, plant establishment, and performance of cultural operations and efficiency of production inputs. Line sowing is beneficial over broadcast as it insures uniform distribution of seeds, placement of seeds at proper depth, better plant stand, easy in cultural operations and also improves drainage. North-south row orientation is generally beneficial over east-west orientation as in the former case there is better interception of light in crop rows.

Seeding depth considerably influences seed germination and emergences. Depending upon the soil moisture conditions the depth of sowing may vary from 5–7.5 cm for ensuring good germination. A row spacing of 20–30 cm and a plant to plant distance of 10 cm is optimum.

Cropping System

Pulses are integral part of the cropping systems and as such they are essentially required to be included in some of the most productive cropping systems for sustaining the soil health and maintaining the productivity. Rabi pulses are grown as main crop, intercrop, mixed crop, relay crop or alley crop depending on the need of the locality. The prominent pulse-based cropping systems which occupy large area, are; chickpea + mustard/barley in Northern India, chickpea + linseed, lentil + linseed and chickpea + safflower in central India and chickpea + coriander in southern India. The management of these systems is complex and needs more attention to obtain higher productivity and better land-use efficiency. Relay cropping of lentil and lathyrus after rice is prevalent in south India, M.P and Bihar. The productivity of these cropping systems can be enhanced by following suitable agronomic practices such as selection of compatible varieties, fertilizer use, weed management, tillage and appropriate plant protection measures.

Water Management

Depending on soil type and climatic conditions, the crop after germination needs two light irrigations which should be given at branching to flowering stage (30–50 DAS) and second at pod filling stage (75 DAS) of the crop. The water logging should be avoided as water logging even for few hours may cause a considerable damage to the crop. Flowering and pod development stages are critical stages for moisture. Drainage is essential. Any of these two irrigation may be skipped if the winter rains coincide with the specified period of irrigation. Rain associated with cloudy weather invites fungal diseases in the crop and the crop requires due care for its protection. The crop may be given an additional light irrigation when frost is expected so that it may not be damaged with frost. In all it needs about 154 mm water. Special precaution should be taken while irrigation a pea crop. Light and uniform irrigation should be given. The soil should have sufficient moisture at the time of sowing so that the seeds may germinate well. Moisture stress at early stage drastically cuts the biological nitrogen-fixation which is ultimately reflected in poor yield. Kasturi (1995) recorded marked decrease in nitrogenise activity and nodulation in field pea due to moisture stress at early vegetative growth stage under New Delhi conditions on sandy loam soils. In North east plains (Faizabad), one irrigation at flowering initiation stage (50 DAS) was found most promising (Yadav *et al.*, 1992) whereas in central India (Jabalpur) 2 irrigation, one each at branching and pod formation were found optimum (Rathi *et al.*, 1995). In NWPZ (Navgaon, Rajasthan) too, branching and pod formation were found most critical.

Manures and Fertilizer Management

The favourable response of fertilizer application has been found in field pea. The optimum dose of nitrogen was found 20–40 kg/ha at many locations of northern India. The field pea response to applied phosphorus various depending on the phosphorus availability in soil and moisture availability (Yadav *et al.*, 1992). In most of the studies the response was observed from 50–100 kg phosphorus/ha (Srivastava and Ahlawat, 1995, Rathi *et al.*, 1995, Yadav *et al.*, 1992). Although an increase in yield to the tune of 25 per cent has also been observed with the application of 20 kg Potash/ha (Ali, 1985). The response to Potash application in general is megre. Sulphur application increased the yield at many locations. Application of 40 kg sulphur gave a response of 332 kg/ha over 20 kg sulphur (AICPIP,

1997) at Varanasi. Amongst micronutrients, significant response to molybdenum application has been observed (Srivastava and Ahlawat (1995). Application of 0.5 kg/ha molybdenum increased the yield by 80 kg/ha.

Being a leguminous crop, the pea does not require high dose of nitrogen. If available, about 15–20 tonnes of organic manures should be incorporated in the soil at the time of land preparation. This should be supplemented with 20–30 kg nitrogen as a starter does per hectare as a basal dressing the time of sowing which can meet plant requirement before the formation of nodules. For good nodulation seed should be treated with proper strain of bacterial culture before sowing.

Weed Control

Weed competition may severely reduce yield of field pea. The degree of yield loss varies from 18 to 90 per cent depending on the growing conditions, crop species and management practices. Heavy weed infestations should be controlled by cultural or mechanical or chemical measures prior to rotating into field pea, and prior to planting. The crop should be kept free from weeds by giving two hoeing preferably with wheel hand hoe at 3 and 5 weeks after sowing. For efficient weed control use stomp 30 EC (Pendimethalin) 2.5 lit/ha has pre-emergence within 2 days of sowing. Dissolve herbicide in 400-500 litre of water and spray it uniform over the entire field. Application of stomp control *Phalaris minor* and broadleaf weeds but not wild oats.

Harvesting, Threshing and Storage

Harvesting at appropriate time is necessary to avoid post-harvest losses. The maturity of different pulse crops varies considerably from region to region even for the same date of planting on account of atmospheric temperature, humidity, photoperiod, profile moisture etc. At maturity, the moisture content of seeds should be around 15 per cent. Seed cracking may occur, if moisture content of seeds drops below 10 per cent due tòp over maturity. Field pea should be harvested when they are fully ripe and threshed after sufficient drying in the sun. Harvesting is generally done by sickles. In some areas, hand pulling at physiological maturity is also practised. Use of combine harvester is rare. The harvested plants are carted to the threshing floor, dried for about a week and trampled under the feet of bullocks or under tractor or by beating with sticks. Seeds are separated from straw and trash by winnower or dropping seeds manually from a height against wind current. Clean seeds should be sundried for 3-4 days to bring moisture content to about 10 per cent seeds should be safely stored in appropriate seed storage bins and fumigated to protect form insects.

Yield

Field pea can yield about 20-25 quintal of grain and about same quantity of straw from one hectare of land. Field pea yield depend upon on the variety and soil conditions.

Diseases

Seed Rot

Seed rots are soil borne fungal diseases caused by *Pythium* spp., *Fusarium* spp. and *Rhizoctonia solani*. Seed is infested shortly after planting and seedlings fail to emerge. Young germinating seedlings are attached at ground level, stem becomes brown and constricted, which results in poor stand of the crop. Treatment of seed with a fungicide can be a cost effective control.

Powdery Mildew

It is a fungal disease caused by *Erysiphe pisi* white powdery growth of fungus mycelium and spores develop on stems, branches leaves and pods. Spray the crop with Karathane 0.1 per cent or Bavistin 0.1 per cent or Sulphur dust 25–30 kg/ha was an old recommendation and is even now very effective though costly.

Wilt

The disease is caused by *Fusarium oxysporum* f. sp. *pisi* rotting of roots and yellowing of the lower most leaves is followed by wilting. Symptoms include a brownish red discoloration of the vascular tissue. The vascular tissue of the root may also be discoloured (reddish streaks) control is best achieved by planting filed pea in a four year rotation with other crops. Treatment of seed with a (1:1) mixture of Benomyl + Thiram in a commercial formulation (1.25 g Benomyl + 1.25 g Thiram kg^{-1}) is advised to eradicate the fungus.

Rust

It is a fungal disease caused by *Uromyces fabae* yellowish, reddish brown spherical raised pustules appear on leaves, stems, branches and pods. Several infected leaves shrivel and fall off. Stem infection results in cracking and breaking of stem which leads to drying and death of plant. Quality of pea is reduced. Early sowing minimizes the disease. Fungicides found effective against the pathogen are Mancozeb (0.2 per cent) and Calixin (0.2 per cent).

Nematodes

Root-Knot Nematode

Root- knot nematode caused by *Meloidogyne* spp is an important nematode of pea. Three species of root-knot nematode, *M.hapla*, *M.incognita* and *M. javanica* parasitize pea roots. The symptoms produced by root- knot nematode on pea include stunting of infected plants, leaf chlorosis, patchy growth and galls/knots. Severe symptoms and even death of heavily infected pea plants have been observed by Singh, (2008) at DLRSS, SKUAST- Jammu (Fig. 1). In general, the total life cycle of the nematode is completed within 30 days. For disease management, use of Carbofuran (Furadan 3G) @ 25-30 kg/ha, however, crop rotation and deep summer ploughing is a very effective control measures for root knot nematode management.

Cyst Nematode

Cyst nematode caused by *Heterodera* spp.is an important nematode of pea. The above ground symptoms include stunting of shoot, yellowing of foliage and patchy growth, however, below ground symptoms the mature female bodies are found attached to roots by their head end embedded almost in the stele. At least two species of cyst forming nematodes *H.goettingiana* and *H. glycines* parasitize pea roots. *H.goettingiana* is the only nematode reported to be a limiting factor in itself in pea production (Hagedorn, 1984).Management practices are used same as root- knot nematode in pea.

Insects

Pea Pod Borer

This pest is more common in north India. *Etiella zinckenella, Lampides bocticus* and *Helicoverpa armigera* damage the crop by feeding on flowers and pods. The young larvae after hatching bore into

Figure 12.1: Root-Knot Nematode Infested Pea Field

the pods and feed inside on developing seeds. For pea pod borer, crop may be sprayed with 0.07 per cent. Endosulfan or Sevin at 10–15 days interval beginning with pod formation.

Aphids

Normally, aphids are considered as minor pest of pea, but sometimes they cause severe damage to the crop both nymphs and adults feed on plant sap and while sucking, inject toxins into plant system. Infested pods become deshaped, withered and malformed. Severe infestation may result in drying up of the affected pods. Aphids also act as vector pea virus. Spray the Rogor 30 EC (Dimethoate) when the attack starts and repeat after 15 days if necessary.

Seed Corn Maggot

Seed corn maggot feeds on sprouting seed or on seedlings. To control this insect, coat the seed with an insecticidal seed treatment.

Leaf Minor

It is widely distributed in north India. Adult females deposit the eggs by puncturing the leaf. Young maggot mines the leaf and feeds within. Infected leaves show numerous zigzag mines or galleries. Normally, it is a minor pest but may become serious.

Vertebrates

Appears highly palatable to and easily damaged by rabbits and ground squirrels.

References

Adsule, R.N. Lawande, K.M. and Kadam, S.S (1989). Pea: In hand book of world food legumes-nutritional chemistry, processing technology and utilization, Vol. II (eds. Salunkhe, DK and Kadam SS), CRC press, Inc. Boca Raton, Florida, pp 215- 251.

Srivastava, R.P. and Srivastava, G.K. (2004). Nutritional quality of some chickpea (*Cicer arietinum*) varities grown under drought condition. Indian Journal of Agricultural Biochemistry 17:15-19.

Kasturi, Koka. (1995). Response of pea to moisture stress and application of P, S and Zn fertilizers.Ph D. Thesis, pp 75-81 IARI, New Delhi.

Yadav, R.P. Chauhan, DVS and Kushwaha, H.S. (1992). Effect of irrigation, phosphorus and row spacing on yield attributing characters of pea (*Pisum sativum*). Indian Journal of Agronomy 37: 617-18.

Rathi, G.S, Sharma, R.S. and Rajak, R.C. (1995). Effect of irrigation schedule and phosphorus on yield attributes and yields of field pea (*Pisum sativum*). Indian Journal of Agronomy 40:82-85.

Srivastava, T.K and Ahlawat, I.P.S (1995). Response of pea to phosphorus, molybdenum and biofertilizers. Indian Journal of Agronomy 40: 630–635.

Ali, M. (1985). Effect of pre sowing seed treatment, foliar nutrition and planting pattern on productivity and water use in chick pea under rainfed conditions Legume Research 80:7–11.

Singh, V.K., Sengar, R.B.S and Singh, R.N (1996).Effect of varieties and sowing dates on disease incidence and productivity of field pea. Indian Journal of Agronomy 41:451-453.

Hill Agriculture: Prospects, Constraints and Mitigations (2013) *Pages* **179–192**
Editor: **Dr. Shahid Ahamad**
Published by: **DAYA PUBLISHING HOUSE, NEW DELHI**

Chapter 13

Underutilized and Underexploited Vegetables Crops in Hilly Tripura

Sukhen Chandra Das[1]* and C. Ravindran[2]
[1]Assistant Professor (Horticulture),
College of Agriculture, Lembucherra, Tripura – 799 210, Tripura
[2]Assistant Professor (Horticulture),
KVK, AC and RI, Madurai – 625 104, Tamil Nadu

Introduction

The crops, which are neither grown commercially on large scale nor traded widely, termed as underutilized crops (UUC). These crops are cultivated, traded and consumed locally. The popularity of these crops varies from crop to crop and locality to locality, which however, can be enhanced to a greater extent through publicity. These are easier to grow and hardy in nature, producing a crop even under adverse soil and climatic conditions. Most of them are very rich sources of vitamins, minerals, and other nutrients such as carbohydrates, proteins and fats. Since, the underutilized crops have a long history of consumption, the local people are aware about their nutritional and medicinal properties. Moreover, these are cheap and readily available. Huge numbers of underutilized and underexploited vegetables and tuber crops have grown in the homestead, fallow, forest areas as well as in the road sides with less care. Those vegetables and tuber crops are highly nutritious and contributing in the household food, nutritional security for tribal and rural peoples. State is a hilly and land locked and stretching latitude 22°56' and 24°32' north and longitude 91°10' and 92°21' east, with total geographical area of about 10,491 sq.km of which 60 per cent is hill with dense forest and 30 per cent is under cultivation. More than 75 per cent of the population either directly or indirectly depends on agriculture. The small and marginal farmers contribute about 90 per cent of total farming community and the average size of land holding is 0.97 ha (1990-91). The diverse agro-climatic conditions, fertile and acidic soil with good depth and abundance rainfall favour the cultivation of various underutilized

* Corresponding Author: E-mail: sukhenchandra@rediffmail.com

and underexploited vegetables and tuber crops. More than 81 species of underutilized and underexploited vegetables grown in the state (Table 13.1) and interior peoples prefers traditional underutilized and underexploited vegetables and tuber crops for their food like bamboo shoot (*Bambusa* sp.), tender leave of elephant food yam (*Amorphophallus paveonifolius*), colocasia or taro and xanthosoma (*Xanthosona sagitifolius*), Greater galangal (*Alpinia galangal*), Tit Begun (*Solanum indicum*), leafy vegetables like fern tender shoots (*Ceratopteris thalictroides*), lai (*Brassica juncea*), amaranth (*Amaranthus* spp), Spiny amaranthus (*Amaranthus spinosus*), Pui (*Basella alba*), Punarnaba (*Boerrhavia diffusa*), Hiencha(*Enhydra flatuans*), Lajalu (*Mimosa* sp), sem (*Dolichos purpureus*), tree bean (*Parkia roxburghii*), jack bean (*Canavalia gladiata*), winged bean (*P. tetragonalobus*), Kakrol (*Monordica cochinsinensis*), Coccinia (*Coccinia indica*), parwal or pointed gourd (*Trichosanthus dioca*), chilli and *Dioscorea* sp., etc. The Tripura is one of the richest reservoirs of genetic variability and diversity of different underutilized and underexploited vegetables and tuber crops, which exist in plant types, morphological and physiological variations, resistance to diseases and pests, adaptability and distribution. It lays emphasis on exploiting the potential usefulness of such a valuable resources, in combating the challenges of food, employment generation and nutritional security to the ever-increasing population. List of different type of underutilized and underexploited vegetables and tuber crops in the state are given below with Table 13.1.

Some of the important underutilized and underexploited vegetables are as follows:

Alpinia galangal (Greater Galangal, Kulanjam)

It is belong to family Zingiberaceae. The plant height 1.2–1.5 m, perennial herb, it possesses short stalk and sheathing bases; leaf sheath giows upward in to a ligule. The leaf blade is lanceolate, ovate (or) oblong with a strong mid rib. The plant is erect and bears the tuberous root and leafy stem; flowers small, branched, 15-20 cm long, dense panicle; rachis densely pubescent; bract small, ovate and concave. The lib is ovate, clawed, emarginated, white; veins lilac; a pair of subulate, red glands at the base of the claw. Stamens arculate shorter than the lip. Ovary glabrous; 1-2 in each ovule. Fruit is orange red in colour, having 3-5 seeds in it. The rhizomes are used as a condiment and source of an essential oil. The flowers are eaten raw and the seeds are used as spice. It is used in various ayurvedic preparations. It is used to control the rheumatism, bronchitis, and cataract. It is also act as stimulant, carminative.The seeds are used for colic diarrhoea and nausea. Rhizome yields an essential oil and in cosmetics and perfumes.

Dioscorea glabra (Pisala Alu)

It is belong to family Dioscoreaceae. The plants are annual (or) biennial climbing to the right, glabrous and prickly at base. The leaves are deeply cordate or sagitate, opposite or lower alternate glaucous or nearly white beneath, petiole long sometimes as long as the leaf blade. The tubers descending deep in to the soil on a long stalk gradually increasing in size, cylindrical and flesh is white. The tubers are eaten as vegetable along with the dry fish.

Solanum ferox

It is belong to family Solanaceae. The plants are small bushy shrubs, stellately tomentose, woolly and prickly. The leaves are ovate with large triangular shallow lobes, prickly and densely woolly with stellate hairs, flowers in very short lateral cymes, calyx penta lobed, densely hirsute hardly enlarges in fruit, corolla white, densely villous outside. Berry is round or oval globose. The fruits are eaten as vegetable.

Table 13.1: List of Wild Edibles or Underutilized and Underexploited Fruits in Tripura

Sl.No.	Botanical Name	Common/Local Name and Family	Other Information Related with Specific Purpose and Medicinal Use
1.	*Emblica officinalis*	Amlaki, Amla, Aonla, Euphorbiaceae	Flowering time Feb–March and fruiting time Nov.–Dec. Chavan Prash, Blackening of hair, Astringent, Cardiac, laxative and liver tonic
2.	*Feronia limonia*	Kaith, Elephant apple, Rutaceae	Flowering time March–April and fruiting time Nov.–Feb. Preparation of Jam, Diuretic, Febrifuge, diarrhea and dysentery
3.	*Manikara achrus*	Khirni, Sapotaceae	Flowering time Feb.–March, Jam and Jelly
4.	*Garcinia pendulata* Roxb	Baikal, Amelvet, Gutiferae	Sep–Oct (Fruiting time), Antihelmintic, Aphrodisiac, Appetizer and cardiac
5.	*Magnifera sylvatica* (Mango)	Aam, Ambi, Anacardeaceae	Jan–Feb (Flowering time), May–June (Fruiting time), Jam, Jelly, Chuteny and pickles
6.	*Tamarindus indica*	Lmli, Tentul, Amlaka, Fabaceae	Used as spice, checks vomiting
7.	*Ficus racemosa*	Country Fig, Moraceae	Feb–March (Flowering time), June–July (Fruiting time), Analgesic
8.	*Carissa carandus*	Karonda, Apocyanaceae	Aug–Sept (Fruiting time), Table purpose and Pickling, Chutney and jelly
9.	Elaeocarpus*floribundes*	Jalpui, Eleocarpaceae	July–Aug (Flowering time), Oct–Nov (Fruiting time), Table purpose and Pickling, Chutney and jelly
10.	*Dillenia indica*	Chalta/Elephantapple	July–Aug (Flowering time), Aug–Sept (Fruiting time), Pickling, Chutney and jelly
11.	*Citrus maxima*	Jambura, Rutaceae	Year round, Pickling and Chutney
12.	*Averhoea carambola*	Kamranga, Oxalidaceae	Year round, Pickling and Chutney
13.	*Averrhoea bilimbi*	Bilimbi, Oxalidaceae	Oxalidaceae, Pickling and Chutney
14.	*Psidium guajava*	Wild Guava, Myrtaceae	Dec–Jan (Flowering time), May–June (Fruiting time), Jam and Jelly
15.	*Aegle marmelos*	Bael, Rutaceae	March–April (Flowering time), Jan–Feb (Fruiting time), Preparation of Sarbath, used in diarrhoea and dysentery
16.	*Artocarpus chaplasa*	Chamal, Moraceae	Jan–Feb (Flowering time), May–June (Fruiting time), Culinary purpose
17.	*A. lakoosha*	Monkey jack, Moraceae	Jan–Feb (Flowering time), May–June (Fruiting time), Culinary purpose
18.	*Annona reticulate*	Annona, Anonnceae	March–April (Fruiting time), Table purpose
19.	A. squamosa	Ata, Anonnceae	March–April (FT), Table purpose
20.	*Phyllanthus acidus*	Harabari, Paramlaki, Euphorbiaceae	Jan–Feb (Flowering time), May–June (Fruiting time), Table purpose, Pickles and Chutney
21.	Passiflora edulis	Passion fruit, Passifloraceae	March–April (Flowering time), June–July (Fruiting time), Squash
22.	*Spordias pinnata*	Amra, Anacardeaceae	Jan–Feb (Flowering time), June–July (Fruiting time), Pickles and Chutney

Contd...

Table 13.1–*Contd...*

Sl.No.	*Botanical Name*	*Common/Local Name and Family*	*Other Information Related with Specific Purpose and Medicinal Use*
23.	*Spondias mombin*	Hog plum, Anacardeaceae	Jan–Feb (Flowering time), June–July (Fruiting time), Pickles and Chutney
24.	*Syzygium scrasoides*	Bhali Jamun, Myrlaceae	March–April (Flowering time), June–July (Fruiting time), Juice, Jam and Jelly
25.	*S. jambos*	Gulab jamun, Myrlaceae	March–April (Flowering time), June–July (Fruiting time), Juice, Jam and Jelly
26.	*S. samarafinds*	Jamrul, Myrlaceae	March–April (Flowering time), June–July (Fruiting time), Juice, Jam and Jelly
27.	*S. cumini*	Kala jam, Myrlaceae	March–April (Flowering time), June–July (Fruiting time), Juice, Jam and Jelly
28.	*Phoenix sylvestris*	Khajur, Palmae	Jan–Feb (Flowering time), May–June (Fruiting time), Toddy extraction, Preparation of "Gur"
29.	*Borassuss blafellifer*	Palmyram, Palmae	Jan–Feb (Flowering time), May–June (Fruiting time), Toddy extraction, Preparation of "Gur"
30.	*Zizyphus funiculosa*	Wild kul, Rhamnaceae	Dec–Jan (Fruiting time), Used by tribals
31.	*Z. oenophila*	Siakul, Rhamnaceae	Dec–Jan (Fruiting time), Table purpose and candy
32.	*Z. mauritiana*	Kul, Rhamnaceae	Dec–Jan (Fruiting time), Table purpose and candy
33.	*Antidesma acuminate*	Sialbuka, Euphorbiaceae	Jan–Feb (Flowering time), May–June (Fruiting time), Table purpose
34.	*Grevia hirsute*	Wild phalsa, Tiliaceae	Dec–Jan (Fruiting time), Table purpose
35.	*G. sapida*	Wild phalsa, Tiliaceae	Dec–Jan (Fruiting time), Table purpose
36.	*Morus australis*	Tut, Moraceae	Jan–Feb (Flowering time), May–June (Fruiting time), Antihelmintic, aphrodisiac and appetizer.

Table 13.2: List of Wild Edible or Underutilized and Underexploited Vegetables Crops in Tripura

Sl.No.	Botanical Name	Common/local Name	Family	Parts Used
1.	*Thumbergia grandiflora*	Nil–late	Acanthaceae	Leaves
2.	*Sagittaria sagittifolia*	–	Alismataceae	Tuber
3.	*Allium tuberosum*	Sugandha piyaj	Alliaceae	Bulb and leaf
4.	*Amaranthus spinosus*	Kanta notya	Amaranthaceae	Leaf and tender shoot
5.	*Amaranthus viridis*	Noteyshak	Amaranthaceae	Leaf and tender shoot
6.	*Amaranthus gangeticus*	Lalshak	Amaranthaceae	Leaf and tender shoot
7.	*Amaranthus lividus*	Danta	Amaranthaceae	Leaf and tender shoot
8.	*Lannea coromandelia*	Jiyal	Anacardeaceae	Leaves
9.	*Centella asiatica*	Thankuni	Apiaceae	Whole parts
10.	*Eryngium foetidum*	Dhandhane	Apiaceae	Leaves
11.	*Xanthosoma atrovirens*	Moulavikachu/Tannia	Araceae	Leave Petioles/Rhizome
12.	*Xanthosoma violaceum*	Dudkachu/Tannia	Araceae	Leave Petioles/Rhizome
13.	*Colocasia esculenta*	Mukhikachu/Eddoe	Araceae	Rhizome
14.	*Alocasia macrorrhiza*	Mankochu/Giant taro	Araceae	Rhizome
15.	*Amorphophalus campanulatus*	Ol	Araceae	Corms
16.	*Colocasia esculenta*	Kochu/Taro	Araceae	Leaf, petiole, corms
17.	*Typhonium trilobatum*		Araceae	Tuber
18.	*Basella rubra*	Puishak (lal)/Indian spinach (red)	Basellacease	Leaf and tender stem
19.	*Basella alba*	Puishak (sabuj)/Indian spinach (green)	Basellacease	Leaf and tender stem
20.	*Begonia barbata*	Begonia	Begoniaceae	Leaf
21.	*Bombax malabarica*	Shimool	Bombaceae	Flower and leaflets
22.	*Bauhinia purpurea*	Deb kanchan	Caesalpiniaceae	Tender leaf, pod and flower
23.	*Drymaria cordata*	–	Caryophyllaceae	Whole plants
24.	*Polycarpon prostratun*	Surata	Caryophyllaceae	Leaves
25.	*Chenopodium album*	Bathua	Chenopodiaceae	Leaves
26.	*Enhydra fluctuans*	Heincha	Compositae	Whole plant
27.	*Ipomoea alba*	Sudhia kalmi	Convolvulaceae	Leaves and tender shoots
28.	*Ipomoea aquatica*	Kalmisak/Water spinach	Convolvulaceae	Leaves and tender shoots
29.	*Ipomoea batatus*	Misti alu/Sweet potato	Convolvualceae	Tubers
30.	*Ipomoea reptans*	Gima kolmi/Kangkong	Convolvualceae	Tubers
31.	*Trichosanthes dioca*	Patal/Parwal/Pointedgourd	Cucurbitaceae	Fruits
32.	*Coccinia indica*	Kundru	Cucurbitaceae	Leaf, stem and fruits
33.	*Trichosanthes anguina*	Chichingga/Snakegourd	Cucurbitaceae	Fruits
34.	*Momordica charantia*	Ucche/Karala/Bittergourd	Cucurbitaceae	Fruits
35.	*Luffa cylindrica*	Dhundul/Sponsegourd	Cucurbitaceae	Fruits

Contd...

Table 13.2–*Contd...*

Sl.No.	Botanical Name	Common/local Name	Family	Parts Used
36.	*Luffa acutagula*	Jhingga/Ribbed gourd	Cucurbitaceae	Fruits
37.	*Cucumis anguina*	Khira/ (short)Cucumber	Cucurbitaceae	Fruits
38.	*Benincasa hispida*	Chal kumda/Wax gourd	Cucurbitaceae	Tender vines and fruit
39.	*Cucurbita maxima syn. Cucurbita moschata*	Misti kumda/Sweet gourd	Cucurbitaceae	Tender vines and fruit
40.	*Melothria heterophylla*	Kudari	Cucurbitaceae	Leaves and tender fruits
	Momordica cochinsinensis	Karkol	Cucurbitaceae	Fruit
41.	*Momordica charantia*	Jangli karela	Cucurbitaceae	Fruit
42.	*Dryopteris filix–mas*	Dhekisak	Cycadaceae	Whole parts
43.	*Eleocharis dulcis*	Sapla/waterchesnut	Cyperaceae	Tuber
44.	*Dioscorea alata*	Matti alu/Greater Yam	Dioscoreaceae	Tuber
45.	*Dioscorea bulbifera*	Gach alu Arial potato	Dioscoreaceae	Tuber
46.	*Dioscorea pentaphylla*	Suralu	Dioscoreaceae	Tuber
47.	*Dioscorea esculenta*	Khamalu	Dioscoreaceae	Tuber
48.	*Antidesma acidum*	Muta, Beloni	Euphorbiaceae	Leaf and fruit
49.	*Manihot esculenta*	Shimul alu/Cassava	Euphorbiaceae	Tubers
50.	*Jatropha curcus*	Bagbherenda	Euphorbiaceae	Tender leaves
51.	*Canavalia glabra*	Makhan shim	Fabaceae	Pods, leaf and bean
52.	*Dolichos lablab*	Sim/Indian Bean	Fabaceae	Seeds and tender pods
53.	*Crotolaria juncea*	Sunhemp	Fabaceae	Flower and tender pods
54.	*Pueraria tuberosa*	Bidali/Indian kudzu	Fabaceae	Tuberous roots
55.	*Perilla fruitescens*	Banjira	Labiatae	Leaves and seeds
56.	*Pachyrrhizus tuberosa*	Shakalu/Yam bean	Leguminoseae	Tubers
57.	*Phaseolus limensis*	Rukuri/Lima bean	Leguminoseae	Bean/Pods
58.	*Phaseolus vulgaris*	Jhar sheem/French bean	Leguminoseae	Bean/Pods
59.	*Vigna sesquipedalis*	Barbati/String bean	Leguminoseae	Bean/Pods
60.	*Sesbania grandiflora*	Bak phul	Leguminaceae	Leaves and flowers
61.	*Psophocarpus tetragonolobus*	Kamranga Sim/ Winged bean	Leguminaceae	Tender pods, seeds and tuber
62.	*Asparagus racemosus*	Satamuli	Liliaceae	Bulb
63.	*Smilax zylancia*	Ramdantan	Liliaceae	Leaves
64.	*Hibiscus sabdariffa*	Lal Makri	Malvaceae	Leaf and fruit
65.	*Abelmoschus manihot*	Ban Kapas	Malvaceae	Tender Fruits
66.	*Entada phaseolus*	Borbi	Mimosaceae	Fruits
67.	*Leucaena leucocephala*	Subabul	Mimosaceae	Tender leaf
68.	*Neptunia prostrata*	Lajalu	Mimosaceae	Whole plant
69.	*Moringa oleifera*	Drumstick/Shajina	Moringaceae	Fruits

Contd...

Table 13.2–*Contd...*

Sl.No.	*Botanical Name*	*Common/local Name*	*Family*	*Parts Used*
70.	*Parkia roxburghii*	Manipuri Sim/Yongchak	Mimosaceae	Tender pods
71.	*Mirabilis jalapa*	Sarpamani	Nyctaginiaceae	Leaves
72.	*Euryale ferox*	Makhna	Nymphaceae	Fruits
73.	*Nelumbo nucifera*	Padma/Lotus	Nymphaceae	Stem and rhizome
74.	*Nymphea pubescens*	Saluk/water lily	Nymphaceae	Stem
75.	*Oxalis corniculata*	Amrul/Indian sorrel	Oxalidaceae	Leaves
76.	*Ceratopteris thalictroides*		Parkeriaceae	Whole parts
77.	*Rumex vesicarius*	Tak palang	Polygoniaceae	Leaves
78.	*Morinda tintoria*	Daruharidra	Rubiaceae	Fruits
79.	*Bacopa monnieri*	Brahmisak	Scrophulariaceae	Leaf
80.	*Melochia corchorifolia*	Tikiokara	Sterculia	Leaves
81.	*Sterculia indica*	Udal	Sterculiaceae	Seed and leaves
82.	*Solanum torvum*	Titbegon big	Solanaceae	Tender Fruits
83.	*Solanum ferox*	Ram begum	Solanaceae	Fruits
84.	*Solanum nigram*	Gurkamai	Solanaceae	Fruits
85.	*Solanum indicum*	Teta Begun	Solanaceae	Fruits
86.	*Premma esculenta*	–	Verbenaceae	Leaves
87.	*Alpinia galanga*	Kulanjajam	Zingiberaceae	Flower

Solanum tarvum **(Tit Begun)**

It is belong to family Solanaceae. The plants are 1.5-2 m height with few branchs, tomentose and sparingly armed. The leaves ovate, shallowly lobed, acuminate, densely stellate, tomentose beneath, particularly prickly on the petiole and on the mid rib beneath. Flowers white, dense, extra-axillary, often cyme. Calyx unarmed, apiculate, corolla stellate, pubescent outside, lobes spreading. The fruits are berry oval in shaps, green and persistent calyx. The tender fruits are edible. The fruits can be dried and used as vegetable during off season.

Basella alba **(Malabar Spinach/Pui Sag)**

It is belong to family Basellaceae. The plant is climbing annual (or) biennial, glabrous, rampant herb with alternate, thick succulent leaves. There are three types of basella those that are red, oval round leaves those with dark green and oval or round shape leaf, and those with dark green and cordate leaves. It is grown for its tender stems and leaves and it is one of the better know substitutes for true spinach. It is suitable for home garden and market growers. Young shoots and leaves (15-30 cm long) can be harvested after 2-3 months of sowing and there after at regular interval for a few months. The leaves are an excellent source of vitamins A and C and contain good amounts of calcium and iron. The dark red juice from the fruite is being used as dye and as an ink.

Trichosanthes dioca **Roxb (Parwal/Batal/Pointed Gourd)**

It is belong family Cucurbitaceae. It is khown by the name of parwal, palwal, or parmal in eastern parts of India and is one of the important vegetables in the state and North-Eastern region. The plant is a perennial, dioecious, and grows as a vine. Roots are tuberous with long tap root system. Veins are pencil thick in size with dark green cordate simple leaves. Folwers are tubular white with 16-19 day's initiation to anthesis time for pistillate flowers and 10-14 days for staminate flowers. Stigma remaines viable for approximately 14 hours and 40-70 per cent of flowers set fruit. Bases on shape, size and striation, fruits can be grouped into 4 categories :(1) long, dark green with white stripes, 10-13cm long. (2) thick, dark green with very pale green stripes, 10-16 cm long. (3) roundish, dark green with white stripe 5-8 cm long, and (4) tapering, green and striped, 5-8 cm long (Singh, 1989). The fruit is the edible part of the plant, which is cooked in various ways either alone or in combination with other vegetables or meats. Pointed gourd is rich in vitamins and minerals contains like 9.0 mg Mg, 2.6 mg Na, 83.0 mg K, 1.1 mg Cu, and 17.0 mg S per 100 g edible part. The leaves and tender shoots are used for preparation of syrup for convalescents. It is easily digestible and good for maintaining healthy heart and brain.

Momordica cochinsinensis Spneng (Kakrol/Kakrol)

It is belong to family Cucurbitaceae. The plants are dioecious and perennial, profusely branching, glabrous, tuberous roots. Leaves ovate, glabrous, acute or obtuse and mucronate at the tendrils filiform, simple glabrous. Male flowers: solitary, peduncle 5-14 cm long, slender, glabrous barcteate at apex, bract sessile, reniform, eglandular, calyx densely villous, Corolla: white colour petals. Female flowers: peduncle 5-15 cm long, minutely barcteate at the base. Ovary ovoid, densely aculeate, fleshy, fruit 10-15 cm long, acute at the apex. Fruits are light green to dark green in colour and it takes 26 days to reach edible maturity. This is one of the popular vegetables grown and consumed by the local people. The fruit contains carbohydrate, fat and calcium 6.4, 0.10 and 1.60 per cent, respectively. The fruits and leaves are used in external application for lumbago, fractures and ulceration. Seeds yield fatty oil, which is employed in paint and varnishes. The unripe fruit act as appetizer and astringent and rich in β- carotene. There are 3 local types of kakrol (Long (15-17.50 cm), medium (12-15cm) and short (8-10 cm) available in Tripura.

Typhonium trilobatum

It is belong to family Araceae. Herbs with sub globose tubers. Leaves are 7.5-20 cm long and as broad, has tately trilobed or sub triplicate, petiole up to 30 cm long. Spathe appearing with the leavesd, 7-15 cm long, constricted with short convolute, persistent tube and ovate, acuminate striate limb, red purple inside obnoxious. Peduncle 2.5 – 10 cm long. Spadix exserated (Vimala and Bala Nambisan, 2005). Whole plant is edible(both foliage and tuber).

Coccinia indica (Kundru/Coccina/Talakuchi)

It is belong to *family* Cucubitaceae. The plants are climbing, profusely branching, with long tuberous roots; tendril simple and rarely bifid. Leaves petoilate, entrire to palmately lobed or angled, with few glistening glands on the midrib towards the base. Flowers are dioecious, solitary; bracts absent, peduncle filiform, striate. Male flowers: calyx companulate, glabrous, 5 lobed, obtuse or subulate, corolla white, companulate, glabrous 5 lobed short and acute; stamens 5(2+2+1) inserted at the base of the calyx tube, filaments connate in to a central coloumn. Female flower: solitary, peduncle slender, calyx and corolla similar as in male flower. Staminode 3, subulate villous at base. The Ovary is tricarpels and parietal placentation. Fruits ovoid to elliptic, 10-12 cm long, 2.5–4.0 cm across, smooth, bright green with white stripes, and bright scarlet red when ripe. Various parts of the plants are used in the indigenous system of medicine in both Ayurvedic and Unani practice. The plant is reported to reduce the amount of sugar in the urine and improve the general metabolism of patients suffering from diabetes. Some parts of the plants particularly the leaves and roots are used by the Ayurvedic and

Unani practitioners in the treatment of conditions similar to diabetics when insulin treatment has failed. The cooked leaves are used externally in eruptions of the skin and also mixed with ghee are applied as ointment to sores. Boiled in ginger oil, they are applied externally in the treatment of ring worm, psoriasis, itch, etc. Tender fruits are used as vegetable and recommended for fever patient. In a preliminary study, the $CHCl_3$ extract of the roots showed a hypo glycaemic effect in fasted *albino rats* (Vaisnab and Gupta, 1995).

Benincasa hispida (Thunb.) (Ash Gourd/chalkumda)

It is belong to family Cucurbitaceae. It is an annual, profusely branching (vine) and climbing by tendrils, softly hair allover. Leaves large, cordate, reniform or orbicular, palmately 5 lobed; petilole eglandular; tendril bifid or trifid. Flower large yellow axillary, solitary; monoecious; bracts absent. Male flowers: calyx companulate, sepals 5, sub foliaceous and serrate above. Petals 5, abovate, slightly connate at the base. Stamens (2+2+1), inserted near mouth of the calyx tube; anthers exserted. Female flowers: calyx and corolla as in the male flower, tricarpallary, syncarpous, inferior ovary and parietal placentation. This crop is cultivated in homestead garden and jhum areas of Tripura. Huge genetic diversity exists in case of jhum cultivated ash gourd. The fruits are harvested at very early stage when it reaches 1-1.50 kg size for culinary purpose.

Vigna unguiculata var. *Sesquipedalis* (Yard Long Bean/Asparagus Bean)

It is belong to leguminaceae. The Plants are trailing, annual climbers attaining a height of 1.5-20 m. The leaves are trifoliate with along petiole. The leaflets are large. Hairy, dark green cordate and subtended by inconspicuous stipels. The flowers are white. Light pink petiole (or) light yellow with their keel incurved and arched upward. The pods are long, smooth, cylindrical and somewhat constricted between. This crop is cultivated mainly for its immature pods. The fruit is 45-100 cm long. Fleshy and inflated when green but shrivels on drying. Immature pods are eaten fresh, cooked frozen (or) canned. The dried seeds can be used as a grain. The seeds may be used as concentrate for farm animals. This crop is used as hay, silage, pasture soil cover and green manure. The fodder is highly palatable to all types of livestock; hence this crop is ideal for integrated farming system. The proximate composition of cowpea is as follows, moisture (12 per cent) protein (24-60 per cent), fat (0.7 per cent) carbohydrate (55-70 per cent), fibre (3.8 per cent) and ash (3.2 per cent).

Enhydra fluctuans Lour (Hiencha)

It is belong to family Compositae. The plants are prostrate rooting at the nodes succulent with short ascending hispid branches. Leaves opposite, sessile, oblong, serrate, glands punctuate beneath. Inflorescence yellow, terminal and axillary, heterogamous involucral bracts 4 foliaceous in two opposite pair, greenish yellow, giving the appearance of a single flower. Ray florets female many sessile, corolla tubular, limb companulate, 5 fid, receptacle convex with palae enclosing flowers. The leaves are eaten as salad and as vegetable. It has the property of blood purification; leaves are laxative, antibilious, demulcent, and useful in skin and nervous disorder (Kumar, 2003).

Parkia roxburghii (Tree Bean)

It is one of the most common multipurpose tree species of Mimosaceae family in the state and other north eastern region of India, its tree commonly grows in every household of hill region. Tree bean is a much-branched legume of medium height (10-12 m) with bipinnate leaves; main rachis with a gland below the lower pair of pinnae, pinnae 8-30 pairs, opposite; leaflets 40-80 pairs close set, sessile, linear oblong, falcate, truncate at base, flowers in dense clavate, long peduncle heads with sterile flowers lower down; bract spathulate, silky pubescent outside, calyx tubular 5 cleft. Corolla pale yellow tubular. Stamens 10 exserted anthers not gland tipped. Pods 25-50 x 3.5 – 4.0 cm. Flowering Season (Oct.–Dec.) and fruiting season (Dec.–April). Pods are formed in clusters of 10-15, each measuring 25-40 cm in length and 2-4 cm in breadth (Kumar *et al.*, 2002). The long tender pods of tree bean are most popular and delicious vegetable in Tripura and other north-eastern state. Based on local preference, the pods are consumed at different stages of maturity, either fresh or processed. The tender and matured beans are used in various dishes. The beans, after scraping the out skin, are sliced into pieces for use indifferent chutney (Iromba) preparations. Matured flowers and young shoots are also used in curries and salad. The leaves are good source of fodder and green manure. The pod contains 15.6 to 18.0 per cent Proteins. Similarly, fat content is 34 per cent in the kernel and ranged from 1-16 per cent in the pod. Oleic, linoleic and linolenic acids made up 63-68 per cent of the total fat in the pods as well as the kernels. The total essential amino-acids amounted in 33, 36 and 39 per cent in the tender, immature and mature pods, respectively, and 42 per cent in the kernels (Longvah and Deosthale, 1998). The oil extract of *Parkia roxburghii* can cause 100 per cent mortality rate of *Aphis craccivora* after 96 h with a concentration of 2 per cent. The tree is also of immense use in local medicines. Decoction of bark, fruit skin and leaf is being used to control diarrhoea and dysentery.

Mimosa prostrate Lamk (Lajalu)

It is belong to family Mimosaceae. An aquatic herb with stout widely creeping or floating stem rooting at the nodes and producting sub erect branches, leaves bipinnate; pinnae 4-6; leaf sensitive, 5-15 pairs, about 8-12 mm long oblong, obtuse glabrous; stipules obliquely cordate. Flower minute in dense heads, the lower replaced by numerous ligulate yellow staminodes; peduncle axillary, ascending 10-30 cm long; bracts small ovate, calyx minute, companulate and corolla 5 lobed. It occurs mainly in tanks and swamp areas. The shoot is used as vegetable. It possesses the property of controlling Jaundice and stomach disorders.

Dolichos purpureus var. *typicus* (Hyacinth Bean, Indian Bean, Sem)

It is to family Fabaceae. The plants are annual, twining vine with purplish stems and alternate, pinnately divided leaves with 3 large leaflets. The flowers are arranged on axillary, erect, long stalked, stiff racemes 30 cm (or) more in length. Pods are variable in length (5-22cm x 1-5 cm), obliquely oblong, often curved and flattened. The beak is also curved with persistant style. The seeds are of different size and colour and have a prominent white hilum along one third of the edge. The seed colour ranges from reddish brown to black with or with out speckling. The young pods and lender beans are used as vegetables. The ripe and dried seeds are consumed as a pulse. The young pods and seeds contain a poisonous glycoside that must be removed by boiling. The haulms, either green or as hay or silage are used as fodder for livestock. It is also grown as a manure and cover crop. The approximate composition of seeds is as follows; water (9.6 per cent), protein (24.9 per cent), fat (0.8 per cent), carbohydrate (60.10 per cent), fibre (1.4 per cent), and ash (3.2 per cent).

Xanthosoma sagittifolium (Tannia)

It is belong to family Araceae. A herbaceous perennial, several large leaves sprout from the main stem, which are sagittate and erect with long, ribbed petioles, inflorescence sprout between the leaves in a spadix, with a white 12 to 15 cm spathe which closes at its base in the form of a spherical chamber and opens at the top into a concave lamina, the spadix is cylindrical, slightly longer than the spathe, with female flower on the lower portion, male flower on the upper portion and sterile flowers in the middle portion. The spadices are rarely fertile and produce few viable seeds. It has a corm or main underground stem in the form of a rhizome from which swollen secondary shoots, or cormels, sprout. The duration is 9 to 11 months, during the first six months the corm and leaves develop and in the last four months, the foliage remains stable and, when it begins to dry, the plants are ready for harvest. The tuber and petiole is edible. The approximate nutrient composition of 100 gms tuber is as follows; Water (65 per cent) Protein (2.0g) Fat (0.3) Carbohydrate (31g), Fibre (1.0g), Calcium (20mg), Phosphorus (47mg) and Ascorbic acid 10mg).

Psoptocarpus telragonalobus (L) D.C (Winged Bean/Khamranga Bean)

It is belong to family Fabaceae. It is grown in the home gardens in Tripura and other north-eastern state. The winged bean plant grows as a vine with climbing stems and leaves, 3-4 m in high. It is a herbaceous perennial; but can be grown generally as an annual Inflorescence raceme, flowers lilac colour peduncle up to 15 cm long, Calyx glabrous teeth shorter than the tube, upper two connate. Corolla: petals are equal in length, keel petal incurved. Stamens: monadelphous, anthers uniform, ovary subsessile. The pods are four sided with characteristic wings, and vary in length from 6-36 cm (upto 50 cm) containing 5-20 seeds in each pod (Sahoo *et al.*, 2002). The globular shaped shining seeds may be white, yellow, brown, black or mottled and vary in weight from 0.06-0.5g each. All parts of the plant, *i.e.*, seeds, flowers, leaves, pods and tuber-like-roots are edible. The plants takes 50 days to flower, young pods are ready to harvest about 20 to 30 days after anthesis. After this winged bean can be used as grain legume. The tubers are harvested 5-12 months after sowing when they reach 2 to 4 cm in diameter. The seeds are used for industrial purpose for commercial exploitation, they are also useful as food when steamed, boiled, fried, roasted, fermented, made in to milk and other methods. The plant also produces underground tubers of varying sizes, rich in carbohydrates (25-30 per cent), proteins (10-15) and vitamin A (300 to 900 IU). Its tender leaves make good sauce and curry. It produces edible immature pods, mature seeds, leaves, flowers, tuberous roots, succulent shoots and edible seed oil. The young tender pods can be stewed, boiled, fried, roasted or made into milk. The seeds are rich in crude proteins (33.70 per cent), fat (17.74) and the minerals like Fe and Ca content 13.78 and 240.11 mg/100 (Chimmad *et al.*, 1999). Flowers have a sweet taste because of the nectar they contain. The tuber-like-roots are eaten after boiling or frying. The plant is a good fodder for cattle.

Canavalia gladiatas (Sword Bean/Jack Bean/Makhan Bean)

It belongs to family fabaceae and cultivated on limited scale in the Tripura. It is a bushy, semi-erect, annual herb, 2 m tall and the tips of its branches tend to twine under shade. Leaves are trifoliate and shortly hairy. Pods are sword shaped, 2-30 cm long and 2-2.5 cm broad. The pods are pendent, ribbed near suture and 8 to 20 seeded. Seeds are white with a brownish hilum extending to half of its length. Young green pods are eaten as a cooked vegetable. The young leaves may be cooked and eaten as a potherb. The dried beans are good source of protein and starch, (27 per cent protein and 42 per cent starch, respectively) with a good balance of amino-acids and some toxic alkaloids of non protein-nitrogen compounds like lectins, canavine, con-canavine, canaline, etc., which are heat labile and hydro soluble. The lectine reacts with the sugar components in intestinal cells causing a disruption in cell structure, which leads to abnormalities in nutrient absorption (Mello *et al.*, 1985). However, the canavine is a major toxic factor, which is present to an extent of 2-3.5 per cent of dry seed (Fearon and Bell, 1955), which also inhibits protein metabolism. Sword bean is used for hiccups, vomiting, abdominal swelling, lumbago (lower back pain) due to kidney deficiency, asthma with sputum, etc.

Amaranthus sp. (*A. tricolour, A. caudatus, A. spinous, A. hyphocondriacus*) (Amaranthaceae)

It is belong to family Amaranthaceae. The plants are herbs, annual; leaves opposite or alternate, simple, entire and estipulate. The inflorescence is racemose type (mostly the flower are being arranged in dense fascicles). Each flower bears a pair of large membranous persistent bracteoles, which are sterile. The flower are bracteates and bracteolate, hermaphrodite, actinomorphic and hypogynous. The seeds are endospermic usually having a polished testa (Mostly black in colour). Amaranth is a multipurpose crop with good potential to use as vegetable, grain and fodder. This crop is rich is iron; β-carotene and it can hgelp in reducing the iron and vitamin A deficiency. The presence of high amount of folic acid can help in increasing the blood haemoglobin. The seed protein of amaranthus is superior that of true cereals because of its high lysine and methionine content.

Brassica campestris (Lai Pata)

It is belong to family Cruciferae. The plants are slender, erect, branched annual herbs about 0.6-1.5 m tail. The leave are auricled, generally pinnate partite. The yellow flowers are borne in a corymbose raceme. The leave of young plants are used as an green or leafy vegetable. The leaves are used as green fodder and occasionally as green manure. Seeds yield edible oil. Seed is also used as an condiment in preparation of pickles and for flavouring curries and vegetables.

Sechium edule (Chow-Chow)

A native of tropical America, Chow-Chow is a very popular vegetable in the state commonly called squash and grows abundantly without much care and attention in high hills and plain land of Tripura and other north-Eastern state. Chow-Chow produces large starchy edible roots in addition to fruits. It is a vigorous, scrambling, tuberous-rooted perennial plant, grown for its starchy, edible fruit and seeds. This climber can spread to fifty feet, producing huge tubers. It looks like a large, green pear, but having a number of deep folds in the skin. Some varieties have smooth skins, while others have dots of prickly spines on the surface. The flesh is crisp and white with a large white oval seed in the center. Chow-Chow is a fruit but most often used as a vegetable. It is often used in the place of potato.

Vicia faba (Broad Bean)

It is a cool-season crop in high altitude areas, grown on a limited extent in the state. Broad bean is an annual legume botanically known as *Vicia faba* L. Plants are erect annuals reaching 60-125 cm in height, very leafy with robust stems. The plant has compound leaves with toothed stipules at the base, and conspicuous white flowers with the wings smeared in black. The fruit is a pod, reaching 10 cm in a length and pubescent. Seeds are 3 cm in length. It is used as green-sheil, the seeds removed from the pod before maturity, or as dry beans. It is the principal protein source for poor people in this region. The protein content of bean seeds is high, amounting to about 20-25 per cent. It is a nitrogen fixing plant, and is not poisonous to humans in the conventional sense, but they cause favism in susceptible individuals, who have a genetically transmitted, male sex-linked deficiency to the enzyme glucose-6-phosphate dehydrogenase. The disease can cause death in severe cases.

Allium sativum (Garlic)

It is grown in Tripura and other north-eastern state. Edible underground stem is the composite bulb made of several smaller bulbs called cloves and is surrounded by a thin white or pinkish sheath. An interesting genotype of garlic having 2-4 cloves, is found in a limited and sporadic manner in state.

Flemingia vestita syn Maughania vestita (Sohphlong)

It is a weak climbing/trailing type, under ground tubers, distributed in the state. It is consumed raw. The tubers of *Flemingia vestita* are rich in iron (2.64 mg), phosphorus (64.06mg) and contain fair amount of protein (2.99 g), calcium (19.77) and carbohydrates (27.02g) (Sarma, 2001).

Amorphophallus spp. (Elephant Foot Yam)

Elephant foot yam is widely grown in Tripura and north-eastern state. Plant is perennial, robust herb, with a big subterraneous depressed globose tuber and erect and solitary leaf. Leaf in adult plants is 1-2.5 cm diameter, 3-partite, each part divided into numerous oblong segments. Flowering generally occurs between May and June when the corms are in 'dormant' stage having shed all vegetative part at the end of the crop season (May to February). The exact conditions that trigger flowering in elephant foot at the starting of monsoon. It is a good source of carbohydrates, rich in minerals and vitamins A and B. The corms of wild types contain crystals of calcium oxalate, on account of which they are highly irritant. In addition, glucose, glactose, rhamnose and xylose are also present as free sugars. It is grown for its corm and used as a vegetable. It is useful in stomachache, a tonic recommended for piles and given as calcium, a restorative in dyspepsia and debility. Its root is used in boils and opthalmia and also as emmenagogue.

Lageneria vulgaris (Bottle Gourd/Lau)

Generally a winter crop, possibly tropical or sub-tropical in origin, rich in carotene, calcium and vitamin C. It is cultivated or grown by direct sowing or transplanting 15 to 20 days old seedlings; crops are ready for harvest within 55-60 days; yield ranges from 35-40 m tons/ha. Bottle gourd is primarily used in making curry with different fishes. Its leaves and vines are also used as vegetables. The tender bottle gourd is used to prepare sweet dishes with sugar and milk.

Cucubrita maxima (Pumpkin/Mistikumda)

Popular summer-rainy season vegetable, good source of carbohydrate, vitamins A and C, and minerals. Pumpkin is thought to be native to Northern and Southern America. It may be grown by sowing seeds directly or transplanting two weeks old seedlings. Fully ripened pumpkins are harvested within 90 days after sowing. The average yield is 20-25 m tons/ha. Leaves, veins, flower and fruits are edible as vegetable. Fruits are used in making curry and different sweet dishes. Ripe whole fruits can be preserved dry for several months.

Momordica charantea (Bitter Gourd/Ucchya or Karala)

Primarily a summer vegetable, possibly a native of India or South-East Africa. The crop is rich in carotene, calcium and iron; can be grown in any type of soil but loamy to sandy loamy soils are preferred for better growth and quality fruit. Seeds are directly sown in the field. Sometimes seedlings are raised and 15-20 days old seedlings are transplanted in the main field during March to May. *Ucchya*, the smaller ones, has no specific local variety but the larger one, *Karala*, has a local variety. The yield of Ucchya is about 5 m tons/ha and *Karala* 15-18 m tons/ha. Bitter gourd is consumed by making curry or frying.

Luffa acutangula (Ribbed Gourd/Jhina)

A minor cucurbitaceous vegetable, a native of South and South-East Asia; rich in carbohydrate and calcium. March-April is the optimum time for cultivation; immature fruits are used as vegetable and are ready for harvest after 6 weeks. Fruits are usually cooked or fried; small fruits are sometimes pickled.

Trichosanthes anguina (Snake Gourd/Chichina)

Smooth, greenish white, sometimes stripped, long, slender and cylindrical, often twisted fruit. It is a summer vegetable; fruit length varies from 30-100 cm. It contains fair amount of vitamins A, C and calcium. Seeds are sown directly in the field during February-June; crop is ready for harvest within 2 months; yield ranges from 10-12 m tons/ha.

Luffa cylindrica (Sponge Gourd/dhundul)

A summer vegetable, is a native of South Africa. It is a good source of vitamins A, C, iron, and calcium. Seedlings are transplanted in March-April; start flowering after 6-7 weeks. The immature fruit is ready for harvest as a vegetable after 8 weeks; yields range from 10 to 15 m tons/ha. Young fruits are used as cooked vegetables.

Hill Agriculture: Prospects, Constraints and Mitigations (2013) *Pages* **193–212**
Editor: **Dr. Shahid Ahamad**
Published by: **DAYA PUBLISHING HOUSE, NEW DELHI**

Chapter 14

Horticulture Scenario in Hilly Tripura: Prospects and Problems

Sukhen Chandra Das[1]* and C. Ravindran[2]
[1]Assistant Professor (Horticulture),
College of Agriculture, Lembucherra – 799 210, Tripura
[2]Assistant Professor (Horticulture),
KVK, AC and RI, Madurai – 625 104, Tamil Nadu

Introduction

Tripura is endowed with vast natural resources. State is covered by picturesque hills and dales, deep and green valleys which have added beauty to its landscape and possesses rich biodiversity, which is under increasing threat from various biotic and abiotic factors. State is a hilly and land locked, located in the South-West extreme corner of the North Eastern region of India and has long international border (nearly 84 per cent) surrounding three sides by Bangladesh and has stretching latitude 22°56' and 24°32' north and longitude 91°10' and 92 °21' east, with total geographical area of about 10,491 sq.km of which 60 per cent is hill with dense forest and 27 per cent is under cultivation. The agro-climatic conditions (humid subtropical), fertile and acidic soil with good depth and abundance rainfall favour the cultivation of various Horticultural crops. State have its own unique combination of living species, habitats and ecosystem, which together makes its Horticulture rich resource. The state is one of the richest reservoir of genetic variability and diversity of different Horticultural crops *i.e.*, various kinds of commercial Fruits including underutilised fruits, different commercial Vegetables including underutilized vegetables, spices, Plantation crops, Ornamental plants including flowers and orchids and also have plenty of medicinal and aromatic plants including indigenous and traditional tribal medicinal plants. State have 88,700 ha area under horticultural crops, from which, cultivation of fruits (36,598 ha), plantation crops (14,614 ha), vegetables (32495

* Corresponding Author: E-mail: sukhenchandra@rediffmail.com

ha), spices (4885 ha) and flowers (108 ha) (Anonymous, 2010). Considerable diversity exists among the Horticultural species including variation in plant type, morphological and physiological characteristics, reaction to disease and pest, adaptability and distribution. Lack of awareness about the potentiality of horticultural crops as commercial crop. Lack of need based systematic cultivation as sustainable Horticulture system/silvi-horti-pastoral system in the need of the hour. Apart from nutritional security, Horticultural crops also provide food, nutrition, livelihood, health care, income generation and poverty alleviation in rural and interior areas of the state.

Climate and Land of Tripura

The State of Tripura enjoys a typical monsoon climate with variations ranging from Sub-tropical to temperate conditions in hilly areas. The rapid changes in topography results in significant climate changes within a short distance. The foot hill plains, sheltered valleys and hill ranges are marked with climatic contrast. The climate of Tripura exhibits a strong seasonal rhythm; the year being divisible into four characteristic seasons, *viz.* (i) Winter (December- February) (ii) Pre-monsoon (March-April) (iii) Monsoon (May – September) and (iv) Post Monsoon (October-November). The monsoon period lasting for about 5 months from may to September in the longest season of the State. The amount of total annual rainfall in the State varies between 1500 mm to 2500 mm. The maximum and minimum temperatures during Winter are 27°C and 13°C and during Summer are 35°C and 24°C respectively. The land of this State is divided into different group like high hills (40 per cent), Tilla with moderate steep slopes (15 per cent), Tilla with gentle moderate slopes (10 per cent), undulating topography (15 per cent) and plain land (20 per cent).

The contribution of agricultural and allied activities to Gross State Domestic product is about 24 per cent and nearly 50-60 per cent of the state's population depends on Agriculture for its livelihood and therefore, this sector has been the single largest provider of employment to the rural people of the state. Among the important horticultural crops Pineapple, orange, Banana, Jackfruit and Cashew Nut grown widely. State also produce True Potato Seed (TPS) and other important vegetables like Cabbage, Cauliflower, Tomato and different Cucurbitaceous crops. The current status of area, production and productivity of important horticultural crops in Table 14.1.

Table 14.1: Area, Production and Productivity of some Important Crops (2009-10)

Sl.No.	Crops	Area (Ha)	Production(MT)	Productivity(MT)
1.	Pineapple	6,476	11,7531	18.15
2.	Jackfruit	7,636	25,1569	32.95
3.	Banana	7,486	10,5615	14.11
4.	Orange	3,162	22,230	7.03
5.	Coconut	5,784	11,384	1.97
6.	Cashew nut	4,387	2345	0.53
7.	Summer vegetables	13,869	16,6177	11.98
8.	Winter vegetables including potato	18,626	28,0741	15.07

Fruits

The soil and agro-climatic condition of Tripura is congenial for growing a variety of fruit crops. Fruits grown in extensive scale in the state are-Orange, Pineapple, Banana, Papaya, Litchi, Jackfruit, lime/lemon. Among the fruits frown in the state, Jackfruit have highest area (7,636 ha), with an annual production (25, 1569 m t) and productivity (32.95 t ha^{-1}), followed by banana, pineapple and citrus Tables 14.1 and 14.5. Beside this many other underexploited and underutilized fruit crops are grown in homestead garden. The different underutilized and underexploited or wild traditional fruits grown in the state they are Carambola *(Averrhoa carambola)*, Elephant Apple *(Dillenia indica)*, Jalpai *(Elaeocarpus floribundus)*, Star Aonla *(Phyllanthus acidus)*, Jack fruit *(Artocarpus heterophyllus)*, Latka *(Baccauea sapida)*, Paniala *(Flacourtia jangomas)*, Bael *(Aegle marmelos)*, Jamun *(Syzygium cumini)*, Karonda *(Carissa carandas)*, Amra *(Spondias pinnata)*, Passion fruit *(Passiflora edulis)*, Satkara *(Citrus macroptera)*, Aonla *(Emblica officinalis)*, Wood apple *(Feronia limonia)*, Ber *(Ziziphus mauritiana)*, Fig *(Ficus carica)*, *Annona* spp, Tamarind *(Tamarindus indica)* and others fruits namely, Jamrul, Bhali Jamun, Chamal, Mankey jack, Bilimbi and Jambura are endowed with food, nutrition and medicinal value. The details list of wild edibles or underutilized and underexploited fruits are also provided in Table 13.2.

Details of the major fruits grown in Tripura are:

Mango

Mango is a popular fruit ranking 4th next to the Jack, banana and pineapple in total annual production and productivity and it provide rich source of vitamins and minerals and is very important for food, nutrition and health security for rural and urban area. The agro-climatic condition favours the cultivation of mango with wild local and hybrids. Mango fruit is one of the commercial fruit crops in Tripura and has high demand for fresh consumption. Amrapalli, Himsagar and Mallika are grown in different parts of Tripura along with other varieties, local and wild type of mangoes. Among the different varieties and hybrids grown in the state, Amrapalli is the leading and excellent passion, which covers 90 per cent area of the state with good performance with respect of quality, size of fruit, test, aroma, flavour, texture and appearance etc. Area, production and productivity under Mango cultivation in Tripura are 4,190 ha, with a annual production 1,2706 mt and 3.03 tones ha^{-1} respectively. Since the Amrapalli is highly suitable for the state because of its good adaptation with the climate and also dwarfness and less canopy area and regular bearer in nature. Fruits are very sweet and have no fibres. Farming community of the state earns lot of money from it. Apart from being used as a table fruit, mango is popular fruit among house wife for making pickles, squashes, syrups and other items. In Tripura it is cultivated as one of the major fruit crops and consume immature and ripen fruits. There is a high potential for increasing area, production and processing as well as conservation of local and wild type of mangos. During May to July the state usually receives high rainfall accompanied with high relative humidity and temperature which deteriorates the keeping quality. Amarapali mango is largely propagated by softwood grafting. The grafting start in the month of April and continues upto September, but maximum success was found in the month of may-June (80-85 per cent success) and August to September (75-80 per cent success), during this month *i.e.*, may-June (80-85 per cent success) and August to September (75-80 per cent success), temperature, relative humidity and rainfall favour the success of softwood grafting. The age of the rootstock 10-12 month old with pencil thickness. In the month of April the success is minimum due to dry weather and during July success is very minimum due to excess rainfall and moisture. Unripe and ripen mango fruits peoples are preparing different kinds of processes product in the season like pickles, juice, squash, jam, and others value added items. For early and uniform ripening fruits are dipped in 250 ppm of ethrel with 52°C hot water are used, as ethrel are natural hormone.

Banana

Banana is the second most important fruit crops next to Jack fruit in Tripura and is cultivated in about 7,416 ha area with an annual production of 96,787 million tones with a productivity of 13.05

tones ha^{-1} which is far below the national productivity. Large numbers of genetic diversity of Banana exists in Tripura. Many local genotype are found and grown widely in hilly area. Local banana genotype has good potential, since, it has been grown for the last several years without any systematic cultivation and management, has resulted, which needs to select a superior genotype. At present most of the Banana genotype are growing and maintain in home gardens by farmers and naturally growing in hill slope. Tripura is considered to be the natural home of Banana and it is reservoir of many local genotypes. Due to favourable agro-climatic conditions, fertile and acidic soil with good depth and abundance rainfall favour the growing of various type of Banana and they are growing wild and semi-wild in the state. Diverse forms of Banana have been growing in hills, slope and plain areas. The available germplasm were evaluated for growth characters. Rich diversity of Banana germplasm was observed across the state. Some local genotypes are Sabri Kela (AAB), Samai Kela or Bangla Kela or Gopi Kela (ABB) and Champa Kela (AAB), Attia kela (BB), Katch Kela (ABB), Ramkala (AA), Kanai Basi (AA) etc are available. Among the all local genotype, Sabri (AAB), Champa (AAB) and Samai(ABB) kala is most popular and excellent genotype with good quality, in respect of test, aroma, flavour texture, size and appearance etc. of fruit for fresh consumption with medium and tall stature plants. Some of the ensete group of wild banana also available in Tripura, the pseudostem sheet of that banana is used as vegetables by Tribles of the state and they are not produce any suckers and that have medicinal value.

Pineapple

Pineapple (*Ananas comosus* L.) is a natural fruit in the state with favourable sub-tropical agro-climatic conditions, fertile slightly acidic soil, with undulating tilla land with varying degrees of slope are highly susceptible to erosion so practices to cheek soil erosion are vital. 'Queen' and Kew ('Smooth Cayenne') are the main cultivars grown in all the four district of Tripura. The area under pineapple cultivation is 5,180 ha with annual production is 1.06 million ton with 'Queen' being the most popular for fresh consumption and occupying the largest area. Pineapple plant has a tendency not to produce uniform flowering even after the attainment of physiological maturity. Pineapple plant generally flowers after the attainment of certain vegetative growth and ripeness–to–flower stage is attained 11-12 months after planting and formation of at least 40 leaves. It is often observed that even after 15-18 months of growth under optimal nutritional and environmental conditions only 50-60 per cent plants come to flowering. Therefore it would be advisable to use growth regulators to regulate the flowering and production of year round Pineapple in Tripura. Recommendation of growth regulators for different months for inducing and regulation of flowering are (i) application of Ethrel at 10 ppm (2.5 ml/100 litres of water)+2 per cent urea+0.04 per cent sodium carbonate for the month of March-May. (ii) application of Ethrel at 25 ppm (6.25 ml/100 litres of water)+2 per cent urea+0.04 per cent sodium carbonate for all months for year round production. 80 per cent of pineapple fruit are harvested within only 2-3 months (May-June- July), resulting in huge market glut all the markets of the state. During May to July the state usually receives high rainfall accompanied with high relative humidity and temperature, which results in rapid deterioration in the keeping quality and because of that the farmers get very less price. In order to over come this problem, Year round production of pineapple are growing by using different grades of suckers and slips (300g, 600g and 900g), staggered planting from April to October at monthly intervals followed by chemical induction by application of ethephon/ ethrel at 25ppm+ 2 per cent urea+ 0.04 per cent sodium carbide at 38-40 leaf stage results in of off season flowering and fruiting, which ultimately provide year round available in the state.

Papaya

Papaya is regarded as the wonder fruit of the tropics and sub tropics. It has got great importance due to its high nutritive value and production potentiality. Owing to the increasing demand for fruits, the area and production have increased during the last few years in the state. Agro-climatic conditions (humid subtropical), fertile and acidic soil with good depth and abundance rainfall favour the cultivation of papaya. Systematic and accurate estimate of area and production are not available in the state. But still papaya is the most important fruit crop in the state and is cultivated in about 1294.30 ha with an annual production of 13187.20 mt with a productivity of 10.19 t ha^{-1} which is far below the national productivity of 31.69 t ha^{-1}. The few genotypes growns in the state are Coorg Honey Dew, Pusa Dwarf, Pusa Majestic, Pusa Nanha, Washington, Surya and local types. The Coorg Honey Dew, Honey Dew, Pusa Nanha, Pusa Majestic and local type are performing good with respect to yield and quality parameters. Since, it has been grown for last several years in backyard garden, many mixed population has developed, which need sanitization and also have scope to choice a superior clone. Low productivity recorded in the state is due to improper management. Most of the orchards are having the problem of water stress from December to March leads to severe moisture stress and heavy rain from May to September causes heavy soil erosion, nutrition loss from top soil and it also offers favourable climate for attract of many disease and insect-pest. The damping off is very serious problem at seedling stage, mosaic, papaya ring spot virus, leaf curl, mealy bug and mites very destructive at different stage of plants. Frost during December–January in the state is another problem for papaya cultivation. The soil across the state is acidic in nature, which hampered the availability of the important nutrients to the plants and affect the yield. Depletion of nutrients by washing of rich top soil and leaching losses are causing nutritional deficiencies. The proper management practices like GAP (Good Agricultural Practices), with moister conservation methods during stress period (December to March) and efficient and timely manage of diseases and pest through environmental friendly approaches will boost the production and productivity of papaya.

Citrus

It is the fourth important fruit crops next to Jackfruit, banana and pineapple and is cultivated with a area of 5,204 ha with an annual production of 32,638 mt with a productivity of 6.27t ha^{-1} which is far below the national productivity. Citrus species has good potential, since, it has been grown for the last several years without any systematic cultivation and management, has resulted, which needs purification and also offers scope to select a superior clone. At present most of the Citrus species are maintain in home gardens by farmers. Majority of the orchards are having the problem of drought stress from November to March leads moisture stress, heavy rainfall from April to September causes heavy soil erosion and nutrients loss and it also offers congenial environment for the attack of so many diseases and insect-pest. The soil is acidic in nature, which hampered the availability of the important macro and micro nutrients to the plants and adversely affect on yield. The farmers of the state seldom use any soil amendments like lime in their orchards. Depletion of nutrients by washing of rich surface soil and leaching by heavy rains are causing nutritional deficiencies. Tripura is considered to be the natural home of many Citrus species and it is reservoir of many Citrus species including mandarin. Due to favourable climatic conditions different species of Citrus are growing wild and semi-wild condition in the state. Diverse forms of Citrus species have been frequently observed growing in lower hills, hill slopes and plain areas. Total 8-9 type of Citrus belonging to different Citrus species are Pummelo (*Citrus gradis*), Kagzi lime (*Citrus aurantifolia*), Elaichi libu (*Citrus limon*), Mandarin(*Citrus reticulate*), Assam lemon(*Citrus limon*), Satkara(*Citrus macroptera*), Pati libu(*Citrus medica*), Mosambi (*Citrus sinensis*) and Sour pummelo (*Citrus megaloxycarpa*) were observed in Tripura and variation has been observed. Rich diversity of Citrus germplasm was observed across the state. Among the different species the Mandarin (*Citrus reticulate*) and some other lime and lemon are grown commercially with good demand. Peoples are preparing different types of value added product from the mandarin, lime and lemon for off season use like juice, squash, pickles and other products etc. Lime and lemon available in the market from April onward and it continue up to September months.

Litchi

Litchi is one of the most relished evergreen tree with spreading branches and dense shining foliage and subtropical fruit with excellent quality, pleasant flavour and attractive colour. The agro-climatic conditions (humid subtropical) favour the cultivation of Litchi fruits in the state. Area under litchi cultivation is 2,762 ha, with an annual production is 16,165 mt and productivity is 5.85 t ha^{-1}. The main varieties grown are Shahi, Swarna Roopa, China, Elachi, Bombai, Late Bedana, Dehradoon, Mazaffarpur, Early Bedana and Rose scented. Out of all the varieties studies in Tripura condition in respect of bearing habit, number of fruits per plants, time of maturity and quality parameters, the varieties Shahi, Mazaffarpur, Swarna Roopa, Bombai and Late Bedana are found in good performance. But Shahi variety showing excellence performance in all aspects like taste, flavour, aroma and other quality parameters with good market demand in Tripura conditions. In Tripura litchi fruits starts flowering in the month of last week of February and mature in the month of end of April and continue up to May month. Litchi fruits available in the market for short period *i.e.,* one and half month. Litchi fruit borers and fruit cracking are severe problems in Tripura. The fruit borer can be control by frequent spraying of insecticide. Fruit cracking is due to water stress condition during the developmental stage of the fruits and can be control by spraying of water during stress condition.

Jackfruit (*Artocarpus heterophyllus* L.)

It is an important fruit crop of humid tropical and near-tropical climatic regions. India is considered to be the native to jack. The tree is monoecious, evergreen and bears the largest edible fruit in the world. In Tripura, the jackfruit is a popular fruit ranking next to the mango, pineapple, banana and orange in total annual production. The agro-climatic condition (humid subtropical), fertile and acidic soil with good depth and abundance rainfall favour the cultivation of jack fruits. They have been exported to neighbouring country like Bangladesh and neighbouring state like Assam. Jack fruit is one of the commercial fruit crops in Tripura and has high local demand. Farming community of state earns lot of money from it, as its cultivation is easy, hardy in nature, producing fruits even under adverse soil and climatic conditions. The succulent and aromatic fruit is eaten fresh. Edible portion (flakes/bulbs) of the fruit is developed from perianth parts. The seeds are also relished when boiled or roasted and eaten out of hand or after soaking in syrup for some time or cooked in dishes. The seeds are also popular ingredients in many culinary preparations. The latex from bark contains a large amount of resins and is often used to plug holes in containers. Jackfruit is a rich source of carbohydrates, carotene, pectin, calories, calcium, potassium and vitamin A. Area under jackfruit cultivation in Tripura is 8,645 ha, with a annual production is 2, 56,280.00 mt. Apart from being used as a table fruit, jack is popular fruit among house wife's for making pickles, sweets and thin round papad. Canned jack fruit, syrup, jam, jelly and candy have also been attempted. Even though it is a minor fruit crop but in Tripura it is cultivated as one of the major fruit crops. People of Tripura consume immature jackfruit as a vegetable and ripen fruit is consumed fresh. There is a high potential for increasing area, production and processing.

Guava

It is another fruit grown in Tripura with small scale as homestead garden. Systematic cultivation is not doing. Some of few varieties and hybrids are grown in Tripura are Lucknow-49, Allahabad Safeda, Baruipur and other local guava with wild small type of are grown in hills and hill slopes. Area under Guava cultivation is only 348 ha, with a annual production is 1636 mt. and productivity is 4.70 t ha^{-1}. At present guava fruits are propagating by air layering, approach grafting and softwood grafting. The success of air layering is 80-90 per cent, approach grafting 90 per cent and softwood grafting 50-60 per cent. Ripen fruits are used for table purpose as well as also used for preparation of jelly.

Vegetables

The state growing major summer vegetables (vandi or okra, pointed gourd, spine gourd, ridge gourd, amaranths, brinjal, cucurbits etc) and winter vegetables (cabbage, cauliflower, tomato, radish, brinjal, bottle gourd, carrot etc). Area, production and productivity of summer vegetables in the state are 13,869 ha, with a annual production 1,66,177 mt and 11.98 t ha^{-1} respectively and winter vegetables are 18,626 ha, with a annual production 28,0741 mt and 15.07 t ha^{-1} respectively in Table 14.1. Besides this their have more than 81 species or types underutilized and underexploited traditional vegetables are grown in rural and interior area as homestead garden for their food and livelihood survival, like bamboo shoot (*Bambusa* sp.), tender leave of elephant food yam (*Amorphophallus paveonifolius*), colocasia or taro and Xanthosoma (*Xanthosona sagitifolius*), Greater galangal (*Alpinia galangal*), Tit Begun (*Solanum indicum*), leafy vegetables like fern tender shoots(*Ceratopteris thalictroides*), lai(*Brassica juncea*), amaranth (*Amaranthus* spp.), Spiny amaranthus(*Amaranthus spinosus*), Pui(*Basella alba*), Punarnaba (*Boerrhavia diffusa*), hiencha (*Enhydra flatuans*), Lajalu (*Mimosa* sp.), sem (*Dolichos purpureus*), tree bean (*Parkia roxburghii*), jack bean(*Canavalia gladiata*), winged bean(*P.tetragonalobus*), Kakrol(*Monordica cochinsinensis*), Coccinia (*Coccinia indica*), parwal or pointed gourd (*Trichosanthus dioca*), chilli and *Dioscorea* sp., etc. The details list of wild edibles or underutilized and underexploited vegetables are also provided in Table 14.7.

Potato

Tripura is one of the leading producer of hybrid TPS (True Potato Seed). More importance has provided for production of hybrid TPS attaching the potato growers to adopt TPS instead of seed tuber for minimizing cost of cultivation. Tripura produced 1162 kg hybrid TPS during last 7 years with approximate valued rupees 1.85 cores, besides production of 60 mt TPS-tuberlets for use as seed. Area under potato grown in the state is 5787 ha, with a annual production is 87214 mt and productivity 15.07 t. 100 grams of True Potato Seed is require to cover one hectare instead of planting 2-2.5 tons of Potato seed tuber. Seed material is diseases free. No cold storage facility is required for storing True Potato Seeds. Transportation cost is very minimum for True Potato seeds compare to seed tuber. True Potato Seeds more resistance to pests and diseases compare to seed tubers. Net profit is more as cost of cultivation is less. Cost of production of potato using TPS is approximately 55 per cent less in comparison to cost of production of potato using seed tuber. Yield obtained upto the 35 MT per hectare.

Mushroom

Mushroom is a very popular food among the tribal peoples of the state. Now a days it is become popular among the non-tribals as it have rich in protein. There are wild mushroom available in the state growing widely in the forest area, which is edible and used by tribal community. Mushroom cultivation generating employment opportunity to the poor farmers, landless labourers, unemployment youths and womens in the state.

Plantation Crops

Cashewnut, Coconut Tea, Rubber and arecanut are the main plantation crops popularly grown in the state. Cashewnut is grown in commercial scale whereas coconut and arecanut are mainly grown in homestead garden. The total area under nut crops 14,314 ha, with annual production of 18,692 mt and productivity 1.31mt/ha (Table 14.2). Other than Cashewnut, Coconut and arecanut, Tea and Rubber are also important economical crop in state.

Table 14.2. Area, Production and Productivity of Nut Crops in Tripura (2008-2009)

Name of the Crops	*Area, Production and Productivity*		
	Area (ha)	*Production (mt)*	*Productivity (mt/ha)*
Coconut	5784.00	8216.00	1.42
Arecanut	4443.00	8354.00	1.88
Cashewnut	4087.00	2122.00	0.52
Total	14314.00	18692.00	1.31
Ginger	1508.00	7932.00	5.26
Turmeric	841.17	5965.00	7.09
Chilly	2278.00	3042.00	1.34
Black Pepper	257.80	106.00	0.41
Total	**4884.97**	**17045.00**	**3.49**

Cashew

Cashew (*Anacardium occidentale* L.) is the commercially important crop in the international trade, which offers considerable scope for enhancing the economy of our country and state by earning more foreign exchange and by generating more employment potential. Tripura having undulating topography (small and medium hillocks) of red-lataritic soil having the opportunity for commercial cultivation of eco-friendly cashew for maintaining sustainable production. Area under cashew cultivation in Tripura is 4,087 ha, with an annual production of 2,122.00 mt with a productivity of 0.52 t ha^{-1}, which is far below the national productivity. The main cashew genotypes evaluated in the state are BLA-39-4, Priyanka (H 1591), Vengurla-4, Ullal-1, Ullal-3, Dhana (H 1608), NRCC Selection-2, BPP-8, Jhargram-1, V.R.I-3 and Bhashkara. Out of all varieties and hybrids studies, the Hybrids BLA-39-4 have found good performance in respect to yields and other quality parameters. The seedling progenies are heterogeneous with respect to flowering, fruiting and other characteristics due to the cross pollinated nature. To overcome this problem, vegetative propagation like softwood grafting are recommended as planting martial *i.e.*, soft-wood grafted plants provide early, uniform and quality yield. The softwood grafting is done in the month of June, July, August, September and October respectively with success rates 65, 68, 60, 55 and 40 per cent respectively and success rates decrease in the cooler month like October, November, December and January respectively. The age of rootstock 45-60 days old. Low productivity recorded in the state is due to improper management and most of the orchards are having the problem of water stress from December to March, leads to severe moister stress and heavy rain from April to October causes heavy soil erosion, nutrition loss from top soil.

Coconut

It is second important nut crop in Tripura. Some of the few varieties and hybrids are grown in the state are East coast tall, West Cost Tall, Sakshigopal etc. Area under coconut cultivation is 5784.00 ha with a annual production 8,216.00 mt and productivity is 1.42 mt/ha.

Arecanut

Arecanut is very economical crop in Tripura. It is growing on slope of the hills and plains with good local demand in the state and it is also cultivated in adjacent state like Assam and Magalaya. The major varieties grown are Mangala, Sri Mangala and Local germplasm. The area under arecanut is 4443ha with a annual production is 8354 mt and productivity is 1.88 mt/ha.

Rubber

Rubber has been identified as one of the thrust areas in the state, in view of its suitability to the terrain and the acceptability amongst the people. Studies have shown that about 100,000 hectares of area in the state can be brought under rubber plantation. The area under rubber cultivation at present is estimated to be about 52,081 hectares, with a annual production 27,276 m t with a market price more than 500 cores. The yield @ 1,500 kg per hectare and the quality of rubber are also comparable to Kerala's plantations. In fact, Tripura has recently been declared the "Second Rubber Capital of India" by the Rubber Board.

Tea

Tripura has been a traditional tea growing State. The agro-climatic conditions suitable for development of tea plantation. The soils are generally fertile, without any major problems of toxicity or deficiencies. The average annual rainfall is about 2100 mm with a fairly even distribution over the year. Tripura has a history of tea plantations going back to 1916 in fact, Tripura is categorised as a traditional tea-growing state - with about 60 Tea Estates and 4,346 small tea growers, with total about 6,000 hectares of area under plantation and producing about 7.5 million kg of tea every year. This makes Tripura the 5th largest, among the 14 tea producing states, after Assam, West Bengal, Tamil Nadu and Kerala. Further, initiative has also been taken by some tea gardens in Tripura to produce "Bio-tea", which is free from any chemical residue. There is a considerable scope to increase the area under tea plantation as well as productivity. The tea currently produced in Tripura is recognised for its good blending qualities. Some plantations (like Ludhah Tea estate, Durgabari Tea estate, Fatikcherra Tea eatate) in the state are going for organic production of tea. There is considerable scope for investment in tea sector in the State.

Spices

Spices are grown in the state traditionally since long. Major spice crops are ginger, turmeric, black pepper and chilli. The area under spices cultivation is 4,884 ha and annual production is 17.04 MT/Ha (Table 14.2).

Floriculture

Floriculture is growing plants for its beautiful flowers. Demand of flowers is increasing day-by-day because change in human attitude towards social and religious values. Flowers are used in almost all the functions due to which its domestic demand is rapidly increasing. Floriculture is emerging as an important sector for income generation in the state and undergone notable transformation in recent years and has become a viable and lucrative commercial farming enterprise.

Realizing the immense potential for commercial floriculture and emerging markets in the other parts of the country, the agro-climatic condition favour the commercial cultivation of flowers in the is gaining popularity among the small and marginal farmers. State started commercial cultivation of orchids, anthurium, gladuolus, tuberose, gerbera, lilies and marigold with the financial help from technology mission or Horticulture mission schemes. The major area of flower cultivation in Bishwalgarh sub-division (Laxmibill and Ranghunathpur village) and Mohanpur sub-division under west Tripura district. Efforts are being made to develop some more villages in flower cultivation with in the vicinity of the city. The numbers of Self-help groups have been come forward to cultivate the commercial flowers. The different type of naturally growing orchids also available in the state.

Table 14.3: Flower Area Grown in Tripura

Name of the Flowers	Area (ha) 2007-2008	Area(ha) 2008-2009
Anthurium	19.3	20.5
Orchids	6.7	31.5
Gerbera	0.6	18.5
Tuberose	3.0	7.0

Medicinal and Aromatic Plants

The agro-climatic conditions of Tripura favour the cultivation of diverse medicinal and aromatic plants and has in abundance, diverse range of herbs, shrubs, trees and vines that have important medicinal value whose healing properties are known to the local healers and practitioner. Tripura being one of richest centre of biodiversity, more than 270 species medicinal plants have been identified for there medicinal uses, which make Tripura a unique position in traditional systems of medicine. The indiscriminate collection of medicinal plants from nature led to scanty and extinction in many cases making them has rare endangered and threatened species. The threats to biodiversity conservation due to deforestation, high populatior growth, urbanization, shifting cultivation, grazing, illegal extraction of medicinal and aromatic plants, forest fires etc. The importance medicinal plants are *Adhatoda Zeylanica, Abrus precatories, Abroma angusta, Achyranthus aspera, Alpinia galanga, Aegle marmelos, Aloe barbadensis, Argemone mexicana, Acorus calamus, Asparagus racemosus, Azadirachta indica, Amaranthus spinosus, Amorphophyllus muelleri, Andrographis paniculata, Aquilaria agallocha, Averrhoa carambola, Bauhinia racemosa, Boerhaavia diffusa, Byayphlium calycinum, Bacopa monnieri, Butea monosperma, Cajanus cajan, Cassia fistula, Centella asiatica, Catharanthus roseus, Calotropis gigantia, Coccimia grandis, Clitoria ternatea, Cissus quadrangularis, Cinnamomun tamala, Cinnamomun zeylanicum, Colocasia esculanta, Costus specious, Crataeva nurvala, Cuscata reflexa, Datura stamonium, Dillenia indica,* Dioscorea (*D.alata, D. floribunda, D. Bulbifera, D. esculenta and D. hamiltonia), Emilia sonchifolia, Enhydra fluctuans, Eryngium foetidum, Leucas aspera, Lawsonia inemis, Heliotrapium indcum, Homalonema aromatica, Hyptis suaveolens, Jatropha carcus, Kalanchoe pinnata, Melastom malabathricum, Mentha arvensis, Mirabilis jalapa, Murraya koenigii, Mucuna pruirens, Oxalis corniculata, Paedaria foetida, Phyllanthus emblica, Phyllanthus amarus, Piper longum, Ocimum sanctum, Ricinus communis, Rauvolfia serpentine, Scoparia dulcis, Sesbania grandiflora, Sida rhombifolia, Sida cordifolia, Solanum indicum, Solanum turvum, Solanum nigrum, Solanum viarum, Solanum xanthocarpum, Terminalia arjuna, Terminalia chebula, Tarminalia bellirica, Tinospora cordifolia, Typhonium trilobatum, Vitex negundo, Withania somnifera,* and others medicinal plants. There are number of wild medicinal plant species are still waiting for identification and documentation in the state. It is highly necessary to document the plants and take efficient steps to conserve them.

Increasing Tend in Horticulture

There has been significant increase over the years in terms of area, production and productivity due to adaptation modern technologies. The increase in productivity in case of fruits, vegetables, floriculture and other horticultural crops. The status of area and production of different crops during 1972, 1978, 1998 and 2009-10 has been presented in Table 14.4.

Table 14.4: Status of Area and Production during 1972, 1978, 1998 and 2009-10

Crops	*Status(1972 to 2009-2010)*							
	1972		*1978*		*1998*		*2009-2010*	
	Area	*Production*	*Area*	*Production*	*Area*	*Production*	*Area*	*Production*
Fruits	9.300	72.20	12.745	139.42	24.23	363.52	36.598	573.82
Plantation crops	3.130	4.75	4.313	6.79	8.260	13.62	14.614	22.32
Spices	1.000	5.00	1.692	7.07	4.000	13.97	4.884	17.04
Vegetables	6.000	30.00	7.592	52.56	12.90	127.78	32.495	446.91
Total	19.430	111.95	26.342	205.84	49.39	518.89	88.591	1060.09

Wild Bamboo Shoot

Bamboo belong to the family Graminae (Poaceae), sub-family Bambusoidae and accounts for 75 genera and 1250 species in the world. Most of the species are grown in the North-Eastern state including Tripura. Tripura is one of the major bamboo producing states in India. Bamboo grows all across the state of Tripura covering over nearly 10 to 15 different species. The common bamboo species of Tripura are *Muli* Bamboo (*Melocanna baccifera*) is a dominant species with over 80 per cent coverage. It is primarily used for incense stick and domestic needs such as gate, fencing and construction. Other species like *Mritinga/Jati(Bambusa tulda), Paura (Bambusa polymorpha), Kanakaich (Thyrsostachys oliver*), Dolu (*Schizostachyum dullooa*), Rupai (*Dendrocalamus baccifera*), Pecha (*Dendrocalamus hamiltonni*), Barak/ Barua (*Bambusa balcooa*), Kailya/Kali (*Gigantochloa andamanica* syn. *Oxytenanthera nigrociliata*), Lata (*Melocalamus compaactiflorus*), Makal (*Bambūsa nutans*), Bari/Jai (*Bambusa vulgaris*) Lathi bans (*Dendrocalamus strictus*), Tetua (*Bambusa jaintiana*) and Bom/Bethua (*Bambusa cacharensis*) with differing properties and more amenable to higher value added products such as handicrafts, furniture and mats are grown in small pockets across the state. Tripura Bamboo Mission (TBM) has a large focus on generating non-muli high value adding bamboo plantations especially as private plantations and homestead plantations. The physico-chemical composition of bamboo shoot shows that the dry matter content, crude protein, fat, carbohydrate, calcium and phosphorus are ranging from 6.5 -11.00, 14.60-27.30, 0.64-2.65,36.10-56.90, 0.73-1.97 and 0.33-0.97 per cent, respectively.

Jhum/Shifting Cultivation and Jhumias

Traditionally, most of the tribal population practiced shifting or jhum cultivation and were termed jhumias. The term "Jhumia" is a generic term used for tribal people dependent on shifting cultivation as the primary source of livelihood. A significant number of persons and families in Tripura continue to depend on forests and jhum cultivation as their main source of livelihood. It showed that in 1961 there were about 25,000 families who were dependent on jhum for their livelihood. By 1978, this number had increased to 46,854 families. By 1987 the estimate was revised to 55,049 families that were more or less dependent on jhum for their survival. In 1999, according to the Department of Tribal Welfare, 51,265 families were dependent on jhum, and majority of them were fully dependent on jhum. The big concentration of jhumia families was in Dhalai and South District of the state (Table 14.5). In 2007, the Forest Department completed a first-ever Census enumeration of hard core shifting cultivators and found 27,278 families (or 1,36, 000 persons) dependent on jhum. In the current circumstances (with reduced years of the jhum cycle), jhum cultivation cannot provide an adequate means of livelihood. Jhum has been continues to be a precarious system of cultivation that yields scarcely enough to survive. The estimated average annual income of a jhumia household was Rs 184 from cultivation; Rs 539 from collection of bamboo, firewood and sungrass; and Rs 1,750 from labour for the Forest Department and forest traders who paid the jhumias minimum prices for the collection of forest produce. A case study conducted in May 2005 in Halodia village of Mungiakami block revealed that one adult worker earned only Rs 1,000 to Rs 1,100 each year from jhum. About 15,000 hectares of area is under Jhum cultivation in Tripura. It is an established fact that Jhuming will continue for many more years to come until the Jhumias could be brought under settled cultivation practices. Government of Tripura provide a package of practices for improving the productivity from the per unit area of existing Jhum crops as well as to encourage them to take up settled cultivation practices. Survey conducted by Tribal Welfare Department, Tripura (1987) and found that total 49800 numbers of hardcore jhumia families, they are fully depend on jhum and 1,465 numbers partly jhumia families, they are partly depend on jhum.

Table 14.5: Sub-division Wise No. of Hardcore Jhumia Families

Sub-Division/District	*Hardcore Jhumias (Nos.)*
Kanchanpur	6739
Kailashahar	198
Dharmanagar	653
Total North District	7590
Longtoraivally	5413
Ambassa	3666
Kamalpur	929
Gandacherra	6041
Total Dhalai District	16049
Amarpur	6607
Sabroom	3914
Belonia	2724
Udaipur	2609
Total South District	15854
Sonamura	875
Bishalgarh	2172
Sadar	2452
Khowai	4808
Total West District	10307
Total State	**49800**
PARTLY JHUMIA	
Sabroom = 516 nos.	
Belonia = 949 nos.	
Total = 1,465 nos.	

Source: Survey by Tribal Welfare Department, 1987.

Major Constraints

Though the Tripura has high potential for the development of horticultural crops, efforts have not been made to develop it as a commercial venture. Factors inhibiting horticultural development in the state are as follows:

Shifting Cultivation

Shifting cultivation also known as jhuming is widely prevalent in North Eastern states of India including Tripura. This jhuming cycle which extended to 15-20 years earlier has now been shortened

Table 14.6: Area and Production of Horticultural Crops for the Year 2009-10 (Area in Hec.; Production in M.T; Productivity in M.T/Hec)

Name of Crops	South District		West District		Dhalai District		North District		State Total		Productivity
	Area	Production	Area	Production	Area	Production	Area	Production	Area	Production	
A: SUMMER VEGETABLE											
Bhindi	324.00	2690.00	790.00	6478.00	140.00	980.00	232.00	1716.00	1486.00	11864.00	7.98
Brinjal	307.00	5103.00	808.00	14312.00	112.00	1362.00	138.00	1874.00	1365.00	22651.00	16.59
Spine guord (Kakral)	167.00	2541.00	346.00	6972.00	29.00	281.00	132.00	1296.00	669.00	11029.35	16.48
Pointed Gourd (Patal)	64.00	415.00	158.00	1608.00	47.00	398.00	24.00	165.00	293.00	2586.00	8.83
Ridge gourd	249.00	2864.00	472.00	7262.00	49.00	401.00	211.00	1790.00	981.00	12317.00	12.56
Bitter gourd	182.00	1482.00	356.00	2743.00	51.50	608.20	139.00	1133.00	728.50	5966.20	8.19
Bottle gourd	11.00	171.00	144.00	2654.00	50.00	667.00	73.00	1098.00	278.00	4590.00	16.51
Sweet gourd	73.00	1213.00	262.00	4086.00	61.00	1054.00	165.00	2453.00	561.00	8806.00	15.70
Ash gourd	40.00	560.00	277.00	3878.00	41.00	492.00	92.00	1242.00	450.00	6172.00	13.72
Colacasia	245.00	4424.00	405.00	5679.00	80.00	897.00	461.00	7426.00	1191.00	18426.00	15.47
Jal kachu	3.00	25.00	118.55	953.00	60.00	556.00	74.00	690.00	255.55	2224.00	8.70
Cucumber	205.00	1878.00	536.00	5446.00	42.00	232.00	85.00	878.00	868.00	8434.00	9.72
Amaranthes	48.00	753.00	725.00	8143.00	125.00	1342.00	164.00	1564.00	1062.00	11802.00	11.11
Raddish	25.00	247.00	181.00	2471.00	52.00	380.00	51.00	363.00	309.00	3461.00	11.20
Cowpea	20.00	206.00	213.00	1315.00	26.00	245.00	115.00	796.00	374.00	2562.00	6.85
Summer Cabbage	65.00	351.00	24.00	128.00	22.00	125.00	20.00	92.00	131.00	696.00	5.31
Summer Cauliflower	66.00	318.00	22.00	102.00	5.00	23.00	6.00	24.00	99.00	311.86	3.15
Others	1242.00	14123.00	682.00	8452.00	508.00	5762.00	336.00	3726.00	2768.00	32063.00	11.58
Total of 'A'	3336.00	39364.00	6519.55	82682.00	1500.50	15805.20	2518.00	28326.00	13869.05	166177.20	11.98

Contd...

Table 14.6–*Contd...*

Name of Crops	South District		West District		Dhalai District		North District		State Total		Productivity
	Area	Production	Area	Production	Area	Production	Area	Production	Area	Production	
B: WINTER VEGETABLE											
Cabbage	400.00	9487.00	903.00	22103.00	177.00	3512.00	291.00	6752.00	1771.00	41854.00	23.63
Cauliflower	337.00	6064.00	815.00	13855.00	138.00	2525.00	271.00	4816.00	1561.00	27260.00	17.46
Brinjal	466.00	5018.00	706.00	7801.00	98.00	701.00	308.00	2775.00	1578.00	16295.00	10.33
Raddish	387.00	4198.00	661.00	10642.00	70.00	712.00	410.00	3516.00	1528.00	19068.00	12.48
Tomato	364.00	9618.00	634.00	12718.00	126.00	2798.00	223.00	4416.00	1347.00	29550.00	21.94
Garden pea	32.00	135.00	158.00	383.00	31.00	104.00	63.00	312.00	284.00	934.00	3.29
Cucumber	22.00	215.00	206.00	981.00	47.00	247.00	61.00	511.00	336.00	1954.00	5.82
Knolkhol	3.00	14.00	62.00	826.00	20.00	70.00	44.00	309.00	129.00	1219.00	9.45
French Bin	12.00	34.00	119.00	296.00	15.00	131.00	35.00	220.00	181.00	681.00	3.76
Carrot	32.00	271.00	64.00	418.00	25.00	398.00	29.00	422.00	150.00	1509.00	10.06
Capsicum	5.00	33.00	23.00	178.00	12.00	66.00	6.00	37.00	46.00	314.00	6.83
Brocoli	1.00	2.50	12.00	31.00	1.00	3.00	4.00	15.00	18.00	51.50	2.86
Chilli	380.00	1915.00	615.00	1891.00	34.00	139.00	10.00	34.00	1039.00	3954.00	3.81
Bottle gourd	117.00	2072.00	325.00	5903.00	52.00	706.00	22.00	208.00	516.00	8889.00	17.23
Beet	13.00	26.00	24.00	38.00	15.00	27.00	3.00	8.00	55.00	99.00	1.80
Others	892.00	11564.00	912.00	11423.00	234.00	2048.00	252.00	2659.00	2290.00	27694.00	12.09
Total of 'B'	3463.00	50666.50	6239.00	89487.00	1095.00	14187.00	2032.00	27010.00	12829.00	181325.50	14.13
Total Vegetable											
Total of A+B	6799.00	90030.50	12758.55	172169.00	2595.50	29992.20	4550.00	55336.00	26698.05	347502.70	13.02
C: NUTS											
Coconut	2238.10	4312.00	2032.00	3772.00	444.28	754.00	1032.00	2546.00	5784.00	11384.00	1.97
Arecanut	1188.00	2479.00	1062.00	1093.00	744.00	1567.00	1449.00	3461.00	4443.00	8600.00	1.94
Cashewnut	3220.00	1418.00	700.00	689.00	239.00	121.00	228.00	117.00	4387.00	2345.00	0.53
Total of 'C'	6646.10	8209.00	3794.00	5554.00	1427.28	2442.00	2709.00	6124.00	14614.00	22329.00	1.53

Contd...

Table 14.6–*Contd...*

Name of Crops	*South District*		*West District*		*Dhalai District*		*North District*		*State Total*		*Productivity*
	Area	*Production*	*Area*	*Production*	*Area*	*Production*	*Area*	*Production*	*Area*	*Production*	
D: FRUITS											
Mango	2080.00	6726.00	1109.00	3421.00	390.00	1011.00	675.00	2012.00	4254.00	13170.00	3.10
Pineapple	1525.00	27518.00	1535.00	28765.00	1196.00	21288.00	2220.00	39960.00	6476.00	117531.00	18.15
Orange	424.00	1725.00	30.00	43.00	561.00	2475.00	2147.00	17987.00	3162.00	22230.00	7.03
Jackfruit	2788.00	75495.00	1355.00	71371.00	1847.00	55601.00	1646.00	49102.00	7636.00	251569.00	32.95
Banana	1395.00	19634.00	3850.00	56854.00	921.00	14121.00	1320.00	15006.00	7486.00	105615.00	14.11
Litchi	938.00	6538.00	525.00	3516.00	489.00	2816.00	810.00	4102.00	2762.00	16972.00	6.14
Lime/Lemon	590.00	3073.00	610.00	4625.00	515.00	2732.00	391.00	1286.00	2106.00	11716.00	5.56
Papaya	398.00	5018.00	320.00	4563.00	268.00	4092.00	343.00	4782.00	1334.00	18455.00	13.83
Sapota	60.00	675.00	22.00	184.00	10.00	97.00	19.00	183.00	111.00	1139.00	10.26
Guava	121.00	591.00	102.00	538.00	36.00	181.00	91.00	417.00	350.00	1727.00	4.93
Minor Fruits	201.00	2506.00	360.00	5087.00	140.00	1892.00	220.00	4216.00	921.00	13701.00	14.88
Total of 'D'	10520.00	149499.00	9818.00	178967.00	6373.00	106306.00	9887.00	139053.00	36598.00	573825.00	15.68
E: SPICES											
Ginger	679.00	4839.00	358.00	896.00	192.00	985.00	279.00	1212.00	1508.00	7932.00	5.26
Turmeric	346.17	2415.00	201.00	1718.00	74.00	407.00	220.00	1425.00	841.17	5965.00	7.09
Chilly	1241.00	1608.00	640.00	927.00	185.00	216.00	212.00	291.00	2278.00	3042.00	1.34
Black Pepper	112.80	17.00	66.00	44.00	5.00	3.00	74.00	42.00	257.80	106.00	0.41
Total of 'E'	2378.97	8879.00	1265.00	3585.00	456.00	1495.40	785.00	2970.00	4884.97	17045.00	3.49
F: POTATO											
Potato	2593.00	44858.00	1513.00	26477.00	664.00	10624.00	1027.00	17457.00	5797.00	99416.00	17.15

to 2-3 years because of increased population pressure on land, decrease in productivity leading to utilization of more area under jhuming. At present many hectare area is under shifting cultivation in the state. This system has caused large-scale deforestation, soil degradation/erosion (removes nutrient rich top soil) and depletion of resource base.

Poor Cultivation Practices and Low Yield

General neglect and non-adoption of scientific cultivation practices are the major constraints for poor return from most of the horticultural crops in the state. Despite conducive environment, the rate of production and growth of all horticultural crops are far below the all India average.

Lack of Desirable Planting Material

The disease free, true to type genuine planting material is absolutely lacking in a number of horticultural crops. It is imperative to produce disease free propagules. Screening of planting materials before its distribution is of utmost importance.

Lack of Marketing Facilities

Due to lack of organized marketing structure in the state, farmers are getting low return compared to the other parts of India, whereas the middleman gets the profit at their expenses. Except the organized tea industry, almost all the commodities. Due to thin primary markets and perishable nature of the products the farmers sell their produce at a throw away prices to the middleman without even getting the opportunity to display them. Transportation of perishable produces is perhaps the most serious constraints in the horticultural development in the state.

Scarcity of Trained Manpower and Extension Support

Dearth of trained manpower and low priority to horticulture in the development plans of states despite high potential are some of the factors responsible for ineffective extension programme. Unlike other states of India like Punjab, Himachal Pradesh, Haryana, Karnataka, Tamil Nadu, Maharashtra etc., where the extension services are very efficient, the state on the other hand is lagging far behind in this aspect. To strengthen this wing not only trained manpower but determined extension activities with full government support are most urgently required.

Land Tenure System or Land Ownership System

- ☆ Land tenure or land ownership system and laws are somewhat problematic in the state and presently it is simplified by the government of Tripura by providing land ownership with different tribes.
- ☆ Lands owned by the chiefs, who allot land among the individual households for shifting cultivation purposes, and
- ☆ Lands owned by individual families.

Since horticultural crops have long gestation period and initial cost of establishment of orchard or plantation is high, it becomes almost impossible for the marginal farmers to go for such ventures without long-term credits from financial institutions like banks. Nationalized banks, do not find it a very favourable investment and are not sure about the recovery of loans because the land tenure system particularly in the tribal belts does not permit land mortgages in favour of lending banks. As the tribal farmers cannot ask for loans individually, intermediaries like District Councils or Village Council can take active part in securing loans for them. Apart from these difficulties, the farmers are not tuned to the idea of considering agriculture as a business proposition and are not accustomed to bank loans. Thus, until and unless this system is changed, the financial investment will not increase to the national level.

Problems of Processing

For a state like this the success of fruits and vegetable growing is closely linked with the success of fruit processing units, because of poor marketing and transport facilities. The processing industry can help in sorting out the problem of proper disposal of perishable commodities. Till today, there are hardly very few cold storage facilities available; few processing units exist but are not functioning up to the desired capacity. Use of appropriate pre and post-harvest practices for horticultural crops is vital for the success of the crops and to provide good return to the growers. Unfortunately this is the weakest spot in the entire state. Value addition should be given top priority for the horticultural crops. Production of oleoresin from ginger, turmeric and chilli using improved techniques as developed by CFTRI, Mysore needs to be tested in the state.

Financial Constraints

The high capital cost involved in establishing orchard/plantation and setting up of equired infrastructure is a serious constraint in the expansion of area under horticultural crops. The situation becomes all the more difficult in view of the large number of small holdings.

Less Expenditure on Research Work

Investments for research on horticulture have always remained low when compared to the large number of crops it covers. This results in poor technological support. The extension system is also weak. The Department of Horticulture has been created in many states but, do not have adequate manpower and infrastructure to address the entire problem of horticulture.

Absence of Insurance Facility

Risk management in horticultural crops is non-existent although crops like onion and potato are covered under the National Agriculture Insurance Scheme. There is a need to cover the risk in case of other horticultural crops in a different manner, perhaps on the basis of potential production coverage instead of average yield. This would encourage higher investment to achieve high productivity.

The other major bottlenecks are as follows:

- ☆ Inadequate facility on conservation and exploitation of horticultural germplasm.
- ☆ Lack of funds and financial support.
- ☆ Remoteness of the state.
- ☆ High rainfall, soil erosion and high rate of leaching of nutrients.
- ☆ Heavy infestation of weeds, insect-pests and diseases.
- ☆ Lack of needed information.
- ☆ Lack of awareness about the potentiality of horticultural crops as commercial crop.
- ☆ Lack of need based research as sustainable agricultural system/silvi-horti-pastoral system in the need of the hour.
- ☆ Weak data base.

Future Thrust

Collection, Characterization and Conservation of Germplasm

There is a need for extensive survey of the region and collection of all the germplasm available in the state. These germplasm should be characterized on morphological and molecular basis and conserved at one place. These germplasm should be utilized in strategic breeding programme involving high yielding national/exotic varieties.

Identification of Area Specific Major Horticultural Crops

There is need to identify important horticultural crops for different areas of the state. The infrastructure facilities for commercial cultivation, marketing, export and processing/value addition for identified crop should be developed. There is also need to identify the high yielding varieties and hybrids available in the country in selected crops for different part of the state.

Hi-tech Horticulture

High-tech horticulture is the deployment of new technology, which is modern, less environment dependent, capital intensive and has the capacity to improve the productivity and quality of horticultural crops. Adoption of this technology in horticultural crops is necessary to ensure the nutritional and health security of future generation. Hi-tech horticulture includes micro-propagation, micro-irrigation, fertigation, protected cultivation (greenhouse/polyhouse cultivation), organic farming, mechanization and use of remote sensing.

Infrastructure for Horticulture

The basic infrastructure facilities like pre-cooling units, packing and grading shed, short and long term cold storage facilities, refrigerated containers, storage and phytosanitary facilities at mandi are lacking in the state. Therefore, there is urgent need to create basic infrastructure facilities to boost the horticulture in the state.

Establishment of Agricultural Technology Information Centre (ATIC)

There is need to establish ATIC in different part of the state. Information through internet connection should be provided regarding demand, supply, price market outlook, knowledge of consumers' preference, marketing channels and practices.

Conduction of On-Farm Trials/Frontline Demonstration (FLD)

The farmers of the state are not aware with the recent technologies of horticulture. Therefore, there is need to conduct demonstration/FLD as much as possible at farmers' field in the identified crops to convince the farmers about the efficacy of measures in enhancing the productivity of identified crops. Apart from this the extension personnel should try to bring the maximum number of farmers to demonstration plots, organize farmers day, fair and yield competition, distribute leaflets/bulletins to the farmers.

Post-Harvest Management and Processing

The state is lacking in trained personnel with sound knowledge of post-harvest management of produce. There is also a need for integrated research for post-harvest handling, packaging, transportation, storage and quality control of perishable commodities. There is also a need to establish processing units to formulate value addition products of excess produce.

Strengthening of Horticultural Farms and Nurseries

The different horticultural farms and nurseries of the state which are meant for supply of disease free, healthy, true to type planting materials of fruits, vegetables and ornamental crops should be strengthened so as to meet the increasing demand for planting materials. The important rootstocks/ mother tree stock should also be maintained at these horticultural farms/nurseries.

Training to Farmers/Extension Functionaries

Non-availability of trained manpower is one of the major problems of the state. As horticulture requires highly skilled personnel for grafting, pruning, orchard management and also for vegetables and ornamental plants. The farmer as well as extension functionaries should be given training from time to time regarding recent advances in horticulture. The entrepreneurship should also be generated by providing training to entrepreneurs for overall development of horticulture in the state.

Protected Cultivation

The state is dominated by hilly areas thus they are highly suitable for protected/off season cultivation of high value crops like tomato and capsicum and flowers/ornamentals crops.

Emphasis on Organic Farming

There is need for research on standardization of doses of different organic fertilizers like FYM, green manure, vermi-compost, neem cake, biofertilizers (Rhizobium, azotobactor, VAM), etc., in different crops. The emphasis should be given on the use of locally available organic materials.

Research on Underutilized Crop

There is need for research on under utilized and under exploited/lesser known horticultural crops for their commercialisation, because these crops are grown at very large scale, in the state. These crop provide lot of food, nutrition and medicinal value in rural area.

Crop Diversification

While giving emphasis on one or two crops, the other crops should also be taken into account to make agriculture sustainable. For example, after ginger and turmeric, which are heavy nutrient feeder, leguminous vegetables like cowpea, pea, etc., should be grown to maintain the fertility of the soil. Similarly, agri-horti-suilvi-pastoral system or multistorey system and agro-forestry will be more successful in hilly areas. The other points which may be taken in to consideration for developing the horticulture industry in the state are:

- ☆ Introduction of export quality horticultural crops suitable for the state. For example black pepper, cashewnut, Banana, cut flowers etc.
- ☆ Technology for the low cost hybrid vegetable seed production.
- ☆ Production of quality seeds (conventional/hybrid) and planting materials.
- ☆ Working out the dynamics of production constraints.
- ☆ Revival of Mandarin cultivation through refinement of rejuvenation practices.
- ☆ Furtherance of research on development of models of multiple cropping including high-density cropping in fruit crops.
- ☆ Production of more quantity of true potato seeds (TPS).
- ☆ Intensification of research on water management including drip system.

- ☆ Improvement of shelf-life and product diversification of banana, pineapple and litchi, citrus, tomato, capsicum, etc.
- ☆ Development of floriculture and establishment of at least one model village in each District of floriculture near urban centre supported with modern sales centre at near by city and linkage with APEDA for export.
- ☆ Establishment of biotech unit (tissue culture lab) for production of disease free planting material in important crops like citrus, banana and ornamental crops etc.

Conclusion

Tripura is blessed with huge plant genetic resources especially horticultural crops. Systematic cultivation Horticultural crop is less. These crops are playing a vital role in providing food, nutrition, health and economic status in the state. There is tremendous scope to popularize these in non-traditional areas and these crops may earn lot of foreign currency in the state in near future. Further, the huge genetic diversity in various horticultural crops has vast scope for collection, conservation and utilization for benefit of people in the state.

References

Anonymous, (2009-10). Annual report. Development of Horticulture. Govt. of Tripura, Tripura.

Pandey. B.P.(1997). Taxonomy of Angiosperms, S. Chand and Company Ltd. New Delhi, pp 497.

Sukhen Chandra Das, Jai Prakash. R.C. Samui and T. Bshwas (2009).Underutilized Fruit Crops in Tripura: Improving Livelihood opportunities for hilly Tribals of Tripura. In Proceeding: International Conference on Horticulture (ICH-2009), Horticulture for Livelihood Security and Economic Growth, Bangalore, India. pp. 1231-1239.

Sukhen Chandra Das, Jai Prakash. R.C. Samui and T. Bshwas (2009).Traditional Medicinal Plants used by Hilly Tripura. In Proceeding: International Conference on Horticulture (ICH-2009), Horticulture for Livelihood Security and Economic Growth, Bangalore, India. pp. 1240-1246.

Sukhen Chandra Das and Jai Prakash (2009). Minor fruits: A Livelihood opportunities for the tribal people of Tripura. Abstract published in 2nd International symposium on Pomegranate and Minor including Mediterranean Fruits(ISPMMF-2009), conducted by International Society for Horticultural Sciences (ISHS), Levuen, Belgium,.June, 23-27, 2009. UAS. Dharwad, India. pp. 67.

Sukhen Chandra Das and Jai Prakash. A. K Deb and T. Bshwas (2009). Studies of Indegenious and Tribals Medicinal Plants in Hilly Tripura. Abstract published in 2nd International Symposium on Medicinal and Nutraceutical plants, conducted International Society for Horticultural Sciences, (ISHS), Levuen, Belgium, November, 25-27, 2009. AIIMS, New Delhi, India. pp. 109.

Hill Agriculture: Prospects, Constraints and Mitigations (2013) *Pages* 213–225
Editor: Dr. Shahid Ahamad
Published by: DAYA PUBLISHING HOUSE, NEW DELHI

Chapter 15

Golden Condiment Saffron in Hill Agriculture

J.N. Srivastava, Upma Dutta and B. Singh
Regional Horticulture Research-Sub Station, Bhaderwah, SKUAST-J

Introduction

Saffron *Crocus sativus*, (Family: *Irridaceae*) is one of the oldest and certainly among the world's most expensive spice, it is known as saffron in English, Kesar in Hindi, Zaffron in Urdu and is popularly called the golden condiment. The plant is native to Southern Europe and its cultivation dates back to 550 A.D. (Kalhan, 12th century). Saffron is a legendary cash crop of Jammu and Kashmir (India). Iran is the largest producer of saffron accounting for more than 65 per cent of the global saffron production. In India it is cultivated in Pulwama, Badgam, Anantnag, Sringar (Kashmir) and Berwar in Kishtwar (Jammu), Bharsa, Joshimath, Doonta, Chaubattia (Uttrakhand) and effort are on for cultivation of saffron in Kinnaur, Lahul, Speethi and Pangi of Himachal Pradesh (Kanan, 2001).

Table 15.1: Area, Production and Productivity of Saffron in the World

Countries	*Area (ha)*	*Production (tons)*	*Productivity (kg/ha)*
Iran	43408	173.0	3.98
Spain	600	5.0	7.84
India	2825	6.46	2.28
Greece	1000	4.30	4.30
Azerbaijan	675	3.70	5.48
Morocco	500	1.00	2.00
Italy	29.4	0.24	6.16

In India most of the saffron production is liminted to the state of Jammu & Kashmir. In Kashmir saffron is cheilfy grown in the Pulwama District (73 per cent) (Pampore, Balhuma, Wayun, Munpur, Mueej, Konibal, Dus, Zundhur, Leptur, Somber, Baras, Ladu and Khrew), Badgam District(Chadura, Nagam, Lasjan, Ompora and Kralpura), Ananatnag District (Zeripur, Srechan, Kaimouh, Santhan and Buch), Srinagar District (Zewan, Zawreh and Ganderbal) and Kishtwar District(Barmer). (Mir, 1992; Zargar, 2001).

In Jammu and Kashmir saffron is cultivated over an area of 5700 ha with about 85 per cent concentration in Pulwama District (Kashmir) and rest in Budgam, Anantnag, Srinagar and Kishtwar (Jammu). More than 1000 farm families are associated with this crop. Saffron is believed to have been introduced into the Kashmir valley by the Persian migrants, who along with saffron, introduced many other plant species, especially the Maple tree known locally as Chinar or Boune and art and crafts such as carpet making, embroidery and papier-mâché products, to name a few (Kafi *et al.*, 2006).

Saffron is unique among spice due to its aroma. It is water soluble and when added to dish, gives a pure and homogenous colour. Saffron is more important in Central Asia and Northern India and is used extensively for rice dishes. Even the North Indian biryanis are relished due to fragrant and aromatic flavour added by the saffron. Indian sweets like kheer, rasmalai, Indian yogurt drink (*lassi*), butter lassi have an everlasting culinary impression due to the saffron added to it. The use of saffron in sweet dishes is famous in desert region of Jodhpur in Rajasthan in the Indian sub continent. The price of Kashmiri saffron is much higher than Iranian saffron (Kalhana, 12th century).

In high doses, saffron exhibits toxic qualities. However, due to its high price, saffron poisoning is very rare. Saffron was used as an ingredient in Ayurvedic medicines by the famous Kashmiri Vaids namely Vegbhatta and Sushtra. Saffron has a variety of uses as a natural colouring and flavouring agent with incredible medicinal and therapeutic attributes.(Zhou *et al.*, 1987) The demand for pure and quality saffron at the global level is anything between 120-130 tonnes annually, to which India is contributing nearly 12-15 tonnes. The entire Indian production comes from Jammu and Kashmir State. World production is not enough to meet the growing global demand. Cost of production in European countries is very high with chances of its horizontal expansion in production comparatively low. It is here that countries like India with tremendous human resource have scope to promote and expand saffron production and become the leading exporters in the global market (Negbi, 1999).

Saffron is widely used as spices and condiment and recently also for other purpose. (Munshi and Zargar, 1989). The ash of saffron is rich in potassium and phosphorus. It possesses a characteristic aromatic odour and a pungent bitter taste. The principal colouring agent of saffron is the glycoside orocin and the bitter substance is the glycoside picroorocin. Saffron is mostly favoured as flavouring and colouring material. It is considered as a stimulant and is used as a sedative. It is also given to promote eruptions in measles. It is used in fevers, melancholia and enlargement of the liver and spleen. It is also an important ingredient of the ayurvedic and unani systems of medicine (Zhou, *et al.*, 1987).

Botany

The plant is a small perennial bulbous herb with narrow, linear radical leaves with globular seed corms which attain the size 1.5-5 cm in diameter or 4-20 g in weight at maturity. The violet or bluish flowers are borne singly on a space. It belongs to family *'Iridiaceae'*. Its propagation is through corms, reproduce annually and gives rise to 3-7 new cormels (Negbi, 1999; Munshi *et al.*, 2001).

Composition

The colouring agent is crocin ($CH_4H_{64}O_{26}$ H_{20}). The clouring compound of Kashmiri saffron is 13-14.5 per cent (Kanan, 2001).

Table 15.2: Main Chemical Constituents of Saffron Growing Soils of Kashmir

Constituents	*Content (per cent)*
Moisture	15.60
Starch and Sugar	13.35
Essential Oil	0.60
Fixed Oil	5.63
Total N- Free Extract	43.64
Crude Fiber	4.48
Ash	4.27

Saffron contains moisture 15.6 per cent, carbohydrate 13.35 per cent, essential oil 0.6 per cent, fixed oil 5.6 per cent, total N-free extract 43.64 per cent, crude fiber 4.48 per cent and ash 4.27 per cent. The ash is rich in potassium and phosphorous. It possesses a characteristics aromatic odour and a pungent bitter taste. The principal colouring agent of saffron is the glycoside orocin and the bitter substance is the glucoside picroorocin. Saffron is mostly flavoured as flavouring and colouring material. It is considered as a stimulant and is used as a sedative it is also given to promote eruptions in measles. It is used in fevers, melancholia and enlargement of the liver and spleen. It is also an important ingredient of the ayurvedic and unani systems of medicine. Its odour (Safranal), taste (picrocin) and pigment (Crocin) components, constituting the spice "Saffron" are located in the red stigmatic lobes of the *C. sativus* flowers (Kanan, 2001; Rees, 1988).

Production Trends

Climate

Saffron prefers cool, dry and sunny site between elevations ranging from 1500 to 2400 m. localities that receives 30 to 40 cm rainfall and covered with snow during winter are suitable for commercial cultivation. Cool and wet climate inhibits flowering but increases the potentialities of mother corms to produce a large number of daughter corms. Heavy rain or frost at the time of flowering is disastrous.(Anonymous, 1988) A warm spring and long autumn are conducive to early flowering (Molafilabi, 1994). The optimal temperature requirement for flower emergence (17ºC) is markedly lower than that for organogenesis. The differences in air temperature explain the different calendar time for saffron flower initiation in different locations. Water availability plays a minor role, if any, on flower formation. The optimal temperature for shoot growth (23-25ºC), has also been shown to be optimal for flower initiation. In view of requirement of critical weather parameter in Kashmir region (Ingram, 1969; Anonymous, 1988)..

Soil

Saffron requires a well-drained loamy soil having a pH range of 7-7.5 free from pebbles. A medium grade light soil with neutral to slightly alkaline reaction is suited for its cultivation. Proper drainage should be ensured to avoid rotting of corms (Ghani, 2001). Soil is main component of saffron farming.

Table 15.3: Status of Primary and Micronutrient Elements in Saffron Growing Soils of Kashmir

Nutrient Element	Range	Mean
Nitrogen (kg ha^{-1})	124.54-336.63	238.82
Phosphorus (kg ha^{-1})	8.00-14.00	10.00
Potassium (kg ha^{-1})	219.52-313.60	277.75
Zinc (ppm)	0.35-1.14	0.64
Copper (ppm)	0.25-0.50	0.35
Manganese (ppm)	8.0-72.0	23.0
Iron (ppm)	6.00-78.0	17.88

Source: Ghani, 2001.

Preparation of Land and Planting

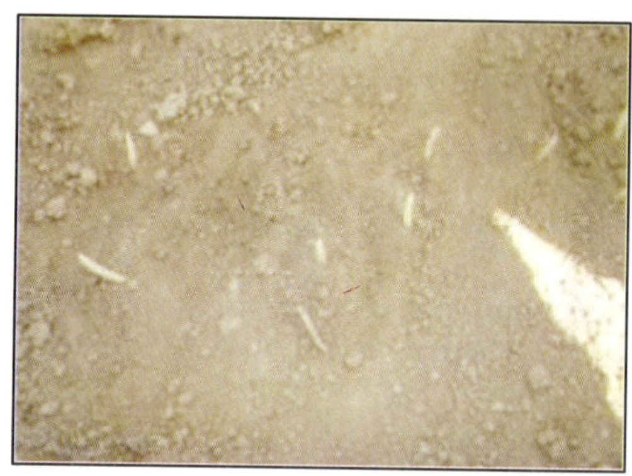

The field is ploughed four to five times to a depth of 30-35 cm fifteen to twenty tons of FYM is incorporated into the soil during land preparation. Furrows are prepared and corms are planted in the furrows at a spacing of 8 to 10 cm (corm to corm) and 15 to 20 cm (row to row). After planting, the furrows are closed and the field is leveled by planker. The field is laid out into 2.5 cm square beds with 30 cm wide and 15 cm deep drainage channels all around (Munshi *et al.*, 2001).

The optimum time of planting ranges from July to first week of August. The corms selected for planting should be 2 cm or more in diameter. Forty to fifty quintal seed corms are required to plant a hectare of land (Mir 1992).

Selection of Corm

The corm required for replanting should above 2.0-5.0 cm in diameter or above 10 g weight Saffron is planted by dormant corms, select disease free and healthy corms, having at least 2.0 cm diameter. The husk dirt etc. adhering to the corms, should be removed, and the corms before planting should be treated with solution of fungicides (Kafi and Showket, 2010).

Cleaning and Sorting of Corms

After sorting productive type of corms followed by removing mother residue and outer most loose covering to identify and isolate injured or disease corms (Munshi *et al.*, 2001).

Seed Treatment

For management of diseases seed corms may be treated / dipped with Carbendazim (50WP) @ 0.1 per cent + Copper oxychloride (50WP) @ 0.3 per cent or Carbendazim @ 0.2 per cent for 10 minutes prior planting.

Time of Planting

The most appropriate time of planting of saffron corms is August to September.

Recommendations to be Followed Strictly

- ✰ Water logging shall be avoided in planting areas.
- ✰ Fields already reported for corm rot shall be avoided for new plantings.

Plant Geometry

This crop requires plant geometry as per the planting cycle for the crops. A suitable plant geometry recommended for 5 years planting cycle requires a spacing 5 cm within plant and 20 cm between rows under a paired row system to harvest 4-5 kg saffron per hectare. Thus, a hectare of land requires 40-80 quintals of seed corms. The seed rate is differ by methods of sowing and spacing.The treatment corms should be planted 15-22 cm deep in the soil (Munshi *et al.*, 2001).

Application of Fertilizers

For higher floral production 40 kg of N, 50 kg P_2O_5 and 30 kg K_2O per hectare should be required. The P_2O_5 and K_2O are given as basal doze, where as N is applied in two equal split doses, the first part at the main hoeing and weeding *i.e.*, first week of September and remaining in the third week of November which is observed to be the active growth period.

Higher floral and cormels production demands 15-20 tones of well decomposed FYM per hectare, the organic manure application is suitable in the month of July- August when monsoon rain are expected and should be mixed during the course of land preparation (Negbi, 1999).

Interculture and Weeding

The planted fields are not disturbed till the month of June. First hoeing is done in the month of June. This hoeing operation provides aeration to the soil and is very important for development of corms. Second hoeing is given in the first week of September *i.e.*, before 30 days of flowering and is accompanied by light dressing of well decomposed farm yard manure @4 tonnes per hectare per year of cycle (In gram 1994).The field is hoed along with drainage channels and the clods are broken and soil is raised from channels to either beds or strips and properly leveled. A third and final light hoeing is given after flowering season is over. These operations are followed every year as the crop remains in the field.

Time of Flowering and Harvesting

Flowering starts in the first week of October and continued up to the first week of November. The flowers are picked daily in the morning after dew disappears as flowers are at full turgor in morning. The *stigmas along with styles* (actual saffron) are trimmed immediately. One kilogram-dried saffron is obtained from the styles and stigmas of about 1, 50,000 flowers. (Nigbi, *et al.*, 1989) Fresh saffron is air dried to moisture content of 8 to 10 per cent, and stored in moisture proof containers. Use of solar dryers and air dryers are recommended for maintaining the better saffron quality. Dryer could be used which takes 7-8 hours only for drying and improve the quality of produce.

Flower Picking

Harvesting saffron includes picking the flowers manually and separating the thread-like stigmas, which is a delicate job also performed by hand. Flower-picking starts as soon as they appear in the field. This should be done on a daily basis because flowers are short-lived and if they are left for a longer period, not only can they get damaged, but the quality of saffron also decreases. Picking flowers generally begins from October-November in Kashmir but this differs in the regions according to the climate variability and time of first irrigation. The flowering period of a field lasts for 15-25 days, starting gradually and reaching a peak from the 7th to 10th day. Flowers are picked early in the morning and before sun rise, while in some areas this is also done in the afternoon (Munshi, 1989).

In Kashmir flowers are not picked up daily but once in four days, before nine O'clock in the morning. This method of flower collecting in Kashmir might cause lost of quality as well as quantity of yield. (Ghani, 2001; Mir, 2001; Munshi *et al.*, 2001).

Separation of Stigmas

The stigmas are to be removed from the picked flowers and this operation should not be delayed until the following day, as it becomes more difficult when the flowers are no longer fresh and in turgid condition.

The blossoms collected daily are brought indoors and put on table or over a cotton sheet/

woolen chaddar round which women workers sit with small earthen pot for collection of stigmas which may be obtained as under:

- The pistil under the corolla is cut with the thumb nail of the left hand and stigma pulled off along with the ovary using finger of the right hand. Alternatively, growers with bumper harvest dry the whole flowers in the sun and the stigmas are picked by hand from the whole dried flowers later.

Drying Techniques

The process of drying has a great effect on the quality and worth of the final product. In Kashmir; saffron is dried by two methods. In the first method the stigma and style are separated from the other parts of the flowers and are then placed directly under sunlight. They are left in the sunlight till the moisture evaporates, and only 10-12 per cent water content remains in the stigma. In accordance with the temperature, it takes 3-5 days to dry the product. In the second method the whole flower is dried under direct sunlight instead of drying only the stigma and style. Subsequently stigma is separated from the flower by hand (Ghani, 2001; Mir, 2001; Molafilabi, 1994).

Flower Characteristics

There are many genetically and environmental factors affecting flower size and ratio of different parts of the flowers. The main objective of plants breeders and growers is to increase the ratio of stigma in the saffron flower. Environmental factors such as physical and chemical properties of the soil time of harvesting the flowers, age of the corms and cultural practices can influence the size of the flowers stigma inside it (Kafi *et al.*, 2006; Mir, 1992). The results of an experiment showed the varied age of the fields had no effect on the number of flowers in a unit weight. On average in each kilogram of fresh saffron flowers in Khorasnan (Iran), there were 2,173 flowers (with the coefficient of variation =9.73 per cent). Table 15.4 shows the weight of the parts of the flowers on the basis when they were fresh or dry. (Molafilabi, 1994)

Table 15.4: District-wise Distribution of Area, Production and Productivity

District	Area (ha.)	Production (qtl.)	Productivity (kg/ha)
Srinagar	210	5.18	2.47
Budgam	439	10.93	2.49
Anantnag	27	0.73	2.67
Pulwama	4987	141.13	2.83
Doda	44	1.55	3.52
Total	5707	159.52	2.80

For producing one kg of dry stigma (superior saffron) 105.4 kg of fresh flowers are required; which is equal to about 230,000flowers. For producing one kg of dry stigma and style (standard saffron) 78.5 kg of fresh flowers are required, which is equal to 17,000 flowers. By taking these results into consideration it becomes quite clear that in 100 g standard saffron of Khorason (Iran) there are 25.5 g of style and 74.5 g of stigma (Kafi, *et al.*, 2006; Mir, 2001; Molafilabi, 1994).

The results for checking the numbers of flowers per kg in three different regions of Kashmir valley showed that the number of flowers per kg varied between 2,680 and 3,840, which indicates that the

Table 15.5: Comparison of Environmental and Managemental Practices for Saffron in Iran (Khorason) and India (Kashmir)

Condition or Practices	*Kashmir*	*Khorasan*
Altitude(m)	>1,500	>1,000
Relative humidity (per cent)	60-85	40-50
Annual precipitation (mm)	600-800	120-210
Type of soil	Lacustrine deposits (Karewas)	Variable soils
Soil organic content (per cent)	1	<1
No. of corms/hill sown	Single corms	3-15
No. of corms/m^2 sown	40-50	150-250
Interspacing (cm)	10-20	25-30
Sowing depth (cm)	10-15	15-20
Corm planting rate (t/ha)	5-7	10-12
Maximum age of the field (years)	5 to 20 years	7-12
Irrigation (No. of times)	Rain, no irrigation	At least 5
Flower yield in first year (kg/ha)	Almost negligible	10-20
Flower yield in second year (kg/ha)	10-20	30-40
Flower yield in third year (kg/ha)	40-50	120-170
Flower yield, third to eight year (kg/ha)	80-150	150-250
Average flower yield (kg/ha/year)	100	175
Pests	Mice	Mice and other rodents
Duration of flowering	20th Oct. -10th Nov.	20th Oct. -10th Nov.
Interval of flower picking (once/day)	3-4	Picked everyday
Time of flower picking in the day	9 am-6 pm	6 am- 10 am
Flower transportation	In baskets	In bags
Mean yield (kg/ha)	3.2	4.7
Marketable part	Stigma	Stigma, fresh flower
Number of flower/kg	2,700-3,000	2,000-2,500
Moisture content at harvesting (per cent)	70	80
Final moisture content (per cent)	Approx.8-10	Approx.7
Labour cost/day (US $)	2-3	3-5
Local price (US$/kg)	580	350
Quality control	No control by Govt. or associations	Controlled by Govt. and associations
Soil utilization after saffron	Vacant or some crops like mustard, Pulses	Wheat, Barley
Method of drying	Direct sunlight	Shade, oven
Five main destinations for export	Spain, France, The USA, The UK, UAE	UAE, Spain, Italy, France, Switzerland

Source: Kafi and Showket, 2010.

saffron flowers of Kashmir are smaller than Khorasan (Iran). There may be two reasons behind these results; firstly the tails of the picked flowers in Khorasan (Iran) are longer, which make the flowers heavier. Secondly, the moisture content of Kashmiri saffron is lower than that of Khorasan (Iran), because the flowers after emergence are kept for few days in the field and the flowers loser their water content thereby becoming lighter in weight, whereas in Khorasan (Iran), the flowers are picked early in the morning, everyday and, therefore, the relative water content of the flowers is high. Comparing the results of Table 15.5 shows that khorasani saffron contains 80 per cent moisture while Kashmiri saffron contains 68.5 per cent moisture. The tests that have been carried out on saffron in Kashmir show that the parts of fresh saffron flowers contain 8 per cent stigma, 2 per cent style, 80 per cent sepals and petals,8 per cent stamen and 2 per cent residue while in the data of Table 15.4, Khorasan (Iran) saffron contains about 3 per cent style (Mir, 1992; Mir, 2001).

Yield

A hectare crop produce during the four year 6-8 kg saffron (dried), or 1.5-2.5 Kg saffron (dried) per annum. Yield during the first year of planting is the lowest.

Grading

The finest quality of saffron "Shahi zafran" is obtained from the red tips of stigma. The remaining part of stigmas and tops of styles constitute saffron of inferior quality and is called 'Mogra'. The ISI recognizes two grades of saffron 'the special' and 'the standard'. (Munshi, 1990). The grading is done according to colour, floral waste content and foreign matter as given in Table 15.6.

Table 15.6

Commercial Grade	*ISI Grade*	*Colour*	*Floral Waste (per cent) Maximum*	*Foreign Matter (per cent) Maximum*
Mongra	Special	Deep red	5	0.5
Lacha	Standard	Light reddish	10	1.0
Guchi	Standard	Bright red	–	–

Pest Management in Saffron

No specific pest responsible to attack saffron except field rats that damages saffron corms. Rat's daily consumed food equal to about 109 times of there body weight. They damaged about 20 times the amount the actually consume by their growing activity. The species of field rats most widely distributed throughout the country which causes damages to crops are *Bandicota bengalensis, Rattus meltada* and *Musa booduga*

Control Measure

Fumigation of Burrows

Close all burrows in the field on first day and mark them. Identity reopened burrows next day and put down tablets of aluminum phosphide (celphos) in each burrow and close them with mud. Recheck the burrow on third day and if any reopened burrow is found, fumigate in the same manner as indicated above.

Baiting

Smear 1kg of crushed grains of wheat, maize, bajra or flour with sarson oil and mixed it with 25 g zinc phosphide. If flour is used, prepare small pellets of gram size. Put 5 to 6 pellets or 1tablespoon of grains in the burrow. These methods of killing rats are effective only when carried out on a large scale covering large contiguous areas.

Diseases and their Management

The difficulties saffron growers have to face normally in the state is not only that the production is less than the investment, but the diseases are also a big problem. Scientifically, this is critical to make proper arrangement for proper diagnosis of several diseases found in saffron. Several diseases are common in saffron such as Corm Rot, Root Rot, Dry Rot, Bacteria Rot, Charcoal Rot, Mosaic etc. (Sud *et al.*, 1999; Mir and Devi, 2003) but corm rot disease is severely found in saffron crop.

Corm Rot

Corm rot has become a major constraint in saffron production. It has been reported to be caused by *Fusarium oxysporum* and *Macrophomina phaseolina*. The disease is characterized by varying degree of stunting, yellowing with reduced number of corms and thereby affecting the yield. The disease can be managed by growing healthy corms. Crop rotation may be followed because the pathogens involved are soil borne. Management of corm rot in saffron, corms must be dipped in 0.2 per cent suspension of bavisitin 50 WP for 30 minutes at the time of sowing or Seed corms may be treated with carbendazim (500WP) @ 0.1 per cent + copper oxychloride (50WP) @ 0.3 per cent before planting in the field (Mir and Devi, 2003; Sedeghi, 1983).

Status

In Jammu and Kashmir around 5700 hectare area is under cultivation of saffron with maximum concentration in District Pulwama (87.0 per cent). However, other areas specially elevated Karewa belt has also been brought under its cultivation in the recent past. Its cultivation has been extended even to District Doda (Berwar in Kishtwar). Approximate annual income to the state through saffron is about 43 crore rupees. There has been a reduction in the production of saffron during the last couple of years. This is being attributed to severe draught, which drastically reduces saffron yield. However, depletion of nutrient status of soil, which are being used without any further enrichment, and non-replacement of seed corms are some of the major causes of low production.

The productivity levels of 2.83 kg/ha is lowest in the world (6-8 kg/ha in Italy and France), suggesting a great scope for vertical expansion of the yield in J&K state (Munshi, 1990). While there is huge domestic demand for saffron, sufficient quantity is also exported to other countries.

Production Constraints

- ☆ Inadequate availability of large sized seed corms
- ☆ Proper crop management
- ☆ Low nutrition of the crop
- ☆ Incidence of corm rot disease
- ☆ Inadequate post harvest handling, processing and marketing
- ☆ Lack of quality standards
- ☆ Inadequate R&D support
- ☆ Non-availability of standard high yielding varieties.

Strategies

Suggestion for the Promotion

- ✩ Supply of high yield varieties seed material.
- ✩ Increase in area and production.
- ✩ Introduce improved post harvest technology.
- ✩ Timely harvest to maximize the yield as it is a sensitive crop.
- ✩ Encourage the use of solar/air driers foe drying saffron.
- ✩ Extending financial assistance for purchase of solar drier/air drier.
- ✩ Educate on better and International packaging requirements.
- ✩ Extending financial assistance for packaging material.
- ✩ Promotional measures for export of saffron-Organizing Buyer-Seller meets in India and delegation of farmers and exporters to other saffron producing countries to understand the requirements of the importers and established better production methods.

Spice Board Strategies and Programmes

- ✩ Spice Board had extended financial assistance of Rs. 1,00,000 to SK University of Agricultural Sciences and Technology, Srinagar for development of corms through Demonstration Plots during 1994-95. Amount was released during September 1994. The amount was not fully spent and there was a balance around of Rs. 27,000.00
- ✩ Spice Board had extended financial assistance to Himachal Pradesh Krishi Vishvavidyalaya, Palampur for Micro Propagation of saffron during 1994-1997.
- ✩ Spice Board had requested SK University of Agricultural Sciences and Technology, Srinagar to continue the programme with a financial outlay of Rs. 1.60 lakh during 1996-97. It is understood that detail of Demonstrations operated are yet to be received. The amount was released on 16.5.1997.
- ✩ Spice Board and Department of Biotechnology have given financial assistance of Rs. 44.72 lakhs to Himachal Pradesh Krishi Vishvavidyalaya, Palampur and Indian Institute of Integrative Medicine (IIIM), Jammu for "Refinement of Micro propagation for Saffron" for commercial use during 1998-2001. Spice Board has released its first installment payment of Rs. 6.61 to Himachal Pradesh Krishi Vishvavidyalaya, Palampur.

Improvement of Local Saffron Cultivation

Selection of Corm

The size of corm used for planting affect flower production. Small sized corms generally tend to produce vegetative growth only. The corms selected for planting should be 2.0 or more in diameter. The seed corms should be free from diseased and cuts or other damages.

Improvement through Tissue Culture

Vegetatively propagated crops can be propagated successfully through tissue culture techniques. Healthy saffron plantation can also be produced through tissue culture. Such plantations are disease free and true to type. Millions of saffron plantlets can be produced in minimum possible time through tissue culture.

Standardization of Management Practices

- ☆ Select healthy and disease free corms having suitable size.
- ☆ Chemical fertilizer may be applied after getting the soil tested both for macro- and micro-nutrients.
- ☆ Introduction of mechanical practices. To avoid huge labour, improvement is needed for sowing of corms through machine. This will help to get more and more area under cultivation.
- ☆ Weeding and hoeing by conventional method requires improvement Mechanization needs to be adopted.
- ☆ The conventional method of sowing corms in Kishtwar is by placement of corms in furrows. This requires improvement in placement; corms can be planted on ridges. This will make weeding/hoeing easier.

References

Anonymous (1988). "Saffron crop" Published by S.K. University of Agricultural Sciences and Technology, Kashmir.

Ghani, M. R. D. (2001). Nutrient status of saffron soil and their management. Proceedings- National Seminar cum workshop, Srinagar, India. pp. 51-54.

Ghani, M.Y. (2001). Saffron diseases and their management. Proceedings -National Seminar cum workshop, Srinagar, India. pp. 107-112.

Ingram, J.S.(1984). Saffron (*Cocus sativus* L.). *Tropical Science*. **11:** 1771-1774

Ingram, J.S. (1969). Saffron (*Cocus sativus* L.). *Tropical Science*. **10 (3):** 177-184.

Kafi, M. and Showket, T. (2010). A Comparative study of Saffron agronomy and production system of Khorasan (Iran) and Kashmir (India). Annual report, Ferdowsi University of Mashhad, Mashhad, Iran.

Kafi, M., Koocheki, A., Rashed Mohassel, M.H., and Nassiri, M. (2006). Saffron, Production and Processing. Science Publisher, New Hampshire, USA.

Kalhana (12th Century). Rajtarangini. An Ancient Conical of Kashmir.

Kanan, S. (2001). Current status of saffron production and export in world. Proceedings -National Seminar cum workshop, Srinagar, India. pp. 67-73.

Masoodi, M.A. (2005). Agriculture in Jammu and Kashmir, pp. 144-151.

Mir, G.H. and Devi, L.S. 2003. Corm rot disease of saffron in Kashmir valley. *Indian Phytopathology*, **56 (1):** 122.

Mir, G.M. (1992). Saffron Agronomy in Kashmir: A Study in Habitate, Economy and Society. Gulshan Publisher Srinagar. (J&K).

Mir, M. A. (2001). Post-harvest handling and processing of saffron. Proceedings- National Seminar cum workshop, Srinagar, India. pp. 75-82.

Molafilabi, A. (1994). Study of the flower component of saffron – Proceeding of second National Congress on Saffron and Medicinal Plants. Scientific and Industrial Research Organization of Iran, Khorasan Institute (Persian).

Munshi, A.M. (1990). Marketing and trade of saffron in India. *A review of J. Agri. Marketing*. **33(1):** 27-31.

Munshi, A.M. (1990). Production and development of saffron in India. *J. Farmer Parliament*. **4(7):** 6-8.

Munshi, A.M. and Zargar, G.H. (1989). Saffron the golden condiment. *J. Ind. Farmer Times*. **7(5):** 15-16

Munshi, A.M., Sindhu, J.S. and Baba, G.H. 1989. Improved cultivation practices for saffron (*Cocus sativus* L.). *J. Ind. Farming*. **39 (3):** 27-30.

Munshi, A.M., Wani, S.A. and Tak, G.M. (2001). Improved cultivation practices for saffron. Proceedings National Seminar cum workshop, Srinagar, India. Pp. 59-64.

Negbi, M. (1999). Saffron Cultivation: Past, Present and Future Prospects, **In:** M. Negbi (ed), Saffron. Harwood Acadmic Publishers, Amsterdam.

Negbi, M., Dagan, B., Dror, A. and Basker, D. (1989). Growth, flowering, vegetative reproduction and dormancy in the saffron (*Cocus sativus* L.). *Israel Journal of Botany*.**38:** 95-113

Nehvi, F.A., Wani, S.A., Dar, S.A., Makhdoomi, M.I., Allie, B.A. and Mir, Z.A. (2007). Biological interventions for enhancing saffron productivity in Kashmir, *Acta Hort. (ISHS)*, 739, pp. 25-32.

Rees, A.R. (1988). Saffron- An expensive plant product. *Plantsman*. **9 (4):** 210-217

Sadeghi, B. (1983). Effect of corm weight on flower initiation of saffron. Scientific and Industrial Research Organization of Iran. Khorasan Institute annual report.

Sud, A.K., Pual, Y.S. and Thakur, B.R.(1999). Corm rot of saffron and its management. *Journal of Mycology and Plant Pathology*, **29 (3):** 380-382.

Zargar, G.H.(2001). Genetic variation in Saffron and Importance of quality seed corm. Proceeding-national seminar cum workshop on the development of Saffron, Srinagar. India-pp. 25-36.

Zhou, Q., Sun, y. and Zhang, X. (1987). Saffron, (*Cocus sativus* L.) *J. Traditional Chinease Med*. **28:** 59-61.

Hill Agriculture: Prospects, Constraints and Mitigations (2013) *Pages* **226–235**
Editor: **Dr. Shahid Ahamad**
Published by: **DAYA PUBLISHING HOUSE, NEW DELHI**

Chapter 16

Saffron Cultivation in Jammu and Kashmir: Problems and Mitigation Strategies

M.A. Khan[1]*, F.A. Nehvi[1], Ajaz A. Lone[1], M.H. Chesti[2], Kamaluddin[3] and P.A. Sofi[2]

[1]Saffron Research Sub-Station, K.D. Research Station, Post Box no.905, SKUAST-K, Shalimar, Srinagar, Jammu and Kashmir
[2]Regional Agricultural Research Station, SKUAST-J, Rajouri, Jammu and Kashmir
[3]Department of Plant Breeding and Genetics, Faculty of Agriculture and Regional Research Station, Wadura, Sopore – 193 201

Introduction

Saffron (*Crocus sativus* L.) is highly valued as a culinary spice for its flavouring and colouring properties (Rios *et al.*, 1996), and is also the subject of ongoing scientific research for its potential medicinal properties. It blooms in autumn and spends a long period of dormancy (aestivation) in the summer. Interest in the impact of saffron carotenoids on human health is also growing due to their high antioxidant capacity (Abdullaev, 2002; Pham *et al.*, 2000; Verma and Bordia, 1998). Saffron (*Crocus sativus*.L) is the legendary crop of Jammu and Kashmir state, belongs to the family irridaceae and genus crocus, of which about 80 species are so far known. Saffron is a product of high commercial value and can be of prime significance in upliftment of rural economy. It is the most expensive spice world over and is produced by drying the long red stigmas of saffron. Kashmir's saffron is known world over as the best saffron due to its intrinsic quality characters, primarily because of its strong colour and flavour as a result of high crocin and safranal content. However, the product even if directly bought from farmers field and made available in the market generally does not conform to the

* Corresponding Author: E-mail: anwarkhan143@gmail.com

norms, mainly due to poor post harvest handling practices being followed by the farmers in the State and lack of proper infrastructural facilities resulting in sharp degeneration in the sensitive quality characteristics.

Area, Production and Productivity

In Jammu and Kashmir state saffron cultivation is presently confined to districts of Pulwama, Budgam, Srinagar and Doda of the State. In Pulwama district, Khrew, Ladoo, Dussu, Sambora, Awantipora and Koil are the major saffron growing areas, accounting for about 75 per cent of the total area under saffron in the state. In district Budgam, the important saffron growing areas include Nagam, Sarwin, Hapatnar, Goplapora, Hyathpora and Chirar-i-Sharief that account for 16.13 per cent of the total area, while in district Srinagar, Zeewan, Khunmoo, Balhama, Sampora and Yachnambal are the major pockets accounting for 6.68 per cent. In Jammu province of the State, saffron is grown only in district Doda in Poochal, Namil, Cherrad, Hullar, Blasia, Gatha, Bandakoota, and Sangrambatta areas, accounting for only 2.5 per cent of the total area in the whole state. More than 16,000 farm families are engaged in saffron cultivation located in 226 villages with 61 per cent of holdings below 0.5 ha, 26 per cent size of holdings between 0.5 -1.0 and 13 per cent having holding size above 1.0 ha.

Table 16.1: Trends in Area, Production and Productivity of Saffron in J&K State

Year	Area (ha)	Production (MT)	Yield (kg/ha)
1996-97	5707	15.95	2.80
1998-99	4116	12.88	3.13
1999-00	3997	7.65	1.89
2000-01	2831	3.59	1.27
2001-02	2713	0.30	0.095
2002-03	2825	6.50	2.28
2003-04	2742	5.15	1.88
2004-05	3143	6.86	2.23
2009-2010	3785	9.46	2.50

Source: Deputy Director, Planning Department, J&K, and Directorates of Agriculture, Jammu and Kashmir Divisions.

The area declined by 52.5 per cent in 2001-02 as compared to that in 1996-97 (Table 16.1). Thereafter it showed slight increase and reached to 3,143 ha (2004-05) and 3,785 ha (2009-2010), which is still 33 per cent less than that in 1996-97 (Table 16.1). The production suffered badly all these years due to several factors. The year 2001-02 was the worst with as low a production as 300 kg for the whole State. This was due to the acute drought situation faced by the State during the summer that year, causing severe soil moisture stress leading to poor plant growth and flowering. This caused a major set back to the prospects of reviving saffron cultivation in the State. The traditional area saw a steady disinterest amongst farmers over the years, leading to their abandoning saffron cultivation, and leaving the land fallow. Consequently, an area of 2882 ha previously under this crop was lying fallow in 2002-03. One redeeming feature, however, observed during the eighties was the gradual extension of saffron cultivation in the non-traditional areas in Kashmir valley, when farmers of Srinagar and Budgam districts started cultivating this crop on a small scale, thus breaking an age-old concept that saffron could be grown only in Karewa lands of 'Pampore' in district Pulwama. This should provide a major platform for working out a long term strategy for area expansion under saffron in the State in the coming years. Recently, much efforts have been made under Projects; A value chain on Kashmir Saffron -NAIP Component-II (2009-2012) and "J&K Saffron Mega Project" Govt. of India (2010-2014) to revolutionize the saffron industry of the state and enhance the productivity level of this precious crop.

Major Constraints Facing Saffron Production in J&K State

Saffron Corms and their Planting

The saffron plant is a triploid, which fails to set seeds, and thus is propagated vegetatively through the use of corms produced by each plant. Indiscriminate use of corms for new planting, irrespective of size or weight, is one of the major factors leading to low plant population and hence low productivity. Khan *et al.* (2011) revealed that initial corm weight affects the production of daughter corms in saffron. No clonal selection has ever been attempted. Hence no single clone of specific characteristics has been in use. The clonal fidelity is grossly compromised by repeated use of the same seed material year after year, leading to the possibilities of somatic mutations occurring without check. Definite evidence is available to suggest genetic variability existing among temporal pollinations being used in saffron growing areas of J&K. Therefore, the need for making clonal selection regularly and multiplying the identified clones for use as seed is an obvious need.

Importance of Planting Cycle

The farmers use very low seed rate which is at least 3 times lesser than the recommended dose required for an optimum plant population. Non-availability of seed corms for new planting in sufficient quantities is another major impediment in improving productivity. Seed corms are not available due to very low rate of multiplication of corms into daughter corms and the time taken by the latter to attain required size for planting. Consequently, the mother corms, once planted are retained in the fields for many years, allowing these to produce daughter corms which continue the production cycle without interruption, though at the cost of declining productivity. The planting cycles are generally of 10 to 12 years duration. Even when seed corms become available when some farmers dig out their fields for new planting, these are sold @ Rs.20,000 to 40,000 per MT, which is quite a high price for small and marginal farmers.

Poor Soil Health

The practice of growing saffron year after year under no supplementation of nutrients through this period has drastically reduced the fertility of the soils in the saffron fields. The organic carbon of these soils has gone down. So have other major nutrients except potassium. Consequently, the size and vigour of the corms produced every season is reduced, directly affecting the crop stand and flowering potential of plants.

Inadequate Moisture

The crop is totally rain-fed, unlike in Iran and Spain where it is irrigated regularly. The State used to receive 600 to 1000 mm rains every year, part of which would be during August to mid-October and then in November, the two critical stages for normal flowering and good crop stand next year. However, in the last several years the weather has become quite erratic. Rains are either scanty or irregular, thus adversely affecting flowering and subsequent plant stand. In 2001-02, the State faced an acute drought situation, with rains having completely failed, thus resulting in the lowest ever recorded saffron production of 0.30 MT for the whole year.

Incidence and Spread of Pests and Diseases

The major biotic stress being faced by saffron since last several years is the corm rot fungal infection, a soil borne disease caused by *Fusarium monliforme* var *intermedium* and an unidentified mycelium of *Basidiomycetous* fungus. The infection generally takes place through an injury in the corms. Disease corms show brown to dark brown sunken and irregular patches below the corm scales, usually located in root and bud regions. In severe cases the entire corm turns into a black powdery mass. Diseased corms produce foliage resembling 'Die-back' like symptoms (Nehvi, 2003). Studies done in Sher-e-Kashmir University of Agricultural Sciences and Technology, Kashmir (SKUAST-K) have shown that over 46 per cent of the soils in the traditional saffron fields are infected with the fungus, and the infection is spreading steadily. Saffron belts of Kashmir are also infested with plant parasitic nematodes, namely *Helicotylenchus* sp; *Tylenchorynchus* sp; *Pratylenchus* sp; *Xiphinema* sp etc. as well as free living mites and thrips. Nematodes inflict direct damage by sucking sap and causing necrosis in roots, stunting, yellowing and reducing the yield. These also predispose saffron corms to corm rotting fungi and enhance infection of corm rot disease, causing heavy production losses. However, no systematic control measures are presently being adopted by the farmers. Rodents and field rats also pose a serious problem.

Un-Scientific Harvesting and Handling Practices

Fresh stigma taken from the Kashmir saffron possess a high percentage of crocin (16.5 per cent to 17 per cent) on moisture free basis, imparting it a superior quality as compared to the material from other countries. However, the post harvest handling practices followed by the farmers for harvesting flowers, separating stigma from flowers, drying, packing, storing and marketing brings down the crocin pigment. An analysis of the farmers' processed samples indicates that the content ranges from 9 to 11.5 per cent. Delay in flower picking from the date of flower opening and the delay in separation of stigma from the flower contribute to the drop in crocin content. Studies conducted by SKUAST (K) have shown that flowers picked on 4th day of emergence give maximum recovery (weight) of stigmas and pistils, called Laccha (0.760g/100 flowers) and crocin content (12 per cent to 13 per cent). Similarly, if the separation of stigmas is completed within 6 to 8 hours after picking, more than 95 per cent of the saffron is recovered, while 24 hours after picking, only 40 to 50 per cent recovery is possible. 72 hours after picking, the entire saffron flower becomes a cake-like mass with 0 per cent recovery (Nehvi *et al.*, 2004).

Poor Post-Harvest Handling of the Crop

Traditional drying practice being followed by the farmers takes about 27 to 52 hours in sun or shade. This leaves the product with a moisture level of nearly 11 per cent as against 8 to 9 per cent required. Such a long drying period results in enzymatic degradation of crocin, thus affecting the quality of the end product. Studies done at the Regional Research Laboratory (RRL), Jammu and SKUAST (K) have shown that "drying at temperature between 35° and 50° C took only 4 to 6 hours with the pigment concentration very close to that found in fresh saffron on moisture-free basis, besides providing a very good texture, colour and bright appearance to the product". It has, therefore, been concluded that the ideal temperature for drying would be 40°C ± 5°C (Nehvi *et al.*, 2005). However, dried saffron is vulnerable to moisture ingress, light and air. Once exposed to these conditions, which often is the case under Indian conditions, hydrolysis of crocin into crocetin sets in, thus reducing the colouring power. Packing of dried product is normally done in polybags and stored under room temperature by a majority of the farmers. This packing is retained till the product is sold to the wholesale traders, which can be between 1 to 6 months and sometimes even a year after drying and packing.

Studies done in SKUAST (K) have shown that saffron packed in 10 gauge polythene bags and stored at ambient temperature in dark over a period of one year loses 56.9 to 70.2 per cent of its crocin content (Mir, 2002).

Quality Parameters

It is well known that good quality saffron, characterized by high crocin and safranal contents, fetches the best price in the international market. Unfortunately, Kashmir saffron suffers severely on quality consideration because of the poor post-harvest handling practices being followed.

Saffron Adulteration

Adulteration in Kashmir saffron is rampant and a most serious malpractice, making it a major constraint in reviving saffron cultivation in the State. It is a common sight to find imported Iranian saffron being mixed with Kashmir saffron and sold as Kashmir saffron to extract a higher price from the consumers/buyers. The traders of the State also allege that students from Iran coming to India for higher studies were allowed to carry 5 kg of saffron each from their home country in their baggage, but not declared to the Indian Customs. The students then sell the saffron to unscrupulous traders in India who use it for adulteration. The students then use the money earned for pursuing their studies. Consequently, the exact figures of import of saffron into India are not known as most of the imports are clandestine in nature. Mixing of pieces of filaments, stamens or petals is also quite common. Saffron samples having only filaments and stamens are marketed as 'white saffron'.

Spurious Saffron

Spurious saffron is available in the open market and sold as genuine stuff to ignorant tourists and pilgrims visiting the State year after year. A study reported by SKUAST (K) showed that out of the samples collected from open market, 52 per cent were genuine saffron, 30 per cent poor grade and 17 per cent totally adulterated. The adulterants detected were tetrazine-dyed starchy fibrous material (probably wheat flour and coconut fruit shell), dyed saffron stamens, and dyed tender adventitious roots of *Salix*. Other adulterants reported are ray florets of marigold, safflower dyed with coal tar dyes *viz.* sunset yellow or matanil yellow; corn silk (stigma of maize), fibres of shredded meat coloured with saffron water, fibrous roots of various grasses, slender roots of willow and coloured nylon fibre. Fats, oils and glycerin are sometimes used to increase the weight of saffron (Mir, 2002).

Quality Standards

As mentioned earlier, saffron produced in Kashmir valley is graded as 'Special' also called 'Mongra' (containing only dried stigmas) and 'Standard' also called 'Laccha' (stigma mixed with style), while the one from Doda district of Jammu Province is known as'Guchhi', equivalent to Laccha, packed in bundles made from 120 to 150 flowers.

At present, there is no mechanism in the country to enforce adoption of quality standards and fixing the price as per the standards each sample of saffron conforms to. There are no state-owned or operated quality evaluation laboratories to carry out regular evaluation and certification of the products being marketed within and outside the country. The Food and Adulteration Act of India (1949), however, lays down specifications (as per the ISI standards) to act as a deterrent against adulteration, which the State could use to nab the adulterators and reduce the menace to save the industry. It has been used very sparingly for saffron, and hence adulteration as a market practice goes unchecked.

Marketing Practices

Saffron marketing in the State and the country as a whole is highly unorganized. It is largely in the hands of private brokers, with a long chain of middlemen linking the grower to the consumer. The main marketing channels are brokers (also called '*Dalals*'), local traders, agents, co-operative societies, government agencies and companies. (Munshi, 2002).A survey of saffron families carried out in Kashmir revealed that 70 per cent of them comprising farmers of large, medium and small land holdings, sell their produce to the brokers (*Dalal*) and 25.71 per cent to sub firms, while only 1.43 per cent sell directly to the wholesalers in Delhi, Amritsar, Mumbai, Kolkata etc., 0.95 per cent sell to the cooperative societies, and 1.91 per cent to the companies. Since the broker is the mainstay of the marketing channel today, there is rampant exploitation of farmers mainly due to the latter's ignorance about the price prevailing in the major trading centres. It was alleged that in some years the unscrupulous traders had even gone to the extent of spreading rumours through the print media about a bumper crop expected in the valley, to force the market prices to crash, and use this ploy to pay a much lower price to the farmers. Interactions held by the author with the farmers during his visit to the saffron villages also confirmed the feeling of being exploited prevailing across the length and breadth of the saffron growing areas. They are prepared to continue growing saffron in the years ahead provided they get a better price for their produce. Ensuring that the farmers receive an optimum share of the market price will be critical to any strategy to be evolved for improving saffron production and productivity in the State, as this enables the farmers to make the required investment in adopting improved technologies, and produce good quality products.

Governmental Support

Saffron development as an agricultural crop falls under the administrative jurisdiction of the Department of Agriculture, Kashmir for the valley area and the Department of Agriculture, Jammu for Jammu division, both operating under separate Directors. Saffron is basically a spice crop, and rightly should be with the two Departments of Horticulture in Jammu and Kashmir. There is no record of any Plan schemes either of the State or Central Governments ever proposed or implemented as a part of any larger scheme for this crop for many decades. In other words, this crop has received no fiscal support all these years despite the crop being indigenous to the State and having a strong cultural influence on the daily life of its people, besides being a good foreign exchange earner. Consequently, the farmers are not able to obtain even the basic required inputs such as fertilizers, plant protection chemicals etc. on subsidized rates as is available to major food and commercial crops of the State. Placement of this crop with the two Agriculture Departments and not with the Departments of Horticulture has deprived it from being considered for support under the most ambitious Central Government scheme of 'Technology Mission for Integrated Development of Horticulture', operating since 2003-04 and meant exclusively for J&K, Himachal Pradesh, Uttarakhand and the 8 states of the North Eastern Region. This scheme includes all the horticultural crops including spices for integrated development covering improving production of seed/planting material, productivity, quality and post-harvest technologies including marketing. Unfortunately, saffron has not been included by the State Government as a priority crop for receiving required support under this scheme.

Research Support

The SKUAST (K) has the primary responsibility of working on this crop to develop location-specific improved packages of practices and evolve long term solutions for the production and post-harvesting problems through a multi-disciplinary approach with the objective of improving its productivity and quality. The SKUAST (K) had taken the required initiative as early as in 1987-88 when it set up a Saffron Research Station at Konibal, in Pampore Tehsil, District Pulwama, right in the heart of the major and traditional saffron belt. However, the station did not function, and remained in limbo till 1997-98 when the work on saffron was taken up on a planned basis. The Station has since done commendable work in close association with the farmers. It has evolved improved soil and crop management and post-harvest technologies, laid On-Farm Trials (OFT) and front-line demonstrations in the farmers' fields besides training selected farmers. However, its operations are severely handicapped due to the lack of availability of sufficient land area at the Station,. This has prevented the University from undertaking extensive field trials and laying out demonstrations for the farmers visiting the Station. No infrastructure facilities are available at the Station for organizing regular training programmes for the extension functionaries and farmers, essential for quicker dissemination of improved technologies. The research efforts are further impeded by lack of regular interactions among the four institutions, namely SKUAST (K), SKUAST (J), CITH, Srinagar and RRL, Jammu to exchange information and research material, undertake collaborative work and avoid overlaps. Hence, the benefit of the work being done is not reaching the farmers in the form of well tested package of practices.

Extension Support

The crop has not had any exclusive extension support from the line Department, or from any NGO or private institution to help the farmer. The Agricultural Extension Officers posted in the four saffron districts are preoccupied with major food and fodder crops being grown in these districts, and have not been detailed for any such work for saffron. Unless there is a dedicated team of extension agents working right in the villages to provide on-the-spot help in adopting improved practices and distribution of inputs, besides providing a link with the research team, improvement in productivity would be difficult to achieve.

Institutional Support

No bank credit facilities are available to the saffron farmers, probably due to low productivity, price fluctuations and risks involved due to the crop's vulnerability to weather fluctuations. Instances of this happened in 2001-02 due to acute drought and the current year (2006) due to sudden fall in day temperature in mid-October just before flowering. Hence, there are no avenues available to the farmers for making investments required for adopting improved technologies. Similarly, institutional support in form of any growers' association or NGO is completely absent, except for the All J&K Saffron Growers Association, Pampore (comprising mostly of saffron traders) which has been in existence since last 20 years. The impact of this association on improving the saffron farmers' lot was not visible or directly ascertainable. This Association has mostly been engaged in trying to protect the interests of the traders and has presented a demand for setting up of a Saffron Market Centre, called a 'Saffron Mandi' in Pampore to the State Government.

The constraints discussed above are all contributing to a large extent to the current state of lowest productivity in the State among all saffron growing countries, besides suffering due to low market quality, thus making saffron cultivation a losing enterprise. During 2004, the average output value per ha, as per reports of SKUAST (K), had been of the order of Rs.24,087.22 against the total input cost of Rs.35067.68, resulting in an input : output (I/O) ratio of 1:0.69, as against ratio of 1:2.49 reported through a study made during *Kharif*, 1998 (Nehvi, *et al.*, 2006). Low I/O ratio of 1:0.69 of recent years clearly suggested that the farmers had suffered over 30 per cent loss on their investment in saffron cultivation in the recent years, which is too alarming to be ignored. Unless these constraints are tackled in a systematic manner, exploiting fully the available improved technologies and backed by infrastructural and policy support, revival of saffron in the State may remain a dream only.

Mitigation Strategies

The question now arises is whether saffron cultivation can be revived in the State, in the face of the constraints it is facing. The above analysis provides a clear answer with an element of firm conviction that the challenges facing the crop are possible to be met and saffron production made quite remunerative, provided a holistic approach is adopted for its development covering all aspects of production and marketing in a Mission-Mode approach. This assumption is based on the fact the State already has a package of improved and user-friendly technologies evolved with proven potential of achieving expected results, backed by political will and commitment of the State government towards the crop. This approach was successfully adopted for the 'Technology Mission on Oilseeds' in late eighties in the entire country. The strategy being proposed would, therefore, be based on the tested modalities used in the Oilseeds Project, which will help to define the broad outlines of the project being proposed for saffron in this Report.

Development of saffron in the State would revolve around the strategy to increase the productivity of crop and improve product quality in the area already under the crop, expand the area further first by covering the lands earlier used for saffron and presently lying fallow, and then expanding in the non-traditional belts to be properly identified in the three Divisions of Kashmir, Jammu and Ladakh. Improvement in productivity and quality would become possible through exploiting the benefits of improved package of practices already available for transfer to be adopted immediately, backed by the governmental support for developing sustainable irrigation sources, supply of inputs etc. Multiplication and distribution of seed corms at very reasonable costs will be a key element of the strategy for adoption of improved technologies and covering new areas. It would need to be taken up systematically, involving both private and public sector farms, introducing for the first time 'off-season multiplication' in the sub tropical plains of the State, and setting up of centralized seed procurement, processing and distribution system for seed corms. The potential of the tissue culture technology to make seed corm multiplication faster and thus cater to the rising demand cannot be exploited immediately, as the technology is still not perfected as yet for mass scale multiplication. However, serious efforts would be put in for taking the research in micro-propagation presently in hand to its logical conclusion of evolving a cost effective and sustainable technology for commercial adoption in the State.

The other major thrust would be on improving post-harvest handling practices by launching awareness campaign among the farmers about the benefits of picking flowers at the right stage and separating stigma and style, both to be completed in shortest possible time. At the same time, popularizing use of solar dryers (of RRL) by one and all for achieving optimum moisture content in the dried product, and preserving the quality during storage would be an integral part of this strategy. This will help to empower the farmers to develop their confidence in their product, and demand an appropriate price from the buyer. Quality based pricing of the product would be introduced through adopting ISO grading, quality evaluation and branding, and enforced through governmental intervention under the proposed Act. This will help in promoting organized marketing in saffron, involvement of saffron grower's forums to be constituted, and setting up of market centre or Mandi for saffron in the State.

The above cited thrusts will call for massive efforts for making all the involved farm families and other stakeholders fully aware of the benefits of improved technologies *vis-à-vis* the harmful effects of the practices presently in use. In other words, transfer of improved technologies including imparting training, undertaking demonstrations in the farmers' fields, and using all established extension methods would be the key to the success of the above approach, and need to be taken up on highest priority. No efforts should be spared for this if the total strategy and the investments proposed are expected to succeed and deliver desired results. On a long term basis, adoption of advanced technologies involving mechanization of digging, planting, sorting and grading, separating stigma from flowers and drying, packing and storing etc would be pursued to achieve much better returns at reduced costs and make saffron production a commercially successful agri-business in the State.

Investments in developing infrastructure and other facilities required will have to come from governmental sources and also from the intervention of international bodies for importing advanced equipment, human resources development, contracting experts as consultants in specific areas etc. Equally important would be a clear commitment to put in place an efficient project implementation system including an effective monitoring and evaluation system as will be outlined in the project proposal.

Summary

In Jammu and Kashmir there is a tremendous human resource base as such saffron industry has an ample scope to expand provided saffron in state is cultivated on scientific lines and replantation in the traditional and non-traditional areas is taken up. Keeping in view a strong association of more than 16000 farm families with this crop directly or indirectly in Jammu and Kashmir efforts are to be made to safeguard the interests of growers by making the industry more profit oriented. Lack of availability of high yielding saffron cultivars, high input cost due to non availability of planting material, lack of mechanization, dry land farming, spreading inoculum of saffron corm rot, post-harvest losses and poor market due to involvement of agents and middlemen are the major causes of saffron decline. In order to increase the productivity and the fertility of saffron soils, there is a need to adopt the integrated nutrient supply and management (INSAM) practices. Introduction of post-harvest technology for flower picking, separation of pistil and quick reduction in moisture have potential to bring about improvement over strenuous traditional practices and returns to the growers from the superior quality product. Most of those practices and much more efforts are underway through NAIP component-II project, "A Value Chain on Kashmir Saffron" (2009-2012) and now a First phase of "Mega Project on Saffron in Jammu and Kashmir" (2010-2014) in the state.

References

Abdullaev F. I. (2002). Cancer chemopreventive and tumoricidal properties of saffron (*Crocus sativus*). *Exp. Biol. Med.* 227: 20–25.

Khan M. A, Sabeena Naseer, Shaheena Nagoo, F.A. Nehvi (2011). Behaviour of Saffron (*Crocus sativus* L.) Corms for Daughter Corm Production. *Journal of Phytology* 3: 47-49.

Mir, M. A. (2002). Post-harvest handling and processing of saffron. Proceedings of the Seminar cum Workshop on the Development of Saffron, 2001. pp. 75-82.

Munshi, A. M. (2002). Marketing and trade mechanism of saffron in Jammu and Kashmir; Proceedings of the Seminar cum Workshop on the Development of Saffron, 2001. pp. 81-91.

Nehvi F.A., Wani S.A., Dar S.A., Makhdoomi M.I., Allie B.A., Mir Z.A.(2006). New Emerging Trends on Production Technology of Saffron. *Proceedings of the Second International Symposiuon Saffron Biology and Technology*, p. 401-408.

Nehvi, F. A. Agarwal, S.G., Verma, M.K., Dar, S. A., Mir, Z. A., Nusrat Nabi. and Allie, B. A. (2004). Technological innovations for saffron production. *Proc. National Symposium on Enhancing Productivity and Sustainability in Hill and Mountain Agroecosystem*. 10-12 August. Dehradun Uttarakhand. p. 58-67.

Nehvi, F. A. (2003). Problems and prospects of Saffron improvement in India. Presented in seminar on Industrial use of Biotechnology 27th September –1 October, 2003. *Islamic Republic of Iran.*

Nehvi, F. A., Agarwal, S.G., Mir, M.A., Dar, S.A., Mir, Z.A. and Nusrat Nabi (2005). Quality drying of Saffron. *SKUAST J Res.*, 7(2): 343-346..

Pham TQ, Cormier F, Farnworth E, Tong VH, Van-Calsteren MR (2000). Antioxidant properties of crocin from *Gardenia jasminoides* Ellis and study of the reactions of crocin with linoleic acid and Crocin with oxygen. *J. Agric. Food Chem.*, 48: 1455–1461.

Rios JL, Recio MC, Ginger RM, Manz S (1996). An update review of saffron and its active constituents. *Phytother. Res.*, 10: 189–193.

Verma SK, Bordia A (1998). Antioxidant property of saffron in man. *Indian J. Med. Sci.*, 52(5): 205-207.

Hill Agriculture: Prospects, Constraints and Mitigations (2013) *Pages* **236–250**
Editor: **Dr. Shahid Ahamad**
Published by: **DAYA PUBLISHING HOUSE, NEW DELHI**

Chapter 17

Minor Millets: Cultivation for Nutritional Security in Hills

***Dinesh Tiwari*[1] *and S.M.A. Rahman*[2]**
[1]*Division of Agronomy, GBPUA&T, Pantnagar), Uttarakhand*
[2]*Krishi Vigyan Kendra, Gwaldam-Chamoli, GBPUA&T, Pantnagar, Uttarakhand*

Hill agro-ecosystem supports almost half of the country's population either directly or indirectly. Minor millets supply sufficient amount of food and nutrients for livelihood in hills since decades. Millets crop can be grown easily in poor soil of hills which occurs due to natural degradation. Millets generally have wide adoptability and can survive even in adverse situations. They can be grown in heavy, light, sandy, alkaline and acidic soils; and can withstand stresses due to lack of moisture and temperature fluctuation. These unique qualities have made them the choice crops to rainfed, tribal and hill agriculture, where crops are limited. Minor millets have been instrumental in imparting food and nutritional security to the people living in tribal and hilly areas. Six minor millets *viz.* finger millet, foxtail millet, barnyard millet, proso millet, little millet and kodo millet are cultivated in India. Among them, finger millet, barnyard millet, foxtail millet and proso millet are cultivated as grain as well as feed purpose in hills or dry land agriculture. Minor millets are rich source of minerals, vitamins and fibre. Traditionally they have been used by the hill farmers as these are hardy in nature and produce reasonable yield even under low fertility or drought conditions. Owing to their nutritional superiority, millets are fast becoming popular in urban area and demand of these crops is likely to increase due to increasing awareness of health of modern people. Minor millets are highly suitable for cultivation in low productivity and drought prone areas especially found in hills. Lack of knowledge about good agronomic practices are the major constraints for boost production of minor millet in hilly areas.

Finger Millet (*Elucine coracana*)

Traditionally various food preparations are made from the flour of finger millet like *roti, milayi* (mixed), *akhadi, kodi phana, jaula, halwa* in hills. Finger millet is a (locally known as ragi/mandua/nagli/kapli) small-seeded annual belonging to family Poaceae (Gramineae). This is a warm weather grass belonging to C-4 group of plant and considered as physiologically efficient. This crop is cultivated up to an altitude of 2100 metres above msl and often grown in diverse soils, climates and harsh environments. Finger millet is an important food and feed crop producing more reliable harvest than many other crops in hill areas. Among millets the productivity is highest in finger millet (around 1,500 kg/ha) followed by sorghum (1,000 kg/ha), pearl millet (750 kg/ha) and other millets (500 kg/ha).

Nutritional Value

Finger millet grains are more nutritious and provide more calcium (344 mg/100g), minerals (2.7 g/100g) and Fibre (3.6 g/100g) per unit grain consumed than wheat and rice. Protein content of finger millet is more than that of rice and has a well balanced amino-acids profile. It is a good source of methionine and lysine (Table 17.1) and is also rich in important vitamins such as thiamine, riboflavin, folic acid and niacin. Govind Ballabh Pant University of Agriculture and Technology, Pantnagar has developed protocol for the preparation of *Laddu* and *Dhokla* from finger millet for diabetic and heart patients that can easily be prepared at home.

Table 17.1: Nutritive Values of Finger Millet as Compared to Wheat and Rice

Nutrient	*Finger Millet*	*Wheat*	*Rice*
Protein (g/100g)	7.3	11.8	6.8
Fat (g/100g)	1.3	1.5	0.5
Energy (K cal)	328	346	345
Fibre(g/100g)	3.6	1.2	0.2
Mineral (g/100g)	2.7	1.5	0.7
Calcium (mg/100g)	344	41	10
Lysine (g/100g protein)	3.5	2.9	3.7
Methionine (g/100g protein)	3.4	1.5	2.4
Cystine (g/100g protein)	2.2	2.2	1.4
Tryptophan (g/100g protein)	1.6	1.1	1.4
Isoleucine (g/100g protein)	6.4	3.3	3.9

Soil

Finger millet can be grown on a wide variety of soils ranging from very poor to very fertile soils. It thrives best on well-drained loam or clay loam soils. It tolerates salinity better than other cereals.

Varieties

A number of high-yielding varieties were evolved and released for cultivation in hill area. The improve varieties have high genetic potential and are able to meet the location specific requirements. This includes varieties having high seed and fodder yield with earliness for hills area. The suitable varieties for hilly areas are given in Table 17.2.

Field Preparation

Thoroughly prepared seed bed free from weeds, imparts better germination and crop growth. Ploughing is advantageous for moisture conservation and weed control. In April or May one deep ploughing with mould board plough followed by one or two harrowing is necessary. For improved farming, light weight Power tiller can be used in valley area of hills. Improved clod crusher is beneficial for breaking the clod.

Table 17.2: Improved Varieties of Finger Millet

Variety	*Average Yield (q/ha.)*	*Character*	*Area of Adoption*
V.L.Mandua 146	20-25	Grain: copper colour, Plant height: 85-90 cm., Maturity: 90-100 days (early mature variety), Resistant to blast disease	Rainfed areas of Uttara-khand and Karnataka
V.L. Mandua 149	22-26	Grain: copper-colour, Plant height: 85-90cm., Maturity: 105-110 days, Highly resistant to Blast disease	Rainfed areas except Tamil Nadu and Aandhra Pradesh
VL Mandua-315	20-25	Grain: light copper colour, Plant height: 90-100cm. Maturity period: 105-110 days, Resistant to Blast disease.	Rainfed areas of Uttara-khand
VL Mandua-324	19-22	Grain: very light copper colour, Plant height: 80-95cm. Maturity period: 105-135 days, 20.43 per cent more fodder than VL 146. More resistant than VL Mandua 146 against spike and greeva blast disease.	Rainfed hilly areas of Uttara-khand
Pant Mandua-3	20	Plant height: 90-95 cm., Maturity period: 95 days, Spikelets medium compact and in curved, Colour of grain light copper, Beneficial for Mandua-Wheat cropping system.	Rainfed hilly areas of Uttara-khand alogwith all over India.
PR 202	30	Medium tall, Plant green, panical medium compact and in curved, grain bold and orange brown, Maturity: 110-115 days	All finger millet growing areas
Muskey 5	20-25	Medium duration, compact earhead.	Recommended for North-eastern region
PRM-1	20-25	Plant height: 100-110 cm., Maturity: 110-115 days, panical medium compact and in curved, Resistant to Blast disease.	Rainfed hilly areas of Uttara-khand
PRM-2	20-25	Plant height: 80-85cm., Maturity period: 95 days, Grain: light copper colour, panical medium compact and in curved, Resistant to Blast and Leaf spot disease. Appropriate for Mandua-Wheat cropping system.	Rainfed hilly areas of Uttara-khand

Seed Rate and Sowing

Maintenance of optimum plant population is an important pre-requisite for getting higher yield under rainfed condition. Higher or lower plant density may reduce the yield. Line sowing ensures better germination, cuts down seed requirement and facilitates intercultural operations as compared to broadcast method of sowing. The recommended spacing for finger millet is 20-25 cm x 10-15 cm. Finger millet seeds are very small (400 seeds/g) and the seed rate of 8-10 kg/ha is recommended in case of line sowing. Seed should be treated with Thiram or Carbendazim/Bavistin at the rate of 2-3 g per kg of seed. The optimum sowing time of finger millet is from second fortnight of May to first fortnight of June upto 1000 m altitude and first fortnight of May at > 1000 m altitude in hill areas. Seed should be sown behind the plough at 2 cm depth in light furrows.

Crop Rotation

Mandua grows in hills in different cropping system. The recommended crop rotation for Mandua alongwith leguminous crops are pea, lentil, soybean and other crops like potato. There are many crop rotation followed in hills:

- Finger millet + Soybean – Oat (Green forage)/Wheat (1 Year)
- Finger millet + Soybean - Wheat (1 Year)
- Finger millet - Lentil (1 Year)
- Finger millet - Pea (1 Year)
- Finger millet + Soybean - Oat - Paddy - Wheat (2 Year)
- Finger millet – Potato (1 Year)

Nutrient Management

Balanced nutrient supply in an integrated manner is important for good harvest of finger millet. Organic manure plays a special role in rainfed farming as it not only supply the nutrients but also improves moisture retaining capacity of the soil. In rainfed areas of hills, generally the finger millet crop is grown in organic condition where the availability of manure is also below the requirement. Farmyard manure or compost @ 10 tones per ha should be applied 2-3 weeks prior to sowing. If well rotten it can be applied at the time of last ploughing before sowing. Finger millet responds well to fertilizer application, especially to N, P and K. In rainfed condition of hills the use of 60 kg N, 30 kg P_2O_5 and 20 kg K_2O per hectare are beneficial. Judicious application of organic manure (farmyard manure, compost or vermicompost) with inorganic fertilizer increase the efficiency. Entire P_2O_5 and K_2O has to be applied at sowing whereas nitrogen has to be applied in two or three split doses depending on moisture availability. In areas of good rainfall and moisture availability, 50 per cent of recommended nitrogen (30 kg) has to be applied at the sowing time and the remaining 50 per cent (30 kg) in 2 equal splits at 25-30 and 40-45 days after sowing is recommended. Seeds treated with *Azospirillum brasiliense* (N-fixing bacfericum) and *Aspergillus awamori* (P-solubilizing fungus) @ 25 g/kg seed is beneficial. If seeds are to be treated with seed-dressing chemicals, treat the seeds first with seed-dressing chemicals and then with biofertilizers at the time of sowing.

Weed Management

Weed management is a difficult task especially during wet rainy season in finger millet. It is essential to control weeds at the initial stage of plant growth and development. The crop should be kept weed free for 30-45 days after sowing. Two to three manual weedings is needed for this purpose. Generally *kutla* in hill areas and *khurpi* in plain are used for this. Sometimes due to continuous rain during the early period of growth, it becomes impossible to enter in the field. For such situations, use of pre or post-emergence herbicides in the field has been found useful. Pre-emergence spray of Pendimethalin (Stomp) 30 EC at the rate of 3.3 litres per hectare in 1000 litres of water within 2-3 days after sowing can shows good control for a wide spectrum of weeds. In assured rainfall and irrigated condition, spraying of 2,4-D sodium salt of 80 per cent at the rate of 0.6 kg/ha in 1000 litres of water per hectare as post-emergence at 30-35 days after sowing effectively control broad-leaved weeds.

Water Management

Finger millet sown during *Kharif* season generally does not require any irrigation. However, at tillering and flowering stages, if rain stops for long spell irrigation would be required to obtain good yield. Finger millet crop does not perform well under water logged conditions. Therefore, proper removal of excess water after rain is also essential.

Disease and Insect Pest Management

The main diseases of finger millet are blast, seedling blights, downy mildew and smut.

Blast

This disease is caused by a fungus (*Pyricularia griseai*). Infection of this disease may occur even in seedling stage with grey-green to yellow lesions forming on leaf blades. In severe condition, spots may coalesce to form large blighted spots and main stem becomes infected and the earheads break over. Grains of infected earheads are shrivelled and become light in weight. It can be controlled by following methods:

1. Use of disease resistant varieties *viz.* VL Mandua 146, VL Mandua 149 etc.
2. Treating seeds with Thiram or Carbendazim @ 2 g/kg seed, 24 hour before sowing.
3. Spray the crop with Carbendazim (0.05 per cent) or Mencozeb (0.2 per cent) or Tricyclazole (0.06 per cent). Repeat the above spray at 50 per cent flowering and again 10 days later to control neck and finger blast.
4. Sowing early in the season and replacing 25 per cent N with farmyard manure/compost/ vermincompost and biofertilizers reduce the blast severity.
5. Avoid use of excessive nitrogen in the field.

Seedling Blight

This disease is caused by a fungus *Cochilobolus noduloeus*. It causes severe damage during continuous rainfall. This disease occurs throughout the life-cycle of finger millet and may infect all parts of the plant. It can be controlled as:

1. Seed treatment with Ceresan or Thiram @ 2 g/kg of seed.
2. Spray the crops with 0.2 per cent Zineb (2 g of Zineb in 1 litre of water) at the time of disease appearance.

Downy Mildew

It is caused by a fungus *Sclerophthora macrospora*. It may cause severe damage in certain areas. Plants are stunted and leaves arise close together in a bunch. For controlling

1. Spray the crops with 0.2 per cent Mancozeb 75 W.P. (2 g in 1 litre of water).
2. Keep the field clean.

Mottle Streak and Streak Virus

The disease is prevalent in Karnataka, Tamil Nadu and Andhra Pradesh. Mottle streak virus is transmitted by 2 species of leaf hoppers, *viz. Cicadulina bipunctella* and *Cicadulina chinai*, and streak virus by *Cicadulina chinai*. It can be controlled by:

1. Application of Carbofuron (10 per cent) granules @ 30-35 kg/ha, 2 days before sowing
2. Spray at seedling crop with Dimethoate 30 EC (0.05 per cent) or Methyl-*o*-demeton (0.05 per cent).
3. Remove the infected plants from the field.

Finger millet is by and large devoid of serious insect pests and considered as ecofriendly crop. However, sporadic incidence of pests such as stem borer, leaf aphid, hairy caterpillar, grass hoppers, army worm and cutworm are noticed.

Pink Stem Borer

Stem-borers sometimes damage the crop severely. The larva bores into stem, resulting in dead heart. It can be controlled by

1. Using Diazinon 5 per cent or Endosulfan 4 per cent granules at the rate of 20 kg per ha.
2. Spraying with Dimethoate (0.05 per cent) or Monocrotophos (0.04 per cent) helps in pest control.

White Grub

Different species of white grubs (major spp. *Anomala dimidiata, Holotrichia seticollis*) are serious pests of finger millets in hills. The adult beetles are attracted to light during night and thereby their population can be gradually controlled through long term use of light mediated insect traps on community basis.

Army Worms and Cut Worms

Army worms and cut worms appear during the early stages of plant growth and continue upto harvest of the crop. The caterpillars cut seedlings at the base of the crop which appears as if grazed by animal. They are active during night and hide themselves under stones and clods during the day. In later stages of plant growth, these insects act as defoliates. Dusting of Malathian 5 per cent @ 24 kg/ha or Quinalphos 1.5 per cent @ 24 kg/ha when the symptoms are noticed can control this insect.

Ear Caterpillars

Ear caterpillars appear at dough stage on ears and persist till the harvest of the crop. The caterpillars chew the maturing seeds and make a fine web out of their casting around half eaten grains. This further attracts saprophytic fungi.

1. Use of Malathion dust 5 per cent @ 24 kg/ha(480 g/Nali) or Quinalphos 1.5 per cent @ 24 kg/ha can control this insect.
2. The incidence of ear caterpillars is negligible on varieties with open or semi-compact ears. So, careful choice of varieties is advocated.

Leaf Aphid

It occurs throughout the corp. growing period. The nymphs and adults suck the sap from tender leaves and stem. They can cause serious damage in the seedling stage upto 30 days. Spraying of Dimethoate (0.05 per cent) or Quinalphos (0.05 per cent) gives effective control.

Bihar Hairy Caterpillars and Grass Hoppers

They feed on leaves leaving only veins of the leaves. Dusting of 2 per cent Methyl parathion at the rate of 25-30 kg/ha in early stages may control both the insects.

Harvesting and Threshing

The crop matures in about 120-135 days (4 to 4 ½ months). Harvesting is generally done in two stages. The earheads are harvested with ordinary sickles and straw is cut close to the ground. Earheads are heaped for three to four days to cure and then threshed with hand or by bullocks. At some places under rainfed conditions, the whole plant with earhead is cut, heaped and then threshed.

Yield

With improved agriculture practices, it is possible to harvest 19-26 quintals of grain and 60 to 80 quintals of fodder/ha. The straw of Mandua (Finger millet) makes nutritious fodder. It can be conserved by putting up in well built stakes.

Foxtail Millet (*Setaria italica*)

Foxtail millet known as Italian millet, German millet and Hay millet is cultivated for straw and grain. Its inflorescence is spike with short side branches bearing spikelets and bristles. The colloquial name is kakun, kangni, navane, tenai, korra and rala. It is one of the oldest crops in China and to some extent in southern and central Asia. Foxtail millet is known for its draught tolerance due to an indispensable crop of vast rainfed areas in semi-arid regions in India. At present, foxtail millet is cultivated in limited area of Andhra Pradesh, Karnataka, Maharashtra, Tamil Nadu, Rajasthan, Madhya Pradesh, Uttarakhand and north eastern states. India (approximately 9 per cent of world production) is the second most important producer of foxtail millet. Foxtail millet is grown in mountainous as well as plain regions. In hills, foxtail millet is cultivated as a border row in the field of rice crop and other crops or as mixed crop with rice in 1:9 ratio. Occasionally, it is cultivated as sole crop. Several local recipes such as roti (bread), jaula, singal, sirol are made from foxtail millet. Jaula made out of foxtail millet is having special medicinal value for patient suffering from chicken pox. Recipes namely dhokla, uppuma and laddu based on foxtail millet in combination with fenugreek seeds and legumes have been formulated. All the food products developed are observed to be effective in reducing blood glucose levels in normal as well as diabeties patient.

Nutritional Value

Foxtail millet grain contains 12.3 per cent protein, 4.7 per cent fat, 60.6 per cent carbohydrates and 3.2 per cent ash. The protein content of foxtail millet is greater than rice and is comparable to wheat. The fat content of foxtail millet is eight and three times more than rice and wheat, respectively.

Soil

Foxtail millet requires a fertile soil for good yield, although it can grow in poor soils. Light soil including red loams, alluvials, and black cotton soils are all suitable for its cultivation but it thrives best in rich, well drained loam soils.

Varieties

A number of varieties with high yield potential have been released for different states as given in Table 17.3.

Field Preparation

Foxtail millet does not require much field preparation. Before the onset of monsoon the field should be harrowed with once with mould board plough. With the onset of monsoon the field should be harrowed or ploughed with a local plough twice in northern India or with blade harrow in south India. Planking should be done for making the field smooth and well levelled.

Seed Rate and Sowing

The seed rate is required 8-10 kilogram per hectare for line sowing and 15 kilogram per hectare for broadcasting. Optimum spacing should be 25-30 cm between rows and 8-10 cm between plants within a row. Optimum time of sowing for rainfed crop is August-September in Tamil Nadu,

July-August in Karnataka, first fortnight of July in Andhra Pradesh and second and third week of July in Maharashtra. In Tamil Nadu, *Kharif* irrigated crop is sown from the beginning of June to end of July and summer irrigated crop in January. For plains of Uttar Pradesh and Bihar, mid June is the optimum time of sowing. For hills of Uttarakhand the optimum time of sowing is second fortnight of May to first fortnight of June at 1000 m altitude and first fortnight of May at more than 1000 m altitude. Generally broadcasting method was followed in hill areas but it is not scientific. Therefore line sowing is recommended for higher germination as well as intercultural operation of field. Sowing should be done behind the plough at 2 cm in shallow furrow.

Table 17.3: Average Grain Yield, Characters and Suitable Area of Improved Varieties

Variety	*Average Grain Yield (Kg/ha)*	*Characters*	*Suitable Area*
Gavari (SR 11)	1500-1600	Tall variety, semi compact panicle tapering at both ends, Small pale green awns/bristles on panicle, 3-4 tillers, bold and cream coloured seed, maturity period-75-80 days	Rajasthan and Meghalaya
Meera (SR-16)	1500-1600	Medium tall variety, compact panicle, pigmented with purple violet colour and bears long stiff violet awns, leaves are broad, erect and glossy, seed is bold and creamish.	Rajasthan and Manipur
PRK-1	1900-2000	Violet pigmentation on stem and leaves and maturity, slender and compact panicle, bristles less ear head.	Mid and high hill areas of Uttarakhand and Nagaland

Nutrient Management

Farmyard manure or compost at the rate of 100 quintal per hectare should be applied 2-3 weeks prior to sowing. In addition, application of 40 kg N, 20 kg P_2O_5 and 20 kg K_2O is recommended for higher yield.

Weed Management

Two hand weedings, 15-20 and 30-35 days after sowing are necessary for effective weed control. Isoproturon 0.75 kg/ha alongwith one hand weeding at 45 days after sowing can control weed effectively.

Cropping System

Generally it is sown as mixed crop with cotton, maize, pigeonpea and blackgram, but sometimes it is raised as a pure crop. As a pure crop the following crop rotation is followed-

- Foxtail millet-linseed
- Foxtail millet-chickpea
- Foxtail millet-barley
- Foxtail millet-fieldpea

Intercropping

Intercropping of foxtail millet with groundnut or foxtail millet with pigeonpea in 2:1 proportion is popular in Andhra Pradesh or foxtail millet with soybean (2:1) is popular in Uttar Pradesh and Uttarakhand.

Sequence Cropping

Relay and sequence cropping systems are also practiced in some areas. In Andhra Pradesh if monsoon is early and favourable, foxtail millet is sown at 45 cm row spacing and *rabi* sorghum is introduced as a relay crop when foxtail millet is near the maturity. In medium deep black soils of Andhra Pradesh, growing 2 crops, *e.g.* foxtail millet and mustard or foxtail millet and green gram, in sequence is profitable than taking single crop of foxtail millet.

Water Management

Foxtail millet sown during *Kharif* season does not require any irrigation. However, if dry spell prevails for longer period, then one or two irrigation has to given for boost yield. Summer crop however, would require two to five irrigations depending upon soil type and climatic conditions. During rainy season, drainage is also essential. Remove excess rain water from the field as it does not withstand water logged conditions.

Disease and Insect Pest Management

The major diseases of foxtail millet are blast, rust and downy mildew (green ear disease). The symptoms and control measures are given below.

Blast

This disease is caused by fungus (*Pyricularia setariae*). Circular spots with light centers are formed in the leaf blade with dark brown margin. For controlling this,

1. Use moderately resistance varieties.
2. Seed treatment with mancozeb @ 2g/kg seed followed by one spray of either carbendazim or mancozeb (0.1 per cent) as per need.

Rust

It is caused by the fungus (*Uromyces setariae-italicae*). Numerous rust pustules full of powdery spore mass are formed on leaves. Colour of pustules is brown at beginning, but turns black at later stage. Foliar spray of wettable sulphur (0.3 per cent) or mancozeb (0.25 per cent) reduces the disease inensity.

Downy Mildew (Green Ear Disease)

It is caused by a fungus (*Sclerospora graminicola*). Affected plants are dwarf and produce excessive tillering from the crown; further branches are produced from auxiliary buds, floral bracts turn into leaf-like malformation giving green ear appearance. Leaves show necrosis and browning followed by splitting and shredding. It can be controlled by

1. Destruction of diseased plant debris.
2. Seed treatment with Ridomil MZ 72 WP or Chlorothalonil (2g/kg seed) followed by one foliar spray of mancozeb (0.2 per cent) is effective in minimizing the disease.

Army worm, cut worm and leaf scrapping beetle appear occasionally in epidemic form. In certain areas shootfly attack, occurs although it is not a regular pest. When pests are noticed, apply dusting of Malathian 5 per cent @ 8-10 kg/ha.

Harvesting and Threshing

The crop matures about in 80-100 days. The crop is harvested when the earheads are dry. Either the whole plant is cut by sickle or only the ears were harvested. After drying for few days, threshing is done with a stone roller or by trampling under the feet of bullocks.

Yield

With improved package of practices 15-18 quintals of grain and 20-40 quintals of straw per hectare can be obtained.

Barnyard Millet (*Echinochloa frumentacea* L.)

Barnyard millet (Madira/Sawan/Kudruvali/Oodalu) is an important crop of dryland agriculture and extensively cultivated in hills for grain as well as fodder production. It is very drought resistant, but also capable of withstanding water logging condition. Easy cultivation, voluminous fodder and assured crop harvest made it a favoured crop for rainfed agriculture. it is cultivated over a wide array of environmental and poor soil health condition. It is generally cultivated in hill slopes and undulating field; where few options exist for crop diversification. Sawan grains are consumed just like rice grains and also used in making rice pudding (*Kheer*).

Nutritional Value

The barnyard millet grain contains about 65 per cent carbohydrate, a high proportion of which is in the form of non-starchy polysaccharide and dietary fibre. This helps in prevention of constipation and facilitates slow release of glucose to the blood stream during digestion and also lowes blood cholesterol. Grains contains 6.2 per cent protein. The digestibility of protein is 40 per cent. Lysine and Isoleucine are the main essential amino-acids in protein extracted from barnyard millet. The grain is eaten mostly by poor classes, but sometimes it is brewed for beer. It is also used as feed for cage birds. The straw makes good fodder for cattle. Green fodder is very much relished by cattle.

Soil

Barnyard millet crops have capacity to with stand drought and flood condition. So it can be grown successfully in heavy and light soil to marsh soil. But light soil is optimum for barnyard millet cultivation. Soil having good water/moisture retention and conservation capacity is good for barnyard millet cultivation.

Varieties

There are several high yielding varieties recommended for specific region.

Field Preparation

The crop is mostly confined to light soil which are ploughed twice before sowing the seed. After onset of monsoon the field is ploughed with soil turning plough. Thereafter two to three harrow was done followed by planks.

Seed Rate and Sowing

A seed rate of 8-10 kg/ha is adequate both for drilling in line sowing and broadcasting. The optimum time of sowing of barnyard millet is last week of June and first week of July. Generally farmers sow this crop by broadcasting method. However this method is not scientific. Seed should be sown in line/furrow. The optimum spacing for this crop is 20-30 cm x 15 cm. A row spacing of 30 cm can facilitate intercultural operation for weed control and other agricultural operation. The depth of seed should not be more than 3-4 cm.

Table 17.4: Average Grain Yield, Characters and Suitable Area of Improved Varieties

Variety	*Duration (Days)*	*Characters*	*Yield Capacity (kg/ha.)*	*Recommended Area*
V.L. Madira 172	80-85	Grain: Grey seed colour; Plant height: 95-100 cm; Maturity: 85-90 days; tolerant to smut and leaf spot.	2200-2500 Fodder yield- 4000-5000 kg/ha	Rainfed areas of Uttrakhand, Uttar Pradesh, Gujarat and Karnataka.
V.L. Madira 181	75-80	-	1400-1700	Uttrakhand, Andhra Pradesh, Jharkhand, Gujarat, Karnataka, Maharashtra, Madhya Pradesh
P.R.J. 1	85-90	Grain: Grey seed colour; Compact ear head, Small awn in spike let; Plant height: 95-100 cm; Maturity: 85-90 days; resistant to smut.	2500	Chhattishgarh, Tamil Nadu, Uttrakhand (Low-Mid-High hills 1000-2000 m)
V.L. Madira 8	80-85	-	1500-2000	Uttarakhand hills plain rainfed area
V.L. Madira 29	80-85	Grain: Grey seed colour; Plant height: 90-95 cm; tolerant to smut and leaf spot.	2200-2500 Fodder yield- 4000-5000 kg/ha	All India except Tamil Nadu and Andhra Pradesh
V.L. Madira 21	80-85	-	2500-3000	Uttrakhand hills and plains rainfed area

Mixed Cropping

It is usually grown as sole crop. Mixed cropping with pigeonpea or short duration pulse crops (urdbean, horsegram) is also practiced. Intercropping of barnyard mille and ricebean in 4:1 proportion is recommended for Uttarakhand state.

Nutrient Management

Barnyard millet crop requires 50 kg nitrogen, 30 kg P_2O_5 and 20 kg K_2O. Half amount of nitrogen should be applied at the time of sowing and remaining half amount at the time of first irrigation. The amount of phosphorus and potassium should be applied at basal dressing as per need. Seed treatment with *Agrobacterium radiobacter* and *Aspergillus awamori* recommended.

Water Management

It does not need any irrigation as it is grown in rainy season.

Weed Management

For control of weeds one hand-weeding in line-sown crop and 2-3 hand weeding in broadcast crop is required. However, howing and weeding become easy in line sown crops.

Crop Rotation

Following crop rotation is observed.

1. Barnyard millet-wheat
2. Barnyard millet-barley
3. Barnyard millet-chickpea
4. Barnyard millet-pea
5. Barnyard millet-lentil
6. Barnyard millet-alsi

Disease and Insect Pest Management

Smut

The crop is infected by 3 different smuts, *viz.* head smut (*Ustilago crus-galli*), grain smut (*Ustilago panici frumentacei*) and kernel smut (*Ustilago paradoxa*). Seed treatment before sowing with Emisan or Carbendazim or Thiram at the rate of 2 g/kg seed is recommended.

Shootfly

It is the serious pest causing significant yield losses. Early sowing with the onset of monsoon is an effective and low-cost method of control for shootfly.

Harvesting and Threshing

Crop generally matures in 2.5 to 3.5 months. This crop is harvested at the end of August or beginning of September. When the crop is ripe, it is cut with sickle and left in the field for two to three days for drying before stacking in the field. Threshing under the cattle feet is usually followed.

Yield

On an average, the grain yield is about 10-15 q/ha and straw 20-30 q/ha, when improved package of practices is followed. The dehusked grain can be cooked as rice or made into flour for porridge or chapaties.

Proso Millet (*Panicum miliaceum* L.)

Proso millet or Cheena is another important minor millet grown in hill areas of India. It is a short duration crop (60-90 days) with low water requirement. It is well adopted to hot summers in tropics and high altitudes as well as rainfed *Kharif* crop in higher hills. In higher hills (> 2000 m), wheat matures in second fortnight of June. Hence proso millet can easily be taken after wheat harvest. Proso millet is an erect annual, 30-100 cm. tall crop with few tillers. The roots are fibrus and shallow. The leaves are linear, slender and the leaf sheath encloses the entire internode. Proso seeds are smaller than pearl millet and grain sorghum. Proso millet seeds are generally oval in shape and about 3 mm long and 2 mm wide and vary in colour from white cream, yellow, red and black to brown. Panicles are usually more or less included at the base and 10-30 cm long. Spikelets are 4.5-5.0 mm long, ovate, acuminate, and strongly many nerved.

Nutritional Value

Proso millet grain contains a high protein content (12.5 per cent), 1.1 per cent fat, 2.2 per cent fibre and 3.4 per cent ash. Grains are used in various ways as cooked grain, flour for making 'Chapatis', parched grains etc. it can also be used in making *kheer*. This makes a good poultry feed also. The straw can be used as good fodder for cattle.

Soil

Proso millet can be grown both in rich and poor soil having variable texture ranging between sandy loam to clay of black cotton soils. Well drained loam or sandy loam soils free from kankar and rich in organic matter are ideal for pros millet cultivation.

Varieties

Very few proso millet varieties are available. Many local varieties with duration of 60-90 days are available for common cultivation. Some of the promising varieties with specific features are given in Table 17.5.

Table 17.5: Promising Proso Millet Varieties and their Specific Features

Variety	*Maturity (Days)*	*Average Grain Yield (kg/ha)*	*Specific Feature*	*Area of Adoption*
GPUP-21	65-75	1500-1600	Plant height: 85-116 cm, drooping arched panicles with bold seed	Uttarakhand, Tamil Nadu and Karnataka
Bhawna	70-75	1200-1500	-	Hills of Uttarakhand
TNAU-63	92-95	2000-2200	It is high grain and fodder yielding variety.	Karnataka

Field Preparation

Soon after harvesting of the previous crop, the field should be ploughed to expose the soil to sun and enable to retain more moisture. With onset of monsoon, the land should be harrowed two to three times and then levelled. Proso millet needs a firm and clean seedbed. It does not respond well to deep ploughing.

Seed Rate and Sowing

Seed rate depends upon the method of sowing. In case of line sowing it requires 10 kg seed/ha and for broadcasting it should be 15 kg seed/ha. As a *Kharif* crop, proso millet should be sown in the first fortnight of July with the onset of monsoon. As a summer crop it should be sown by the middle of April. During summer, it would be desirable to sow this crop as soon as the harvesting of the *Rabi* crop is over. Proso millet can be sown by broadcasting or line sowing in furrows with 3-4 cm deep. Row to row distance should be kept 25 cm and plant to plant is 10 cm. Line sowing ensures better germination, cuts down seed requirement and facilitates intercultural operations compared to broadcast sowing.

Nutrient Management

Prosomillet, being a short duration crop requires relatively less amount of nutrients compared to other cereals. If organic manure is available, it may be added to the soil about a month before sowing at the rate of 15-20 tonnes per hectare. Organic manure retains moisture in soil during crop duration and also provides micro-nutrient. For a good crop, fertilizer recommendation under irrigated condition is 40 kg nitrogen, 20 kg P_2O_5 and 20 kg K_2O per hectare. Under rainfed condition, fertilizer dose is reduced to half of the irrigated crop (20, 10 and 10 kg/ha of N, P_2O_5 and K_2O, respectively). Apply half of the nitrogen and whole amount of phosphorus and potash as a basal dose at the time of sowing. The remaining half of nitrogen should be applied at the time of first irrigation.

Water Management

Proso millet sown during *Kharif* season, generally does not require any irrigation. However, at tillering stage, if dry spell prevails for longer period, then one irrigation must be given for higher production.

Weed Control

To minimize loss of soil moisture and nutrients, the field should be kept weed-free upto 35 days' stage, which produce higher yield. Two weedings at an interval of 15-20 days can control it easily.

Disease and Insect Pest Management

Head Smut

Head smut (*Sphacelotheca destruens*) is a common disease of pros millet. The affected panicles become elongated and thickened. The smut masses rupture before harvest. This is a seed borne disease. It can be controlled by treating seeds with Thiram or Ceresan at the rate of 2.5 g per kg of seed or hot water treatment (soaking seeds in hot water at 55 degree for 7-12 minutes).

Bacterial Streak

Sometimes bacterial streak is also noticed. It can be controlled by seed treatment with 5 per cent magnesium arsenate at the rate of 1 g of the chemical in 1 kg of seed.

Rust

It is caused by a fungus. Brown postules are seen on leaves. This disease hinders photosynthesis and causes considerable loss in yield. It can be controlled to some extent by spraying 0.2 per cent solution of Mancozeb 75 Wettable Powder.

Shootfly

Among insects, shootfly is important for this crop. This insect damages the crop at the seedling stage. It can be controlled as:

1. Apply 15 kg of phorate (Thimate) 10 per cent granules per hectare in the soil at the time of field preparation.
2. Shootfly can also be controlled by spraying the crop with 0.15 per cent Endosulfan. Two sprays at an interval of 15 days is enough to control shootfly.

Harvesting and Threshing

Proso millet becomes ready for harvest after 65-75 days of sowing for most of the varieties. Harvest the crop when it is about to mature. The seeds in the tip of upper heads ripe and shatter before the lower seeds mature. Therefore, the crop should be harvested when about two-thirds of seeds are ripe. Crop of proso millet is threshed with hand or by bullocks.

Yield

With improved package of practices it is possible to harvest 20-25 quintal per hectare of grain and 50-60 quintals of straw per hectare.

Conclusion

In hilly areas, minor milllets provide access to sufficient food for livelihood at the household level. However, due to less aggressive research activity and low priority coupled with negligence by policy makers, these crops remained untouched. Realizing the nutritional values of these grains, they are now considered as nutri-cereals. The minor millets, which are rich in micro-nutrients can contribute effectively to make the diets more balanced, hence play an important role in combating silent hunger. For obtaining higher productivity in hill areas there is a need to popularize new varieties suitable to the specific region and make it a part of sustainable hill farming system. Use of inorganic manure along with organic manure and developing value addition in millets should be given first priority in this regard.

Hill Agriculture: Prospects, Constraints and Mitigations (2013) *Pages* **251–269**
Editor: **Dr. Shahid Ahamad**
Published by: **DAYA PUBLISHING HOUSE, NEW DELHI**

Chapter 18

Bio-Suppression of Insect Pests in Hill Agri-Ecosystem

Kuldeep Srivastava, Devinder Sharma and J.N. Srivastava
Division of Entomology, FOA,
SKUAST-J, Jammu – 180 009

Over 800 million people in the world go hungry each year. The World Food Summit held in June 2002 announced an international alliance against hunger to renew commitments to halve this number by 2015. Poverty alleviation and sustained agricultural production are key strategies. Since the vast majority of the hungry and those living in poverty are in rural areas and dependent on agricultural or natural resources for their livelihoods, attention has focused on how to improve subsistence farming systems, especially in South Asia. The challenges in terms of loss of livelihood security and fall in the commodity prices in the agriculture sector are some of the apprehensions being raised at various fora, especially civil society organizations. By 2015 the region would be food deficient by 16 million tonnes and by 2030 the magnitude would be of the order of 26 million tones (Anonymous, 2011). The quality of food available in the region is also a great concern as 40 per cent of world's malnourished, almost 330 million people, are in South Asia. This challenge becomes much more pressing given that 28 per cent of the GDP in the region comes from agriculture and more than 70 per cent of the population is dependent on this sector. Prospects of raising agricultural production through area expansion and application of existing technologies appear to be severely constrained. Land frontiers are closing down, and there is little, if any, scope to bring additional land under cultivation. Green revolution technologies have now been widely adopted, and the process of diminishing returns to additional input usage has set in. Concurrently, agricultural production continues to be constrained by a number of biotic and abiotic factors. For instance, insect pests, diseases and weeds cause considerable damage to potential agricultural production. Evidences indicate that pests cause 25 per cent loss in rice, 5-10 per cent in wheat, 30 per cent in pulses, 35 per cent in oilseeds, 20 per cent in sugarcane and 50 per cent in cotton (Dhaliwal and Arora, 1996). It has been estimated that crop losses due to insect-pests and disease are over Rs. 29,000 crore. Rejection of agricultural exports upto Rs. 1,000 crore per annum is reported because of very high pesticides residue contents crops. The losses though cannot be eliminated altogether, these can be reduced. Until recently, chemical pesticides were increasingly relied upon to limit the production losses. The pollution of environment, loss of biodiversity, toxic residue in food, development of pest resurgence, out break, hazardous to consumers and elimination of natural enemies from crop ecosystem are well illustration in the recent years. This is the fact that 1 per cent of pesticides are reaching to the target pest and rest of 99 per cent is reaching to the non-target sectors. Now, the policy makers have to think the alternative measure of pest control. IPM is the only resource to conserve biodiversity comprising beneficial pollinators, parasites and predators, to obtain high yields of crops through superior pest control. Biological control is one of the most important and sustainable approach in IPM programme, and therefore we have discussed important insect-pests of hill agri-ecosystem and their management through bio-agents within following heads.

Insect-Pests in Rice Ecosystem

Rice, as the single most important human energy source, feeds about half of the world's population, so it was vital that predicted 1980's Asian famines were averted by the development of Green Revolution high yielding rice and technology between 1965 and 1975. Great advances were made in irrigated rice which provided over 72 per cent of total yield and will remain outstandingly important; with ecologically comparable rainfed lowland rice, it produces more than 92 per cent of the world's crop, of which over 9 1 per cent is grown in Asia. Rice is the dominant staple food in the developing world. More than 90 per cent of the world's rice is produced and consumed in Asia. Water is a prominent factor in the irrigated rice system making it different from other agricultural productions. The biodiversity of irrigated rice is higher than in many natural ecosystems (Siggsgard, 2000). Unlike wheat rice is more prone to insect-pests damage than other pests. More than 100 insect-pests have been reported in rice in which stem borer, Leaf folder, Brown plant hopper and White backed plant hoppers are major ones (Table 18.1).

Table 18.1: Important Insect-Pests of Rice

Common Name	*Scientific Name*	*Damage*
Leaf folder	*Cnaophalocrosis Medinalis*	Folds leaves and remains inside, scraping the green tissues between the veins making the leaves white and papery and can give scorched appearance on drying up.
Brown plant hopper	*Nilaparvata lugens*	Plants become yellow and die. Insects congregate in large numbers, causing hopper burn in circular patches. Hopperburn frequently appears uniformly over large areas.
White backed plant hopper	*Sogatella furcifera*	Insects suck sap causing reduced vigor, stunting, yellowing of leaves, delayed tillering and grain formation.

Natural Enemies Complex of Rice Insect Pests

The rice ecosystem is bestowed with a lot of natural enemies; naturally occurring biological control has a potential role to play in the management of rice fields of tropical south and south east Asia and there is a need to emphasize the impact of indigenous natural enemies as an essential part of IPM. Conservation of the natural enemy fauna in situ for suppressing the pest population seems to be a very good alternative. More than 100 species of natural enemies have been recorded from each major group of insect pest *viz.*, stem borers, foliage feeders and sapsuckers attacking paddy crop and it may be due to the presence of these natural enemies that out of more than 100 insect species recorded on this crop, only about 20 cause economic damage (Kumar and Patil, 2006). However, it is observed

Egg Masses of Yellow Rice Stem Borer

Larva of Rice Leaf Folder, *Cnaphalocrocis medinalis*

Moths of Rice Stem Borer, *Scirpophaga incertulas*

Stem Borer Inside Rice Stem

that inundative release of egg parasite, *Trichogramma japonicum* that *T. chilonis* is really an effective approach to decrease the Yellow stem borer or Leaf folder incidence. It is also observed that a very high count of *Trichogramma viride* and a few bacterial antagonists are effective against fungal pests. The important NEs of leaf folder included egg parasitoids, *Trichogramma* spp. and larval parasitoids, *Apanteles* spp. and *Macrocentrus* sp. Inundative releases of *T. japonicum* at 10 adults/m^2, 4-8 times during the season reduced leaf folder damage by 12 to 60 per cent. Among natural enemies of stem borers egg parasitoids, *Tetrastichus schoenobii* Ferriere, *Telenomus rowani* (Gahan), *Telenomus dignoides* Nixon and *Trichogramma chilonis* Ishii are predominant (Manjunath, 1991). In case of leaf and plant hoppers, predators are more important regulating factors. The important ones are the myrid bug, coccinellid beetle, staphyllinid beetles, valid bug, spiders, water strider, damsel and dragonflies. The presence of 3 predatory spiders/hill checked the population of brown plant hopper and white backed planthopper. Releases of mired bug @100 bugs or 50-75 eggs/m2 at 10 day intervals have been found effective for the control of brown plant hopper. Thus biocontrol agents (predators and parasitoids) form an important component of BIPM and reduce the pest population from 40-60 per cent under certain situation.

Management of Rice Pests

The damage caused by the paddy steam borer (*Scirpophaga incertulas*), which is the most important pest, becomes evident only by the presence of symptoms-dead hearts during the vegetative phase and white heads, after the ear heads are formed. However, the adults lay their eggs on the leaves and these can be controlled effective by release of *Trichogramma japonicum*.

To be effective, release of *Trichogramma* should coincide with period when eggs of the pests are present on the leaves. Normally is very difficult to locate the tiny eggs on paddy leaves. This can be overcome by using pheromone traps/lures. It is recommended that 5 traps per hectare and lures changed once in 15-20 days. The males that are attracted are caught in the trap. The adult male catches indicate the presence of female moths that are in the process of laying eggs. Therefore, release of *Trichogramma* can be commended soon after catches of male moths are noticed in the traps. Release of

Inundative Release of *Tricho card* Released in the Rice Field

Wolf Spiders, *Lycosa pseudoannulata* Feeding on Rice Stem Borer Moth, *Scirpophaga incertulas*

5 cards of *Trichogramma japonicum* each containing 20,000 parasitoids per hectare, to be repeated at 15-20 day intervals, depending on the level of infection, as determined by trap catches is advocated. The releases of *Trichogramma chilonis* are useful in controlling leaf roller eggs. Targeting the key pest (stem borer) by non-chemical methods would encourage the build up of natural enemies such as parasitoids and predators, including spiders in the paddy fields. This in turn would results in a reduction in the population build up of sucking pests such as brown plant hopper and caterpillars that feed on the leaves of the crop.

Table 18.2: Recommendation of Bioagents

Pest Complex	*Bioagents*	*Recommendation*
Insect Pests	*Trichogramma japonicum*	5 cards/ha/release; 3 release at 20 days interval's
Scirpophaga incertulas	*Trichogramma chilonis*	5 cards/ha/release; 3 release at 20 days interval's
(Yellow stem borer)		
Cnaphalocrosis medinalis		
(Leaf folder)		

Insect-Pests in Sugarcane Ecosystem

Sugarcane is a major commercial crop in India and is cultivated in nearly 3.5 million hectares with an annual cane production of about 205 million tones. The present cane yield per hectare is 60.7 tones with an average sugar recovery of 10.2 per cent. One of the reasons for low cane yield and sugar productivity is the damage caused by several species of pests. Among the different tools for the management of sugarcane pests, biological control holds a pivotal position. This is facilitated by the uninterrupted cropping pattern and favourable canopy structure of the crop offering conducive micro-climatic conditions for the natural enemies. In addition, because of the compensating ability of the plant and the consequent high level of the economic threshold for many of the pests, together with minimal use of insecticides in the sugarcane ecosystem, biological control is practical.

Table 18.3: Species of Natural Enemies Reported on Important Insect-Pests

Insect-Pests	*Number of Species of*		
	Parasites	*Predators*	*Pathogens*
Shoot borer	36	5	10
Top borer	61	"	5
Stalk borer	31	"	2
Internode borer	17	15	3
Pyrilla	13	12	5
White flies	8	11	1
Scale insect	17	6	1
White grubs	2	2@	4

Some of these natural enemies, occur throughout the year, while others occur during certain seasons and decimate pest populations. Their activity together with the influence of abiotic factors regulate pest populations below economic injury level on several occasions.

Table 18.4: Extent of Mortality Caused by Bio-agents

Insect-Pest	*Bio-agents*	*Extent of Kill (per cent)*
Shoot borer	*Trichogramma* sp.	4 – 5
	Telenomus spp.	15 – 22
	Apanteles flavipes	4 – 14
	Sturmiopsis inferens	6.5 – 22.2
	Stenobracon deesae	5 – 7
	Egg predators	13.4
	Granulosis virus	2.6 – 14.6
Top borer	*Trichogramma* sp.	4 – 5
	Telenomus spp	5 – 85.5
	A. flavipes	4 – 14
	Elasmus zehntneri	5 – 15
	Rhaconotus scirpophagae	7 – 12.2
	Isotima javensis	5 – 40
Stalk borer	*A. flavipes*	10 – 14
	Campyloneurus mutator	60 – 78
	S. inferens	20 – 50

Contd...

Table 18.4–*Contd...*

Insect-Pest	*Bio-agents*	*Extent of Kill (per cent)*
Internode borer	*Trichogramma* spp.	2 – 78.5
	Telenomus spp.	15 – 90
	A. flavipes	4 – 14
	Rhaconotus sp. nr. Signipennis	25
	Egg predators	9.4 – 22.4
	Granulosis virus	3.8 – 31.5
Pyrilla	Egg parasites	2.7 – 97.6
	Epiricania melanoleuca	15 – 100

Management through Bio-agents

As revealed from Table 18.4, large number of parasites, predators and pathogenic micro-organisms occur in the sugarcane ecosystem and we can use as fallows.

Shoot Borer

1. Application of granulosis virus at $10^6 – 10^7$ inclusion bodies/ml along with a surfactant like teepol or sandovit 0.05 per cent on 30th day. If necessary the application may be done at fortnightly intervals.
2. Sequential release of 125 gravid females of the tachinid parasite, *S. inferens* per ha. from 30th to 50th day of planting.
3. Weekly release of 50,000 adults of *Trichogramma chilonis* per ha. (when planting is done during autumn or special season).

Top Borer

Inoculative release of 125 females of *I. javensis* per ha in fields showing more than 10 per cent top borer infestation. This technique is useful only in tropical India.

Internode Borer

Release of *T. chilonis* @ 2,50,000 parasites/ha in phases during the different stages of crop growth *i.e.*, 25,000 parasites per ha during 4th and 9th months and 50,000 per ha during 5th, 6th, 7th and 8th months.

Pyrilla

Release of 8,000 to 10,000 cocoons or 8 to 10 lakh eggs of *E. melanoleuca* per ha.

Scale Insect

Release of at least 1500 beetles of *Pharoscymnus horni* or *Chilocorus nigritus* per ha at the first appearance of the pest.

Predators and Rains

A recent technique is to mix the eggs parasitized on different days and prepare the cards which ensure sequential emergence of parasites in the field for several days.

The mated females of *S. inferens* and *I. javensis* can be released by holding the container open and slowly moving in all directions in the field. For effective control of pyrilla, leaf bits bearing 2 to 3 egg masses or 5 to 7 cocoons of *E. melanoleuca* can be stapled to the underside of sugarcane leaf at several places in the infested field.

The granulosis virus can be sprayed in the evening hours on the leaf whorls and stem preferably using high volume sprayer.

Biological Control of Sugarcane Pyrilla

Eggs of *Pyrilla perpusilla*

Eggs and Adult of *Epiricanea melanoleuca*

Epiricanea melanoleuca* on *Pyrilla

Insect-Pests in Pulse Ecosystem

A variety of insect pests infest pulses and the annual yield loss is estimated to be 20 per cent in pigeonpea, 15 per cent in chickpea and 30 per cent in urdbean and mungbean. On an average 2.5 to 3.0 million tonnes of pulses are lost annually due to pests (Ali, 1998). Pulses are valuables to many pests and about 250 species of insects belonging to 10 orders and 100 families are reported to attack this crop. The major pests in pulse ecosystem include Pod fly, Gram pod borer, legume pod borer, stem flies, white flies, Bihar hairy caterpillar, sucking bugs and thrips (Tables 18.5 and 18.6). Important Natural Enemies of important insect pests attacking pulse crops are :

Parasitoids

- ✰ Tachinids: *Carcelia* spp., *Goniophthalmus hali* and *Sturmiopsis inferens*
- ✰ Braconids: *Apanteles* spp., *Chelonus* spp and *Cotesia* spp.
- ✰ Ichneunonids: *Campoletis* spp. and *Eriborus* spp.
- ✰ Scelionids: *Telenomus* spp.

Predators

- ✰ Coccinellids: *Chilomenes sexmaculata* and *Coccinella septumpunctata*
- ✰ Lygaeid: *Geocoris* sp.
- ✰ Pentatomids: *Cantheconidea* spp.
- ✰ Eumenids: *Delta* spp.
- ✰ Formicid: *Camponotus* spp.
- ✰ Vespids: *Polistis, Vespa* and *Ropalidia*
- ✰ Chrysopids: *Chrysopa* and *Chrysoperla* sp.
- ✰ Arachnids: Spiders
- ✰ Birds: *Acridotheres tristis, Bubulcus ibis, Discrusus adsimilis, Passer domesticus* and *Sterna aurantia.*

Table 18.5: Insect Pests and Natural Enemies of Pulse Crops

Insect Pests	*Natural Enemies*
Gram pod borer	*Trichogramma* spp., *Chelonus blackburnii*
Blue butterfly	Egg parasitoid *Trichogramma* sp.Larva parasitoids *Aploymia* sp., *Hyperencyrtus lycoonephila, Listrodromus crassipes.*
Plume moth	Larval parasitoids, *Apanteles paludicolae, Diadegma* sp., *Phanerotona hendecasisella, Tropimeris monodon*
Spotted pod borer	larval parasitoids *Bracon greeni* and *Phanerotoma hendecasisella*
Spiny pod borer	*Larval parasitiod : Bracon* sp., *Tetrastichus* sp., *Bracon hebetor, Phanerotoma* sp. and *P. lenedecasisella.*
Field bean pod borer	larvae parasitoids, *Microbracon lefroyi, Bracon hebetor, B. brevicornis, Carcelia evolans.*
Pea Pod fly	*Euderus lividus, Eurytoma* sp., *Euderus agromyzae*
Pea Stem fly	*Euderus lividus, Eurytoma* sp., *Euderus agromyzae*

Table 18.6: Major Insect Pests and their Management in Pulse Ecosystem

Pest Complex	*Control Measures*	*Recommendation*
Prophylactic Measures Aphids, jassids, thrips, white flies and early larval stages of bollworm	300 ppm. Azadirachtin or 1500 ppm. Azadirachtin	5 ml/litre (1.25 Litre/ha)
Insect Pests *Helicoverpa armigera*	Pheromone traps	5 traps/ha. (Distance between traps is 30 meters)
	Lures	20/ha. (Lures to be changed in every 15 days)- 4 changes
	Nuclear Polyhedrosis Virus (NPV)	250 LE/ha/spray
	Fungal pathogen – *Beauveria bassiana* or *Neumarea riley*	
	*Trichogramma chilonis***	5 cards/ha/release; 3 release at every 20 days gap
	Chilonus blackburni/*Bracon brevicornis* or *Telonomus heliothidae* (larval parasitoids)	@ 2000 adults/ha at 15 days inteakes.
	Brachymeria sp. (Pupsl parasitoid)	
	Chrysoperla carnea	2500 eggs/ha/release
Aphids, thrips, mites and whiteflies	*Chrysoperla carnea*	2500 eggs/ha/release
Rodenticides Rodents	Roban cake 0.005 per cent	600 g/ha

* *Chrysoperla* are effective against both early stage larvae of *Helicoverpa* and sucking pests.

Insect-Pests in Vegetable Ecosystem

India is the second largest vegetables producing countries next to China and occupies 6.09 million ha area with an annual production of 84.8 million tones. Majority of Indians are vegetarian, with a per capita consumption 135 g per day as against the recommended 300 g per day. We are still lagging

Lacewing Adult (*Chrysoperla carnea*)

V / VI instar Larva of *Helicoverpa armigera*

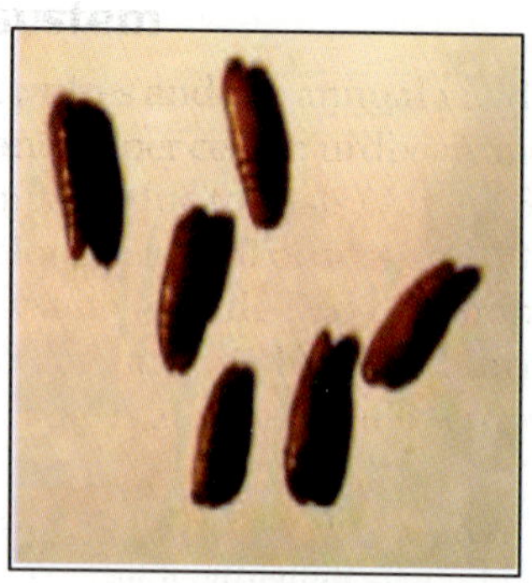

Pupae of *H. armigera*

***Helicoverpa armigera* Adult**

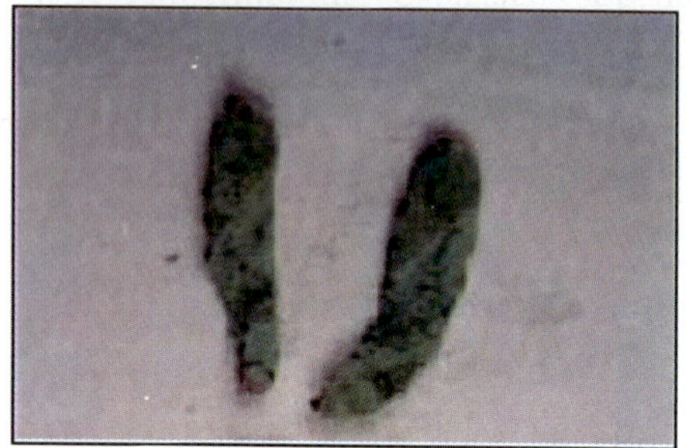

***Nomuraea rileyi*, Fungus Parastitic on *Helicoverpa armigera* Larvae**

***Helicoverpa armigera* Larva Infected by NPV**

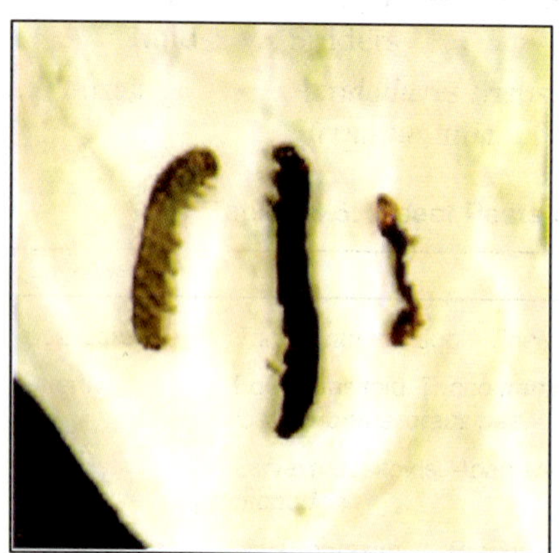

Stages of Infection of *H. armigera* Larvae by *Bt*

***Chrysoperla carnea* Grub Feeding on Aphids**

behind in terms of recommended diet level. In near future, there is a need of around 56 million tons of food to feed our 1.3 billion Indian population expected by the year 2020. Our natural resources are scarce and day by day they are shrinking and reducing due to the expanding urbanization and deforestation. In this context, to feed our burgeoning population we must have to look after the vertical expansion through increase in the intensity of vegetable crops; diversification of crops and through mixed and intercrops. These will only the solution to meet the requirements of growing demands in future.

Table 18.6: Major Insect-Pest of Vegetable Crops

Crop	*Pests*	
	Common Name	*Scientific Name*
Tomato	Fruit borer	*Helicoverpa armigera*
Brinjal	Shoot and fruit borer	*Leucinodes orbonalis*
	Stem borer	*Euzophera perticella*
	Epilachna beetle	*Henosepilachna viginti octopunctata*
	Gall midge	*Asphondylia sp.*
	Aphids	*Myzus persicae*
	Leaf webber	*Psara bipunctalis*
	Grey weevil	*Myllocerus bubfasciatus*
	Spider mite	*Tetranychus lideni*
Okra	Fruit borer	*Helicoverpa armigera*
		Earias vittella
	Leaf hopper	*Amrasca biguttula biguttula*
	Aphids	*Aphis gossypii*
	Stem fly	*Melanogromyza hibisci*
	Spider mite	*Tetranychus ludeni*
Cabbage	Diamondback moth	*Plutella xylostella*
	Leaf webber	*Crocidolomia bionotalis*
	Leaf caterpillars	*Spodoptera litura*
		Trichoplusia ni
	Cabbage borer	*Hellula undalis*
	Cabbage butterfly	*Pieris brassicae*
	Mustard sawfly	*Athalia proxima*
	Aphids	*Lipaphis erysimi*
		Myzus persicae
		Brevicoryne brassicae
Cucurbitaceous crops	Borer	*Diaphania indica*
	Caterpillar	*Sphenarches caffer*
	Semilooper	*Plusia*
	Red pumpkin in beetle	*Aulacophora foveicollis*
	Fruit fly	*Bactrocera cucurbitae*
	Spider mite	*Tetranychus cinnabarinus*
Amaranthus	Leaf caterpillar	*Hymenia recurvalis*
	Weevil	*Lixus truncatulus*
Chillies	Fruit borer	*Helicoverpa armigera*
	Thrips	*Scirtothrips dorsalis*
	Aphid	*Myzus persicae*
Onion	Thrips	*Thrips tabaci*
Field beans	Pod borer	*Adisura atkinsoni*
	Mite	*Tetranychus urticae*

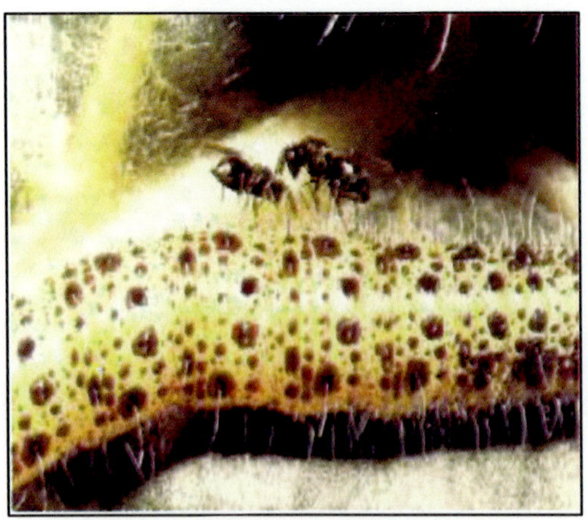

Adults of *Apanteles* sp. Ovipositing on *Pieris brassicae*

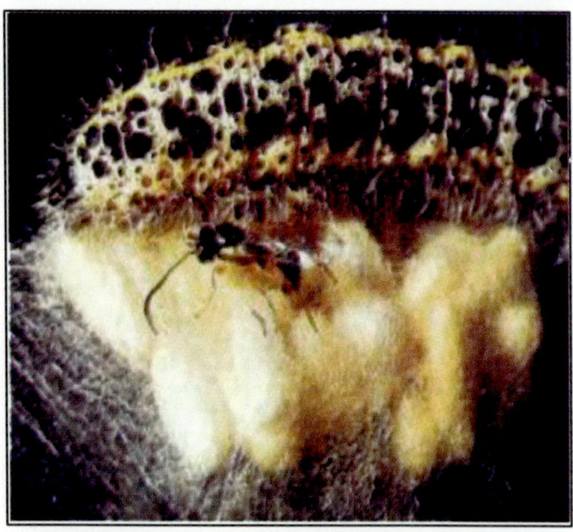

Cocoons of *Apanteles* sp. from Parasitised Larvae of *Pieris brassicae*. An Adult is Emerging from a Cocoon

Spider, Predating on *Spodoptera litura* Larva

***Coccinella septumpunctata* Adults Feeding on Aphids**

Trichogramma chilonis* Parasitizing the Egg of *Helicoverpa armigera

Ichneumonid **Wasp Parasitizing Larvae**

Ichneumonid **Wasp Ovipositing in DBM**

Hymenopteran **Wasp Ovipositng in Potato Tuber Moth**

The major limiting factor, include the extensive crop devastations due to increased pest menace. The insect pests inflict overall crop losses to the tune of 40 per cent in vegetable production. To mitigate the increased demand and limited productivity, high yielding varieties and hybrid vegetable seeds have been introduced in recent past. This has resulted in dramatic changes in insect pest scenario leading to minor pests assuming the status of major, like serpentine leafminer of tomato, *Liriomyza trifolii; Hellula undalis* in cabbage and red spider mites in okra and cucurbits, and have become more serious in recent years. Vegetable farmers by and large depend on chemical pesticides to counter the problem of insect pests. It accounts for 13-14 per cent of total pesticides consumption, as against 2.6 per cent of cropped area. Excessive and indiscriminate use of pesticides has triggered several ecological consequences like destruction of natural enemy fauna, effect on non target organisms, residues in consumable food commodities including packaged mineral water and ultimately resistance to the pesticides. Hence, for higher productivity and production of the vegetable crops, it is vital to adopt the Integrated Pest Management (IPM) strategy with a minimum use of pesticides, which are considered as one of the major factors for the burgeoning pest problem.

The diamondback moth, a worldwide pest of crucifers, such as cabbage, is resistant to several insecticides and is often the most numerous caterpillar attacking crucifers six species of parasites attacked diamondback moth in the study. Although parasitism by the major species (*Diadegma insulare*) was as high as 95 per cent in some plots, parasitism by this insect was not increased significantly in intercrop or trap crop treatments. Parasitism by the second most common species (*Microplitis plutellae*), however, was increased in the sweet alyssum-safflower-sweet alyssum-cabbage treatment. The lady beetle *Coleomegilla maculate* was the most common predator. Its numbers were considerably higher on cabbage in treatments that contained com in the pollen-shedding stage. This predator was probably attracted to both aphids and pollen in the corn-intercrop treatments because these are important food sources.

Table 18.7: Management of Vegetable Insect-Pests through Bio-agents

Pest Complex	*Bio-agents*	*Recommendation*
Fungal pathogens Soil borne fungal diseases	*Trichoderma viride* for seed dressing	4-6 g talc based NIPROT/1 kg. Seed (Minimum packing is 100 g)
	Trichoderma spp. For nursery bed treatment.	Drench beds with NIPROT @ 5 g per litre of water/40 m area
	Trichoderma spp. For seedling root dip treatment	100 g NIPROT in 7.5 litre water. Dip seedling roots for 10 minutes
	Trichoderma spp. For direct soil application after FYM treatment	2.5 kg/ha
Prophylactic Measures Aphids, jassids, thrips, white flies and early stages of caterpillar pests.	300 ppm. *Azadirachtin* or 1500 ppm. *Azadirachtin*	5 ml/litre (1.25 Litre/ha)
Insects Pests *Spodoptera litura*	Pheromone traps Lures Nuclear Polyhedrosis Virus (NPV)	5 traps/ha. (Distance between traps is 30 meters) 20/ha. 4 changes (Lures to be changed in every 15 days) 250 LE/ha/spray
Helicoverpa armigera	Pheromone traps Lures Nuclear Polyhedrosis Virus (NPV) *Trichogramma chilonis* *Chrysoperla carnea*	5 traps/ha. 20/ha. (4 changes) 250 LE/ha/spray 5 cards/ha/release; 3 release at every 20 days gap 2500 eggs/ha/release
Aphids, thrips, mites and whiteflies	*Chrysoperla carnea*	2500 eggs/ha/release
Rodenticides Rodents	Roban cake 0.005 per cent	600 g/ha

* *Chrysoperla* is effective against both early stage larvae of *Helicoverpa* and sucking pest

In case of vegetable pests, exotic tachinid parasitoid, *Eucelatoria bryani* Sabrosky has established on tomato fruit borer, *Helicoverpa armigera* (Hubner). Inundative releases of egg parasitoid, *Trichogramma brasiliense* (Ashmead) @ 4000 adults/ha/week for six weeks suppressed the attack of fruit borer on tomato. Four exotic parasitoids, *Orgilus jenniae* Marsh, *Chelonus kellieae* Marsh, *Chelonus blackburni* Cameron and *Apanteles subandinus* Blanchard introduced against potato tuber moth, *C. Blackburni* have been recovered from Karnataka and others from Maharashtra.

The indegenous parasitoid, *Apanteles plutellae* (Kurdyumov) effectively checked the population of diamondback moth on cabbage in Gujarat and Karnataka under favourable environmental conditions. In the hill regions of Tamil Nadu, *Diadegma semiclausum* Hellen exercises 70 per cent parasitism of diamondback moth in winter months.

Phytophagous mites on okra and brinjal can be effectively checked by augmentative releases of phytoseiid mites, *Phytoseiulus persimilis* Athias-Hanriot and *Amblyseius tetranychivorus* (Gupta) @ 10 adults plant^{-1}.

Insect-Pests in Fruit Ecosystem

The success rate of introduced natural enemies in pest suppression is very high in orchard ecosystem. Most of the natural enemies today used for the pest suppression have been introduced from their native regions.

Table 18.8: Major Pests of Orchard Ecosystems and their Management

Pests	*Natural Enemies*	*Recommendations*
	MANGO	
Mango leaf webber *Orthaga euadrusalis* (Lepidoptera: Pyralidae)	*Bacillus thuringiensis* var. *Thuringiensis* or *Bt* var. kurstaki	@ 1 kg/ha.
Diaspine Scale *Aspidiotus destructor* Signoret	*Chilocorus nigrita* Fabr.	
Mango Scale *Chloropulvinaria polygonata* (Ckll.)	*Cryptolaemus montrouzieri* Mulsant	20-25 beetles/infested tree
Mango Coccid *Rastrocococcus iceryoides* (Green)	*Cryptolaemus montrouzieri* Mulsant	20-25 beetles/infested tree
Mango mealy bug *Drosicha stebbingi* (Green)	*Coccinella septempunctata* L.	
	C. transversalis F.	
	Menochilus sexmaculatus (F.)	
	Chrysopa carnea (Steph)	
	Beauveria bassiana (bals.) vuill.	
	Rodolia fumida	
Mango hopper, Idioscopus Letheiry	*Coccinella septempunctata* L., *C. transversalis* F.	
	Menochilus sexmaculatus (F.)	
	Araneus sinhagadensis Tikader	
	Beauveria bassiana (bals.) vuill; *Verticillium lecanini* (Zimm.); *Aspergillus flavus* Link	
Mango shoot borer, *Chlumetia transversa* Walker	*Goryphus* sp.; *Barcon greeni* Ashmead, *Meteorus* sp., *Araneus singhagadensis*, *Stagodyphus sarasinorum*; *Serratia marcescens* Bizio.	
Oriental fruitfly, *Dacus dorsalis* (Hend.)	*Bracon fletcheri* Silv.; *Opius incisi* Silv.; *O. watersi* (Full.); *O. fletcheri* Silv; *Dirhinus luzonensis* Roh.	
Mango shoot gall psylla, *Apsylla cistellata* Buckton	Micromus timidus Hagea; *Bacca pulcherrima*	

Contd...

Table 18.8–*Contd...*

Pests	*Natural Enemies*	*Recommendations*
	CITRUS	
Citrus butterflies	*Trichogramma chilonis* Ishii	50,000 adults/ha
	Bt. var. kurstaki	A spray of *Bacillus thuringiensis* var. kurstaki (0.5 kg/ha)
Citrus leaf miner *Phyllocnistis citrella*	*Mallada boninensis*	@20-50/tree
Citrus blackfly *Aleurocanthus woglumi*	*Eretmocerus serius* Silvestri	@ 2000 parasitoids/tree
	Amitus hesperdium Silvestri	
	Encarsia clypealis Silvestri, *E. opulenta* Silvestri, *E. citrifoliai*	
	Chrysoperla sp.	
	Pullus xerampelonus	
	Aschersonia sp.	
	Cladosporium sp.	
	Verticillium lecanii	
Dialeurodes citri Ashmead	*Encarsia lahorensis* Howard	@ 2000 parasitoids/tree
California red scale, *Aonidiella aurantii* Maskell	*Aphytis melinus* DeBach	@ 2000 parasitoids/tree
	Chilocorus nigrita Fabricius	@ 10 beetles/tree
Florida red scale*Chrysomphalus aonidum* Linnaeus	*Chilocorus nigrita* Fabricius	
	Aphytis spp.	
Purple scale *Lepidosaphes beckii*	*Aphytis lepidosaphes* Compere	
Citrus mealy bug Planococcus citri Risso	*Coccidoxenoides peregrina* Timberlake	@ 2000 parasitoids/tree
	Leptomastix dactylopii Howard	10 beetle/infested plant
	Cryptolaemus montrouzieri	
	GRAPE	
Grape mealy bug, *Maconellicoccus hirsutus* (Green)	*Allotropa japonicum* Ashmead	60,000 parasitoids/ha.
	Anagyrus dactylopii Howard	60,000 parasitoids/ha.
	Cryptolaemus montrouzieri Mulsant	600-2500 per ha
	Scymnus coccivora Ayyar	600-2500 per ha
Leaf roller, *Sylepta lunalis* Guien	Apanteles clita Nixon (larval parasitoid) and *Cardiochiles fulvus* Cameron (prepupal parasitoid)	
	GUAVA	
Aphid, *Aphis gossypii* Glover	*Cheilomenes sexmaculata* Fabricius	50 grubs of beetle/plant
	Mallada boninensis (Okomato)	50 larvae/plant
Green shield Scale *Chloropulvinaria psidii* (Maskell)		
Mealy bug, *ferrisia virgata* (Ckll.)	*Blepyrus insularis* Cameron	60,000 parasitoids/ha

Contd

Table 18.8–*Contd...*

Pests	*Natural Enemies*	*Recommendations*
	APPLE	
San Jose Scale *Quadraspidiotus perniciosus*, Apple	*Encarsia perniciosi* (Hymenoptera)	@ 2000/scale infested tree once in spring
	Aphytis proclia (Hymenoptera)	- do-
	Chilocorus bijugus (Coleoptera)	@ 20/50 per infested apple plant.
Wooly aphis, *Eriosoma longigerum*, Apple	*Aphelinus mali* (Hymenoptera)	Twigs with parasitized aphids placed in wire gauge cages
	Coccinella septumpunctata (Coleoptera)	@ 20/50/infested plant in April
Codling Moth, *Cydia pomonella*, Apple	*Trichogramma embryophagum* (Hymenoptera)	@ 2000 adults/plant

Conclusion

Hill agriculture is comparatively more vulnerable to insect pest infestation due to the occurrence of varied climatic conditions. Locational specificity in hills is determined by factors like altitudinal differences, steepness of slopes, directions of mountain and valley etc. Micro-climatic specificity of some hilly pockets results in development of hot spot for certain pests. Because of rich organic matter, light textured soil and major area (about 90 per cent) under rainfed condition, white grubs have become key pest of the region. In addition, the intensive agricultural technologies in recent past have added to pest related problems. Intensive vegetable farming especially in valley areas of hills has witnessed increased pest build up during recent past and the management of pests has been mostly relied on the chemical means, which in turn poses several ill effects such as killing of natural enemies, environmental pollution, health hazard etc. One of the strongest alternatives for the management of the insect pests would be integrated pest management with emphasis on biological control using potential insect parasitoids and microbial pathogens. Consideration of the biological and ecological needs of natural enemies is critical for the success of any biological control effort. In practice, effective biological control may require that several tactics be combined to achieve effective pest management. Biological control should be exploited as a method for the control of pest arthropods in hill agro-ecosystem. As summarized by Tauber *et al.* (1985), research needs in biological control "form a continuum from the very basic to the applied.."

References

Ali, M. (1998). Research, Development and management for production of pulses.In : IPM System in Agriculture.Volume 4. Pulses, R.K.Upadhyay, K.G.Mukerji and R.L.Rajak (Eds.) Aditya Books Private Limited, New Delhi. pp1–40.

Anonymous (2011). Crop prospects and food situation. http://www.fao.org/worldfoodsituation/wfs-home/en/

Atwal, A.S. and Singh, B. (1990). Pest population and assessment of crop losses, Indian Council of Agricultural Research, New Delhi. 131 p.

Bellows, T. S. and Fisher, T. V. (1999). Handbook of biological control. Academic Press, New York. pp1015.

Chiri, A.A. and Legner, E.F. (1988). Field applications of host searching kairomones to enhance parasitisation of the pink bollworm (Lepidoptera: Gelechidae). *Journal of Economic Entomology*, 96(2): 254-255.

DeBach, P. (1974). *Biological Control by Natural Enemies*. Cambridge University Press, London. 323 pp.

Dhaliwal, G.S. and Arora, R. (2001). Integrated pest management Ludhiana, Kalyani Publishers. 427p.

Dhaliwal, G.S. and R. Arora.(1996). Principles of insect management. Commonwealth Publishers, New Delhi.pp350.

FAO. (1967). Report of the first session of the FAO panel of experts on Integrated Pest Control, Rome, Italy, Sept. 18-22, 1967, 19 p.

FAO. (1980). Research Summary. Integrated pest management. EPA-600/8-80-044. 28 pp.

Flint, M. L. and Van den Bosch, R. (1981).Introduction to integrated pest management. Plenum Press, New York, 240 p.

Garg, D.K.(1992). Insect pests of agricultural crops in UP hills and their management. Almora VPKAS, 25 p.

Hawtin, G. (1998). Conservation of agro-biodiversity for tropical agriculture. In: Crop Productivity and Sustainability: Shaping the Future, edited by V.L.Chopra, R.B. Singh and A. Varma. Oxford and IBH, New Delhi. pp. 917-925.

Hoy, M. A.(1994). *Insect Molecular Genetics*. Academic Press, San Diego, CA.546 pp.

Jayaraj, S. (1992). Integrated pest management. In: Pest management and pesticides : Indian Scenario. Edited by B.V.David, Chennai, Namrutha Publications. pp 7-16.

Johnson, M. W. and Tabashnik, B. E. (1994). Laboratory selection for pesticides resistance in natural enemies. In : *Application of Genetics to Arthropods of Biological Control Significance.* Narang, K. S., Bartlett, A. C. and Faust, R. M. (Eds.). CRC Press, Boca Raton.pp91-105.

Kumar, V. and Patil, B. V. (2006). Natural Enemy Complex of Paddy Insect Pests in Tungabhadra Project Area of Karnataka. *Karnataka J. Agric. Sci.*,19(3): 544-548.

Paroda, R.S. and P. Kumar. (2000). Food production and demand situations in South Asia. *Agricultural Economics Research Review.* 13(4): 1-25.

Pedigo, L.P.(1989). Entomology and Pest Management. Macmillan Pub. Co., Inc.N.Y. 646 pp.

Poehling, H.M. (1989). Selective application strategies for insecticides in agricultural crops pp 151-175. In Jepseon, P.C (ed). *Pesticides and Nontarget invetebrates.* Intercept, Wimborne, U.K.

Ridgway, R. L. and S. B. Vinson. (1977). *Biological Control by Augmentation of Natural Enemies*. Plenum Press, New York. 480 pp.

Sachan, J.N.(1992).Present status of *Helicoverpa armigera* in pulses and strategies for its management.In: Sachan, J.N.(ed.) *Helicoverpa* Management. Current and future strategies.Proc. of first National Workshop held at Directorate of Pulses, Kanpur, 30–31 August, 1990.

Shankar, U., Kumar, D. and priya, S. (2008). Vegetable Pest Management Guide for Farmers. International Book Distribution Company, Lucknow, U. P. 181 p.

Sigsgaar, L. (2000). Early season natural biological control of insect pests in rice by spiders - and some factors in the management of the cropping system that may affect this control. *European Arachnology*. pp. 57-64.

Sithanatham, G., M.Balasubramanian and S.Chelliah. (1981).Cotrol of *Heliothis armigera* on red gram with NPV and Insecticides. Madras Agric.J. 68:417–420.

Smith, R. F. and Reynolds, H.T. (1966). Principles, definitions and scope of integrated pest control. Proc. FAO Symposium on Integrated Pest Control 1: 11-17.

Smith S. M. 1996. Biological control with Trcihogrammatids : Advances, successes and potential of their use. *Annual Review of Entomology*. 115 : 375-406.

Srivastava, K.P. (1993). Applied Entomology-II, Ludhiana, Kalyani Publishers, 424 p.

Stern, V. M., Smith, R.F., van den Bosch, R and Hagen, K.S.(1959). The integrated control concept. *Hilgardia*, 29: 81-101

Thomas, M. (1990). Diversification of arable ecosystem to control natural enemies of cereal aphids. *Game Conservancy Review*. 21: 68-69.

Thomas, M.B., Wratten, S.D. and Sotherton, (1991). Creation of island habitats in farmland to manipulate populations of beneficial arthropods: predator densities and emigration. *Journal of Applied Ecology* 28: 906-917.

Van Driesche, R. G. and T. S. Bellows, J.R. (1996). *Biological control*. Chapman and Hall, New York.

William, I. S., Jones, T. H. and Hartley, S. B. (2001). The role of resources and natural enemies in determining the distribution of an insect hervobore population. *Ecological Entomology*. 26 : 204-11.

Yazdani, S.S. and Agarwal, M.L. (1997). Elements of insect ecology. Narosa Publishing House, New Delhi. 210 p.

Hill Agriculture: Prospects, Constraints and Mitigations (2013) *Pages* **270–298**
Editor: **Dr. Shahid Ahamad**
Published by: **DAYA PUBLISHING HOUSE, NEW DELHI**

Chapter 19

Insect Pest Management of Vegetables Crops in Hill Agri-Ecosystem

Kuldeep Srivastava[1], *Devinder Sharma*[1], *P.K. Rai*[2] *and Manoj Kumar*[3]

[1]*Department of Entomology, FOA, SKUAST-J, Chatha – 180 009*
[2]*Krishi Vigyan Kendra (SKUAST-J), Jammu*
[3]*Division of Vegetable Science, Sher-e-Kashmir University of Agricultural Sciences and Technology, Jammu*

Hill agriculture is comparatively more vulnerable to insect pest infestation due to the occurrence of varied climatic conditions. Locational specificity in hills is determined by factors like altitudinal differences, steepness of slopes, directions of mountain and valley etc. Microclimatic specificity of some hilly pockets results in development of hot spot for certain insect pests. Because of rich organic matter, light textured soil and major area (about 90 per cent) under rainfed condition, cut worms, white grubs, termites etc. have become key pest of the region. In addition, the intensive agricultural technologies in recent past have added to pest related problems. Intensive vegetable farming especially in valley areas of hills has witnessed increased pest build up during recent past and the management of pests has been mostly relied on the chemical means, which in turn poses several ill effects such as killing of natural enemies, environmental pollution, health hazard etc. and therefore, Indian Government is pressing hard for organic farming. The major challenge in the event of this farming would be the management of insect pests and diseases. Hence, for higher productivity and production of the crops, it is vital to adopt the Integrated Pest Management (IPM) strategy with a minimum use of pesticides, which are considered as one of the major factors for the burgeoning pest problem.

India is the second largest vegetables producing countries next to China and occupies 6.09 million ha area with an annual production of 84.8 million tones. Majority of Indians are vegetarian, with a per capita consumption 135 g per day as against the recommended 300 g per day. India is still lagging behind in terms of recommended diet level. In near future, there is a need of around 56 million tons of food to feed our 1.3 billion Indian population expected by the year 2020. Our natural resources are scarce and day by day they are shrinking and reducing due to the expanding urbanization and deforestation. In this context, to feed our burgeoning population we must have to look after the vertical expansion through increase in the intensity of vegetable crops; diversification of crops and through mixed and intercrops. These will only the solution to meet the requirements of growing demands in future.

The major limiting factor, include the extensive crop devastations due to increased pest menace. The insect pests inflict overall crop losses to the tune of 40 per cent in vegetable production. Yield losses due to major insect pests and their relative crops in vegetables are portrayed in Table 19.1. To mitigate the increased demand and limited productivity, high yielding varieties and hybrid vegetable seeds have been introduced in recent past. This has resulted in dramatic changes in insect pest scenario leading to minor pests assuming the status of major, like serpentine leafminer of tomato, *Liriomyza trifolii*; *Hellula undalis* in cabbage and red spider mites in okra and cucurbits, and have become more serious in recent years. Vegetable farmers by and large depend on chemical pesticides to counter the problem of insect pests. It accounts for 13-14 per cent of total pesticides consumption, as against 2.6 per cent of cropped area. Excessive and indiscriminate use of pesticides has triggered several ecological consequences like destruction of natural enemy fauna, effect on non target organisms, residues in consumable food commodities including packaged mineral water and ultimately resistance to the pesticides. Hence, for higher productivity and production of the vegetable crops, it is vital to adopt the Integrated Pest Management (IPM) strategy with a minimum use of pesticides, which are considered as one of the major factors for the burgeoning pest problem.

Table 19.1: Major Insect Pests of Various Vegetables in North West Himalayas

Pest	*Nature of Damage*	*Crops*
Helicoverpa armigera	Defoliation and fruit boring by larvae	Tomato, pea, onion
Leucinodes orbonalis	Larvae feed on shoots and fruits	Brinjal
Aulacophora foveicollis	Grubs feeds on roots and under ground parts of stems and adults feeds on leaves, flower buds, flower etc.	Cucurbitaceous vegetable
Trichoplusia orichalcea	Caterpillar feeds on leaves and flowers	Cabbage, cauliflower
Plutella xyllostella	Caterpillar feeds on leaves	Cabbage, Cauliflower
Agrotis segetumA.ipsilon	Larvae feed on root of the plants	Brinjal, capsicum, chillies, tomato, cole crops etc.
Dacus sp.	Maggot feeds on fruits	Squash, bottle gourd
Lampides boeticus	Pod boring by larvae	Pea
Phytomyza horticola	Maggots feeds on leaf tissues by making zig-zag tunnel	Pea
Brevicoryne brassicae	Nymphs and adults suck the sap from plants	Cabbage, cauliflower
Myzus persicae	As above	Potato
Chauliops choprai	As above	French bean
Platypria hystrix	Adult and grub feed on leaves and make typical white patches	French bean
Bagrada sp., *Nezara viridula*	Adults and nymphs suck the sap from plants	Cabbage, cauliflower, French bean
Anomala dimidiata *Holotrichia* sp. and other species of white grubs	Grubs are root feeding and adults are defoliators	All vegetables
Mylabris pustulata and other species of *Mylabris*	Adults feed on all parts of plants	French bean, okra, potato

Pest Management Strategies

Integrated pest management was practiced knowingly or unknowingly by human being since time immemorial. However, IPM as an organized concept was evolved since its introduction in the late 1950s, when the focus was on combining suitable methods to limit pests in a crop, within the framework of sustainable agricultural development. The concept of IPM has been defined in different ways by different workers. Stern *et al.* (1959) defined applied pest control, which combines and integrates biological and chemical control. Chemical control is used as necessary and in a manner, which is least disruptive to biological control. Integrated control may make use of naturally occurring biological control as well as biological control affected by manipulated or induced biotic agents. Smith and Reynolds (1966) defined integrated pest control as a pest population management system that utilizes all suitable techniques in a compatible manner to reduce pest populations and maintain them at levels

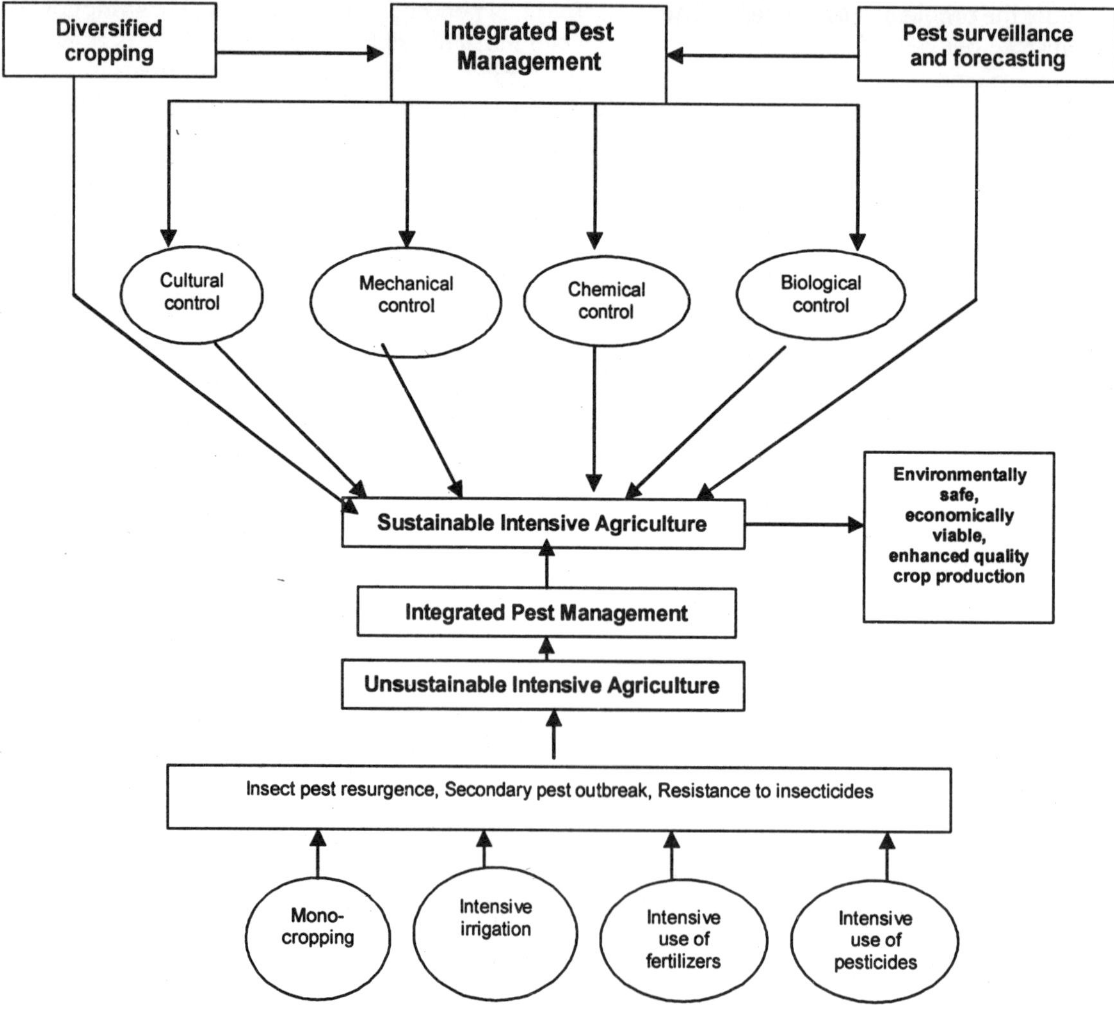

Figure 19.1: Diagrammatic Representation of Integrated Insect Pest Management

below those causing economic injury. FAO, 1967 defines integrated control as a pest management system that in the context of the associated environment and the population dynamics of the pest species, utilizes all suitable techniques and methods in as compatible manner as possible and maintains the pest populations at levels below those causing economic injury. While FAO, 1980 defines it as an interdisciplinary approach incorporating the judicious application of the most efficient methods of maintaining pest populations at tolerable levels.

Ideally, an integrated pest management program considers all available pest control actions, including no action, and evaluates the potential interaction among various control tactics, cultural practices, weather, other pests, and the crop to be protected (Flint and Van den Bosch, 1981). Pedigo, 1989 defined integrated control as control of pests that emphasizes selective use of insecticides so as to conserve natural enemies in the agro-ecosystem. Allen and Rajotte, 1990 defined IPM as a systematic approach to crop protection that uses increased information and improved decision-making paradigms to reduce purchased inputs and improve economic, social, and environmental conditions on the farm and in society. Moreover, the concept emphasizes the integration of pest suppression technologies that include biological, chemical, legal, and cultural controls.

Different pest management strategies or practices are being used by the farmers to reduce the pest infestation. The most important tool for evaluating an insect problem is its economic threshold. The economic threshold is the pest density at which management action should be taken to prevent an increasing pest population from doing economic damage to the crop. The economic threshold takes into consideration the market value of the crop, injury and damage, insect density, lag time before implementation occurs, and cost of management. This threshold information is important in distinguishing the difference between minor plant injury and true economic damage. Economic threshold information is available for some, but not all insect pests. When economic threshold information is not available, other expertise must be used to determine damage potential of a given pest to a crop. Although the concept of the economic threshold can be applied to the use of many tactics, it is primarily used in association with insecticidal control. The most common strategies used for insect control are chemical and cultural practices. There are different insecticides available in the market with different modes of action. Insecticides have the ability to inflict high mortality quickly, and are the most common and effective form of rescue treatment. These chemicals vary in toxicity to humans or other animals, and also in their potential impact on the environment.

Use of resistant crop varieties is another useful practice in managing insect pests. By using resistant varieties, other insect control measures may not be needed because the crop is less likely to be economically injured by the insect pests. Biological control is the use of any biological agent to reduce insect pest populations. Predators, parasitoids, and pathogens are common and effective biocontrol agents. Biological control agents may occur naturally in the cropping environment or may be introduced as needed. This practice usually has low environmental impact but often requires a long time before natural enemy populations are established and control is achieved.

The IPM principle does not preclude the chemical pesticide use, instead it uses it as one of the tools in the management package, to be used prudently integrated with other tools. Thus, the concept of integrated pest management of insects contains three basic elements. These include: maintaining insect populations below levels that cause economic damage, the use of multiple tactics to manage insect populations, and the conservation of environmental quality.

Sustainable Agriculture and IPM

Sustainable agriculture is a system of agriculture that is ecologically sound, economically viable and socially acceptable in the short as well as long term. It avoids specific set of farming practices, rather represents the goal of developing a food production system that: i) yields plentiful, affordable, high-quality food and other agricultural products, ii) does not deplete or damage natural resources (such as germplasm base, soil, water, wildlife etc.), iii) promotes the health of the environment, iv) supports a broad base and diversity of farms and the health of rural communities.

IPM and sustainable agriculture share the goal of developing agricultural systems that are ecologically and economically sound. IPM may be considered as a key component of a sustainable agriculture system. The hill ecosystem is considered as very fragile and the conventional agriculture is followed in most part of the hilly region. However, there is a bright scope for organic farming in the hills due to less usage of inorganic chemicals. Primary strategy in organic farming for the management of insect pests and diseases is the prevention through proactive strategies which include healthy, biologically active soils (increasing belowground diversity), habitat for beneficial organisms (increasing above ground diversity) and appropriate plant cultivars and maintenance of crop diversity. Hence, ensures minimal losses from the pests.

Integrated Pest Management (IPM)

Integrated pest management is a systems approach to pest control that combines biological, cultural, and other alternatives to chemical control with the judicious use of pesticides. The objective of IPM is to maintain pest levels below economically damaging levels while minimizing harmful effects of pest control on human health and environmental resources. IPM is a dynamic and constantly evolving system in which all the suitable control tactics and available surveillance and forecasting information are combined into a holistic management programme delivered to the farmer at requisite interval as part of the sustainable crop production technologies. In another words, IPM aims at combining all available methods or tools of insect pest control in a judicious manner that minimizes insecticide use and disturbance to the ecosystem but becomes a multidisciplinary one.

Integrated pest management helps the farmers to reduce the dependence on chemical pesticides and ecological deterioration. IPM includes all possible and available techniques like cultural, physical, mechanical, biological control agents, host plant resistance, regulatory methods and all conventional non-chemical and chemical methods of pest control. Indian agriculture, which is going through a transition phase, is slowly but surely adopting the ways and means of pest management for long term sustainable agriculture.

Use of an IPM program enables the farming community to:

- Maintain effective and economical pest control over the long term
- Minimize pesticide usage/pressure/load on various field crops
- Minimize risks to human health and the environment
- Reduce the risk of the development of pesticide resistant, pest resurgence and secondary pest outbreak in pest populations.

Factors Influencing Pest Population in Hills

The insect pest population is drastically influenced by any change in the environmental factors. The environmental factors are broadly grouped into abiotic and biotic factors.

Abiotic Factors

The population density of pest fluctuates with the prevailing weather conditions such as temperature, moisture, light, wind, edaphic factors etc. Since, insects are poikilothermic animals *i.e.*, they do not posses temperature regulation system of their body, their body temperature, more or less, follows the surrounding temperature. Exposure to temperature beyond the favourable range, whether low or high, may retard growth and development of the insects or may even cause their death. The species adapted to live in temperate climate, hibernate in a particular developmental stage. A constant supply of moisture is essential for metabolic reactions as well as for the dissolution and transport of salts. The water content in insects varies from less than 50 per cent to more than 90 per cent of the total body weight. Excessive moisture can prove harmful to the insect population by affecting normal development and by encouraging disease-causing micro-organisms. Edaphic factors such as structure, texture and composition of soil, drainage, topography etc. also affect the distribution and abundance of the insects. Example, the cut worm, *Agrotis ipsilon* larva live in light texture soil in which they move around freely in response to daily or seasonal temperature and moisture changes. Major topographic factors that affect biography of insects include height of mountain chains, steepness of slope, directions of mountains and valleys and exposure of the slope. These factors also affect the climate of an area, thus influences the distribution of certain insects. However, their impact on the status of the major insect pests has yet to be quantified. Mountain ranges also act as physical barriers to the spread of insects (Yazdani and Agarwal, 1997).

Biotic Factors

The growth and development of insects is largely influenced by biotic factors such as natural enemies and food. The natural enemies of insect pest population include predators, parasitoids and disease causing microorganisms such as fungi, bacteria, viruses etc. The natural enemies are also influenced by various environmental factors such as weather and hyper parasitism. Weather is known to determine the effectiveness of the natural enemies. Example, in the control of apple wooly aphid with *Aphelinus mali*, the prevalent temperature in spring would determine the effectiveness of the parasitoid. Similarly, quality and quantity of food play important role in insect's survival, longevity, distribution, reproduction, speed of development etc. (Dhaliwal and Arora, 2001; Atwal and Singh, 1990).

Hill Agriculture Dynamics and Insect Pest Problem

With the increase in the productivity and production of the agricultural crops, by intervention of high yielding varieties, agro-chemical, irrigation, monoculture etc., there have been tremendous changes in the pest scenario of the crops (Dhaliwal and Arora, 2001, Atwal and Singh, 1990, Anonymous, 1992-2004). Agronomic recommendations for exploitation of full potential of HYVs has certainly helped in enhancing crop production but at times it has provided congenial condition for many insect pests. Close spacing of paddy has resulted in the increased population of BPH, gall midge and leaf folder. In North West Himalayan Region, cultivation of spring rice is the most common practice which provides availability of the rice crop for a quite longer duration in the field, hence promotes pest build up in the region. Routine crop rotation in some cases may provide congenial environment for insect pest multiplication. Cultivation of off-season vegetables in hills provides food for many insect pests round the year. The tobacco caterpillar, *Spodoptera litura* and tomato fruit borer, *Helicoverpa armigera* were found to damage cabbage, cauliflower, broccoli, tomato, chilli, capsicum etc. round the year. Destruction of forests for cultivation of agricultural crops led to the migration of insect pests of forest, mainly white grubs to agricultural fields for egg laying and subsequent damage to the crops. So far, authors have recorded about 40 species of white grubs from the region. Among which, many of them are considered as major pests of agricultural crops.

Human preference for certain agricultural crops like rice, wheat, maize prompted farmers to grow more of these crops. Despite the rich biodiversity in agriculturally important crops, wheat, maize and rice accounts for 60 per cent of plant derived calories in human diet (Hawtin, 1998). In hill regions, some of the less preferred crops like finger millet and barnyard millet are now gaining momentum because of their nutritive value. Monoculture of certain crops is considered as one of the major factor for increased insect pest problem. Mono-cropping of vegetable crops especially in the valley areas prompted the chances of outbreaks of many insect pests.

Nutritional requirement of the crops are mostly met with the application of FYM in traditional farming of the hills. However, application of compost, which is not well decomposed, in the rainfed conditions also helps in increasing the white grub problem in the region.

Modern agriculture has undoubtedly contributed a lot in increasing the crop productivity as well as crop production all over the country, but at the same time it has provided congenial conditions for multiplication of many of the insect pests. Therefore, before advocating any practice for crop production it has to be thoroughly tested from insect pest point of view.

Status of Major Vegetable Insect Pests

Vegetables are the prime source of vitamins and minerals and also the important source of income generation among the farming communities. The commercial cultivation of these vegetables is *four times more remunerative* than the cereals and fetches higher income and employment to the rural people. Among the various vegetables that are produced, the summer vegetables *viz.*, okra, tomato, brinjal, chillies, and cucurbits especially pumpkin, ash gourd, bitter gourd, snake gourd, ridge gourd, cucumber etc. are the important ones. The major constraints in the production of these crops is the ravages caused by various insect pests and diseases which not only result in the low production of vegetables but also drastically impair the quality of these, rendering them unfit for human consumption, reducing marketable yield and thereby posing a serious threat to farming communities. Insect pests plague everyone who has ever engaged in raising vegetables, from kitchen gardener to researcher horticulturists and commercial vegetable growers. Among the various insect pests that attack these vegetable crops, the most important ones are tomato fruit borer, *Helicoverpa armigera* Hubner; whiteflies, *Bemisia tabaci* Gennadius; tomato leaf minor, *Lariomyza trifolii*; tobacco caterpillar, *Spodoptera litura* Fabricius; Fruit flies, *Bactrocera cucurbitae*, Red pumpkin beetle, *Aulacophora foevicollis*, Cutworm, *Agrotis ipsilon* and mites (especially red and yellow mites). These pests render 50 to 80 per cent losses under severe infestation causing drastic reduction in the marketable yields.

Vegetables have enough potential under hill agro-ecosystem especially in valley areas wherever irrigation facility is available. Major vegetables grown in hills are brinjal, cabbage, cauliflower, onion, pea, capsicum, french bean, chillies, squash etc. besides several off season vegetables such as toma o, capsicum, cauliflower etc. Heavy infestations by the insect pests have been observed on veget ble crops due to prevailing congenial conditions for the growth of the insects in the hilly areas. Crop wise important insect pests of NWHR are discussed hereunder along with available management strategies. (Anonymous, 1992-2004, Anonymous, 2002, Garg, 1992, Bhatt and Sushil, 2004).

Table 19.2: Insect Pests of Cole Crops

Common Name	*Scientific Name*	*Order and Family*	*Status*
	Lepidopteran insect pest complex		
Cabbage caterpillar (large and small)	*Pieris brassicae, Pieris rapae*	Lepidoptera: Pieridae	Major/Minor
Tobacco caterpillar	*Spodoptera litura, S. exigua*	Lepidoptera: Noctuidae	Major/Minor
Cabbage semilooper	*Trichoplusia ni*	Lepidoptera: Noctuidae	Major
Cabbage head borer	*Hellula undalis*	Lepidoptera: Pyralidae	Major
Cabbage borer	*Helicoverpa armigera*	Lepidoptera: Noctuidae	Major
Diamondback moth	*Plutella xylostella*	Lepidoptera: Plutellidae	Major
Tussock moth	*Orgyia* spp.	Lepidoptera: Lymantridae	Minor
	Coleopteran insects		
Flea beetles	*Phyllotreta cruciferae*	Coleoptera: Chrysomelidae	Minor
	Monolepta signata	Coleoptera: Chrysomelidae	Minor
Dipteran insects			
Cabbage maggot	*Delia antique*	Diptera: Anthomyiidae	Minor
Cabbage leaf miner	*Liriomyza trifolii*	Diptera: Agromyzidae	Minor
Fruit fly	*Bactrocera tau*	Diptera: Tephritidae	Major
	Homopteran insects		
Aphids	*Bravicoryne brassicae* *Myzus persicae* *Lipaphis erysimi*	Homoptera: Aphididae	Major
	Hemipteran insects		
Painted bug	*Bagrada cruciferarum*	Hemiptera: Pentatomidae	Minor
Spined bug	*Cletus* spp.	Hemiptera: Coreidae	Minor
	Orthropteran insects		
Grasshoppers	*Chistocera gregaria*	Oethroptera: Acrididae	Minor
Katydids	*Microcentum retinerve*	Oethroptera: Tettigonidae	Minor
	Molluscs		
Snails	*Achantia fulica*		Minor

Table 19.3: Yield Losses Due to Major Insect Pests in Important Vegetable Crops

Major Crops and their Relative Insect Pests	*Yield Losses (per cent)*
Chillies	
Chilli thrips (*Scirtothrips dorsalis*)	12-90
Chilli mites (*Polyphagotarsonemus latus*)	34
Brinjal	
Fruit and shoot borer (*Leucinodes orbonalis*)	11-93

Contd...

Table 19.3–*Contd...*

Major Crops and their Relative Insect Pests	*Yield Losses (per cent)*
Tomato	
Fruit borer (*Helicoverpa armigera*)	24-65
White flies (*Bemesia tabaci*)	30
Okra/Lady's finger	
Shoot and fruit borer (*Earias vitella*)	23-54
Jassids/Leafhoppers (*Amrasca bigutulla bigutulla*)	54-66
White flies (*Bemesia tabaci*)	54
Fruit borer (*Helicoverpa armigera*)	22
Cucurbit crops	
Red pumpkin beetle (*Aulacophora foevecollis*)	30-50
Cucurbits mites (Red mites)	20-25
Cut worms (*Agrotis ipsilon*)	NA
Fruit flies (*Bactrocera cucurbitae*)	
Bitter gourd	60-80
Cucumber	20-39
Muskmelon	76-100
Snake gourd	63
Sponge gourd	50

Table 19.4: Economic Threshold Level (ETL) for Major Vegetable Insect Pests

Insect Pests	*Economic Threshold Level (ETL)*
Chilli thrips (*Scirtothrips dorsalis*)	2 thrips/leaf
Chilli mites (*Polyphagotarsonemus latus*)	1 mite per leaf
Brinal Fruit and shoot borer, BFSB (*Leucinodes orbonalis*)	0.5 per cent shoot and fruit damage 0.91 per cent on shoots and 0.67 per cent on fruits
Brinjal white fly (*Bemesia tabaci*)	5-10 flies per leaf
Tomato fruit borer (*Helicoverpa armigera*)	8 eggs in 15 plants or one larva per plant
Cut worms (*Agrotis ipsilon*) on potato	One larvae per ten plants or 2 larvae/metre row
Bhendi fruit borer (*Earias vitella*)	5.3 per cent infestation of fruits
Okra Jassids/Leafhoppers (*Amrasca bigutulla bigutulla*)	4-5 nymphs per plant.
Onion thrips (*Thrips tabaci*)	13-15 thrips per plant 15 days after transplanting

Cruciferous Crops

Cabbage Butterfly (*Pieris brassicae* and *P. rapae*)

It is also known a imported cabbage worm. Two species of cabbage butterflies *i.e.*, large butterfly, *Pieris brassicae* and small butterfly, *P. rapae* have been recorded and found damaging the cole crops in Jammu province. They are serious threats to the crops grown for seed purpose especially in hills.

Tabacco Caterpillar (*Spodoptera litura* and *S. exigua*)

Spodoptera litura has a wide host range and economic insect pest of cole crops but it is usually kept under damaging levels by controls targeted to lepidopteran pests of cole crop. Its population rises in the spring coincides with abundance of other vegetable crops production in Jammu. Many weeds plant also serve as hosts, including bathua, choulai and parthenium. The young larvae feed gregariously and scrape the leaves and later on it may completely defoliate the leaves. Since adults can readily invade a field from nearby crops or weeds, monitoring the crop twice a week for Spodoptera damage and their presence is recommended.

Cabbage Semilooper (*Trichoplusia ni, T. orichalsea*)

As the name indicates, caterpillars are pale green in colour and moved with remarkable loop movement. It is polyphagous insect and defoliates a wide variety of crops including cole crops. The insect remains active and found throughout the winter months and their peak populations are observed in September-October and again during February-March in Jammu province. Its larvae damage the crops by chewing irregular holes in leaves. The feeding damage and excrement left behind on heads by semilooper caterpillar make cabbage and cauliflower unmarketable.

Cabbage Head Borer (*Hellula undalis* and *Helicoverpa armigera*)

Cabbage head borer is seen sporadically but its attack coincides with the head/curd initiation stage and a single larvae is responsible to damage the entire head/curd. The early instar may feed on the growing point, and bore inside the head and damaging the head and excrements which prevents head formation. When fully grown, larvae pupate inside the head, or on stems, or in soil. Helicoverpa armigera larvae have also been recorded at several places in Jammu region as head borer of cabbage. It may be due to absence of preferred host in winter months or some environmental and behavioural change.

Diamondback Moth (*Plutella xylostella*)

Diamondback moth became the main and universal pest of cole crops and it remains one of the most serious problems for the intensive cole crops growing areas. Plants at all stages of growth may be attacked by this pest. In Jammu province, diamondback moth is recorded on late season cole crops and on crucifers *i.e.,* from February to April-May.

Tussock Moth (*Orgyia* spp.)

It is polyphagous pests on various crops and seen sporadically on cole crops. It has also been found defoliating the leaves of cole crops.

Flea Beetles (*Phyllotreta cruciferae* and *Monolepta signata*)

Flea beetles are common insects in cole crops, and can cause damage by retarding growth of crops. The larva live in the soil and feed upon the roots of the host plants. The adult feed on the leaves of young plants leading to a characteristic 'shoot hole' appearance. Severe infestation results in loss of plant vigour, wilting and death. The stem and flowers may also be attacked.

Cabbage Leaf Miner (*Liriomyza* spp.)

It is one of the polyphagous pests causing serious damage to several crops including cole crops. Damage is done by mining into leaves and petiole by the larvae. Due to loss of chlorophyll containing cells, the photosynthetic activity of the plants is adversely reduced.

Cabbage Aphids (*Bravicoryne brassicae, Myzus persicae* and *Aphis gossypii*)

Aphids are small, soft-bodied, slow-moving insects. They are often found in large colonies on the undersurface of leaves. They are the major pests of cole crops and if not controlled in time it may cause serious damage to the crops. It appears in November and remain active till April. Cloudy and moist weather favours rapid multiplication of this pest. Both nymphs and adults are similar in appearances and suck the sap constantly from the leaves which results into wilting, stunting and leaves distortion in the plants.

Bagrada Bug (*Bagarada cruciferum*)

This is mainly a pests of crucifers but can not cause significant damage to the cole crops. The nymphs and adults suck the sap causing the patchy whitish growth on leaves.

Solanaceous Crops

The one of the important and economic group of vegetables including brinjal (*Solanum melongena*), tomato (*Lycopersicon esculentum*) and chillies (*Capsicum annum*). These crops are extensively cultivated in South East Asia and in Indian sub- continent for their taste and nutritive value. Tomato is a very popular constituent of Indian diet. These popular crops are devastated by an array of insect pests, however the major losses are caused by fruit borers (*Helicoverpa armigera* and *Spodoptera litura*), serpentine leaf miner (*Liriomyza trifoli*), tomato fruit fly (*Bactrocera tau*), white flies (*Bemesia tabaci*), tomato thrips (*Frankliniella* spp.), aphids (*Aphis gossypii*), red mites (*Tetranychus* spp.) and tomato mealy bug (*Phenacoccus solenopsis*), the later five being the vector also for transmitting the viral diseases in tomato crop. Due to certain aberration in climatic change, some insect pests which were earlier considered as minor pests have now assumed the status of major pests. These include serpentine leaf miner, fruit fly, mealy bugs, red mites, thrips and white flies. Besides these insects, stink bugs are also recorded causing damage to the tomato fruits.

Tomato

Tomato Fruit Borer, *Helicoverpa armigera* Hubner

Helicoverpa is a very serious insect pest of several vegetable crops and has been reported on more than 200 host plants in India. Caterpillars are the immature stage of the insect which causes the considerable damage to the crops in various stages. In the pre-fruiting stage, the caterpillar feeds on the tender foliage including leaves, flowers, buds and even sometimes they make holes in the tender twigs which reflects the perforated look to the crops. After fruiting, the larvae bore large, clear, circular holes into fruits and feed on the pulp. Its name has been changed according to its host. Its larvae are stoutly built and robust. In tomato, it is responsible to destroy many fruits and make them unfit for the human consumption. Its larvae have peculiar feeding habit on host crop. The larvae thrusts its half of the body into the fruits and feeds the inner contents and rendering them unfit for the human consumption. Specific phenomenon cannibalism is found among its larvae. A great colour variation have also been observed among its larvae which scientists opined that it is due to its substrate/host food consumption but the recent studies proved that it is due to their genomic make up. It causes damage up to the tune of 50 to 80 per cent if not intervened at right time.

Tobacco Caterpillar, *Spodoptera litura* Fabricius

Spodoptera litura has also wide host range and economic insect pest of tomato but it is usually kept under damaging levels by management practices targeted to Helicoverpa in tomato crop. The young larvae feed gregariously and scrape the leaves and later on it may completely defoliate the leaves. The insect is active during the entire year except severe cold months and females can lay up to 600 eggs each, usually in groups of about 100. Eggs are laid on the underside of lower leaves and are covered with fuzzy, white creamy scales. Under warm conditions, eggs hatch within two to three days. The neonate larvae feed from one to three weeks, in groups at younger stage and scattered on the plants when larger.

White Fly, *Bemisia tabaci*

Whiteflies are minute sized whitish yellow adults and nymphs and they suck sap from the leaves of the host plants and reduce the vitality of crops. They often congregate and can be seen on the underside of leaves. They also act like vector for transmitting the virus in tomato and disease is known as tomato leaf curl virus. Severe stunting of plant with downward rolling, crinkling of leaves and severe chlorosis of newly formed leaves takes place. Older leaves become leathery and brittle in severe infestation of white flies.

Serpentine Leaf Miner, *Liriomyza trifoli*

Leaf miners cause heavy leaf mining on upper or lower side of the leaves. Mines starts from the margins of leaves and progress towards the base/centre of leaf. Mines are initially narrow and gradually enlarge often twisting through leaf which causes stippling of young seedlings. Severe leaf mining slows down the plant growth and accelerates the leaf drop.

Tomato Aphids, *Aphis gossypii*

Tomato aphids also feed on cucurbits, potato, okra, brinjal, chilli and peach etc. Both adults and nymphs suck sap from the underside of the young leaves in the early growth stage of tomato leading to curling and yellowing of leaves. They also secrete honeydew on which black sooty mould develops which inhibits the photosynthetic activity of the leaves. Besides this, they also acts as vector and transmit mosaic viruses in tomatoes.

Tomato Fruit Fly, *Bactrocera tau* and *Bactrocera cucrbitae*

Fruit flies have been found infesting the tomato fruits, growing shoots of brinjal and curds of cauliflower in and around the fields of Jammu region. The adult female flies have been found puncturing the tomato fruits even in unripe condition. This behaviour was confirmed through rearing in laboratory condition. Females lay eggs by inserting their ovipositor into the fruits. The eggs hatch within 7-10 days. The newly hatched maggots start mining inside the fruits which become rotten due to secondary infection of bacterial. After attaining maturity, the maggots come out from the rotten fruit and pupate in soil. The adult emergence takes 7-10 days.

Tomato Thrips, *Frankliniella* spp.

Several species of thrips also infest tomatoes including western flower thrips, flower thrip and melon thrips. These are tiny, slender insects that may vary in colour from yellow to dark brown or black. Nymphs and adults usually prefer to feed in flowers but also feed in flower, leaf buds and on leaves. Thrips feeding in blossoms may cause flower abortion and blossom drop, or fruit may not develop properly and become deformed. Feeding on foliage may cause a silvering of foliage. Crops are at greatest risk during flowering and fruit set stage. Eggs inserted in fruit causes dimpling, and the infested area may appear white. Thrips are also vectors of tomato spotted wilt virus (TSWV). Infected plants have dark lesions on the foliage and fruit show characteristic halo markings.

Red Mites : Tomato Red Mite, *Tetranychus* spp.

Nymphs and adults of these mites lacerate the leaves from the lower surface and suck sap resulting in the production of white patches at feeding sites between main veins. In case of severe attack, intense webbing occurs that invites trapping of dust particles giving dusty and purplish appearance of leaves. The infested leaves dry up and fall off prematurely. The growth of the plant is hindered and fruit setting is greatly reduced.

Tomato Mealy Bugs, *Phenacoccus solenopsis*

Mealybugs are found infesting on leaves, stems and fruits of tomato, brinjal and okra in Jammu region during experimentation. They are covered with white wax, which makes them difficult to control. Both nymphs and adults suck the sap from leaves causing withering and yellowing of leaves. Fruit may drop prematurely on crop plants. Heavy infestation can cause defoliation and even death of the plants. The excess sap is excreted as honeydew which attracts ants and further development of sooty mould.

Brinjal

Brinjal (*Solanum melongena*), a native of India, is grown almost round the year in Indian sub-continent. It is an important source of vegetable in South-East Asian countries and has high nutritive and commercial value. It assumes a special significance among vegetables as it gives higher returns over investment in short span of time to the farmers. However, this crop is invaded by several insect pests. The major insect pests include shoot and fruit borer, *Leucinodes orbonalis;* Hadda beetle, *Epilachna viginctioctopunctata;* White flies, *Bemesia tabaci* (acts like a vector for little leaf of brinjal); and aphids, *Aphis gossypii*.

Brinjal Shoot and Fruit Borer, *Leucinodes orbonalis*

Brinjal/Eggplant shoot and fruit borer (ESFB), *Leucinodes orbonalis* Guen, is one of the major constraints contributing up to 70 per cent loss in marketable yield. The pest is active throughout the year on spring, summer, rainy and autumn season crops. The adult moth is a cottony white with black spots on thorax and abdomen. It measures 20-22 mm across the spread of wings. Eggs are laid singly on ventral surface of leaves, shoots, flower buds and occasionally on fruits. Larva/caterpillar is the only damaging stage. In early stage, the larva bore into tender shoots and growing points as a result the infested shoots droop down (withers) and eventually dry up. The young larvae are creamy white in colour and turn pinkish when full grown and bore into the fruits and close the entry point with their excreta and feed within. Its larval stage is the concealed phase so it is very difficult to control. There are 5 stages are found in the larvae. One larva is responsible to damage at least 4-6 fruits. The full grown larva comes out from the fruit with making a big exit hole and pupates in tough silken cocoon among the fallen leaves. They feed inside the internal contents of the fruit. The infested fruit is secondarily infested by the bacterial attack so the fruits become unfit for the human consumption. The extent of damage is reported to be around 50- 64 per cent in severe condition.

Hadda Beetle, *Epilachna viginitioctopunctata* and *E. dodecastigma*

E. viginitioctopunctata is the most common and destructive one especially to brinjal, bitter gouard, bottle gourd, sponge gourd, potato and tomato. The adults being strong fliers and infest wide areas during their peak activity period. Hadda grubs feed gregariously on the underside of leaves. Both grubs and adults feed by scrapping chlorophyll from epidermal layers of leaves, the affected leaves get skeletonised and gradually dry away.

Jassids, *Amrasca biguttula biguttula*

The adults are wedge shaped, about 2 mm long and pale green in colour. The front wings have a black spot on their posterior parts. The nymphs are wingless, pale green in colour and are founding large numbers on the lower surface of leaves. They have a characteristic way of quickly walking diagonally.

Both adults and nymphs suck plant sap from underside of the leaves. The attacked leaves curl upwards along the tips and margins, turn yellow and finally dry up.

Little Leaf of Brinjal

This has become the serious problem in Jammu province since last 2 years. This Mycoplasma disease has been attributed by the two sources *viz.*, contaminated seed materials and by the sap feeding of white flies/leaf hoppers and *Cestius phycitis*. The leaves become very small, narrow, soft, smooth and yellowish in colour in the affected plants and the entire plant seems to be bushy in appearance. Newly formed leaves further reduced in size and internodes shortened. Flowers deformed and withers leading to the sterility in the crops. The plant doesn't bear any fruit in this case, if fruiting takes place, they become very hard.

Chilli

India is one of the largest producers of chilli, *Capcicum annuum*. It is preferred for pungency and an important constituent of Indian vegetarian and non vegetarian preparations. The major pests of chilli include thrips, *Scirtothrips dorsalis*; mites, *Polyphagotarsonemus latus*; fruit borer, *Spodoptera litura* and *Heliothis armigera*.

Chilli Mites *(Polyphagotarsonemus latus)*

Mites attack has become a serious problem in the entire country. Chilli crops are attacked by yellow mite, *Polyphagotarsonemus latus*. This mite is very small, measuring about 0.15 mm and is not visible to the naked eyes. Adult females lay eggs on the under surface of leaves, around midrib and adjoining veins, on buds and sometimes even on closed leaf buds. Mite affected leaves of the chilli plant turn towards underside and look like an inverted boat. As young and adult mites attack the tender leaves, almost all new leaves curled downward and lower surface appears silvery whereas, the upper surface of leaves look more dark green as compared to healthy ones. Further the plant growth is checked and the internodes become shorter.

Thrips (*Scirtothrips dorsalis*):

Besides mites, thrips also attack the chilli crop all over the country. Thrips also suck the plant sap from the leaves and the affected leaves turn towards upper side, internodes becomes long and look like normal boat shaped. Both the nymphs and adults of chilli thrips lacerate the leaf and imbibe the oozing sap. Sometimes even buds and flowers are also attacked. Tender leaves and growing shoots are prefereed for feeding, while the older leaves are rarely attacked. The infested leaves start curling and crumbling and ultimately shed. Whereas buds become brittle and drop down. Chief symptom are curling of leaves upward (normal boat shaped), small leaf size and shorter internodal space with stunting growth of plants, no fruiting or small deformed fruiting in infected plants takes place.

White Fly, *Bemicia tabaci*

It is vector of chilli leaf curl virus. It is a major limitation in successful cultivation of both hot and bell pepper. The most characteristic symptoms are the curling of leaves, their small size, shortened internode and general dwarfing of plant which assumes a bushy appearance. Leaves are of pale yellow colour and roll downwards. Fruiting is stopped, if fruits are formed, they are small and deformed. The disease is transmitted by white fly. The vector spreads maximum disease during the months of August to October.

Cucurbits

Cutworms (*Agrotis ipsilon)*

It is a polyphagous insect pest which feeds mainly in seedling stages of many vegetables, potato, maize and other crops. Caterpillar is the only damaging stage of this pests.

Brinjal Hadda Beetle

Tomato Yellow Leaf Curl Virus Transmitted by Whitefly

Cabbage Semilooper

Tomato Aphid

Cabbage Worm

Cabbage Butterfly on Cabbage

Brinjal Fruit and Shoot Borer

Whitefly and Jassids on Okra

full grown in 5-7 days and pupates within the leaves galleries. It completes one generation in 14-15 days and passes through several overlapping generations during the period of their activity.

Pea Pod Borer Complex

Etiella zinckenella are widely distributed pest of pea. The pinkish larvae feed on the floral parts and seeds of younger pods. The damaged pod has a large emergence hole made by pupating larvae.

Maruca vitrata

Moths are medium sized with conspicuous transverse white elongated markings on forewing. The eggs are laid in small batches on flower buds which hatch within 2-3 days. The newly hatched larvae may found in groups and damages several flowers. The larvae feed on any part of flower or foliage but the later instars causes' damage to the pods by boring.

Pest Management Strategies in Vegetable Crops

Pest Management in Vegetable Nursery

Nursery raising is an important practice of vegetable growing. Vegetable crops are comparatively more susceptible to various biotic and abiotic stresses responsible for different diseases and soil insects pests. Many soil and seed borne pathogens are responsible for heavy losses of the vegetable crop nursery in which pre and post emergence damping off is very serious and can cause loss up to 70 per cent of seedlings.

Protected Nursery Raising Technology

Farming community of India may not be able to use all kinds of protected technology due to their very high initial, running and maintenance cost, but some protected technologies are simple, suitable and highly profitable under various agro-climatic conditions of India, which can be adopted by Indian farmers for production of different vegetable crops nursery. These protected production technologies helps in minimizing the insect pest and viral and mycoplasmal diseases under nursery conditions especially in tomato, chilli, brinjal etc. Now a day, ballooning plastic materials are being used in developed countries which rises with the growth of crops to protect it from insect pest problems.

Cultural Control Practices

Strategies in Integrated Pest Management of Vegetables

Cultural practices are the manipulation of the cropping environment to reduce rates of pest increase and damage. There are many different cultural practices that can help to reduce pest impact. Some of the more common techniques include improving plant vigor, tillage practices that disrupt the insect's life cycle and destroy crop residue, changing planting dates to minimize insect impact, and crop rotations that include non-susceptible crops. Some cultural controls may also increase or decrease pests' natural enemy populations. Sound cultural control methods often become stable farming practices that serve multiple purposes.

Crop Selection and Cultural Practices

It needs careful consideration with sufficient planning in advance. This component does not require the additional attention of the farmers and more sustainable and eco-safe in long run. Routine agronomic practices like selection of varieties, intercultural operation, sowing/planting date, trap cropping/intercropping should be suitably modified to make uncongenial condition for insect pests and to enhance the activity of natural enemies.

Sowing/Planting Time

A little bit adjustment in sowing and planting date helps to avoid the critical stage of the crop from insect pest attack. Careful consideration of sowing/planting date in vegetable reduced the attack of red pumpkin beetle, fruit fly, shoot and fruit borer. Early planting of cucurbits in November escapes the attack of red pumpkin beetle whereas flowering beyond October in bitter gourd suffers less from fruitfly. Sowing of okra during second week of June retain less population of borers, there by incurred maximum healthy yield whereas July planted brinjal face the ravages of shoot and fruit borer. Thus, synchronization of most susceptible stage of the crop with the inactive period of insect reduces the infestation and chemical intervention.

Tillage

Summer ploughing is an effective practice to destroy the soil dwelling inactive stages of insect. Deep ploughing of the field after the harvest reduces the activity of fruitfly, red pumpkin beetle and cut worm as these insects remain in the soil in earthen cocoon to complete the dormant stage of their life-cycle.

Intercropping, Trap Crops and Deterrent Crops

Intercropping of crops with main crops breaks the standard mono-cropping and limits the infestation from the pest. Diverse nature of plant not only obstructs the adults from egg laying but also the release of volatile allelo-chemicals from a particular crop deters the adult insect from damaging the other. Above all such planting combination enhances the activity of predators and parasites also. Trap crop must be distinctly attractive to the pest than the main crop. It provides protection either by preventing the pest from reaching the main crop or by concentrating them in certain part of the field where they can be economically destroyed. Mustard as a trap crop and coriander and tomato as deterrent crop along with cabbage has been successfully utilized for the control of diamondback moth, aphid and leaf webber. African marigold functions as good source to trap the adults of *Helicoverpa armigera* besides it also attracts the adults of leafminer for egg laying on the leaves. Maize plants applied with bait spray trap and kill fruit fly adults when sown in combination with bitter gourd.

- ✰ Removal and complete destruction of crop residue to check migration of pests from one crop to the other.
- ✰ Deep ploughing should be performed to kill eggs and grubs of red pumpkin beetle and other pests harbouring in the soil.
- ✰ Use of resistant/tolerant varieties.
- ✰ Balanced use of fertilizers and micro-nutrients.
- ✰ Trap crop of marigold after 16 lines of tomato can be transplanted to avoid *H. armigera* attack on main crop (tomato).

Mechanical Control Practices

- ✰ Collection and destruction of eggs and immature stages of insect pests wherever feasible.
- ✰ Removal and destruction of brinjal fruits attacked by *L. orbonalis* and tomato fruits by *H. armigera.*
- ✰ Use light trap/pheromone traps for mass trapping/monitoring of major pests.

Aphids on Cauliflower

DBM on Cabbage

***Helicoverpa* on Peas**

***Helicoverpa* on Okra**

***Helicoverpa* on Tomato**

Mites on Okra (Lady's Finger)

Mites on Okra (Lady's Finger)

Biological Control

Vegetables ecosystem is bestowed with a large number of natural enemies comprising of predators, parasites and parasitoids causing damage to different stages of the insect to varying extent. Among the egg parasitoids *Trichogramma* spp. has been utilized to some extent for control of tomato fruit borer. Inundative release of 2,50,000 parasitised eggs has been found to control the fruit borer. Another predator, which needs immediate attention, is *Chrysoperla carnea*. This is an effective predator for control of aphid, jassid and eggs of some lepidopterous borers, when the first instar larvae are released

@50,000/ha. Mass rearing of this predator needs to be streamlined for economising its use as biocontrol agent. Besides, improvement of field release technique to maintain optimum potential is most important. The strains having potency to survive in toxic environment will fit well into the Integrated Pest Management system.

Recommendations

- ☆ Use of *Trichogramma* spp. @50,000 insects/ha/7-10 days interval from initiation of flowering against important lepidopterous pests is recommended. Parasitoids like *Trathala flavoorbitalis*, *Pristomerus testaceus*, *Bracon* sp., against *L.orbonalis*, *Campoletis chlorideae* against *H.armigera* should be encouraged.
- ☆ Spraying of HaNPV 250 LE with 1 kg jaggary and 0.1 per cent tepol in 800 litres water/ha may be sprayed at 10 days intervals for 3 times for control of *H.armigera* on tomato.
- ☆ Spraying of *Bacillus thuringiensis* formulation @1 kg/ha may be sprayed for the management of *P. xyllostella* and other lepidopterous pests.

Microbial Control

Microbial pesticides as a component of Integrated Pest Management are cheaper than pesticides in a long run. They are also proved to be eco-friendly and compatible with natural enemies existing in the ecosystem. *Bacillus thuringiensis (Bt)* is the most extensively used biocontrol agent in vegetables. It has been widely applied against *Earias vittella* and *Helicoverpa armigera*. Because of its host specificity, use of nuclear polyhedrosis virus (NPV) is limited to *H. armigera*, *Spodoptera litura*. Lack of commercial production has suppressed its popularity and is used in limited areas. Application of HaNPV and SlNPV @250 larval equivalent (LE).

Potential Natural Enemies Fauna in Tomato Crops

- ☆ Coccinellids beetle were recorded to regulate the mealy bugs in field conditions.
- ☆ Predatory bugs, *Eocanthecona* spp. were found predating on the Helicoverpa and Spodoptera larvae in the field conditions.
- ☆ *Telenomus* spp. of parasitoids were found parasitizing the egg masses of *Spodoptera litura* and their egg masses have also been recovered from tomato fields.
- ☆ Another abundant larval parasitoids, pupae of *Campoletus chloridae* were collected from tomato fields and brought to the laboratory for emergence of adults to ascertain the *Campoletus* species.
- ☆ Black drongo or king krow were also observed predating the fruit borers larvae from tomato field. To increase the visit of black drongo, bird perches should be installed in the fields.
- ☆ Spiders are generalist predators of insects and various type of spider species were noticed in tomato crops to predates upon small larvae.

Judicious Selection of Insecticides

Due to the easy accessibility and powerful tool as ready reckoner action against the target organism, chemical control still remains first choice of farmers. But indiscriminate use of insecticides let the problems of resistance, resurgence of whitefly and red spider mite have now threatens the widespread use of synthetic insecticides. However, these insecticides can be judiciously used considering certain aspect like waiting period of chemical, economic threshold level and initiation of control measures through pheromone/light trap catches. Residue of major group of insecticides has been detected to considerable extent in okra, brinjal and tomato. To avoid the persistent residues, vegetables with less picking interval should be sprayed with insecticide having less waiting period (Table 19.3).

Table 19.5: Waiting/Safe Periods of Insecticides on Vegetable Crops

Sl.No.	Insecticides	Crops	Waiting Period (Days)
1.	Lindane 20 EC, 1.3 D	Okra	6
		Onion	9
2.	Malathion 50 EC	Brinjal	1-4
		Okra	3-4
		Tomato	5
		Onion	6
3.	Dichlorvos (DDVP) 76 EC	Okra, Brinjal	7
4.	Dimethoate 30 EC	Toria, Rough surface Fruits and Vegetable	10
		Smooth surface vegetable and fruits	5-7
5.	Methyl parathion 50 EC	Smooth surface vegetable and fruits	4-6
		Rough surface fruits and vegetable	5-8
6.	Carbaryl 50 WP	OkraBrinjal, cabbage and tomato	121
7.	Cypermethrin 10 EC	Okra	4
8.	Fenvalerate 20 EC	Okra	9
9.	Decamethrin 2.8 EC	Okra	1

For control of *H. armigera* and other major pests, deltamethrin (2.8EC) 0.5 ml/lit or may be sprayed. For control of *L. orbonalis*, neem seed kernel extract 4 per cent or carbosalfan (25EC) 2ml/lit or cartap hydrochloride (50WP) 2gm/lit water spraying or 5 per cent savin mixed with ash may be applied. For control of fruitfly on cucurbitaceous plants, malathion (50EC) 20ml with 200 gms of sugar or jaggary in 20 litres of water may be sprayed.For aphid control, use of 0.05 per cent malathion or 0.03 per cent DDVP or 0.03 per cent endosulfan or 4 per cent NSKE was found effective.

Future Projection

Rich biodiversity and fast changing cropping pattern toward commercial agriculture invite vibrant monitoring and surveillance system. Due to emphasis on organic farming in the region, there is a need to stress upon exploitation of potent local bioagents and botanicals against major insect pests, besides conservation and augmentation of exotic as well as native natural enemies. In addition, already existing proven IPM technologies available in different parts of the country need to be validated under hill condition. As the hills posses varied climatic conditions due to several factors like topography, steepness of slope, variations in soil structures, face of mountain etc., niche based IPM module has to be worked out to ensure better effectivity of the various management tools. Of late, a greater emphasis has been on cultivation of tea, medicinal plants, vegetables, flowers etc. especially in Uttarakhand hills. In the event of shift in cropping system, particularly perennial crops, a critical vigilance would be of paramount importance in order to prevent emerging pest problems. Overall general awareness about IPM among the inhabitants coupled with government and institutional support would be necessary in the geographically challenged region of the country.

Conclusion

Insect pests have been the major constraint in vegetable crop production in the country in journal and particular in hills. To combat the burgeoning problem, relying on the pesticides, is not desirable in the present day of global consciousness towards health hazards and environmental safety. One of the strongest alternatives for the management of the insect pests would be integrated pest management with emphasis on biological control using potential insect parasitoids and microbial pathogens

References

Allen, W. A. and Rajotte, E.G. (1990). *Ann. Rev. Entomology*. **35**: 379-397.

Anonymous, (1992-2004). *Annual report.*, Almora, VPKAS.

Anonymous,(2002). Technological options for improving agricultural productivity in the North-Western Hill, Almora, VPKAS, 70 p.

Atwal, A.S. and Singh, B. (1990). Pest population and assessment of crop losses, Indian Council of Agricultural Research, New Delhi. 131 p.

Bhatt, J.C. and Sushil, S.N. (Eds) (2004). A compendium on diseases, insects and weeds of rice, maize and wheat crops in Indian Himalayas.

Dhaliwal, G.S. and Arora, R. (2001). Integrated pest management, Ludhiana, Kalyani Publishers. 427p.

FAO. (1967). Report of the first session of the FAO panel of experts on Integrated Pest Control, Rome, Italy, Sept. 18-22, 1967, 19 p.

FAO. (1980). Research Summary. Integrated pest management. EPA-600/8-80-044. 28 pp.

Flint, M. L. and Van den Bosch, R. (1981). Introduction to integrated pest management. Plenum Press, New York, 240 p.

Garg, D.K.(1992). Insect pests of agricultural crops in UP hills and their management. Almora VPKAS, 25 p.

Hawtin, G. (1998). Conservation of agro-biodiversity for tropical agriculture. In: Crop Productivity and Sustainability: Shaping the Future, edited by V.L.Chopra, R.B.Singh and A. Varma. Oxford and IBH, New Delhi. pp. 917-925.

Jayaraj, S. (1992). Integrated pest management. In: Pest management and pesticides : Indian Scenario. Edited by B.V.David, Chennai, Namrutha Publications. pp 7-16.

Monobrullah, Md. and Shankar, U. (2008). Sub-lethal effects of *Splt*MNPV infection on developmental stages of *Spodoptera litura* (Lepidoptera: Noctuidae). *Biocontrol Science and Technology*, 18(**4**):431-437.

Pedigo, L.P. (1989). Entomology and Pest Management. Macmillan Pub. Co., Inc.N.Y. 646 pp.

Shankar, U., Kumar, D. and Priya, S. (2008). Vegetable Pest Management Guide for Farmers, International Book Distribution Co., Lucknow. 181 pp.

Shankar, U., Kumar, D., Singh, S. K. and Gupta, S. (2009). Pest complex of tomato and their management. Technical Bulletin No. 2, SKUAST-J, Jammu.

Smith, R. F. and Reynolds, H.T. 1966. Principles, definitions and scope of integrated pest control. Proc. FAO Symposium on Integrated Pest Control 1: 11-17.

Srivastava, K.P. (1993). Applied Entomology-II, Ludhiana, Kalyani Publishers, 424 p.

Stern, V. M., Smith, R.F., van den Bosch, R and Hagen, K.S. (1959). The integrated control concept. *Hilgardia*, 29: 81-101.

Yazdani, S.S. and Agarwal, M.L. (1997). Elements of insect ecology. Narosa Publishing House, New Delhi. 210 p.

Hill Agriculture: Prospects, Constraints and Mitigations (2013) *Pages* **299–313**
Editor: **Dr. Shahid Ahamad**
Published by: **DAYA PUBLISHING HOUSE, NEW DELHI**

Chapter 20

Management of Insect Pest Complex of Maize in Hill Agro-Ecosystem

Hafeez Ahmad and Devinder Sharma
Division of Entomology,
Faculty of Agriculture, Sher-e-Kashmir University of Agricultural Sciences and Technology,
Main Campus, Chatha, Jammu – 180 009

Maize or :

(*Zea mays* L.) is a plant belonging to the family of grasses (Poaceae). Itis cultivated globally being one of the most important cereal crops worldwide. Maize is not only an important human nutrient, but also a basic element of animal feed and raw material for manufacture of many industrial products. The products include corn starch, maltodextrins, corn oil, corn syrup and products of fermentation and distillation industries. It is also being recently used as biofuel. Maize is a versatile crop grown over a range of agro-climatic zones. In fact the suitability of maize to diverse environments is unmatched by any other crop. It is grown from 58° N to 40° S, from below sea level to altitudes higher than 3000 m, and in areas with 250 mm to more than 5000 mm of rainfall per year (Dowswell *et al.*, 1996) and with a growing cycle ranging from 3 to 13 months (CIMMYT, 2000). However the major maize production areas are located in temperate regions of the globe. The United States, China, Brazil and Mexico account for 70 per cent of global production. India has 5 per cent of corn acreage and contributes 2 per cent of world production.

Maize is the third most important food grain in India after wheat and rice. In India, about 28 per cent of maize produced is used for food purpose, about 11 per cent as livestock feed, 48 per cent as poultry feed, 12 per cent in wet milling industry (for example starch and oil production) and 1 per cent as seed (AICRP on Maize, 2007). A major shift in global cereal demand is underway: by 2020, demand for maize in developing countries will surpass the demand for both wheat and rice. This shift will be reflected in a 50 per cent increase in global maize demand from its 1995 level of 558 million tons to 837 million tons by 2020. Maize requirements in the developing world alone will increase from 282 million tons in 1995 to 504 million tons in 2020 (IFPRI, 2000). In the last one decade, it as registered the highest growth rate among all food grains including wheat and rice because of newly emerging food habits as well as rice enhanced industrial requirements. The challenge of meeting this unprecedented demand for maize is daunting, especially for the developing world and its poor and subsistence farmers.

Hill Agriculture Dynamics and Insect Pest Problem

With the increase in the productivity and production of the agricultural crops, by intervention of high yielding varieties, agro-chemical, irrigation, monoculture etc., there have been tremendous changes in the pest scenario of the crops (Dhaliwal and Arora, 2001, Atwal and Singh, 1990, Anonymous, 1992-2004). Agronomic recommendations for exploitation of full potential of HYVs has certainly helped in enhancing crop production but at times it has provided congenial condition for many insect pests. Close spacing of paddy has resulted in the increased population of BPH, gall midge and leaf folder. In North West Himalayan Region, cultivation of spring rice is the most common practice which provides availability of the rice crop for a quite longer duration in the field, hence promotes pest build up in the region. Routine crop rotation in some cases may provide congenial environment for insect pest multiplication. Cultivation of off-season vegetables in hills provides food for many insect pests round the year. The tobacco caterpillar, *Spodoptera litura* and tomato fruit borer, *Helicoverpa armigera* were found to damage cabbage, cauliflower, broccoli, tomato, chilli, capsicum etc. round the year. Destruction of forests for cultivation of agricultural crops led to the migration of insect pests of forest, mainly white grubs to agricultural fields for egg laying and subsequent damage to the crops. So far, authors have recorded about 40 species of white grubs from the region. Among which, many of them are considered as major pests of agricultural crops.

Human preference for certain agricultural crops like rice, wheat, maize prompted farmers to grow more of these crops. Despite the rich biodiversity in agriculturally important crops, wheat, maize and rice accounts for 60 per cent of plant derived calories in human diet (Hawtin, 1998). In NWHR, some of the less preferred crops like finger millet and barnyard millet are now gaining momentum because of their nutritive value. Monoculture of certain crops is considered as one of the major factor for increased insect pest problem. Mono-cropping of vegetable crops especially in the valley areas prompted the chances of outbreaks of many insect pests. Nutritional requirement of the crops are mostly met with the application of FYM in traditional farming of the hills. However, application of compost, which is not well decomposed, in the rainfed conditions also helps in increasing the white grub problem in the region. Modern agriculture has undoubtedly contributed a lot in increasing the crop productivity as well as crop production all over the country, but at the same time it has provided congenial conditions for multiplication of many of the insect pests. Therefore, before advocating any practice for crop production it has to be thoroughly tested from insect pest point of view.

Insect Pest Complex in Maize

Maize is one of the major staple food crops in North India and over 70 per cent of the maize produced is used as food. The crop possesses great genetic diversity and is grown across varied agro-ecological zones. In India Maize (*Zea mays* L.) crop is grown in about 6.90 m ha area with production of about 17 million tones (2006-07). Its average productivity (2.5 tonnes/ha) is very low in comparison to its potential. Maize is an important staple food of people living in hilly and sub-mountainous regions of Himalayas in Jammu division, occupying an area of 221.93 thousand hectares with an average productivity of 17.33 q ha^{-1} (Anonymous, 2007). However, the average yield is far below as compared to the world's average yield 38q ha^{-1}. Most important reason for its low productivity is improper management of insect pests' infestation. About 139 insect-pests cause varying degree of damage to maize crop. However, only about a dozen of these insect-pests were 5 per cent but has reached up to 25 per cent. Arthropod pests are among the key factors contributing to low yield of maize. Globally losses due to insect pests of maize are reported to be approximately 9 per cent, which is equivalent to 52 million MT and worth approximately $5.7 billion annually. Despite the annual investment of US $35,000 for the application of there million metric tonnes of pesticides, plus the use of various biological and other non-chemical controls worldwide, global crop losses remain a matter of concern (Pimentel, 2009).

In spite of the wide spread use of synthetic pesticides and other control measures, the losses due to insect and mite pests increased in post-green revolution era (Oerke *et al.*, 1994) than in pre-green revolution era (Cramer, 1967). Worldwide total pre harvest losses for post-green revolution era (1988 through 1990) period value at US $90 billion for eight principal food and cash crops (barley, coffee, cotton, maize, potato, rice, soybean and wheat) (Benedict, 2003) (Table 20.1). Oerke (2006) updated the loss data for major food and cash crops for the period 2001-03. The actual losses due to various pests have been estimated as 26-29 per cent for soybean, wheat and cotton, and 31, 37 and 40 per cent for maize, rice and potatoes, respectively (Table 20.2). Thus, there is a decline in global crop losses due to various pests from 42.1 per cent during 1988-90 to 32.1 per cent during 2001-03. The corresponding decline in loss due to animal pests, weeds and pathogens was from 15.6 to 10.8, 13.2 to 8.8 and 13.3 to 12.5 per cent, respectively. Losses due to insect pests in Indian agriculture have been estimated from time to time (Dhaliwal *et al.*, 2010) (Table 20.3). In general, the losses in post-green revolution era (Dhaliwal *et al.*, 2004) have shown an increasing trend than in the pre-green revolution era (Pradhan, 1964). Overall, the losses increased from 7.2 per cent in early 1960s to 23.3 per cent in early 2000s (Table 20.4). The maximum increase in loss occurred in cotton (18.0 to 50.0 per cent), followed y other crops like sorghum and millets (3.5 to 30.0 per cent), maize (5.0 to 25.0 per cent) and oilseeds (other than groundnut) (5.0 to 25.0 per cent). There has been a paradigm shift in the crop management scenario of Indian agriculture since the beginning of this century.

Table 20.1: Worldwide Crop Losses (per cent) Due to Insect and Mite Pests during Pre- and Post-Green Revolution Era

Crop	Pre-Green Revolution (1965) (1)		Post-Green Revolution (1988-90) (2)		Changes in Losses (2-1)	
	World	India	World	India	World	India
Maize	13.00	5.00	14.50	25.00	+1.50	+20.00
Rice	27.50	10.00	20.70	25.00	-6.80	+15.00
Wheat	5.10	3.00	9.30	5.00	+4.30	+2.00

Table 20.2: Current Global Losses (per cent) Due to Various Categories of Pests in Major Crops

Crop	Animal Pests	Weeds	Pathogens	Viruses	Total
Maize	9.60	10.50	8.50	2.70	31.30
Rice	15.10	10.20	10.80	1.40	37.50
Wheat	7.90	7.70	10.20	2.40	28.20

Of the 130 insect-pests that affect maize crop, stem borers, shoofly, armyworm, jassids, thrips, white ants, pyrilla, grasshoppers, grey weevil, hairy caterpillars, root worms, earworms and leaf miner are more serious, though the spectrum varies in different agro-ecological regions. Most of the research efforts have gone into breeding for resistance to European corn borer (*Ostrinia nubilalis*), a pest of maize in the USA and Europe. Fall armyworm (*Spodoptera frugiperda*) is another very important pest in tropical and subtropical areas. In India, spotted stem borer (*Chilo partellus*) is the most serious pest. The major stalk-borer (*Chilo partellus*) is the major pest throughout the country, particularly during the *Kharif* season.

Table 20.3: Estimation of Crop Losses Caused by Insect Pests to Major Agricultural Crops in India

Crop	*Actual Production (million tonnes)*	*Approximate Estimated Loss in Yield*		*Hypothetical Production in the Absence of Losses (Million tonnes)*	*Monetary Value of Estimated Loss (Million rupees)*
		Percentage	*Total (Million tonnes)*		
Maize	19.00	20.00	4.80	23.80	29,450
Rice	96.70	25.00	32.20	128.90	2,40,138
Wheat	78.60	5.00	4.10	82.70	41,368

* Production and minimum support price (MSP) fixed by Government of India for 2007-08, are adapted from Anonymous (2010).

Table 20.4: Insect Pests of Maize (*Zea mays* L.)

Common Name	*Scientific Name*	*Family*	*Order*
Wireworm	*Melanotus communis* (Gyllenhal)	*Elateridae*	Coleoptera
Stem borer	*Chilo partellus* (Swinhoe)	Crambidea	Lepidoptera
Cutworm	*Agrotis ipsilon* (Hufnagal)	Noctuidea	Lepidoptera
White grub	*Holotrichia* sp	Melolonthidae	Coleoptera
Grass hopper	*Hieroglyphus nigrorepletus* Bol. *Chrotogonus* spp., *Ceracris* sp	Acrididae	Orthoptera
Aphid	*Rhopalosiphum maidis*	Apididae	Hemiptera
Jassid	*Zyginidia manaliensis*	Cicadellidae	Hemiptera
Blister beetle	*Mylabrus pustulata* Thunberg	Meliodae	Coleoptera
Stink bug	*Nezara virudula* (Linn.)	Pentatomidae	Hemiptera
Hairy caterpillar	*Amsacta moorei* (Butler)	Arctiidae	Lepidoptera
Red cotton bug	*Dysdercus koenigi* (Fab.)	Pyrrhocoreidae	Hemiptera
Pyrilla	*Pyrilla perpesula* (Walker)	Fulgoridae	Hemiptera
Corn ear worm	*Heliothis armigera*	Noctuidea	Lepidoptera
Tobacco caterpillar	*Spodoptera litura* (Fab.)	Noctuidea	Lepidoptera
Flower eating beetle	*Protaetia alboquattata* Vig.	Cetonidae	Coleoptera
Snail	*Helix pomatia*		Stylomatophora
Termites	*Odontotermes obesus* Ramb. *Microtermes obesi* Holm.	Termitideae	Isoptera
Hairy caterpillar	*Amasacta moori* Butter*Diacrisia oblique*	Arctiidae	Lepidoptera

A variety of insect pests attack maize crop right from seedling to maturity. These insect pests includes: lepidopterous pests (borers, cutworms, armyworms, earworms, and grain moths) are the most damaging world over, followed by the coleoptera (root worms, wireworms, grubs, grain borers, and weevils). Next in importance is the group of insects that serve as carriers (vectors) for disease agents or pathogens, among which leafhoppers and aphids are of prime importance. Maize cultivation and yield realization are highly affected by the degree of insect-pest management. For realizing potential yield, particularly from HYVs, under precision management strategy holds the 'key' to success. To ensure good yields, insect-pest management programme should be implemented unscrupulously

Table 20.5: Succession of Insect Pest Complex on Maize

Scientific Name	*Peak Period*	*Crop Stage*	*Status*
Melanotus communis (Gyllenhal)	Ist F.N. Jun.	Emergence -6th leaf	Minor
Chilo partellus (Swinhoe)	Jul.	Tasseling	Major
Agrotis ipsilon (Hufnagal)	Ist F.N. Jun.	Emergence -6th leaf	Major
Holotrichia sp.	Ist F.N. Jul.	Tasseling	Major
Ceracris sp.	Ist F.N. Sep.	Miliking	Minor
Rhopalosiphum maidis	Aug.	Siliking	Majoı
Zyginidia manaliensis	IInd F.N. Sep.	Dent	Minor
Mylabris pustulata	Ist F.N. Aug.	Siliking	Minor
Nezara virudula (Linn.)	IInd F.N. Sep.	Dent	Minor
Amsacta moorei (Butler)	IInd F.N. Sep.	Dough – Dent	Minor
Dysdercus koenigi (Fab.)	Ist F.N. Sep.	Miliking	Minor
Pyrilla perpesula (Walker)	Aug.	Siliking	Minor
Heliothis armigera	Sep.	Dent	Minor
Spodoptera litura (Fab.)	IInd F. N. Sep	Dough – Dent	Minor
Protaetia alboquattata Vig.	2nd F.N Aug	Miliking	Minor
Helix pomatia	Ist F.N. Aug.	Siliking	Minor
Canis aureus Linnaeus	Ist F.N. Sep.	Miliking	Minor

In India the crop is being attacked by about 139 species of insect pests with varying degree of damage to the maize crop. However, only about a dozen of these are quite serious (Siddiqui and Marwaha, 1993). Seventeen species in two families (Pyralidae and Noctuidae) have been found to attack maize in various parts of Africa. However, *Chilo partellus* (Swinhoe), *Chilo orichalcociliellus* Strand, *Busseola fusca* Fuller, *Sesamia calamistis* Hampson, and *Eldana saccharina* Walker are of great importance. The yield losses caused by stem borers to maize vary widely in different regions and range from 25-40 per cent according to the pest population density and phenological stage of the crop at infestation (Khan *et al.*, 1997a). Among them maize stem borer, *C. partellus* is most dominant contributing 90-95 per cent of the total damage in *kharif* season (Jalali and Singh, 2002). This pest is principal limiting factors in achieving the increased production of maize in tropical and subtropical regions especially in the northern India.

The changing scenario of insect pest problems in agriculture as a consequence of green revolution technology has been well documented (Dhaliwal and Arora, 2006; Dhaliwal and Koul, 2010). There has been further shift in the status of several insect pests after the introduction of transgenic crops and the current scenario of climate change. The pest scenario in cotton ecosystem is changing fast and is assailed by multitude of pests as it evolves through various production levels. Various species of mealy bugs have started appearing in serious proportions on field crops, vegetables, fruits and ornamentals (Tanwar *et al.*, 2007). In fact, mealy bugs have become indicator insects for the current ecosystem alterations due to slow changes in climate during the period from 2002 to 2005. *Maconellicoccus hirsutus* (Green) causes extensive damage to ornamentals, though its host range extends to 76 families and over 200 genera, including maize (Tanwar *et al.*, 2007; Rajendran, 2009).

Factors Influencing Pest Population in Hills

The insect pest population is drastically influenced by any change in the environmental factors. The environmental factors are broadly grouped into abiotic and biotic factors.

Abiotic Factors

The population density of pest fluctuates with the prevailing weather conditions such as temperature, moisture, light, wind, edaphic factors etc. Since, insects are poikilothermic animals *i.e.* they do not posses temperature regulation system of their body, their body temperature, more or less, follows the surrounding temperature. Exposure to temperature beyond the favourable range, whether low or high, may retard growth and development of the insects or may even cause their death. The species adapted to live in temperate climate, hibernate in a particular developmental stage. A constant supply of moisture is essential for metabolic reactions as well as for the dissolution and transport of salts. The water content in insects varies from less than 50 per cent to more than 90 per cent of the total body weight. Excessive moisture can prove harmful to the insect population by affecting normal development and by encouraging disease-causing microorganisms. Edaphic factors such as structure, texture and composition of soil, drainage, topography etc. also affect the distribution and abundance of the insects. Example, the cut worm, *Agrotis ipsilon* larva live in light texture soil in which they move around freely in response to daily or seasonal temperature and moisture changes. Major topographic factors that affect biography of insects include height of mountain chains, steepness of slope, directions of mountains and valleys and exposure of the slope. These factors also affect the climate of an area, thus influences the distribution of certain insects. However, their impact on the status of the major insect pests has yet to be quantified. Mountain ranges also act as physical barriers to the spread of insects (Yazdani and Agarwal, 1997).

Biotic Factors

The growth and development of insects is largely influenced by biotic factors such as natural enemies and food. The natural enemies of insect pest population include predators, parasitoids and disease causing microorganisms such as fungi, bacteria, viruses etc. The natural enemies are also influenced by various environmental factors such as weather and hyper parasitism. Weather is known to determine the effectiveness of the natural enemies. Example, in the control of apple wooly aphid with *Aphelinus mali*, the prevalent temperature in spring would determine the effectiveness of the parasitoid. Similarly, quality and quantity of food play important role in insect's survival, longevity, distribution, reproduction, speed of development etc. (Dhaliwal and Arora, 2001, Atwal and Singh, 1990).

Management Strategies

Ecological Management

Agricultural systems are designed to alter a given eco system to increase the flow of energy to man. With in this scope, crop monocultures are extreme examples of environmental simplifications and specialized management. Although highly productive and efficient, these systems have been criticized because of their genetic and horticultural uniformity, and the resulting continuous pest susceptibility. Grow only recommended hybrids and composites, which are relatively less susceptible to insect pests. Early sown crop during *kharif* is less susceptible to borers. Crop rotation with non host crops reduces pests infestation. Recommended plant density should be maintained. Burning of stubbles and trash, which harbour borer and act as source of infestation for the next crop. Plants for fodder should be stored in chaffed or cut condition to kill burrowed insects. The fodder should be fed to the cattle before the start of the next season. Use of higher seed rate to compensate for the young infested plants that are pulled out and destroyed. This practice is much helpful in cutworm and stem borer prevalent area. In severe cutworm attacked areas, several smooth holes should be dug all over the effective field so that the caterpillars moving about in the night fall in them and being cannibalistic kill each other. Heaps of green grasses may be kept at suitable places in the field in the evening so that the caterpillars seek shelter below them. The heaps along with the caterpillars may then be destroyed in the morning. Crop rotation should be followed in disease and insect prone areas.

Resistant Cultivars

World wide damage due to insect pest attack accounts for 15 per cent crop losses, despite the use of insecticides which represents over US $100 billion (Krattiger, 1997). The annual cost of insect control itself amounts to US $8 billion, thus warranting immediate and economical control measures. The crop varieties developed in the past 30 years were high yielders, but unfortunately had poor storage characteristics and were highly susceptible to pest damage. Insect-pests are capable of evolving to biotypes that can adapt to new situations, for instance they overcome the effect of toxic materials or bypass natural or artificial plant resistance, which further confounds the problem (Roush and McKenzie, 1987). Under these circumstances, provision of food to the rapidly expanding population has always been a challenge facing mankind. This problem is more acute in the tropics and sub-tropics, where the climate provides a highly conducive environment for a wide range of insects and necessitates massive efforts to suppress the population densities of different pests in order to achieve an adequate supply of food. In developing countries, the problem of competition from insect-pests is further complicated with a rapid annual increase in human population (2.5-3.0 percentage) against a meager 1.0 per cent increase in food production. In order to feed the ever expanding population, crop protection plays a vital and integral role in the present day agricultural production to minimise yield losses. Currently, the crop protection practice in agricultural systems relies exclusively on the use of agrochemicals, although a few specific cases do exist, where inherent varietal resistance and biological control have been successfully employed. The exclusive use of chemical pesticides not only results in rapid build-up of resistance to such compounds, but their non-selectivity affects the balance between pests and natural predators, and is generally in favour of pests (Metcalf, 1986). Therefore, an integrated pest management (IPM) programme, comprising a combination of practices including the judicious use of pesticides, crop rotation, field sanitation and above all exploitation of inherently resistant plant varieties would provide the best option (Meiners and Elden, 1978). The last option includes the use of transgenic crops, expressing foreign insecticidal genes which could make a significant contribution to sustainable agriculture and thus could be an integral component of IPM (Boulter, 1993). The production of transgenic crops has seen rapid advances during the last decade with the commercial introduction of *Bt* transgenics, but the major concern with these crops has been the development of resistance by pest and public acceptability.

Table 20.6: Improved Cultivars of Maize

Hybrid Maize	*Composite Maize Varieties*
Hybrid AH-58 (PEHM-3)	Pusa composite -3
Hybrid AH-421 (PEHM-5)	Pusa composite -4
Ganga-4	Vijay
Ganga-5	Kissan
Ganga Safed-2	Vikram
Deccan	Agati-76
Hi-Starch	Shakti
Himalayan	Rattan
NMH 909 (Raja)	
NMH 360 (Krishana)	
NMH 589 (Suwarna)	
NMH 234 (Kamdehnu)	

Host plant resistance provides an alternative option to overcome insect pests under intensive cultivation system. The use of insect resistant varieties has been recognized to be of immense value in IPM programme. This necessitated identifying resistant germplasm through varietal screening. Therefore incorporating resistant genes in agronomical sound cultivars of maize is a good practice against the pests.

Habitat Management

Conservation of Natural Enemies

Interplanted ryegrass (*Lolium multiflorum* Lambert) in seed maize (*Zea mays* L.) fields to reduce the temperature of the soil surface, increasing survival of augmentatively released *Trichogramma brassicae* Bezdenko Non-crop vegetation may be favored by natural enemies as oviposition sites. It has been observed that *Coleomegilla maculata* (Coleoptera: Coccinellidae) lays more eggs on a native weed, *Acalypha ostryaefolia* Ridell than the sweet corn (*Zea mays* L.) crop, even though the plant supported few prey. Larvae then disperse from the weed and climb maize plants. Maize plots bordered by *A. ostryaefolia* contained significantly more *C. maculata* than did plots without a border. Other examples of ground covers or intercrops influencing natural enemy density, activity or impact include carabids in maize, Work with *Eriborus terebrans* (Gravenhorst), the primary parasitoid of *Ostrinia nubilalis* (Hubner) in maize in Michigan showed that females were most frequently captured in maize fields close to wooded field edges and that parasitism of *O. nubilalis* was higher in these areas. Laboratory, greenhouse, and field cage studies showed that adults lived longer when provided with sugar and that longevity increased at 258C versus 358C. The wasps were also shown to be more active on hotter days, which could result in their leaving crop habitats to find shelter in adjacent wooded areas. It was suggested that wooded edges benefited *E. terebrans* by providing both a source of adult food (nectar and honeydew) and access to a moderated microclimate. Intentionally established "insectary hedgerows" containing a diversity of shrubs and herbaceous plants to provide continuous sources of pollen, nectar, and shelter have been studied. Marking studies showed that natural enemies utilize these habitats and disperse into adjacent crops. Up to 47 per cent of the marked *Hyposoter* wasps and 23 per cent of lady beetles were found up to 75 m into the crop.

Recently it has been observed that use of semiochemicals may hold potential to manipulate an agroecosystem in a "push-pull" or stimulodeterrent diversionary strategy. Kahn *et al.*, studied lepidopteran stem borers and the parasitoid *Cotesia sesamiae* (Cameron) in Africa. In this study, the grass *Melinis minutiflora* Beauv. produced volatiles that repel female stem borers and attract the foraging female parasitoids. Intercropping maize with this grass led to reduced infestation by the stem borer and increased rates of parasitism compared with a maize monoculture. some new technologies such as transgenic maize expressing *Bacillus thuringiensis* Berliner toxins may appear to reduce disturbance by eliminating pesticide treatments, but may in fact represent a more pervasive disturbance through the potential for cascading multitrophic level impacts.

Combining Deterrents and Attractants/Stimulants

Several authors have suggested that the efficacy of a deterrent-based method may be increased if used in combination with another method that attracts the pest to a nonvalued resource in a stimulo-deterrent diversion or push-pull strategy. The combination of methods might overcome such difficulties with deterrents as desensitization and untreated areas of the resource. The stimulo-deterrent strategy (SDD) was conceived for insect herbivores but is applicable to any pest and any resource type. To date, it has only been tested against a small number of herbivores in small field or glasshouse trials and has yet to be used commercially. The onion fly *D. antiqua* can be deterred from laying eggs on seedling onions by cinnamaldehyde and stimulated to lay eggs on worthless cull onion bulbs that are planted in the same field. In large-scale field trials using microencapsulated cinnamaldehyde as the deterrent and cull onions as the stimulant, the SDD strategy appeared to give good protection for the first few weeks following application. Control subsequently broke down due to the short lifespan of the deterrent formulation. Moths of the genus *Heliothis* were deterred from ovipositing on cotton by azadirachtin and stimulated to oviposit on pigeon pea or maize crops.

In plant-based systems, naturally generated plant stimuli can be exploited using vegetation diversification, including intercropping and trap cropping. Push stimuli can be delivered by intercropping with nonhost plants that have repellent or deterrent attributes appropriate to the target pest. Intercropping reduces pest density in crops, principally by disrupting host location through reducing the visual apparency of the host plant (48), by repellent or deterrent semiochemicals in the nonhosts, or both (75, 78). Enhanced natural enemy abundance in diversified systems may also lead to increased herbivore mortality through predation and parasitism. Molasses grass (*Melinis minutiflora*) and silverleaf desmodium (*Desmodium uncinatum*), which release repellent HIPVs, are used as intercrops in a push pull strategy for maize in Africa.

Push-Pull Trap Cropping

The concept of trap cropping fits into the ecological framework of habitat manipulation of an agro-ecosystem for the purpose of pest management. Trap crops have been defined as "plant stands grown to attract insects or other organisms like nematodes to protect target crops from pest attack, preventing the pests from reaching the crop or concentrating them in a certain part of the field where they can be economically destroyed". The fundamental tenet of this definition involves differential pest preference between plant species, the plants that function as trap crops and those to be protected. The push-pull or "stimulo-deterrent diversion" strategy is based on a combination of a trap crop (pull component) with a repellent intercrop (push component). The trap crop attracts the insect pest and, combined with the repellent intercrop, diverts the insect pest away from the main crop. A push-pull strategy based on using either sweet sorghum, Napier or Sudan grass as a trap crop planted around the main maize crop, and either desmodium or molasses grass planted within the field as a repellent intercrop, has greatly increased the effectiveness of trap cropping for stem borers (*Chilo partellus* Swinhoe). Stem borers are the most important biotic constraint to corn production, and the push-pull strategy has allowed small farmers to control them while managing various parasitic weed species in the genus *Striga*. In addition, the use of molasses grass as a repellent intercrop enhances stem borer parasitoid abundance, thereby improving stem borer control. The push-pull strategy could also be combined with Bt corn to reduce the potential evolution of *Bt* resistance in stem borers.

Predators and parasitoids can make valuable contributions to biological control in IPM and many are commercially available for inundative release. Their use as a population-reducing component in push-pull strategies has been limited so far, but parasitiods of thrips have been tested to contribute to the population reduction of stem borers in maize. The importance of population reduction by natural enemies in push-pull strategies is likely to increase in the future as strategies for their behavioral manipulation are developed. Thousands of farmers in east Africa are now using push-pull strategies to protect their maize crop. The strategies involve the combined use of intercrops and trap crops, using plants that are appropriate for the farmers and that also exploit natural enemies. These plants were selected following trials in Kenya of potential host and non host plants. Stem borers are repelled from the crops by repellent nonhost intercrops, particularly molasses grass (*M. minutiflora*), silver leaf *Desmodium* (*D. uncinatum*), or Greenleaf desmodium (*D. intortum*) (push), and are concentrated on attractive trap plants, primarily Napier grass (*Pennisetum purpureum*) or Sudan grass (*Sorghum vulgare sudanense*) (pull). Molasses grass, when intercropped with maize, not only reduced stem borer infestation, but also increased parasitism by *Cotesia sesamiae*. Coupled gas chromatography-electroantennography of stem borers with volatiles from molasses grass showed attractive compounds similar to those found from maize but, in addition, identified five other compounds including (*E*)-β-ocimene and (*E*)-4,8-dimethyl-1,3,7- nonatriene. These had already been identified from herbivore-damaged plants and were repellent to stem borers in oviposition assays. Desmodium intercrops also produce these compounds, together with large amounts of other sesquiterpenes, and furthermore, when intercropped with maize or sorghum, suppress the parasitic African witchweed (*Striga hermonthica*), a significant yield constraint of arable land in the savannah region. A trap crop of Sudan grass also increased the efficiency of stem borer natural enemies. Although stem borers oviposit heavily on Napier grass, it produces a gummy substance that restricts larval development, causing few to survive. Six host volatiles were attractive to gravid stem borers: octanal, nonanal, naphthalene, 4-allylanisole, eugenol, and (*R*, S)-linalool. Recent studies have indicated that the differential preference of moths between maize and sorghum and Napier grass trap crops is related to a large burst of four electro-physiologically active green leaf volatiles released from the trap crop plants within the first hour of the scotophase, the time at which most oviposition occurs. The push-pull strategy has contributed to increased crop yields and livestock production, resulting in a significant impact on food security in the region. However, wherever these approaches are developed for the specific needs of local farmers, it is essential that the scientific basis of the modified systems is elucidated.

Biological Control Practices

Various living organisms lived in perfect harmony and balance with each other in different ecosystems before chemical agriculture and chemical pest control came into the picture. Today, we see a kind of agricultural development which is based on monoculture and extensive use of chemicals. This has led to the emergence of several pests which has in turn led to the failure of crops. The number of pests developing resistance to pesticides over the years has been increasing at a very alarming rate. The pesticides have proven to be extremely toxic and have led to a number of side effects: impact on public health, toxic residues in food and disturbance of local ecosystems. Use of such chemicals has also led to the eradication of natural predators such as spiders which normally keep pest populations in balance.

Biological control with beneficial insects makes dollars and sense. Costs of sprays, scheduling sprays when workers are not present, managing residue and resistance problems, particularly the resistant insect pests can be avoided. Difficulty of achieving satisfactory spray coverage by itself is incentive enough to seek more sustainable systems of pest management. Unsprayed corn fields host an army of beneficial arthropods, a valuable free source of natural pest control best nurtured, but all too easily and often destroyed by harsh pesticide regimens. In hill agro-ecosystem stem borers, cutworms, blister beetle, armyworms, aphids and other potential pests seldom pose major problems on maize in diversified organic farms where the indigenous pest-fighting natural enemies are preserved. Among the hundreds of beneficial species commonly devouring corn pests are green and brown lacewings, pirate bugs, big-eyed bugs, assassin bugs, damsel bugs, spined soldier bugs, Staphylinid rove beetles, Carabid ground beetles, Collops beetles, lady beetles, six-spotted thrips, Tachinid flies, Phytoseiid mites, spiders and several dozen parasitic wasp species, including Trichogramma.

Minimizing crop losses and optimizing crop production through biological control is dependent upon maximizing the diversity and distribution of the above plant species and resident arthropod wild life. To counter the forces acting against natural enemies of pests of corn in conventional farming systems it is necessary to manipulate these enemies as much as the crop is managed, both must be farmed together. Slight changes in farming can overcome some of the cultural practices taken for granted under conventional chemical farming that form habitats for biological controls. Crop rotation, hedging, refuge management can make a difference in the behaviour of both pests and the beneficials that attack them. Legume green manure crops add to the diversity of organisms provided by resident weeds. When managed by periodically, alternately cutting every other border, they keep the weeds from going to seed and further increase the diversity of composting organisms. This process adds humus to the soil and fosters the food chains that feeds the sets of beneficials in the biological control of insect pests and diseases.

Table 20.7: Natural Enemies of Maize Stem Borer, *Chilo partellus* on Maize Crop (J&K)

Natural Enemy	*Stage Attacked*	*Order*	*Family*
Predators			
Lycosa sp.	Larval	Araneae	Lycosidae
Orius tantillus	Larval	Heteroptera	Anthocoridae
Mantis religiosa	Larval/Adult	Dictyoptera	Mantidae
Coccinella spp.	Egg	Coleoptera	Coccinellidae
Forficula sp.	Egg and Larval	Dermaptera	Forficulidae
Parasitiods			
Cotesia sp.	Larval	Hymenoptera	Braconidae
Trichogramma sp.	Egg	Hymenoptera	Trichogrammitidae
Apanteles sp.	Larval	Hymenoptera	Braconidae

Source: Ahad, I. 2008.

Many natural enemies have been reported to attack *C. partelllus*. Among the parasitoids reported, *Trichogramma* spp. as egg parasitoids and *Aphanteles favipes* and *T. Chilonis* and larval parasitoids are important. *T. cllilonis* has been found parasitising egg masses of *C. partelllus* upto 70 per cent. *A. flavipes* is the most dominant and most widely distributed larval parasitoid of *C. partellus* in India with a parasitisation ranging from 5.3 - 42.8 per cent. Biological control agents such as *Trichogramma minutum* as egg parasitoid, *Apanteles flavipes* and *Bracon brevicornis* as larval parasitoids against *C. partellus* and *Tetrastichus nyemitawus*, *Callitula bipartitius* against *A. soccata* larvae should be encouraged. Coccinellid beetles were also found feeding on early stage larvae and eggs of stem borer and aphids. Entomopathogenic fungus *Fusarium pallidoroseum* cause epizootics in aphids. Armyworm caterpillars can be managed through Bt and NPV. Two applications of NPV at 10 days interval at 250 LE/ha along with crude sugar 2.5 kg + cotton seed kernel powder 250 g on the ear heads to reduce the larval population of *Helicoverpa* sp.

Table 20.8: Examples of Natural Enemies Used in Mass Release

Crop	Pest	Biotic Agent	Dosage	Application Method
Maize	Stalk borer, *Chilo partellus*	*T. chilonis*	2 release of @ 1 Lakh/ha at weekly intervals	Stapling parasitized card at bits on lower surface of leaves
		Bt. kurstaki	0.5kg/ha for proper distribution of spores, 2 g of spore dust (containing 200 million spores) is deposited with a spacing of 10 feet	Spraying with knapsack sprayer
	White grub	*Bacillus popilliae*	0.5kg/ha	Spraying with knapsack sprayer
	Agrotis sp.	*Steinernema carpocapsae*	5 billion infective juveniles/ha, on detection of cutworm larvae	Apply in the soil through irrigation water

Chemical Control Practices

Clothianidin, a new neonicotinoid insecticide for seed treatment, has been extensively tested against major insect pests of maize, canola and other crops. The compound is highly root systemic and enters the transpiration stream through the roots of newly germinating seedlings and developed plants. Pests become intoxicated mainly through ingestion of protected plant tissues and stop feeding immediately. When used as seed treatment at 0.125-0.5 mg a.i./kernel, it has been reported to be effective against corn rootworm (*Diabrotica* spp.), wireworm (*Melanotus* spp.), seed corn maggot (*Hylemya platura* [*Delia platura*]), flea beetle (*Chaetocnema pulicaria*), chinch bug (*Blissus leucopterus*), white grub (*Lachnosterna implicita* [*Phyllophaga implicita*]), Southern green stink bug (*Nezara viridula*) grape colaspis (*Colaspis brunnea*), and cutworm (*Agrotis ypsilon* [*Agrotis ipsilon*]), similar to chlorpyrifos, soil-applied tefluthrin and fipronil. The compound has good activity against variety of pests resulted in a significant increase in yield up to 17.6 per cent on the average compared to the control plots. Soil treatment by mixing carbofuran 3G @ 20 kg/ha in the soil at time of last ploughing has been found to be very effective for cutworm control. Spray the crop with 2-3 weeks after sowing or as soon as the borer injury to the leaves is noticed with any of the synthetic pyrethroids using 125 L of water/ha, fenvalerate @ 100ml/ha, cypermthrin 10EC @ 100ml/ha, deltamethrin 2.8 EC @ 200ml/ha. Usually no additional spray or Granular application is required after the spray. Spraying the crop with 1 kg of carbaryl 50 per cent WP or 500ml dichlorvos 100EC/ha or endosulfan35 EC @ 0.07 per cent for other insect pests. Seed treatment with imidacloprid @ 4g a.i./kg seed or thiamethoxam @ 1g a.i./kg seed followed by whorl application of carbofuran 3G @ 10 kg/ha effectively reduce the stem borer infestation.

Table 20.9: Economic Threshold Level of Major Insect-Pests

Crop	Insect Pests		ET level
	Common Name	Scientific Name	
Maize	Stem borer	*Chilo partellus*	5-10 per cent infestation
	Shoot fly	*Atherigona soccata*	5-10 per cent dead hearts
	Earworm	*H. armigera*	25-30 per cent damage to cobs
	Cut worm	*A. ipsilon*	1 larva/tillers
	Blister beetle	*M.pustulata*	10 per cent infestation
	Aphids	*S.graminum*	Appearance of honeydew on 50 per cent plants

Biotech Interventions in Maize

Maize has been genetically engineered to insert agronomically desirable traits *i.e.* incorporation of a gene that codes for the Bacillus thuringiensis (*Bt*) toxin, protecting plants from insect pests and resistance to herbicides. Both pest resistance and herbicide tolerant genes/events have also been stacked. As of now several events of the two traits have been approved in different countries and are being extensively cultivated. In early 2009, there were nine different event of GM maize in the varieties cultivated globally and three additional maize events have been authorized in at least one country worldwide, but not yet commercialized anywhere (Table 20.10) (Stein and Rodriguez-Cerezo, 2009).

Table 20.10: Commercial GM Maize and GM Maize in the Commercial Pipeline Worldwide

Developer	Product Name	Event Name/Genes	Trait
Commercialized maize events			
Monsanto	Yield Gard Corn borer	MON 810	Insect resistance (to Lepidopteran)
Monsanto	Yied Gard Rootworm	MON 863	Insect resistance (to Coleopteran)
Monsanto	Yield Gard VT	MON 88017	Insect resistance (to Coleopteran)
Dow Agro Sciences and pioneer Hi-bred	Herculex 1	1507	Insect resistance (to Lepidopteran)
Dow Agro Sciences and pioneer Hi-bred	Herculex RW	59112	Insect resistance (to Coleopteran)
Syngenta	Agrisure CB	Bt 11	Insect resistance (to Lepidopteran)
Syngenta	Agrisure RW	MIR604	Insect resistance (to Coleopteran)
Maize events authorized in at least one country but not yet commercialized anywhere			
Monsanto	Yield Gard VTPRO	MON 89034	Insect resistance (to Lepidopteran)
GM maize in the regulatory pipeline worldwide			
Syngenta	Agrisure Viptera	MIR162	Insect resistance (to Lepidopteran)
n/a (China)	n/a	Cry 1A	Insect resistance
GM maize in the advanced R&D pipeline worldwide			
Pioneer Hi-Bred	Optimum Acre Max 1	n/a	Insect resistance (to Coleopteran)
n/a (India)	n/a	Cry 1Ac + cp4epsp4	Insect resistance

On a global basis, in 2008, genetically engineered maize occupied 37.3 million hectares equivalent to 24 per cent of the global maize area of 157 million hectares (James, 2008). There have been a substantial increase in the deployment of stacked traits of *Bt* and herbicide tolerance, particularly in USA. The triple gene products in GM maize, featuring two *Bt* genes (one to control the European corn borer complex and the other to control root worm) and one herbicide trait continued to grow in adoption in USA in 2008. The area occupied by maize with insect resistant and herbicide tolerant genes and the two characters stacked together is as follows. In total maize was grown in 17 countries worldwide. Major countries growing GE maize are USA, Argentina, Canada, Brazil and South Africa. Other countries include Uruguay, Philippines, Chile, Egypt, Honduras and seven EU countries.

Future Projection

Rich biodiversity and fast changing cropping pattern toward commercial agriculture invite vibrant monitoring and surveillance system. Due to emphasis on organic farming in the region, there is a need to stress upon exploitation of potent local bioagents and botanicals against major insect pests, besides conservation and augmentation of exotic as well as native natural enemies. In addition, already existing proven IPM technologies available in different parts of the country need to be validated under hill condition. As the hills posses varied climatic conditions due to several factors like topography, steepness of slope, variations in soil structures, face of mountain etc., niche based IPM module has to be worked out to ensure better effectivity of the various management tools. Of late, a greater emphasis has been on cultivation of tea, medicinal plants, vegetables, flowers etc. especially in Uttarakhand hills. In the event of shift in cropping system, particularly perennial crops, a critical vigilance would be of paramount importance in order to prevent emerging pest problems. Overall general awareness about IPM among the inhabitants coupled with government and institutional support would be necessary in the geographically challenged region of the country.

Conclusion

Insect pests have been the major constraint in agricultural crop production in the hilly tracts of NW Himalayan region of the country. To combat the burgeoning problem, relying on the pesticides, is not desirable in the present day of global consciousness towards health hazards and environmental safety. Moreover, the newly formed Uttaranchal state of the region is all set to be declared as organic state and the major hurdle in the event of organic farming is the management of insect pests and diseases. One of the strongest alternatives for the management of the insect pests would be integrated pest management with emphasis on biological control using potential insect parasitoids and microbial pathogens.

References

Ahad, I. (2008). Pest complex and their sucession on maize with special reference to management of stem borer (*Chilo partellus* Swinhoe). Ph.D. Thesis. SKUAST-Jammu. pp 85.

Alejandro, O. C. (1987). *Insect Pests of Maize: A Guide for Field Identification*. Mexico, D.F.: CIMMYT.

Altieri M. A., Francis, C. A., Choonhoven, A. V. S. and Doll, J. D. (1978). A review of insect prevalence in maize (*Zea mays* L.) and bean (*Phaseolus vulgaris* L.) polycultural systems. *Field Crops Research*, 1 : 33 - 49.

Anonymous (2011). *Biology of maize. Series of Crop Specific Biology Documents*. Department of Biotechnology, Ministry of Science and Technology, Government of India. pp 24.

Bhatt, J.C. and Sushil, S.N. (Eds) (2004). A compendium on diseases, insects and weeds of rice, maize and wheat crops in Indian Himalayas. 118 p.

Cardwell, K. F., Schulthess, F., Ndemah, R. and Ngoko, Z (1997). A systems approach to assess crop health and maize yield losses due to pests and diseases in Cameroon. *Agriculture, Ecosystems and Environment.* 65(1):33-47

Cook, S. M., Khan, Z. R. and Pickett, J. A. (2007). The Use of Push-Pull Strategies in Integrated Pest Management. *Annu. Rev. Entomol.* 52:375-400.

Landis, D. A., Wratten, S. D. and Gurr, G. M. (2000). Habitat management to conserve natural enemies of arthropod pests in agriculture. *Annu. Rev. Entomol.* 2000. 45:175–201.

Nair, M. R. G. K. (1995). *Insects and Mites of Crops in India*. ICAR publication, New Delhi. pp 407.

Penagos, D. I., Magallanes, I., Valle, J., Cisneros, J., Marti, A. M., Goulson, D., Chapman, J. W. and Caballero, P. (2003). Effect of weeds on insect pests of maize and their natural enemies in Southern Mexico. *International Journal of Pest Management*. 49(2) : 155–161.

Shelton, A. M. and Perez, F. B. R. (2006). Concepts and applications of trap cropping in pest management. *Annu. Rev. Entomol.* 51:285–308.

Siddiqui, K. H. and Marwaha, K. K. (1993). *The Vistas of Maize Entomology in India*. Kalyani Publishers, Ludhiana. pp250.

Hill Agriculture: Prospects, Constraints and Mitigations (2013) *Pages* **314–330**
Editor: **Dr. Shahid Ahamad**
Published by: **DAYA PUBLISHING HOUSE, NEW DELHI**

Chapter 21

Aphids in Agro-ecosystems and their Management

Mohd Abas Shah and Akhtar Ali Khan
Division of Entomology,
Sher-e-Kashmir University of Agricultural Sciences and Technology of Kashmir,
Shalimar Campus, Srinagar – 191 121, Jammu and Kashmir, India

Introduction

Aphids (Homoptera: Aphididae) are minute pear-shaped, soft-bodied insects. They may be winged (alate) or wingless (apterous). Most of the aphids reproduce sexually and develop through gradual metamorphosis (eggs, nymphs and winged or wingless adults) but also through a process called parthenogenesis. The result being shortening of generation time, overlapping generations, increased reproductive potential and rate of development of resistance to insecticides. Their reproductive potential, salivary secretions, and ability to transmit viral diseases make them the most potent and worldwide enemies of many crops, with the greatest number of species in the temperate regions. Out of 4702 aphid species so far known world over (Remaudiere and Remaudiere, 1997) about 1015 species occur in oriental region (Agarwala and Ghosh, 1984). Of course, about 653 species belonging to 208 genera represent Indian aphididae.

Aphids cause direct damage by sucking the cell sap from the plants and indirect damage by transmitting diseases from one plant to another. Host alternation is common in aphids. Often, aphids use one host plant as the primary host for reproduction and another plant as a secondary host for parthenogenetic reproduction. The small size, biology and behaviour have made the aphids as one of the most successful group of insects which infest all kinds of vegetables, fruits, ornamental plants and field crop and cause economic damage (Atwal, 1976). Aphids play an important role as vectors in spreading of plant diseases, particularly viral diseases. The cotton aphid is known to transmit over 50 plant viruses and green peach aphid, over 100 (Kennedy *et al.*, 1962). Viruses are carried by migrating aphids which may travels as far as 1300 km. Yield losses caused by aphids have direct crop yield reductions ranging from 10-50 per cent and indirect from 20-80 per cent. Aphids constitute one of the most important groups of phytophagous insects because of their polymorphism, host alternating, migration, heteroecious behavior, disease transmission and reproductive habits. They are remarkable for their role as largest group of insect vector of plant viral diseases (Ghosh and Singh, 2000). These insects are considered as obligatory parasites of plants.

Host Range

Over 250 species of the superfamily Aphidoidea feed on agricultural and horticultural crops throughout the world (Blackman and Eastop 1984). Pest aphids tend to have a wider host range than economically unimportant species (Table 21.1). The green peach aphid, *Myzus persicae,* has an extremely wide host range of over 100 plants including a wide variety of vegetable and ornamental crops (Baker 1982). Most aphid species are relatively host-plant specific, although they may occasionally be found colonizing certain plants or 'reserve hosts' with which they are normally not associated. The Aphidoidea are predominately a north temperate group (North America, Europe, Central Asia).

Table 21.1: Important Aphids and their Host Plants

Sl.No.	*Aphid*	*Scientific Name*	*Host Plants*
1.	Bean aphid	*Aphis fabae*	Artichoke, asparagus, bean, carrot, corn, lettuce, parsnip, rhubarb, spinach, squash, Euonymous, *Viburnum* spp.
2.	Cabbage aphid	*Brevicoryne brassicae*	Broccoli, Brussels sprouts, cabbage, collards, kale, kohlrabi, and radish.
3.	Corn leaf aphid	*Rhopalosiphum maidis*	Corn, paddy, wheat.
4.	Green peach aphid	*Myzus persicae*	Beet, celery, cole crops, cucurbits, lettuce, pepper, potato, spinach, tomato, cherry and peach
5.	Melon aphid	*Aphis gossypii*	Asparagus, bean, beet, celery, cucurbits, okra, spinach.
6.	Pea aphid	*Acyrthosiphon pisum*	Peas, alfalfa.
7.	Potato aphid	*Macrosiphum euphorbiae*	Asparagus, bean, corn, cucurbits, eggplant, lettuce, pepper, potato, sweet potato, tomato, rose.
8.	Turnip aphid	*Lipaphis erysimi*	Mustard, radish, turnip, other crucifers
9.	Cabbage root aphid/Poplar petiole gal aphid	*Pemphigus populitraus-versus*	Cruciferous crops/poplar trees.
10.	Soybean aphid	*Aphis glycines*	Soybean, pulses
			Fruits Aphids
1.	Almond leaf curl aphid	*Brachycaudus amygdalins*	*Prunus* spp, tobbacco
2.	Leaf costing plum/ peach aphid	*B. Lelichrysi*	*Prunus* spp, *Brassica* spp, *Solanum* spp
3.	Wooly apple aphid	*Eriosoma lanigerium*	Apple, pear, wild apple, other members of family rosaceae
5.	Citrus aphid	*Toxoptera citricidus*	Citrus, members of family rutaceae and rosaceae.
6.	Coconut aphid	*Ceratrphis variables*	Coconut, *Areca* sp.
7.	Banana aphid	*Pentatonia nigronervosa*	Banana, cardamom, *Calocasia* sp.

Contd...

Table 21.1–*Contd...*

Sl.No.	Aphid	Scientific Name	Host Plants
			Oilseed crops
1.	Mustard aphid	*Lipaphis erysimi*	*Brassica* spp., sugarbeet
2.	Groundnut aphid	*Aphis craccivora*	Grour.dnut, pea, beans, safflower
3.	Green peach aphid	*Myzus persicae*	*Brassica* spp.
4.	Safflower aphid	*Uroleucon compositae*	Safflower, zinnia, calendula, *Blumea* spp.
5.	Wooly aphid	*Ceratovacuna lanigera*	Sugarcane
			Cereals
1.		*Sitobion avenue*	*Triticum* spp., barley, oats, maize, tobacco
2.		*Rhopalosiphum padi*	Oats, paddy, wheat, maize, grasses.
3.		*Metopolosiplum dirhodum*	*Triticum* spp., other grasses
4.		*Macrosophum miscanthi*	Oat, wheat, rice, *Brassica* spp.
			Ornamentals
1.	Rose aphid	*Macrosiphum rosaeiformis*	*Rosa* spp., pea, *Secale cereale*
			Landscape trees
1.	The linden aphid	*Eucallipterus tiliae*	Landscape trees
2.	Little leaf linden	*Tilia chordata*	Landscape trees

Source: Modified after Raychaudhuri, 1983.

Biology and Life-cycle

The life history of aphids is complex and involves sexual and asexual (parthenogenetic) modes of reproduction (Moran, 1992). In the sexual mode, the life stages consist of the egg stage, four nympal instars and adult. In contrast, the parthenogenic females give birth to live progeny. Both winged (alate) and wingless (apterous) individuals and generations exist. Sexual and parthenogenetic reproduction alternate in the life-cycle (cyclical parthenogenecity).

Most aphids require regular alternation of plant hosts where sexuals mate and lay eggs on the primary host for several generations and give rise to a generation of alate morphs that move to the secondary host, which is usually a distinct plant species. On the secondary host, the parthenogenic mode of reproduction is initiated and used for subsequent generations. Some aphids have lost or can facultatively lose, the sexual life cycle. However, some aphid species don't change their hosts, and thrive on single species of plant or group of closely related species.

Sexual forms usually appear in the fall and oviposit overwintering eggs, the eggs hatch in spring just as new growth on plants is developing. These aphids are all wingless females, each of which without fertilization bears succeeding generations of as many as 100 or more live wingless females. Incredibly, the aphids are actually born pregnant- even before birth; the female's own daughters are developing within her (paedogenesis). This strategy of bearing live young without fertilization accounts for the extreme rapidity with which aphid populations can grow, making control difficult. Young plant shoots can quickly be completely covered with a teaming colony of aphids.

As the colony grows it may begin to run out of room or food may become limited. When this occurs, the aphids adapt by producing winged females which can colonise surrounding plants. Most work on aphid flight behaviour has examined trivial flights- the plant to plant flight of an aphid searching for a suitable host (Dixon, 1998). Trivial flight is horizontal and for short distance and always within the atmospheric boundary layer (Kring, 1972). In contrast, migratory flight, from birth colony to a new host, bypassing potential targets nearby is a means of dispersal to geographically distant locations (Hardie, 1993). As migratory aphids are carried by the wind, more emphasis has been laid on long distance migration of aphids ranging to few hundred kilometres (Hardrie *et al.*, 1985; Riley *et al.*, 1995). However, there are few reports where aphids were found to migrate 1000 Km from the possible source (Elton, 1925; Kring; 1972).

The sexual males and females are produced late in the fall. After mating the females lay one more overwintering eggs. Protected in plant crevices, these eggs withstand inclement weather to hatch in the spring, repeating the cycle. Aphid species found on indoor plants or in tropical climates don't need to produce eggs, and thus males, whose only function is fertilization, may never appear (Figure 21.1).

Among all the abiotic environmental factors, temperature markedly increases the proliferation rate of aphids and temperature is often seen as the most important factor for pest population development (Wennergren and Land, 1993).The optimum temperature for reproduction and development of aphids is around 20°C. Temperature changes affect not only the pest but also its host plant and natural enemies, for example, low temperature can reduce aphid population growth but simultaneously the low temperature reduces the number of natural enemies or their activity and the aphids can thereby reach similar densities in hot and cold years (Skirvin, Perry and Harrington, 1997). Rainfall and wind are also often reported as being of importance for pest densities. Aphid pest attacks fluctuate widely from year to year because of changes in the weather pattern and because of the activity of naturally occurring parasitoids and predators.

Damage and Severity of Losses

Aphids are serious pests of plants. They damage the crops both directly and indirectly. The feeding of adults and nymphs harms plants directly, while indirect damage can result from the secretion of honeydew, transmission of plant diseases especially viral, and the contamination of harvested crop. Major damage is manifested by sucking of nutrients and transmission of diseases.

Aphids have piercing and sucking mouthparts. They insert their needle like mouthparts in plant tissues and remove plant juices, thus robbing the plant of nutrients. Aphids can feed on all plant parts, but most prefer young shoots and leaves are they are more susceptible to their attack. Aphids become pest when their feeding begins to affect plant health and this condition is dictated by the aphid number.

The plant may gradually weaken as fewer nutrients become available to it. Aphids can also cause stunting and distortion of plants, further weakening the plant and suggesting some sort of biochemical interaction and sometimes chlorosis as well. Plant distortion also compounds the control problems because aphids are somewhat protected from contact insecticides inside the curled leaves. Several aphid species are also known to cause gall formation by hypertrophy and hyperplasia of the cells under the influence of aphids. Several types of galls are produced by the gall-producing aphids like simple leaf gall, leaf pouch gall, shoot axis gall, stem gall etc.

Most aphids excrete a sweet, sticky substance called honeydew from their cornicles as they feed. The sugar protein mixture can cover the surface particularly leaves and serve as a food source for ants,

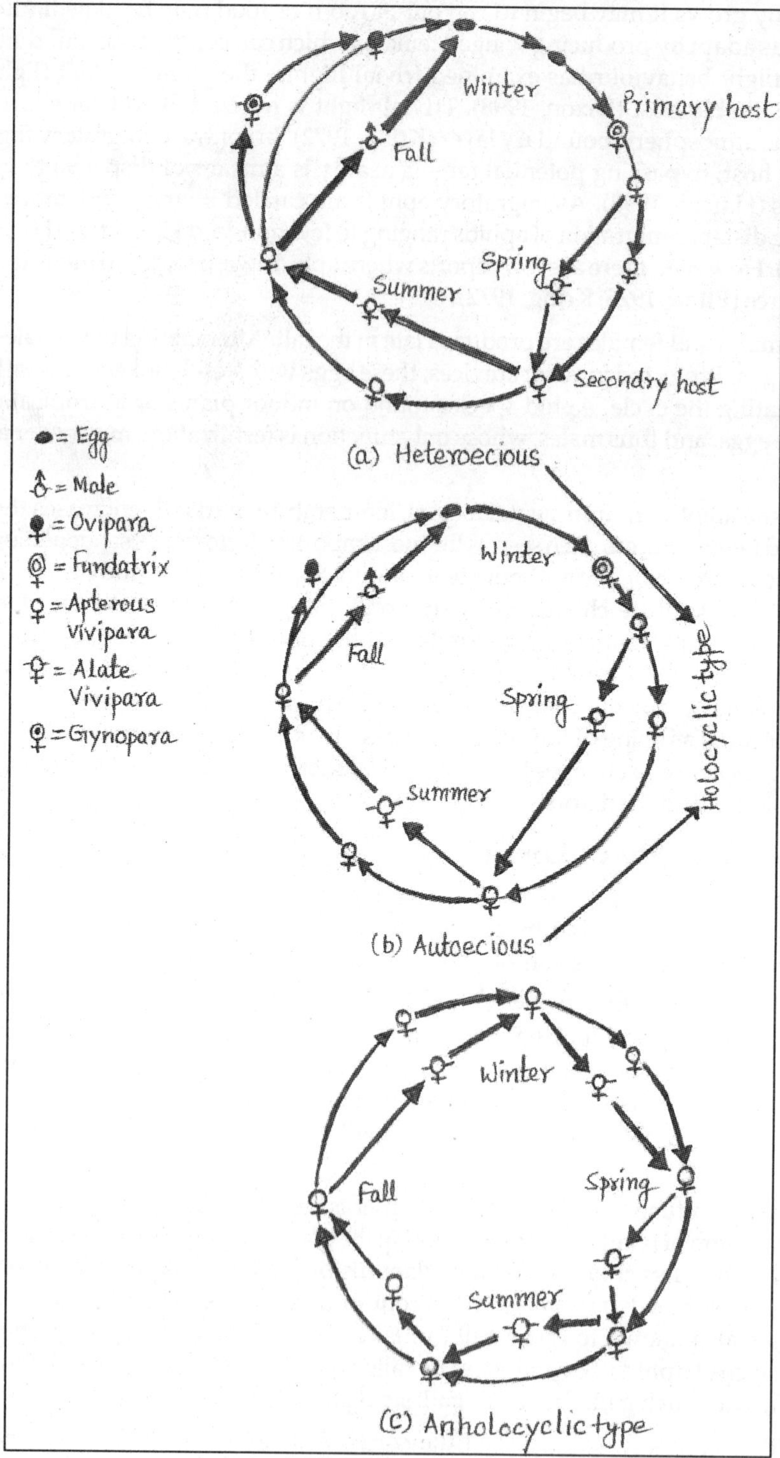

Figure 21.1: The Life-cycle of Aphids

bees, flies etc. Some of the insects attracted prey on the aphids but others *e.g.* ants may attack the beneficial predators. Thus controlling ants is an important part of aphid reduction. Honey dew also serves as a medium for sooty mould fungus (saprophytic) which appears as a black film growing over it. With low pest density, honey dew and sooty mould merely contaminate the crop and reduce its marketability (particularly vegetables). With heavy pest pressure, sooty mould can become thick enough to block sunlight which reduces photosynthesis and affects yield. Honey dew produced by the aphids and the sooty mould which develops on it are nuisances on landscape trees, affecting their aesthetic value.

Aphids may introduce pathogens as they feed. Sometimes the transmission of disease is incidental; the pathogen simply hitches a ride on the aphid while other pathogens use aphids as an important part of the disease cycle. Transmission of diseases is of primary concern with aphids in many crops. The role of aphids in disease transmission and the diseases they transfer shall be discussed further in the following sections.

Various workers have reported appreciable crop yield losses inflicted by different aphid species to various crops throughout the world. Direct yield losses ranging from 10-50 per cent and indirect yield reduction up to 80 per cent have been documented. Tradan and Milivoj (1999) reported direct crop yield reductions ranging from 10-50 per cent and indirect from 20-80 per cent in cereals. Voss *et al.* (1997) have also reported yield losses of 19 and 31 per cent in wheat crop at the boot stage and 14-15 and 20 per cent during anthesis stage of plant growth. Aphids are among the important key pests affecting quality and quantity of the produce of vegetables and oilseed crops, besides cotton and various fruit and plantation crops.

Table 21.2: The Nature and Extent of Losses Caused by Aphids on Selected Crops

Crop	*Aphid*	*Type of Damage*	*Average per cent Loss*	*Reference*
Winter wheat	*Sitobion avenae*	Direct damage	12.59-7	George and Gair, 1979, Rabbinge *et al.*, 1981
	Metopolophium dihodum	Direct damage	13.0	Prew *et al.*, 1983
Winter barley	*Rhopalosiphum padi*	Virus vector	0-86	Doodson and Saunders, 1970
Spring barley	*Metopolophiumdi rhodum*	Direct damage	8.8	George, 1974
Sugar beet	*Myzus persicae*	Virus vector	6.5	Hull, 1963
Potatoes	*Macrosiphum euphorbiae* *Myzus persicae*	Direct damage	5.74.4	Kolbe, 1970; Southall and Sly, 1976
Field beans	*Aphis fabae*	Direct damage	46.316.4	Way, 1967 Gould and Graham, 1969
Peas	*Acyrthosiphon pisum*	Direct damage	15.88.0	Maiteki and Lamb, 1985

Source: Modified after Tatchell, G.M (1989).

Aphids as Vectors

A vector is an organism capable of transmitting pathogens from one host to another. A number of plant diseases caused by viruses, phytoplasmas, bacteria and fungi are transmitted by insects. Majority of plant diseases are transmitted by insects and a few by other arthropods like mites and only a small percentage by nematodes. Homopteran insects (aphids and leafhoppers) alone are known to transmit about 90 per cent of the plant diseases. Most plant viruses are transmitted by aphids than by any other group (Swenson, 1968) and it's likely that aphids do more damage by transmitting viruses than by removing the sap of plants many a times.

Of around 850 described plant virus species, about a half is transmitted by aphids, a third by leafhoppers and the rest by other organisms. The main aphid vectors are *Myzus persicae*, *Aphis gossypii* and *Aphis craccivora*. Viruses transmitted by aphids can be classified according to the way in which the aphid carries them: the stylet borne and circulatine viruses (Kennedy *et al.*, 1962). A few viruses are intermediate in their characteristics.

The stylet borne or non-persistent viruses are believed to be transmitted as contaminants of the mouthparts, hence the transmission is mechanical. The vector is able to acquire the virus from a disease source and transmit to a healthy plant by feeding for a few seconds. Optimal period is approximately 30 seconds. These viruses can't persist hunger within the vectors, which transmit them soon after feeding on infected plant but the ability to transmit fresh infection soon disappears after insect feeds on healthy plants. The vectors usually begin to lose ability to transmit viruses within 2 minutes. Aphids transmit majority of viruses by this method. Sometimes the viruses are carried in the anterior region of the gut of the vector, where they multiply to a certain extent. Vectors don't remain ineffective after a moult, presumably because the viruses are lost when the forget intima is shed. This is the semi persistent mode of disease transmission.

In circulative or circulatine-cumulatine or persistent mode of virus transmission, the aphids feed on the diseased plant (acquisition feed), require sometime after acquisition feed to transmit the virus (latent or incubation period), feed on healthy plant (inoculation feed) and in the process transmit the viruses acquired earlier and may remain infective for the rest of its life. The acquired viruses pass through the midgut wall to the salivary glands from where they can infect new hosts. In such mode of transmission, the insect vector has to feed on the source of virus for comparatively longer periods to acquire infectivity. The incubation period may vary from several hours to 10-20 days during which the vector is incapable of virus transmission.

Virus transmission is closely related to the feeding behaviour of the aphids. While feeding, the plant tissue is punctured by a number of forward and backward movements of the inner pair of stylets. During the forward movements, the fluid flows into them and during the backward movements the saliva is ejected. Generally an insect injects by feeding on any part of the plant, but in some cases the virus is only found in the phloem and has to be injected into the phloem, the movement of which is perhaps controlled by the PH gradient between the mesophyll and the phloem.

Irrespective of the mode of transmission, virus-vector relationship is highly specific. Generally one type of viral disease is transmitted by insects belonging to a particular group, for example mosaics by aphids. The relationship is sometimes more specific, for example cabbage ring spot is transmitted only by *Myzus persicae* and not by *Myzus ornatus*. A vector can also acquire and transmit more than one virus or a virus may be transmitted by many vectors for example the cotton aphid is known to transmit over 50 plant viruses and green peach aphid, over 100 (Kennedy *et al.*, 1962), and the onion yellow dwarf is known to be transmitted by 60 insect vectors.

Some of the economically important plant viruses transmitted by different aphid species are listed in Table 21.3.

Table 21.3: Aphid Vectors of Plant Viruses

Sl.No.	Insect Vector(s)	Varies(es)	Host(s)
1.	*Aphis craccivora* Koch	Alfalfa mosaic	Alfalfa
		Cowpea mosaic	Cowpea
		Onion yellow dwarf	Onion
		Papaya mosaic	Papaya.
2.	*Acyrthosiphon pisum* (Harris)	Alfalfa mosaic	Alfalfa
		Bean mosaic	Beans
		Onion yellow dwarf	Onion
		Pea mosaic	Peas
		Soybean mosaic	Beans, Peas.
3.	*Aphis fabae* Scopolis	Alfalfa mosaic	Alfalfa
		Beet yellow	Spinach, Sugarbeet
		Dahlia mosaic	Calendula, Dahlia, Zinnia
		Onion yellow dwarf	Onion
		Pea mosaic	Peas.
4.	*Aphis gossypii* Glover	Alfalfa mosaic	Alfalfa, Potato
		Chilli mosaic	Chillies
		Cucumber mosaic	Cucumber
		Dahlia mosaic	Dahlia, Zinnia
		Lettuce mosaic	Lettuce, Sweet pea
		Papaya mosaic	Papaya
		Sugarcane mosaic	Sugarcane
5.	*Aphis rumicis*	Bean Common mosaic	Beans
		Pea mosaic	Broadbean, Garden pea, Sweet pea
6.	*Brevicoryne brassicae* L.	Cauliflower mosaic	Cabbage, cauliflower
7.	*Macrosiphum euphorbiae* (Thomas)	Alfalfa mosaic	Alfalfa
		Lettuce mosaic lettuce mosaic	Garden pea, lettuce, sweet pea
		Potato virus Y	Potato, Tobacco, Tomato
8.	*M. granarium*, *M. miscanthi*	Barley yellow dwarf	Barley, oat, rye, wheat
9.	*Myzus persicae* (Sulzer)	Alfalfa mosaic	Alfalfa
		Beet yellow	Spunaic, Sugarbeet
		Cabbage yellow ring spot	Cabbage
		Cauliflower mosaic	Cauliflower
		Cucumber mosaic	Cauliflower
		Dahlia mosaic	Cucumber

Contd...

Table 21.3–*Contd...*

Sl.No.	*Insect Vector(s)*	*Varies(es)*	*Host(s)*
		Lettuce mosaic	Calendula, Dahlia, Zinnia Garden pea, lettuce, sweet pea
		Onion yellow dwarf	Onion
		Pea mosaic	Broad bean, Garden pea, Sweet pea
		Potato virus Y	Potato, Tobacco, Tomato
		Soybean mosaic	Soybean
		Sugarcane mosaic	Maize, Sorghum, Sugarcane
10.	*Pentalonia nigronervosa* Conquerel	Banana bunchy top	Banana
		Cardamom mosaic	Cardamom
11.	*Rhopalosiphum maidis* (Fitch)	Barley yellow dwarf	Barley, eye, oat, wheat
		Onion yellow dwarf Sugarcane mosaic	Onion
			Maize, Sugarcane, Sorghum
12.	*Schizaphis graminum* (Rondani)	Sugarcane mosaic	Maize, Sorghum, Sugarcane

Source: Modified after Dhaliwal (2006).

The indirect crop loss caused by aphids by virus transmission is closely related to weather. Dixon (1978) put down the following account, "Most aphids that transmit viruses are host alternating and the trees and shrubs on which these aphids normally overwinter do not usually harbour viruses of agricultural crops. However outbreaks of viral diseases in agricultural crops are associated with mild winters and springs and this phenomenon is more pronounced in temperate zones. The reason for this is that not all the individuals of some host alternating aphids return to their primary host in the autumn; some remain in their herbaceous hosts all through the year, without entering an egg stage during the winter. In mild winters the survival rate of these forms is high, and in spring they multiply early and migrate. The herbaceous plants on which these aphids overwinter are often infected with virus. Thus in years when alate aphids arrive very early on the crops they have overwintered on herbaceous hosts and are likely to be infected with virus, but the aphids which arrive later come from their woody hosts and are usually virus free."

As far as the adaptive significance of virus-aphid relationship is concerned, the relationship is mutually beneficial since the aphid is the means of dispersal of the virus and the virus improves the quality of the plant as food for the aphid. Since the amino-nitrogen content of the phloem sap of mature plants or parts of plants is generally low, thus the rate of growth and fecundity of aphids is restricted. Virus infection of plants results in increased concentration of free amino-acids, as a result aphids develop faster and give birth to more young on virus infected plants. Therefore, it would appear as Kennedy (1951) suggested that at least in the short term aphids have much more to gain from transmitting plant viruses.

Management of Aphids

Scouting and Economic Thresholds

Aphid populations vary from year to year, plant to plant and with the growing season so regular inspection is an important step to determine if the aphid problem is escalating.

Measures for aphid control should be initiated only if the population swings over the economic threshold level. Moderate aphid populations usually don't cause noticeable or economic damage as different crops can tolerate a certain amount of damage inflicted by aphids besides various natural enemies take care of small aphid populations. However, this concept doesn't hold good if the aphid species is responsible for the transmission of virus (es) that may prove dreadful for the crop. In the later case, few winged aphids can cause appreciable damage.

Scouting or monitoring must be carried out to determine if aphid populations are reaching the economic threshold. Plants should be visually inspected for signs of an aphid infestation, especially the plant parts prone to aphid problems like underside of leaves, stems and new growth like buds. Also watch for honey dew, sooty mould and cast skins. Sometimes the aphids are difficult to see on plants with fine foliage, such plants should be held over a white piece of paper and tapped to dislodge aphids: if any. It is recommended that most vegetables should be scouted for aphids and other pests per week throughout the growing season. Progress of aphids can be determined by the aphid/leaf count method (Ramon, 1985) or sometimes aphid per unit area of branch depending upon the crop plant and the aphid species. Monitoring of alate aphid populations is conducted using yellow pan traps ($50\times30\times8$ cm^3) with sides sloping outward. Traps are filled with fresh water and a few drops of detergent added to break surface tension and prevent aphids from escaping (Raman, 1985).

It is important to determine whether the aphid population is increasing or decreasing before taking management decisions. In general the conditions that favour an increase in aphid population are potassium deficient soils, cool temperatures, absence of beneficials, plants under drought stress, plants treated with higher levels of nitrogen etc. Heavy rains and beneficial insects may reduce low or moderate populations. If most of the aphids are winged or nearing this stage, they leave the plant or may be the field. Under such condition, an insecticide may not be needed because the population will rapidly decline. Similarly, a high proportion of parasitized aphids in the field is indicative of a decline in population. Economic threshold levels of some of the important aphid pests for various crops of local importance are given in Table 21.3.

Cultural and Physical Control

Cultural practices can be helpful in preventing aphid infestation or reducing their severity. Crops should be produced in a manner that encourages rapid, vigorous growth. Planting of susceptible crops on or near an infested site should be avoided. If it is necessary to have aphid attractive plants or tree in landscape designing, place them where aphid presence and honeydew production won't become an unbearable nuisance. Also planting of susceptible cultivars or crops should be avoided on downwind side of infested area as the winged adult aphids are usually blown downwind. Alternate host plants like weeds in the vicinity of crop should be destroyed before planting. Similarly the crop residues and culls left in the field after a crop is harvested should be destroyed as they may help aphids carry over to next crop or provide refuges to produce aphids that infest nearby crops. Heavily infested areas should be kept fallow. On similar lines, crop rotation and mixed/inter cropping system involving non host crops or resistant varieties have shown encouraging results in decreasing aphid infestation in the next season. When possible, planting time should be adjusted so that the crop escapes the environmental conditions most favourable to aphid development.

Table 21.3: Economic Threshold Levels (ETL) of some Important Aphid Pests of Crops

Sl.No.	*Aphid*	*Crop*	*ETL*
1.	Wheat aphid, *Macrosiphum miscanthi*	Wheat	5 aphids/ear (at ear head stage)
2.	Mustard aphid, *Lipahis erysimi*	Brassica crops	50-60 aphids/10 cm terminal portion of shoot or 40-50 per cent plants are infested by aphids
3.	Green peach aphid, *Myzus persicae*	Sunflower	48 aphid/cm^2 of shoot or 5 per cent galls
4.	Groundnut aphid, *Aphis craccivora*	Groundnut Bean	50-100 aphids per plant10 per cent affected parts
5.	Cotton aphid, *Aphis gossypii*	Cotton	Appearance of honeydew on 50 per cent plants.
6.	Soybean aphid, *Aphis glycines*	Soybean	250 aphids/plant
7.	Apple woolly aphid, *Eriosoma lanigerum*	Apple	
8.	Cabbage Aphid, *Brevicoryne brassicae*	Cole crops	50-60 aphids/10 cm terminal portion of shoot or 40-50 per cent plants are infested by aphids

Plants that receive levels of nutrients, water and light adequate for the species will thrive, usually. Improper levels produce stress on the plants. Air pollution and physical damage also contribute to plant stress. Reports suggest that an excess of nitrogen can also stress plants- particularly indoors plants under environmental stress will usually exhibit increased disease and pest problems. Thus proper crop husbandry is a prerequisite for the combat against pestiferous aphids.

Physical control measures include the use of mulches that reflect light and heat. Silver reflective mulches and row covers have successfully been used to repel aphids from plants. Mulches help plants get off to a healthy start, and are effective until expanded foliage covers the reflective surface. Various types of reflective mulches have been worked out suiting to various cropping systems. Aphid populations near to ETL can be reduced by washing the terminals of sturdy plants such as shrubs and trees with strong water sprays. If a regular water wash doesn't work, adding a soap or detergent can help dislodge aphids by reducing surface tension.

Biological Control

Were it not for the natural control that limits the aphid population, the whole biosphere would soon be overwhelmed by them. In the majority of landscapes, natural control does an adequate job of keeping aphid population under check. Important agents of natural control are the natural enemies of aphids and the climate. Aphids have many natural enemies, including many species of predators, parasitoids and pathogens. When aphid populations are small, predators and parasitoids can frequently eliminate them. Diseases also can eliminate aphid populations, but tend to have their greatest impact only after the aphid population has greatly exceeded an acceptable level but the epizootics are more at the beck and call of climatic conditions. Even under moderate pest pressure natural enemies can frequently eliminate aphids if given enough time. Since natural enemy populations usually somewhat lag in time behind the aphid populations or may be less in number and in extreme cases altogether absent, there may be periods in the year when the aphids appear to be out of control, but sooner or later the natural enemies can catch up and restore control. The decision whether to allow the beneficials to work or to go for insecticidal control is influenced by the stage of the crop and the efficacy and speed of control by natural enemies.

In the order of importance, predators are currently the most abundant and have the ability to delay and suppress aphid outbreaks. The most common predators of aphids are adult and larval lady beetles, green lace wing larvae, and syrphid fly larvae. Lady beetle adults and larvae ear as many as 200 aphids per day and as many as 2400 in their life span. Among the most common lady beetles are *Hippodamia convergens, Harmonia axyridis, H. eucharis, Coccinella septempuctata, Adalia tetraspilota, Coleomegilla maculate, Adalia decempunctata* etc. The predation potential of various coccinellid predators on various aphid species has been worked out by different workers (Khan and Mir, 2008, Khan 2009, Khan, 2010) under laboratory conditions. Larvae of green lace wings (*Chrysoperla carnea, C. rufilabris, Chrysopa* spp.) are also efficient predators, consuming as many as 100 aphids per day (Mushtaq and Khan, 2010). Among the syrphids, *Metasyrphus* spp., *Syrphus* spp., *Scaeva pyrastri* etc are prominent in different agro-ecosystems.

Hymenopterans are perhaps the most important parasitiods (egg) of aphids' *e.g. Aphidius* spp, *Aphelinus* spp. The adult wasps are tiny and seldom seen, but their presence is readily observed. Aphids that have been parasitized swell up until they are almost spherical in shape and turn brassy brown in due course of time. These swollen, hardened aphids are called mummies. The parasitoid develops inside the aphid, then chews a circular hole in the abdomen of the aphid and emerges. Parasitism can increase rapidly in an aphid population often from about 10 per cent to more than 90 per cent in about a week. Unfortunately this often occurs too late to be of commercial benefit, and aphid mummies are a crop contamination. However the predators and parasitiods can be introduced into or augmented in landscape as a means to control aphid populations.

Aphids are susceptible to fungal diseases when it is warm and humid. Entire colonies of aphids can be killed when conditions are right. Important ones are *Beauveria bassiana, Verticillium lecanii* etc. Potential pathogens may include some bacteria and viruses.

Although aphids are generally considered to be completely helpless (Imms, 1947), their defensive and escape mechanisms have been observed in several cases. The aphids show various defence responses, *e.g.* kicking movements, movements of the body, pulling free the appendages seized by the predator and waxing the predaceous coccinellids with an oily liquid appearing on the tip of siphunculi. Moreover the aphids may escape by walking away or dropping off the plant. The intensity of response varies with the size and stage of the predator. Aphids with longer appendages are generally more difficult for the predators to capture than the more compact and tenacious aphids with short appendages (Kaddou, 1960) *e.g.* older larvae of *Adalia bipunctata* succeed in capturing 90-100 per cent of first instar and 60-70 per cent of adult *Myzus persicae* but only 0-50 per cent of the much larger *Acyrthosiphon pisum* (Klinguaf, 1967). The aphids may apparently succeed in escape because the prey perceives the predator at a distance of several centimetres, while the predators perceive the prey only by contact. Thus the aphid attains a time advantage and may use this for escape. Some aphids are protected against their predators by attendant ants which protect the whole area from invasion of other insects.

Insecticidal Control

When aphid populations exceed acceptable levels and beneficial insect activity is low or rapid control is needed, insecticides become a necessary part of aphid management. But this should be the last resort in any pest control program owing to various problems arising because of insecticidal application. In general the older broad spectrum insecticides (organophosphates, carbamates, organochlorines) kill more pests, work, faster and cost less, mostly contact in action.

Newer insecticides are generally more expensive and more selective, hence fit well in an IPM programmer. Another advantage of these insecticides is that they are systemic or locally systemic.

This means that they penetrate leaf tissues or roots and move within the plant. Thus, thorough spray coverage is not as critical as it is with contact insecticides. This is particularly beneficial late in the in the season when coverage of some crops is very difficult because of crop structure. Systemic insecticides also work better on aphid infested plants whose structure may protect aphids from contact insecticides. Also some root systemic insecticides control aphids from four weeks to all season. They are used at planting to prevent aphid damage when the infestation potential and production history justify their use. Thus systemic insecticides are beyond doubt a better option in insecticidal control of aphids.

Optimal aphid control and yield protection depends on insecticide selection, timing of the application and method of application. Besides insecticide resistance management in aphids and environmental safety particularly safety of beneficials and bees are other important concerns.

Timing of insecticide application is situation dependant and depends on the period of activity of the aphid. Sometime the insecticides are soil incorporated when the aphid infestation is sure during the crop growth season. At times, insecticides are sprayed in the off season to kill the overwintering forms *e.g.* in case of apple. Aphids like green apple aphid, apple grain aphid and rosy apple aphid overwinter as eggs on twigs and bask crevices of apple trees. A delayed dormant oil application between green-tip and half inch green controls newly hatched aphids. Failure to use a properly timed delayed dormant oil application may require additional seasonal treatments.

Aphids can be effectively killed with either ground or aerial application; however the following three elements of insecticide coverage are required for optimum control: increased application pressure, increased carrier (water) per unit area and small droplet size, especially in fields with dense crop canopy. Complete coverage of the crop is essential for optimum aphid control especially because most aphids feed on underside of leaves. If coverage is poor or an insecticide doesn't give effective control, then the remaining aphids will reproduce and the population will reach the economic threshold again. Generally recommended insecticides for field application include Endosulfan, Dimethoate, Methyl-*o*-demotion, Cruiser, Gaucho, Monocrotophos, Acetamiprid, Acephate, Hydrocarbon oils, insecticidal soaps etc.

The soil incorporated insecticides are systemic hence they spread throughout the plant system and have a tendency to concentrate in the actively growing areas in the plant. These are generally effective in the vegetative stage however because of limited residual activity; they fail to protect older, reproductive stages of plants. Commonly used insecticides for soil incorporation or seed treatment include thiamethoxam, imidacloprid, carbofuron etc. Sometimes the insecticides are tank mixed with herbicides to reduce costs *e.g.* in case of soybean. Similarly the aphid insecticides are mixed fungicides and sprayed provided the chemicals mixed are compatible with each other. Indoor sprays for aphid infested house plants are usually pump sprays containing insecticidal soap (Potassium salts of fatty acids) or pyrethroids.

Other factors to consider while selecting an insecticide are the pre-harvest interval and re-entry interval. The former is particularly important for vegetables as they usually involve multiple harvests. Hence insecticides with minimum residual life are preferred.

As already mentioned, insecticidal control should be the last resort in pest management. Major problems associated with indiscriminate use of broad-spectrum insecticides is destruction of natural enemies leading to problems such as secondary pest outbreak and pest resurgence, and honey bee toxicity besides environmental pollution and contamination of food commodities. Thus the selective insecticides should always be preferred. In a safe aphid management strategy involving synthetic insecticides, insecticidal soaps and pysethroids get the top most priority. Similarly spraying of crops at bloom poses a serious threat to honey bees, the major pollinators of crop plants. When there is concern about honey bees, pyrethroids are the better insecticide choice.

Aphids stand 4th or 5th in the list of insect groups which have developed resistance to a wide variety of insecticides. It is due to their faster rate of breeding (more number of generations per year) and parthenogenecity. Aphids reproducing parthenogenetically produce offsprings similar to themselves (clones); thus aphids that survive an insecticide are likely to produce aphids capable of surviving the same or similar insecticides rotation, using insecticides with different mode of action and switch over to novel methods of pest management is important.

Management of Vector Aphids

From the viewpoint of viral disease control, there are two strategies for vector aphid control.

Direct Strategies: to prevent virus acquisition and transmission by vector aphids by

1. Preventing the vector aphids from landing (repellents, mulches),
2. Preventing the landing aphids from probing (insecticides, feeding deterrents, fleece, nets),
3. Preventing probing aphids from transmitting viruses (mineral oil) and
4. Preventing the landing aphids from settling (alarm pheromone).

Indirect Strategies: to reduce the population density of vector aphids (insecticides, insect natural enemies). None of the tactics mentioned above is able to protect plants completely against virus infection when used alone, except for certain fleeces. Multiple tactics must be used, to build integrated pest management programs (IPM). Pinese *et al.* (1994) evaluated two insecticides, two mineral oil sprays and different-coloured reflective mulches, alone and in combination, for the control of PRSV in zucchini (*Cucurbita pepo*) in Australia. A treatment strategy incorporating the repellent qualities of reflective silver mulches against itinerant aphid vectors, plus the protectant qualities of non-phytotoxic mineral oils and deltamethrin sprays to reduce the number of vectors on crops, provided the best control options. It may be possible to apply a similar strategy successfully to other crops susceptible to viruses.

Majority of management tactics have already been discussed, the rest are being discussed here. Repellents are the chemicals that prevent insect damage to plants by rendering them unattractive, unpalatable or offensive. Physical barriers may also act as repellents while as the chemicals that affect the tactile, olfactory or gustatory receptors of insects are the common ones. (-)-Polygodial isolated from water-pepper (*Polygonum hydropiper*), is a sesquiterpenoid which is highly repellent to aphids (Asakawa *et al.*, 1988, Griffiths *et al.*, 1989). Under conditions of high pest and disease pressure, three two-weekly treatments of this compound at 50 g/ha increased the yield by over 1 mt/ha against barley yellow dwarf virus (BYDV), transmitted by the aphid *Rhopalosiphum padi*. This is similar to the control achieved by the conventional broad-spectrum insecticide cypermethrin (Pickett *et al.*, 1992). Feeding deterents or antifeedents are the chemicals which prevent feeding of insects on a treated surface without impairing their appetite and gustatory receptors or driving them away from the potential food. Azadirachtin, the triterpenoid, isolated from the neem tree (*Azadirachta indica*), reduced probing activity by *R. padi* and *Sitobion avenae* on winter barley treated with concentrations of <500 ppm. The effect lasted for at least four days after application (West and Mordue, 1992). The reduction in probing activity would diminish the probability of BYDV transmission by these aphids.

(E)-betafarnesene has been reported as one of the alarm pheromones effective against aphids. Under normal conditions it is not effective as control tactic because it makes the aphids more active, and might increase transmission of viruses. However, a spray application of carbamate or organophosphate, at one-tenth of the recommended dose, in combination with the pheromone, was found successful in controlling aphids on iceberg lettuce (Ester *et al.*, 1993). It thus seems to be possible to diminish virus transmission by aphids by using an insecticide in combination with the pheromone. Compounds derived from the alarm pheromone can affect aphid settling and feeding, and decrease virus spread both in the laboratory and in the field (Dawson *et al.*, 1988, Griffiths *et al.*, 1989). An alternative way of using alarm pheromone is to disrupt effective alarm communication between aphids, by the presence of a high background level of synthetic alarm pheromone. Under these conditions, predators may become more effective in feeding on aphids. However, this hypothesis was not supported by the experimental results of El- Agamy and Haynes (1992) when they worked with the predator *Navis americoferus*. The potential role of habituation or sensory adaptation is well worth further investigation.

References

Agarwala, B. K. and A.K.Ghosh. (1984). A checklist of Aphididae of India.Rev. Zool. Surv. India. Occ. Paper No. 50, 1-71.

Aphididae) to an insecticide and a predator in the presence of synthetic aphid alarm pheromone. *Journal of Economic Entomology* 85: 794-798.

Asakawa, Y., G.W. Dawson, D.C. Griffiths, M. Jalali-Naini, J.-Y. Lallemand *et al.* (1988). Activity of drimance antifeedants and relatėd compounds against aphids, and comparative biological effects and chemical reactivity of (-) - and (+) - polygodial. *Journal Chemical Ecology* 14: 1845-1855.

Atwal, A. S. (1976). Agricultural pests of India and Southeast Asia. Kalyani Publishers Delhi. 502 pp.

Baker, J. R. (1982). Insects and related pests of flowers and foliage plants. The North Carolina Agricultural Extension Service. 75 pp.

Dawson, G.W., D.C. Griffiths, J.A. Pickett, R.T. Plumb, C. M. Woodcock and Z.N. Zhang. (1988). Structaure/activity studies on aphid alarm pheromone derivatives and their field use against transmission of barley yellow dwarf virus. *Pesticide Science* 22: 17-30.

Dhaliwal, G. S. (2006). An outline of Entomology. Kalyani Publisher, India.

Dixon, A.F.G. (1978) Biology of Aphids. Edward Arnold (Publishers) Limited, London.

Dixon, A.F.G. (1998). Aphid Ecology. Chapman and Hall, London, pp 300.

E1-Agamy, F.M. and K.F. Haynes (1992). Susceptibility of the pea aphid (Homoptera: Aphididae).

Elton, C.S. (1925). The dispersal of insects to Spitsbergen Trans. R. Entomol. Soc. Lon. 1925: 289-299.

Ester, A., J. Gut, A.M. van Oosten and H.C.H. Pijnenburg (1993). Controlling aphids iniceberg lettuce by alarm pheromone in combination with an insecticide. *Journal of Applied Entomology* 115: 432-440.

Ghosh, L.K. and R. Singh (2000). Biodiversity of Indian insects with special reference to aphids (Homoptera: Aphididae).*J. Aphidol.* 14:113-123.

Gillot, C. (2005) Entomology. Springer, Dordrecht, the Netherlands.

Griffiths, D.C., J.A. Pickett, L.E. Smart and C.M. Woodcock (1989). Use of insect antifeedants against aphid vectors of plant virus disease. *Pesticide Science* 27: 269- 276.

Hardie, J. (1993). Flight behavior in migrating insects. Journal of Agricultural Entomology. 10: 239-245.

Hardie, L.K., Irwin, M.E.s and Scott, R.W. (1985). Conceptual approach to modeling aphid migration, Eds. The movement and dispersal of agriculturally important biotic agents. pp. 541-582.

Imms, A.D. (1947). Insect natural history. London.

Kaddou, I.K. (1960). The feeding behaviour of Hippodamia 5-signata (Kirby) larvae. University of California. Publications in Entomology, 16:181-230.

Kennedy, J.S.(1951), Nature, London 168: 890-5

Kennedy, J.S., Day, M.F. and Eastop, V.F. (1962). A conspectus of aphids as vectors of plant viruses. Commonwealth Institute of Entomology. London, 114 p.

Khan, A. A. (2009). Functional response of *Adalia tetraspilota* (Hope) (Coleoptera: ccinellidae) on cabbage aphid, *Brevicoryne brassicae* (L.) (Homoptera: Aphididae). *Journal of Biological Control*, **23**: 243–48.

Khan, A. A. and Mir, R. A. (2008). Functional response of fourpredaceous coccinellids *Adalia tetraspilota* (Hope), *Coccinella septempunctata* (L.), *Calvia punctata* (Mulsant) and *ippodamia variegata* (Goeze) feeding on the green apple aphid, *Aphis pomi* De Geer (Homoptera: Aphididae). *Journal of Biological Control*, **22**: 291–298.

Khan. A. A. (2010). Stage-specific functional response of predaceous ladybird beetle, *Harmonia eucharis* (Mulsant) (Coleoptera: Coccinellidae) to green apple aphid, *Aphis pomi* De Geer (Hemiptera: Aphididae). *Journal of Biological Control.* 24 (3): 222–226,

Klingauf, F. (1967). Abwehr- und Meideraltionen von Coccinellidenlarven (Coleoptera) in Bodenfallen auf Fitchenstandorten. Ent. Nachr., 13: 145-148.

Kmetijstvo. 33(3): 119-128.

Kring, J.B.(1972). Flight Behavior of Aphids. Annual Review of Entomology. 17: 461-492.

Moran, N.A. (1992). The evolution of aphid life-cycles. *Annual Review of Entomology*. 37: 321-348.

Pickett, J.A., L.T. Wadhams and C.M. Woodcock (1992). The chemical ecology of aphids. *Annual Review of Entomology* 37: 67-90.

Pierr.e, J.S. and Dedryver, C.A.(1995). Un model de prevision des pullullation du puceron Sitobion avenae sur ble d'hiver. Phytoma. Juin 1985.

Pinese, B., A.T. Lisle, M.D. Ramsey, K.H. Halfpapp and S. de Faveri (1994). Control of aphid-borne papaya ringspot potyvirus in zucchini marrow (*Cucurbita pepo*) with reflective mulches and mineral oil-insecticide sprays. *International Journal of Pest Management* 40: 81-87.

Raychaudhuri, D. N. (1983). Food – plant catalogue of Indian aphididae. Aphidological society of India, Gorakhpur. Grafic Printall, Culcutta (India). 188pp.

Remaundiere, G and Remaundiere, M. (1997). *Catalogue of the world;s Aphididae*. Institut National de la Remaundiere agronomique, Peris. pp 473.

Riley, J.R.D., Reynolds, R., Mukhopadhyay, S., Ghosh, M.R. and Sarkar, J.R. (1995). LongDistance migration of aphids and other small insects in north east India. European. *Journal of Entomology.* 92: 639-653.

Skirvin, D.J., Perry, J.N. and Harrington, R. (1997). A model describing the population dynamics of *Sitobion avenae* and *Coccinella sepfumpunctata*. Ecobgical Modeling 96: 29-39.

Swenson, K.G. (1968). Annual Review of Phytopathology, 6, 351-74.

T. Mushtaq and A. A. Khan (2010). Functional and aggregational response of *Chrysoperla carnea* (Stephens) (Neuroptera: Chrysopidae) to different densities of *Bravicoryne brassicae* (Homopetera: Aphididae). *Journal of Biological Control,24:28-34*

Tatchell, G.M. (1989). An estimate of the potential economic losses to some crops due to aphids in Britian. *Crop protection*, **8**: 25-29.

Traden, S. and Milivoj, L. (1999). The cereal aphid (Sitobion avenae F.), wheat pest. Sodobno.

Voss, T.S., Kieckhefer, R.W., fuller, B.W., Meleod, M.J. and Beck, D.A. (1997). Yield loss inmaturing spring wheat caused by cereal aphids (Homptera: Aphididae) under laboratory conditions. *Journal of Economic Entomology*. 90(5): 1346-1350.

Wennergren, U. and Landin, J. (1993). Population growth and structure in a variable environment I. Aphids and temperature variation. Oecologica. 93: 394-405.

West, A.J. and A.J. Mordue (1992). The influence of azadirachtin on the feeding behaviour of cereal aphids and slugs. *Entomologia Experimentalis et Applicata* 62: 75-79.

Hill Agriculture: Prospects, Constraints and Mitigations (2013) *Pages* **331–337**
Editor: **Dr. Shahid Ahamad**
Published by: **DAYA PUBLISHING HOUSE, NEW DELHI**

Chapter 22

Nematode Pests of Mushrooms and their Management

V.K. Singh[1] and Rajanish Kumar Singh[2]
[1]*Division of Plant Pathology, Sher-e-Kashmir University of Agricultural Sciences and Technology, DLRSS, Dhiansar – 181 133, Jammu*
[2]*Department of Horticulture, U.P College, Varanasi, U.P*

Mushrooms are a group of fungi some of which are edible and rich in nutritive value. From time immemorial, people have been using mushrooms for food. Edible mushrooms are cultivated and used in various ways as food. It is useful food in modern diet. Of the edible mushrooms, about 20 types are being cultivated but only three types of mushrooms are promising in India, namely the white button mushroom (*Agaricus bisporus*), Oyster or 'dhingri' (*Pleurotus sajor caju*) and Chinese or paddy straw mushroom (*Volvariella volvacea*). In India, a major share of 85 per cent is contributed by button mushroom, though specialty mushrooms have greater scope in the country (Tewari, 2005).Recently, milky mushroom (*Calocybe indica* (P&C), gained popularity and is cultivated commerciality in tropical region of the country. Substrates and cultivation technology vary with different species of mushrooms. Mushrooms are grown in almost all parts of India and the annual production is approximately 40,000 t. Several species of naturally occurring edible mushrooms are being harvested and consumed locally in tribal and interior forest areas. However, commercially only white button and oyster mushrooms are popular. They are being cultivated under semi-scientific conditions (use of partially sterilized media and casing soil, use of implements, container etc. seldom treated with formalin, etc.) with limited inputs. Other biotic stresses, such as fungi, bacteria, insects and nematodes, cause heavy losses. Hence, the commercial productivity is much less than that of other countries. Button mushrooms are generally highly susceptible to nematode infection while oyster mushrooms are relatively resistant. Of these, the most dangerous are nematodes which once entered into the crop cannot be eradicated without completely destroying the crop. Their presence in the beds simply means very poor yields or total crop failures.

Edible mushrooms are extremely delicate and are highly susceptible to various pests and pathogens, which may attack the crop at any stage, right from mixing of substrate ingredients to cropping. The severity of problem increases manifold when cultivation is done under unhygienic conditions in makeshift houses lacking controlled conditions (Kumar, 2007; Kumar *et al.*, 2007). Various nematode pests cause significant damage to mushrooms under such unhygienic conditions. Although several nematode control methods have been worked out, under Indian conditions maintenance of hygienic production conditions, prophylactic (sterilized production substrates, containers, implements, gloves etc.) and use of plant products (Leaves and non-edible oil cakes etc.) were more successful in minimizing nematode damage to mushrooms and maximizing the yields. In this review the important mushrooms nematodes and their management have been discussed.

Cultivation of Mushrooms

Substrates and cultivation technology vary with different species of mushrooms. Modern cultivation techniques for *A. bisporus*, which is cultivated on composts prepared from manure and wheat or rice straw are reviewed by Gaze (1985). After 'composting' (the microbial degradation of straw) the substrate is inoculated with mushroom mycelium (spawn) and covered with a layer of soil or 'casing' material (moss-peat and chalk) to induce fruiting. The first mushrooms may be harvested 15-20 days after casing and then they appear in batches (flushes) at intervals of 7-15 days. However, mushroom production is continuous through out the year.

Nematodes of Mushrooms

Parasitic and saprophytic species of nematodes associated with commercial mushroom growing are routine inhabitants of most agricultural soils. Parasitic nematodes possess needle like mouth parts capable of drawing out the contents of a mushroom cell. The mouth of the saprophytic nematodes is built for chewing. Saprophytic (free-living) nematodes ingest bacteria, protozoa and fungal spores, but do not attack the mushroom mycelium. Large populations of saprophytic nematodes (*Rhabditis* sp.) may produce by products which adversely affect mushroom spawn growth, but the extent of this is still to be proven. One fact regarding nematodes is certain – if present in the compost, they either survived Phase II pasteurization or over-cropped in the woodwork of the growing room. Pasteurization of compost and wood is a keystone in mushroom-farm hygiene, so the presence of nematodes suggests a failure in sanitation procedures. According to feeding nature of the nematodes associated with mushroom are categorized into several groups *i.e.*, mycophagous (fungal feeder), saprozoic, predatory, animal parasites and plant parasite etc. Mostly fungal feeder nematode species representing two important orders (*Aphelenchida* and *Tylenchida*) have been reported to be harmfully associated with mushroom cultivation from different parts of the world. Several important species of the nematodes belonging to four genera (*Aphelenchoides, Aphelenchus, Paraphelenchus* and *Seinura*) fall in order *Aphelenchida*, whereas *Ditylenchus* is in *Tylenchida*. In which, mycophagous or fungal feeder and saprozoic nematodes are very important damaging nematodes (Sharma, 2002).

The primary symptom associated with the parasitic nematodes *Aphelenchoides composticola* and *Ditylenchus myceliophagus* is a degeneration of the mushrooms mycelium in the compost. If nematodes are present in the compost at spawning, the spawn's mycelium will grow slowly and degenerate; mushrooms will not form. Mycelium in affected areas is completely destroyed and as the compost decomposes, it turns black and a medicinal odour is detectable. Surface of nematode-infested areas sink and take on an irregular appearance as a result of compost decomposition. Infested areas enlarge as the nematodes migrate to new feeding sites. Myceliophagous nematodes are highly destructive pests and are known to incur 41-100 per cent damage in button mushrooms, depending upon the nematode species involved, their initial population and mushroom stages at the time of infestation (Sharma *et al.*, 1984; Khanna, 1991; Khanna 1993; Khanna and Kumar, 2005; Kumar *et al.*, 2007; Kumar *et al.*, 2008; Kumar *et al.*, 2009.

Ditylenchus myceliophagus

It is widely distributed nematodes species present in almost all mushroom growing belts of the world. Damage to mushroom mycelium by *Ditylenchus* sp. was first recorded by Lambert *et al.* (1949). Before the description of *D. myceliophagus*, *Ditylenchus* nematodes in mushrooms were usually identified as *D. destructor* (Hussey *et al.*, 1969). Although both *D. destructor* and *D. myceliophagus* are capable of destroying mushroom mycelium, the later is more pathogenic (Goodey, 1960). Several species have been identified from mushroom cultivation, but only *D. myceliophagus* is considered economically important in commercial culture of *A. bisporus*. In India, this species has been found to be distributed in mushroom farm of Himachal Pradesh, J&K, and Punjab (Thapa *et al.*, 1981). *Ditylenchus myceliophagus* is capable of feeding and multiplying on a range of fungi found in mushrooms substrates (Choleva, 1973). It completes a generation in 26 days at 18°C but the optimum temperature for development on *A. bisporus* is 25°C (Cayrol, 1967; Evans and Fisher, 1969). Sex ratio depends on the stage of population development sub-optimal nutrition results in a greater frequency of males. Females lay about 50 eggs (Cayrol, 1970a).All juveniles stages of *D. myceliophagus* are capable anhydrobiosis (Cayrol, 1970b). Reactivated females are often more fecund than those that have not been in anhydrobiosis (Cayrol, 1970b). When *D. myceliophagus* has 'over-colonized' its substrate, it emerges from it, moving in water films and aggregating as these dry out. Nematodes are most commonly spread by passive dispersal, in draining water and on contaminated implements.

Symptoms

First sign of infestation by *D. myceliophagus* is yield reduction and crop failure (Hesling, 1974). Mycelium is sparse in the affected areas which become sunken, soggy and foul-smelling, and rise in pH value. At this stage, aggregations of nematodes may be seen on or beneath the trays of compost. Pathogenicity to *A. bisporus* is affected by temperature; the higher temperature the sooner damage to mycelium becomes apparent (Cairns, 1954). Standard pasteurization and hygiene techniques have much reduced the incidence of crop losses due to *D. myceliophagus*, but occasional destruction of crops still occurs. The earlier the nematodes is introduced into the crop the greater is its effect on production and when 3 nematodes were present initially at the time of spawning then can be 30 per cent yield loss of mushroom. (Goodey, 1960). Arrold and Blake (1968) showed that the addition of 20, 100 or 300 *D. myceliophagus* 100 g^{-1} of compost, at casing, reduced the total yield of mushrooms by 50 per cent,68 per cent and 75 per cent, respectively.

Aphelenchoides composticola

Several *Aphelenchoides* spp. are known to be associated with *A.bisporus* cultivation, however, *A. composticola* is economically important throughout the world. Populations of *A. swarupi* and/or *Aphelenchus avenae* were isolated from mushroom substrate samples collected from different localities of Himachal Pradesh. *Aphelenchoides composticola* feeds on a wide variety of fungi (Grewal, 1990). When cultured on *A. bisporus*, a generation is completed in 10 days at 18°C but the optimum temperature for reproduction is 23-25°C (Cayrol, 1967b; Arrold and Blake, 1967). *Aphelenchoides composticola* can survive starvation, freezing and slow desiccation. Populations in mushroom beds frequently, reach 10^4 -10^5 nematodes g^{-1} of compost and result in aggregations on trays and other structures and in contamination of drainage water. Survival may also be enhanced by other fungy that commonly occur in mushroom substrates and support large nematode populations (Cayrol, 1970c; Grewal, 1990).

Symptoms

Symptoms are very similar to those described for *D. myceliophagus*. In commercial mushroom growing, the first sign of an infestation by *A. composticola* is crop reduction, usually in patches, where the surface compost sinks and the underlying substrate becomes soggy, with a characteristic foul smell (Hesling, 1977). Sunken patches may be seen covered with a grey mould consisting of the fruiting bodies of nematophagous fungi (*Arthrobotrys* spp.). *Aphelenchoides composticola* is a major factor limiting mushroom production in India (Sharma and seth, 1986) and has also caused significant crop losses through out Europe, China, Australlia, New Zealand and the USA. Arrold and Blake (1968) showed that the introduction of 1, 10 or 50 *A. composticola* 100^{-1}g of compost at spawning caused 26 per cent, 30 per cent and 42 per cent losses in mushroom yields, respectively. Similar introductions at casing resulted in 14 per cent, 18 per cent and 21 per cent losses.

Other Mycophagous Nematodes

Aphelenchoides sacchari has been recorded from mushroom farms in Europe (Janowicz, 1978) and is an important pest in India (Sharma *et al.*, 1981). Many other species of *Ditylenchus* and *Aphelenchoides* are associated infrequently with *A. bisporus* culture. In addition; *Aphelenchus avenae* has been recorded in several countries of the world. *Paraphelenchus myceliophthorus* was described from mushrooms in England (Goodey, 1958) and also found in Bulgaria (Choleva, 1969).The damage potential of *A. swarupi* and *A. avenae* in white button mushrooms was recorded by Kumar *et al.*, 2008 and Kumar *et al.*, 2009 has worked on population levels of *A. swarupi* and *A. avenae* in the some mushrooms. Aphelenchida contributes the major share with four Myceliophagous genera, twenty species of which have been associated with mushroom cultivation (Seth and Sharma, 1986; Bajaj and Walia, 1999).

Saprozoic Nematodes

Among saprobes, *Rhaditids* are most commonly found in mushroom compost. The most commonly present genera are *Rhabditis Caenorhabditis, Caphalobus, Panagrolaimus, Diplogaster* and *Acrobeloides*. Generally mixed populations of some of these genera occur in the beds. The feeding approaches of these nematodes are unable to feed upon the mycelium directly, and whenever are considered to be non pathogenic. But several workers have reported that when these nematodes present in hugs amount, these can highly damage the mycelium of mushroom (Cairus and Thomas, 1950). *Caenorhabditis elegans* has been isolated from mushroom substrates in several countries. Studies on the bionomics and effect of this nematode on *A. bisporus* have confirmed the disease status of this species. Grewal and Richardson (1991) showed that losses by *Caenorhabditis elegans* in mushroom yield upto 26 per cent. Life-cycle may be rapid taking 2.5 days at 25°C 4.5 days at 15°C (Grewal, 1991 a).

Nematode Management

Chemical Management

Nematicides to be used for nematode control should be non toxic and or non persistent, easy in application and economically cheap. Methyl bromide is effective but only rarely used to control of nematodes in casing soil because having a hazards nature. Formaldehyde is applied to clean up peat and also as a general disinfectant of woodwork and equipment. The nematicides to be used for nematode control should be non-toxic and/or non persistent, easy in application and economically cheap.

Biological Management

Biological control is a good source for the control of nematodes. Nematode trapping fungus, *Arthrobotrys robusta* has been developed as a biological control agent for mycophagous nematodes in France. When added to compost at spawning, even in the absence of nematodes, it stimulates mycelial growth and improves cropping. Similar effects were achieved by Matskevich *et al.* (1978) with *A. oligospora*. Insect pathogenic bacterium, *Bacillus thuringiensis*, may have potential for the control of *Ditylenchus myceliophagus* nematodes (Cayrol, 1974).

Cultural Management

Maintenance of a strict level of hygiene in and around mushroom houses restricts the dispersal of nematodes. In India, products from plants such as *Azadirachta indica*, *Cannabis sativa* and *Ricinus communis* are not only nematicidal to *A. composticola* and *A.sacchari* but also increase mushroom yields by up to 19 per cent (Grewal, 1989).The incorporation of ground leaves and seeds of these plants is regarded as a safe and cheap alternative to treatment with commercial nematicides (Sharma and Seth, 1986).

Physical Management

Use of heat has been the most successful method of nematode control in mushroom cultivation. During pasteurization, the temperature of compost is maintained at 55-60°C for up to eight hours (Turner, 1988); such treatment is lethal to most nematodes (Hesling, 1979b). Effective circulation of heated air and open packing of trays of compost help to ensure adequate and uniform temperatures (Wuest, 1977). *Ditylenchus myceliophagus* in an anhydrobiotic state may survive within the wood trays, in the absence of mushroom crops, and be resistant to heat treatment. Its use has been particularly advised during composting and casing process as well as at the end of cropping (Sharma and Seth, 1986).Casing materials such as moss- peat, spent- compost or field soil are generally not sterilized, in order to preserve sporophore inducing microflora (Eager, 1972); alternative treatments such as solarization (Grewal and Grewal, 1988) and irradiation (Wood and Goodey, 1957) have been tested.

Conclusion

Nematode management in mushroom is to be dealt in a different way because of its unique growing pattern, its short cropping duration and fresh sporophore consumption immediately after harvest. Due to some what specific method of cultivation, mushroom beds are generally not distributed after casing and this hides the early symptoms of nematode attack. Cleanliness should be maintained inside the surroundings of mushroom farms so that nematodes, insects, and other diseases are avoided. Economic losses due to decline in yields although generally very high are not estimated for most of the myceliophagous nematodes infesting mushroom in India.

References

Arrold, N.P. and Blake, C. D. (1968). Effects of the nematodes, *Ditylenchus myceliophagus* and *Aphelenchoides composticola* on the yield of the cultivated mushroom. *Annals of Applied Biology* 61: 161-166.

Bajaj, H.K. and Walia, K.K (1999). Three new species of Aphelenchoides *Fisher*. 1894 (Aphelenchina: Nematoda) from India alongwith a compendium of Indian species. *Indian Journal of Nematology* 29: 154-170.

Cairns, E.J. (1954). Effect of temperature upon pathogenicity of the mushroom spawn nematode, *Ditylenchus* sp. Mushroom Science (1953) II, 164-167.

Cayrol, J.C. (1967 b). Etude du cycle evolutifd, *Aphelenchoides composticola*. *Nematologica* 13: 23-32.

Cayrol, J.C. (1970a). Etude Imsemination chez le nematode mycophage *Ditylenchus myceliophagus* (J.B. Goodey, 1958). *Annals de Zoologie Ecologie Animals* 2: 327-337.

Cayrol, J.C. (1974). Action of *Bacillus thuringiensis* toxins on the mycophagous nematode *Ditylenchus myceliophagus*. In: Simposio Internacional (XII) de Nematologica, Sociedad Europea de Nematologos,1-7 September, 1974, Granada, Spain.21-22.

Choleva, B. (1969). Studies on the sources of nematode infection in mushroom houses.*Rastenievudni Nauki* 6,179-186.

Eger, G. (1972). Experiments and comments on the action of bacteria on sporophore initiation in *Agaricus bisporus*. Mushroom Science (1971) VII, 719-725.

Evance, A.A.F.and Fisher, J.M. (1969). Development and structure of populations of *Ditylenchus myceliophagus* as affected by temperature. *Nematologica*. 15:395- 402.

Gaze, R.H. (1985).Cultivation systems and their evolution.In: Flegg, P.BSpencer, and D.M. and Wood.D.A. (eds), The biology and technology of the cultivated mushroom. John Wiley and Sons, Chichester. Pp.23-41.

Goodey, J.B. (1958). *Paraphelenchus myceliophthorus*. *Nematologica*. 3: 1-5.

Goodey, J.B. (1960). *Annals of Applied Biology*. 48:655

Grewal, P.C. and Grewal, S.K. (1988). Solar heat pasteurization of casing soil and chicken manure for mushroom cultivation. Mushroom Journal for the Tropics.8: 35-45.

Grewal, P.C.(1989). Effects of leaf–matter incorporation on *Alphelenchoides composticola*, mycofloral composition, mushroom compost quality and yield of *Agaricus bisporus*. *Annals of Applied Biology*.115: 299-312.

Grewal, P.C.(1990). Reproduction of the nematode *Alphelenchoides composticola* cultivated mushrooms and common weed moulds. Revue de Nematologie. 13:117-119.

Grewal, P.C. and Richardson, P.N. (1991). Effects of *Caenorhabditis elegans* on yield and quality of the cultivated mushroom *Agaricus bisporus*. *Annals of Applied Biology*. 118: 381-394.

Janowicz, K. (1978). Zerowanie i szkodliwose *Aphelenchoides sacchari* Hooper (Nematoda, Aphelenchoididae) dla pieczarki. *Rocznic Nauk Rolniczych*, E7, 209- 220.

Khanna, A.S. (1991). Yield losses in white button mushrooms due to nematodes infestation. *Nematologia Mediterranea* 19: 35-36.

Khanna, A.S. (1993). Estimation of yield losses in white button mushrooms due to nemic infestation. *Current Nematology* 4: 173-176.

Khanna, A.S. and Kumar, S. (2005). Effect of myceliophagous nematodes on flush pattern and yield of *Agaricus bisporus* (Lange) Imbach. *Indian Journal of Mushrooms* 23: 34-36.

Kumar, S. (2007). Faunsitic studies on cultivated edible mushrooms and bio-management of their nematodes pests. Ph. D. Thesis, Department of Entomology and Apiculture, Dr. Y.S. Parmar University of Horticulture and Forestry, Nauni, Solan (HP), India. pp. 157.

Kumar, S., Khanna, A.S. and Chandel, Y.S. (2007). Effect of population levels *of Aphelenchoides swarupi* and *Aphelenchus avenae* inoculated at spawning on mycelial growth of mushrooms and nematode multiplication *Nematologia Mediterranea* 35: 155-163.

Kumar, S., Khanna, A.S. and Chandel, Y.S. (2007a). Susceptibility of mushrooms to *Aphelenchoides swarupi* and *Aphelenchus avenae*. *Nematologia Mediterranea 35*: 205-211.

Kumar, S., Khanna, A.S. and Chandel, Y.S. (2007). Effect of population levels of *Aphelenchoides swarupi* and *Aphelenchus avenae* inoculated at spawning on mycelial growth of mushrooms and nematode multiplication. *Nematologia Mediterranea* 35: 155-163.

Kumar, S., Khanna, A.S. and Chandel, Y.S. (2007a). Susceptibility of mushrooms to *Aphelenchoides swarupi* and *Aphelenchus avenae*. *Nematologia Mediterranea* 35: 205-211.

Kumar, S., Khanna, A.S. and Chandel, Y.S. (2008). Damage potential of *Aphelenchoides swarupi* and *Ahelenchus avenae* in white button mushroom, *Agaricus bisporus*. *Indian J. Nematology* 38: 146-153.

Kumar, S., Khanna, A.S., Verma, A.K., Chandel, Y.S. and Jandaik, S. (2009). *Indian J. Nematology* 39: 14-24.

Lambert, E.B. Steiner, and Drechsler, L. (1949). The "Cephalothecium" disease of cultivated mushrooms caused by a nematode (*Ditylenchus* sp.) evidenced by surface development of predacious fungi. *Plant Disease Reporter* 33: 252-253.

Matskevich, N.V., Kasovets, V.S., Upadhyay, V.B. Kurt, L.A. and Teplyakova, T.V.(1978).The possibility of using nematophagous fungi in the control of some plant nematodes under enclosed conditions.

Sharma, N.K., Thapa. C.D. and Nath, D. (1981). Pathogenicity and identity of myceliophagous nematodes infesting *Agaricus bisporus*.In: Himachal Pradesh (India). *Indian Journal of Nematology* 11: 230-231

Sharma, N.K. Thapa, C.D. and Damanjeet. (1984). Estimation of loss in mushroom yield due to myceliophagous nematode, *Aphelenchoides sacchari. Indian journal of Nematology* 14: 121-124.

Sharma N.K. and Seth, A. (1986). Nematode problems in mushrooms: losses, symptomatology and management. In: Swarup, G. and Dasgupta, D.R. (eds), Plant parasitic nematodes of India, Problems and progress. IARI, New Delhi, India, pp. 384 -399.

Tewari, R.P. (2005). Mushrooms, their role in nature and society. In: Frontiers of mushroom Biotechnology (eds.Rai R.D., Upadhyay R.C. and Sharma S.R.). NRCM, Solan, India. pp. 1-8.

Thapa, C.D., Munjal, R.C. and Seth, P.K. (1977). The mushroom nematodes and their control. *Indian Journal of Mushrooms* 3: 1-11.

Thapa, C.D., Sharma, N.K. and Nath, A. (1981). Mushroom nematodes and their source of infestation in mushroom beds in India. *Mushroom Science* 11: 443-448.

Thapa, C.D. and Sharma, N.K. (1987). Effect of temperature on pathogenicity and multiplication of two myceliophagous nematodes. *Indian Mushroom science* 2: 167-168.

Turner, S.J. (1988). Effects of compost production on nematode survival. *Mushroom Journal* 183: 505-507.

Wood, F.C. and Goodey, J.B. (1957). Effects of gamma-ray irradiction nematode infesting cultivated mushroom beds.*Nature* 180: 760-761.

Wuest, P.J. (1997). Managing air during phase II composting and the nematode nemosis. *Mushroom News* 25: 7-16

Hill Agriculture: Prospects, Constraints and Mitigations (2013) *Pages* **338–349**
Editor: **Dr. Shahid Ahamad**
Published by: **DAYA PUBLISHING HOUSE, NEW DELHI**

Chapter 23

Nematode Constraints in Pulse Production in Hill Agriculture

Virendra Kumar Singh

Division of Plant Pathology,
S.K.University of Agricultural Sciences and Technology,
DLRSS, Dhiansar – 181 133, Jammu

Plant parasitic nematodes are a limiting factor in agricultural production in many parts of the world. Plants are constantly under attack by a variety of pathogens including bacteria, viruses, fungi, nematodes and insects. The percentage of crop losses caused by plant pathogens, insect pests and weeds world wide has been estimated to 42 per cent according for $500 billion worth of damage. The estimated overall annual yield loss of world's major crops due to damage by plant parasitic nematodes has been reported to the extent of 12.3 per cent (Sasser and Freckman, 1987). Estimated losses due to nematodes attack on different cultivated crops all over the world by FAO are around 400 million dollars. Annual crop losses in India is Rs. 242.1 billion. A large number of plant parasitic nematodes have been found associated with pulse crops in all over India but pathogenicity has not been proved for all. Plant parasitic nematodes are one of the most unwanted organisms in agricultural soils. They parasitize the roots of our carefully nurtured crops, reduce their ability to produce yields, make them weak and perhaps vulnerable to many pests and diseases. They reduce the crop yields that may equal those associated with more severe and easily identifiable plant diseases and we do not even realize their presence. Damage to crops by nematodes often goes unnoticed or attributed to other causes such as lack of fertility, deficient soil moisture. Sandy and warm soils found in India and other developing countries in the arid zone are very favourable for nematode infection, especially in irrigated areas which are used for crop production continuously (Taylor, 1967). The detail work have been done under pulses improvement program at ICRISAT by Sharma (1985).Time to time several workers have reported various type of nematodes on pulses which cause severe damage to crops. These nematodes have associated with pulses such as *Heterodera* spp., *Meloidogyne* spp., *Rotylenchulus* spp., *Hoplolaimus* spp., *Pratylenchus* spp., *Helicotylenchus* spp., *Tylenchorhynchus* spp. etc.

Root-Knot Nematodes (*Meloidogyne* spp.)

Root-knot nematodes are important pests of pulse crops such as Pea, cowpea, bean, black gram, green gram, chickpea, lentil, pigeonpea and several other leguminous crops. India about 12 species have been reported to occur. But, of all these species, predominating populations encountered are of *M. incognita*, *M. javanica*, *M. arenaria* and *M. hapla*. Bio-chemical differences in protein and enzyme content among species have also been used for taxonomic identification. These four species account for more than 95 per cent of the root-knot nematode population world over (Sasser and Carter, 1985). The above ground symptoms: Root knot nematode damage results in poor growth, a decline in quality and yield of the crop and reduced resistance to other stresses (drought and other diseases). A high level of root-knot nematode damage can lead to total crop loss.

Nematode damaged roots do not utilize water and fertilizers as effectively leading to additional losses for the grower. Below ground symptoms: The most distinctive nematode symptoms on roots are galls caused by *Meloidogyne* spp. They are small, individual, bead like or fusiform swellings in some roots. In other plants, galls may be massive lumps of fleshy tissue more than 1 inch in diameter containing dozens of nematodes. Other root swellings must not be mistaken for root-knot galls. Nitrogen fixing bacteria cause swellings on the roots of most legumes. These swellings called nodules are easily distinguished from root-knot galls by differences in how they are attached to the root and their contents. Nitrogen nodules are loosely attached to the root, and can generally be very easily removed, root knot galls originated from infection at the center of root, so they are an integral part of the root whose removal requires tearing the cortex apart. In addition, fresh nodules should have a milky pink to brown liquid inside them, while root-knot galls have firmer tissues and contain female root-knot nematodes inside the gall tissues, near the fibrous vascular tissues of the root. The male and female root-knot nematodes are easily distinguishable morphologically. The males are wormlike and about 1.2 to 1.5 mm long by 30 to 36 µm in diameter. The females are pear shaped and about 0.40 to 1.30 mm long by 0.27 to 0.75 mm wide. The life cycle includes egg, juvenile and adult stages. A life cycle is completed in 25 days at 27°C, but it takes longer at lower or higher temperatures. The female lays about 400-500 eggs per in a gelatinous matrix secreted by rectal glands. The root knot nematode, *M.incognita* on pea and bean was reported first time from farmer's field of Jammu (Singh and Satpriya, 2008) and also *M.incognita* on Lentil (Singh *et al.*, 2008). A mixed infestation of *M.incognita* and *M. javanica* reduced seed yield 39 per cent (Honda and Mishra, 1989). Various types of work *viz.*, penetration, development, biology and pathogenicity have studied of *M.incognita* on pea (Singh, 2009 and Singh, 2010).

Cyst Nematodes (*Heterodera* spp.)

The nematodes belonging to the genus Heterodera are called cyst nematodes because at the time of death the mature females get converted into a brown cyst in which the eggs are held. Various species of *Heterodera* are popularly known as the cyst nematodes. The above ground symptoms resemble these associated with root damage and include stunting of shoot, yellowing of foliage and reduced size of various shoot parts. An experienced observer can often see cyst nematode, *Heterodera* spp. on the roots of their hosts without magnification. The young adult females are visible as tiny white colour. After a female cyst nematode dies, her white body wall is tanned to a tough brown capsule containing several hundred eggs. The severity depends upon density of nematodes attacking roots. The mature female bodies are found attached to roots by their head end embedded almost in the stele. The site of feeding is modified into a syncitium similar to that found in case of *Meloidogyne*. On the surface of infected roots, white to brown bodies of females can be discerned with naked eyes. The intensity of body colour depends upon the maturity stage of the young female or cyst. In case of cyst nematode matrix with eggs may also be found attached to the posterior region of the female body. The male is wormlike, about 1.3 mm long by 30 to 40 µm in diameter. Fully developed females are lemon shaped, 0.6 to 0.8 mm in length and 0.3 to 0.5 mm in diameter. Approximately 21-30 days is required for the completion of life cycle of this nematode. Several workers have studies the effect of initial inoculum level of *Heterodera* spp. on pulse crops (Sharma and Sethi, 1976; Dalal and Bhatti, 1989). Several types of work such as penetration, development, volume, biomass, biology and pathogenicity have studied of *Heterodera cajani* on pigeonpea (Singh and Singh, 1988; Singh, 1992; Singh and Singh, 1997; Singh and Singh, 1998; Singh and Singh, 1999).

Lesion Nematodes (*Pratylenchus* spp.)

Root Lesion nematodes occur in all parts of the world. Assessment of exact loss by lesion nematodes has not been made possible under field conditions due to presence of mixed population of the nematode in the field. Lesion nematodes have wide host range which can affect the selections of crop used to control the nematode in crop rotation sequences. Soil type tillage operations have also been reported to affect lesion nematodes population dynamics. The plants show chlorosis, stunting and general lack of vigour resulting into wilt. The plant form patches or zones in the field. The roots show necrosis and lesions which become ideal for infection of other microorganisms. The presence of small brown to black lesions on the root surface is the most important symptoms or damage produced by the lesion nematode. Both male and female of these nematodes are wormlike, 0.4 to 0.7 mm long and 20 to 25 µm in diameter. They are migratory endo parasitic nematodes. The life cycle of the various species of *Pratylenchus* is completed within 45 to 65 days. Temperature, soil type, moisture and tillage operations are important environment factors which greatly affect the development and reproduction of nematode species as well as disease development.

Reniform Nematodes (*Rotylenchulus* reniformis)

Reniform nematode is pest of a large number of pulses in India, it is considered as pest of great significance after root-knot nematode. The nematode is a semi-endoparasite and remains attached to roots. Infected plants show stunting in growth with reduced and discolored root system. Damage during pre and post emergence of seedlings leads to reduction in germination and crop stand. The total duration of life-cycle is about 4-5 weeks under optimum conditions. Soil pH is also an important factor affecting reproduction of the nematode. Soil moisture and temperature have profound influence on infectivity, penetration and the biology of the nematode. There was significant reduction in plant height, shoot and root weight of mash at 1000 or more juveniles/500 g soil.

Lance Nematodes (*Hoplolaimus* spp.)

Lance nematodes are ectoparasites. Some times they feed at a particular site for a long time with nearly half of the body inside the rot system (sedentary ectoparasite). In many cases, juveniles of the lance nematode completely enter the cortical tissue (endoparasite). *Hoplolaimus* spp. multiplies slowly in comparison to endoparasitic nematodes but they inflict significant crop damage at a lower level of infection. Several workers have reported significant reduction by the *Hoplolaimus* spp. on various crops in the different parts of India (Haider *et al.*, 1978, Gaur and Mishra, 1981; Das, 1982; Mulk and Jairajpuri, 1975; Sharma, 1985). The availability of feeder roots and temperature are important factors for population build up of this nematode (Haider and Nath, 1992).

Cultural Management

The cultural practices include crop rotation, fallowing, flooding, sanitation, ploughing during summer season, mulching, organic manure, spacing of plants in the field, time of sowing, resistant varieties etc.One of the most effective eco-friendly management practices is the use of proper combination of cultural practices. The affectivity of cultural method depends upon the proper knowledge of life history, population dynamics host range of the plant parasitic nematodes infesting crops. Nematode management through inter cropping and crop rotation is based on the fact that some species of nematodes are able to feed and multiply only on host crop.Rotation is a very old practices for reducing soil borne problems. Rotation to non-host crops may cause many of those pests to cease reproduction and allow natural mortality factors to reduce their numbers. Flooding or following may be used to help reduce numbers of nematode pests. The period of flooding appears to vary with several factors such as kind of soil, season etc. Juveniles (larvae) are more easily killed by flooding than eggs. The period of flooding needs to be worked out for each condition. Fallow periods in cropping sequences can also reduce nematode populations. The summer ploughing 2-3 times during hottest period of the year help to expose nematodes to the drying action of sun and wind and reduce the population. Haque and Prasad (1982) reported drastic reduction of plant parasitic nematodes and significantly increased the yield with three deep ploughing. Singh and Singh (1999) studied the different inoculum levels of *H. cajani* on pigeonpea. They found that various growth parameters were decreased in different inoculum levels such as 2000, 1000, 500 and 100 juveniles/pot in comparison to without inoculated pot. The intensity of symptoms was directly related to the inoculum levels. In general, the entry of the nematode in roots of pigeonpea was higher at higher levels of inoculum. The nematode multiplication was universally proportional to inoculum density size of the egg sac was more at lower inoculum with higher number of eggs/egg sac. Singh and Singh (1995) investigated development and pathogenicity of *Heterodera cajani* on susceptible and moderately susceptible cultivars of pigeonpea. The penetration by second stage juveniles in roots of ICPL-87, susceptible cultivar was earlier *i.e.* within 24 hours as compared to moderately susceptible cultivar PDA –I where penetration occurred only after 24 hours. The different stages of *H. cajani* were found earlier in susceptible cultivar than the moderately susceptible one. All the stages of the nematode persisted longer in cv. PDA-I than ICPL-87. Number of eggs/egg sac was also lower in the moderately susceptible cultivar as compared to susceptible cultivar of pigeonpea. Reduction in plant growth characters was more in susceptible cultivar ICPL-87 than the moderately susceptible cultivar PDA-I to nematode infection. Multiplication of *H. cajani* was higher in ICPL-87 than in PDA-I as higher number of cysts, egg sacs, eggs and juveniles/pot. Sanitation terms covers a wide range of cultural practices, including weed control, crop residue destruction and disinfestations of farm equipment before moving it from heavily infested fields to uninfected fields. In monocultures, eliminating the weed hosts can be important in reducing the populations of plant parasitic nematodes. Soil temperature plays crucial role in the activities of plant parasitic nematodes, the time during which crop is planted is important (Ayaub, 1980). Singh and Singh (1995) studied the effect of time of inoculum on growth of pigeonpea and biology of *Heterodera cajani*. The pathogenic effect of the nematode on the plant decreased with increasing time of inoculation after germination of the seed. The pathogenic effect was minimum in 28th day old plant inoculated with nematode. The multiplication of the nematode was maximum in 7 day old plant inoculated followed by 14 day, 21 day. Minimum multiplication was recorded in 28th day plant and than 0 day. The development of the nematode was faster in 7 day old inoculated plants while the persistence of their different stages of the nematode was minimum. Most pathogenic nematodes are inactive at lower temperature. Manipulation of date of sowing of chickpea escaped root-knot nematode, lesion nematode and raniforam nematode was reported by (Gaur *et al.*, 1979, Anon, 1999).

The addition of inorganic fertilizers alone without organic manure usually increases the nematode population and disease intensity. NH_4-N reduces the disease incidence while NO_3-N may increase the same. Particular forms as well as dose and proportion of NPK may also reduce or increase the incidence of disease. Use of tolerant/resistant varieties is most practical approach for the management of nematode diseases. Crop cultivars resistant to phytonematodes can be the most useful and cheapest means of nematode control for the small-scale farmers.

Nematicidal plants with roots containing nematicidal substances have been investigated. These toxic substances reduce the population level of some nematode species. African marigolds (*Tagetes* spp.), asparagus, crotalaria, mustard and several cruciferous plants have been reported to produce toxic substances. Soil amendments with green manure, compost, oil cakes of neem, mahua, mustard, groundnut, cotton, linseed, karanj and saw dust etc. have been found to reduce nematode populations. Neem, karanj and groundnut cakes incorporated into soil at the rate of 1,500 kg/ha give good control of plant parasitic nematodes and could be practiced wherever possible. Apart from encouraging the multiplication of natural enemies like nematode trapping fungi, the decomposition products of these organic amendments are toxic to nematodes. Use of organic manures is of great value since the decomposition products and promotion of natural enemies decrease nematode populations (Singh and Sitaramaiah, 1973; Singh and Singh, 1991; Singh and Singh, 1992; Alam, 1990; Haseeb *et al.*, 1984 a; Mankau,1963;Mishra and Prasad,1974; Singh and Singh, 2001; Singh, 2004; Singh, 2008; Ononuja and Kpodobi, 2008) reported that among three cakes used *i.e.* neem cake, linseed cake and mustard cake each @ 5q/ha neem cake reduced the population *viz.*, cysts, males, egg-sacs, eggs and Juveniles per pot of *Heterodera cajani* on pigeonpea, more as compared to others. Haque *et al.* (1997) reported significant reduction in infections of *M. incognita* on mungbean, when soil amended with *Calotropis procera* leaves @ 0.5 and 1.0 per cent (w/w). Patel and Patel (1992) reported significant increase in plant growth with 73 per cent reduction in *Rotylenchlus reniformis* population infecting pigeonpea, when treated with poultry manure @ 0.720 g/pot with 800 g soil followed by press mud @ 1.198g/pot. Singh (2008) reported significant reduction infection of *M. incognita* on pea, when soil amended with various organic amendments @ 2.5 and 5.0 (w/w).The amount of oil cakes or any organic matter to be incorporated depends on various factors like, soil type and texture, crop to be planted, the predominant nematode fauna of the soil and the amount of soil moisture present in the soil. In trap crops when grown in infested soils, the nematodes penetrate into the root system and start multiplying. Before the nematodes complete its life cycle the plants are uprooted and destroyed. Applications of botanicals are easy, environmentally safe and having no phytotoxic effect on crops. The inhibition of root-knot development may be due to the accumulation of toxic by products of decomposition (Alam *et al.*, 1979) to increased phenolic contents resulting in host resistance (Alam *et al.*, 1979, 1980), or to changed physical and chemical properties of soil inimical to the nematodes (Ahmad *et al.*, 1972). One of the most economical and effective ways to control plant parasitic nematodes is the growing of nematode resistant plant cultivars. The identification of a dominant *Mi* gene importing resistance to the root-knot nematodes and its linkage to an acid phosphates gene is a major break through in this area of research. Several varieties of pulse crops were listed to see their resistance against plant parasitic nematodes.

Biological Management

Application of natural enemies of plant parasitic nematodes for controlling nematode population is an essential component of eco-friendly management. Nematologists in all over the world are working very hard to identify and lean to manipulate natural enemies of nematodes so they can be used as biological control agents.Nematodes have many natural enemies including fungi, bacteria and predacious nematodes. Certain fungi capture and kill nematodes in the soil. *Arthrobotrys* spp., *Dactylaria* spp., *Dactylella* spp., *Catenaria* spp., and *Trichothecium* spp., are the genera most commonly represented. Siddiqui and Mahmood (1996) reported significant reduction infection of the *H. cajani* and

Fusarium udum on pigeonpea, when applied bioagents. Kumar and Prabhu (2008) reported significant increase in yield with 52-73 per cent reduction in cyst population infecting pigeonpea. When used *Trichoderma harzianum* and *Pochonia chlamydosporia*. The biological control of root knot nematode on tomato under green house condition by using predaceous fungi has been reported by Singh *et al.* (2001). Some fungi capture nematode by adhesion, but many employ specialized devices that include networks of adhesive branches, stalked adhesive knobs, non constricting rings and constricting rings. The surface of the nematode is penetrated and the fungus hyphae grow throughout the nematode body, digesting and absorbing its contents. Under favourable conditions, large numbers of nematodes may be captured, and killed especially by those fungi that form adhesive net works or hyphal loops. Wide range of opportunities fungi parasitizes eggs and cysts of nematodes. *Trichoderma harzianum, Trichoderma virens, Aspergillus niger, Paecilomytes lilacinus, Verticillium chlamydosporium* are found promising biocontrol agents. Now mycorrhiza is not restricted to its use only as biofertilizers, its potential role in the biological control of plant parasitic nematodes is reported by many workers. Sikora (1979) found that prior presence of VAM fungi *Glomus mosseae* has resulted into an increase in plant resistance against *Meloidogyne* spp.

A bacterial parasite of nematodes *Pasteuria penetrans*, has received much attention and research effort in recent years, *P. penetrans* is probably the most specific obligate parasite of nematodes, with a life cycle remarkably well adapted to parasitism of certain phytonematodes. It directly parasitizes juvenile nematode, thus affects penetration and reproduction. Seed bacterization, soil drenching and bare root dip application with *Pseudomonas fluroscenes, Bacillus subtilis, B. polymyxa* effectively controls plant parasitic nematodes was also reported by many workers. Gogoi and Neog (2003) reported *pasteuria* treated plots significantly root-knot nematode population on green gram, cowpea and french bean. Among the predatory nematodes, monarchs may be proved efficient predators because of stronger predatory potential, high rate of predacity and high strike rate (Bilgrami, 1998).

Chemical Management

Nematicides are chemicals used to control plant parasitic nematodes. Two major groups of nematicides are distinguished by the manner in which they spread through the soil. Soil fumigants are gases in cylinders or liquids which spread as gases from the point where they are infected into the soil. Non fumigant nematicides include a variety of water soluble compounds which are applied to the soil as liquid or granular formulations. Most belong to the carbamate or organophosphate families of pesticides. These distributions in the soil depend on physical mixing during application and moving in solution in soil water. There is no perfect nematicides for all purposes. Nematicides vary in their effectiveness against different kinds of nematodes, ease to handling, cost effect on other classes of pests (weeds, disease organisms, insects) behavior in different soils, toxicity to different plants and availability. The performance of the nematicides will depend on soil conditions, temperatures and rainfall. A yield benefit is not guaranteed and nematicides are expensive.

The older nematicides are mostly fumigants which are applied in soil. These are seldom recommended because of their hazardous nature and high toxicity to non target organisms. Recently, large number of non fumigant and systemic nematicides are available which are safe on plant. DBCP (nemagon) has been found very effective for standing crops against root-knot nematodes but its use has been suspended due to adverse effect on human beings. Carbofuran 3G (furadon), phorate 10G' (thimet), fenamiphos (nemacur), fensulphothion (dasanit) have been recommended for a variety of crops. Application of some granular nematicides such as phorate, aldicarb and carbofuran @2.5 mg a.i/kg soil (5.0 kg a.i/ha.) reduced the entry of *H. cajani* in roots of pigeonpea. The development of the nematode was also delayed by all these nematicides. Maximum adverse effect on penetration and development of *Heterodera cajani* was produced by aldicarb followed by carbofuran and phorate. All the nematicides delayed egg laying and reduced the number of eggs/egg sac was found in aldicarb. All nematicides decreased the final population of nematode. The browning of cysts was also delayed by all nematicides (Singh and Singh, 1998). Singh (2008) reported significant reduction in root - knot nematode, *M.incognita* population of pigeonpea, when soil amended with carbofuran and phorate nematicides @1.0 and 1.5kg/ha. However, growth of the pigeonpea was maximum in both nematicides at both doses of each as in comparing to control. The dose and method of application would vary with crop. Chemicals and Phytoalexins that have implications in the nematode control (Prasad and Swarup, 1986), synthetic pheromones to disrupt nematode reproduction Bone (1987), and chemicals that directly interfere with nematode chemoreceptors Janson (1987) should be exploited for nematode population management. Prasad (1990) studied the control by seed pelleted with carbofuran, aldicarb, sulfone @ 3 and 6 per cent w/w. Application of carbosulfan nematicides reduced root-knot nematodes in pulses and seed treatment was more effective than soil application.

Regulatory Management

Numerous attempts have been made to prevent the introduction of nematodes into countries or provinces by means of quarantine. Quarantines are established by legislative action in parliament, etc., and usually give quarantine authorities power to make and enforce regulations to accomplish the purpose. Such regulations usually prohibit bringing infected seeds into protected areas where similar crops might become infected.

Physical Management

Physical means of nematode control includes heat treatment of soil, solar drying, steam sterilization, hot water treatment and soil solarization. Soil solarization with transparent polyethylene sheet has been attended as a means of raising the soil temperature to lethal levels to control soil pathogens (Sharma and Nene, 1990) Soil solarization, a method of pasteurization can effectively suppress most species of nematode along with other microbes and weeds under field conditions (Rao and Krishappa, 1995, Kumar *et al.*, 1993). Generally, sheets of 50-100 µm thickness are most suitable for raising the soil temperature. The use of transparent polyethylene had yielded better results than black sheets, since transparent sheet transmit most of the incident radiation to soil (Mehrer, 1979, Mehrer and Katan, 1980). Mutagenic treatment with EMS and MH etc. were found effective for inducing resistance in black gram (Routaray *et al.*, 2001). Additions of salt, sugar, charcoal etc. create osmotic stress on nematodes and can be used for controlling nematodes.

Integrated Management

The integrated nematode management is based upon the system approach, follows location specific principle and is environment specific (Pankaj and Sharma, 1998). Utilization of the best combination of available management strategies for the pest complex at hand (nematodes, insect pests, disease organisms, weeds etc.) constitutes an integrated crop protection system. Resistant cultivars, crop rotation, pesticides, and sanitary and cultural practices can all be employed to the best possible advantage. An integrated management strategy prevents the excessive build up any single nematode, insect, or disease population and minimizes the development of pest resistance to any single tactic. Integrated pest management systems require flexibility and depend upon the specific pest problem and locally available management options. A fixed set of recommendations may keep a pest complex in check for a limited period of time, but as the pest population shifts, recommendations will have to change also. Therefore, system development takes into account many factors including the species and race of pests present, the availability of resistant host plants, the longevity of the pest, and the crops, cropping systems, and climate of the geographical region. The end result is a management strategy tailored to fit the unique circumstances of each pest situation.

Application of Molecular Biology and Biotechnology

Advances in molecular biology and biotechnology provide means to develop novel strategies for nematode management. Nematode diagnostics is essential for success of any nematode management programme even more when the control method is highly specific to the control species/pathotypes/race. A number of new techniques for analyzing nucleic acids, proteins, carbohydrates and lipids can be helpful in the identification of pests of those allozyme; monoclonal antibody and DNA based systems are most well developed for nematodes. Polyacrylamide gel electrophoresis (PAGE) has mainly focused on cyst and root-knot nematodes (Davies, 1977). PCR techniques or RFLP analysis used as supplementary tools wherever necessary. Nematode resistant transgenic plants can be designed by various approaches. The simple method is to introduce resistant genes effective against plant parasitic nematodes from wild species to commercial cultivars.

Conclusion

Nematodes parasitize the roots of our carefully nurtured crops, reduce their ability to produce yield, make them weak and perhaps vulnerable to many pests and diseases. There is a need to reduce these avoidable yield losses by developing new environment friendly, economically acceptable and ecologically based management strategies as the current options become ineffective or unacceptable. Crop resistance offer to be of great importance in this directivity. In addition, an intensive search of plant germplasm collection for natural resistance to nematodes and their races is also required. An in depth understanding is needed of the molecular basis of how and why plants are susceptible to nematodes. New progress is being made in studying changes in gene expression during the infection of plants in nematode host interactions where feeding sites are formed. The research on identifying the promising biocontrol agents, their mass production, application techniques and their behavior in the soil under varying agro climatic conditions need to be intensified.

References

Ahmad, R., Khan, A.M and Saxena, S.K. (1972). Changes resulting from amending the soil with oil cakes and analysis of oil cakes (Abstract) Proc. 59th Ses. Indian Sci. Con.Calcutta. Part III, pp. 164.

Alam, M.M., Kahn, A.M., and Saxena, S.K. (1979). Mechanism of control of plant parasitic nematodes as a result of the application of organic amendments to the soil. V. role of phenolic compounds. *Indian J. Nematol.* 9: 136-142.

Alam, M.M. (1990). Control of plant parasitic nematodes with organic amendments and nematicides in nurseries of annual plants. *J. Bangladesh Acad. Sci.* 14:107-113.

Anonymous (1999). All India Coordinated Research Project on "Plant parasitic nematodes with integrated approach for their control, ICAR Biennial report (1997-99), C.C.S. Haryana Agricultural University, Hissar.

Ayaub, S.M. (1980). Plant Nematology: An Agricultural Training and Nema aid Publication Sacramento, C.A. P-195.

Bandyopadhyay, P. Kumar, D. Singh, V.K. and Singh, K.P. (2001). Eco-friendly management of root-knot nematode of tomato by *Arthrobotrys oligospora* and *Dactylaria brochopaga*. *Indian J. Nematol.* 31 (2): 153-156.

Bilgrami, A.L. (1981). Predatory nematodes and protozoans as biopesticides of plant parasitic nematodes. In plant nematode management a biological approach, (Ed.) by P.C. Trivedi. pp. 4-23.

Bone, I.W. (1987) Pheromone communications in nematodes. Pages 147 – 152 In vistas on Nematology. A commemoration of the twenty fifth anniversary of the Society of Nematologists. (Veech, J.A. and Dickson, D.W., eds.), Hyattsville, Maryland, U.S.A. Society of Nematologists.

Dalal, M.R. and Bhati, D.S. (1989). Pathogenicity of *Heterodera cajani* in mungbean and cluster bean as affected by presence or absence of Rhizobium. *Indian J. Nematol.* 19: 153-158.

Gaur, H.S., Mishra, S.D. and. Sud, U.C (1979). Effect of date of sowing on the rotation between the population density of root-knot nematode, *M. incognita* and the growth of three varieties of chickpea. *Indian J. Nematol.* 9: 152-159

Gaur, H.S., Mishra, S.D. (1981). Pathogenicity of the lance nematode, *Hoplolaimus indicus* to cotton. *Indian J. Nematol.* (1): 87-88.

Gogoi, B.B and Neog, P.P. (2005).Efficacy of *Pasteuria penetrans* against *Meloidogyne incognita* on green gram under field conditions. *Indian J. Nematol.* 35: 80-81.

Haider, M.G. and Nath, R.P. (1992). *Curr. Nematol.* 3: 173-177.

Haider, M.G., Nath, R.P., Prasad S.S. (1978). Studies on the lance nematode *Hoplolaimus indicus I*-pathogenicity and histopathogenesis on maize. *Indian J.Nematol.* 8 (1): 9-12.

Haseeb, A., Alam, M.M. and Khan, A.M (1984a) Control of plant parasitic nematodes with chopped plant leaves. *Indian J. Plant Pathol.* 2: 180-181.

Haque, M.M. and Prasad, (1982). Effect of agronomical practices and DD application to nematodes and crop yield. *Indian J. Nematol.* 13: 126-129.

Haque, S.E., Sultana, V., Abid, M., Ara, J. and Ghaffer, A. (1997) Use of *Calotropis procera* and microbial antagonists in the control of *Meloidogyne* root knot on Mungbean. *Pak. J. Phytopathol.* 9: 108-110.

Jansson, H.B. (1987). Receptors and recognition in nematodes. Pages 153-158 in vistas on Nematology. A commemoration of the twenty fifth anniversary of the Society of Nematologists. (Veech, J.A. and Dickson, D.W., eds.), Hyattsville, Maryland, U.S.A. Society of Nematologists.

Kumar, B.N., Yaduraju, T., Ahuja, K.N. and Prasad, D.P. (1993). Effect of soil solarization on weeds and nematodes under tropical Indian conditions. *Weed Research,* 33: 423-429.

Kumar, S. and Prabhu, S. (2008). Biological control of *Heterodera cajani* in pigeonpea by *Trichoderma harziaum* and *Pochonia chlamydosporia. Indian J.Nematol.*38 (1): 65-67

Katan, J. (1981). Solar heating (solarization) of soil for control of soil borne pests. *Ann.Rev. Phytopathol.* 19:211-236.

Mehrer, Y. (1979). Prediction of soil temperature of a soil mulched with transparent polyethylene. *J. Appl. Meteorol.* 18: 1263-1267.

Mehrer, Y. and Katan, J. (1980). Prediction of soil temperature under polythene mulch. *Hass.* 60: 1384-1387.

Mankau, R. (1963). Effect of organic soil amendments on nematode population. *Phytopathology*. 53: 881-882.

Mishra, S.D and Prasad, S. K (1974). Effect of soil amendments on nematodes and crop yields. *Indian J. Nematol*. 4:1-19.

Mulk, M.M. and Jairajpuri, M.S. (1975). Nematodes of leguminous crops in India III three new species of *Hoplolaimus*. *Indian J. Nematol*. 5 (1): 1-8.

Ononuju, C.C and Kpadobi, U.C. (2008). Effect of plant leaves on the control of *Meloidogyne incognita* in soybean. *Indian J. Nematol*. 38 (1): 1-4.

Pankaj, and Sharma, H.K. (1998). IPM strategies for nematode management. In: *Potential IPM tactics* (eds. D. Prasad and R.D. Gautam) Westville Publ. House, New Delhi, pp. 131-145.

Patel, R.G. and Patel, D.J. (1992). Comparative efficacy of some organic amendments and granular nematicides in the control of *Rotylenchulus reniformis* infecting pigeonpea. *Indian J. Nematol*. 22: 14-18.

Prasad, D. (1990). Control of root-knot nematode *Meloidogyne arenaria* on groundnut by chemical seed pelleting. *Curr. Nematol*. 1:31-34.

Prasad, D. and Yaduraju, N.T, (1993). Solarization of filed soil for effective management of nematodes and weeds. *National Conference on Ecofriendly Approaches*, Dec. 20-22, 67, (Abstr.)

Rao, M.S. and Reddy, P.P (1992). Prospects of management of root-knot nematode on tomato through the integration of bio control agent and botanicals. Seminar on current trends in diagnosis and management of plant diseases. IIHR. Bangalore, pp. 10.

Rao, V.K. and Krishappa, K. (1995). Soil solarization for the control of soil borne pathogen complexes with special reference to *Meloidogyne incognita* and *Fusarium oxysporium*. *Indian Phytopathol*. 48: 300-303.

Routaray, B.N., Dash, S.B. and Mishra, R.C. (2001). Effectiveness of mutagenic treatment in induction of resistance to *M. incognita* infection in black gram. National Congress on Centenary of Nematology in India: Appraisal and future plans 5-7 Dec, 2001, I.A.R.I., New Delhi, PP. 154-155 (Abstract)

Sasser, J.N. and Carter, C.C. (1985). An advanced treatise on *Meloidogyne*. Vol I: Biology and Control. North Carolina State University, Graphics pp.1-422.

Sasser, J.N. and Freckman, D.W. (1987). A World perspective on Nematology. In: vistas on Nematology (eds. Veech, J.A. and Diskson, D.W.) Pub. by Society of Nematologist, U.S.A pp. 7-14

Sharma, S.B. (1985A). Nematode diseases of chickpea and pigeonpea. Pulse pathology progress report No. 43 Patancheru, A.P. (India) International Crops Research Institute for the Semi Arid Tropics pp. 103.

Sharma, S.B. and Nene, Y.L. (1985). Effect of presowing solarization on plant parasitic nematodes in chickpea and pigeonpea fields. *Indian J. Nematol*. 15: 277-278.

Sharma, S.B. and Nene, Y.L. (1990). Effects of soil solarization on nematodes parasitic to chickpea and pigeonpea. *J. Nematol*. 22: 658-664.

Sharma, V.K. and Sethi, C.L. (1976). Interrelationship between *Meloidogyne incognita, Heterodera cajani* and *Rhizobium* spp. on cowpea. *Indian J. Nematol*. 6 (2): 117-123

Sikora, R.A. (1979). Predisposition to *Meloidogyne* infection by the endotropic mycorrhizal fungus *Glomus mosseae*. In root-knot nematode Meloidogyne spp. systematic biology and control (eds.F.Lamberti and C.Taylor Academic Press London) pp. 399-404.

Singh, R.S. and K. Sitaramaiah (1973). Res. Bull. No. 6, G.B. Pant University of Agril. Sci. and Tech., pp 1-289.

Singh, K.P. and Singh, V.K. (1988). Effect of time of inoculation with standard inoculum of *Heterodera cajani* on the manifestation of symptoms in pigeonpea. *Indian Phytopathol*. 41: 279 (Abstract).

Singh, K.P. and Singh, V.K. (1991). Nematicidal natures of Arjun bark *Terminelia arjuna* on cyst nematode, *Heterodera cajani* of pigeon pea. *New Agric*. 2(1) 77-78.

Singh, V.K. (1992). Biology of *Heterodera cajani* associated with pigeonpea roots with emphasis on pathogenesis. *Ph.D Thesis*, B. H.U., Varanasi (India).

Singh, K.P. and Singh, V.K. (1992). *Terminelia arjuna* leaf powder reduces population density of *Heterodera cajani*. International pigeonpea newsletter 16:17-18.

Singh, V.K. and Singh, K.P. (1995). Effect of time of inoculation on growth of pigeonpea and biology of *Heterodera cajani*. *Curr. Nematol*. 6 (2): 107-115.

Singh, K.P. and Singh, V.K. (1997). Studies on volume and biomass of *Heterodera cajani* on susceptible and moderately susceptible cultivars of pigeonpea. *Int. Tropical Plant Diseases*. 13: 237-244.

Singh, V.K. and Singh, K.P. (1998). Effect of granular nematicides on the biology of *Heterodera cajani* of pigeonpea. *Indian J. Nematol*. 28 (2).168-173.

Singh, V.K. and Singh, K.P (1998). Penetration and development of *Heterodera cajani* on pigeonpea with reference to its biology. *Acta Botanica Indica* 26: 47-53.

Singh, V.K. and Singh, K.P. (1999). Effect of different inoculum levels of *Heterodera cajani* on biology of nematode and plant growth of pigeonpea. *Curr. Nematol*. 10 (1): 1-12.

Singh, V.K. and Singh, K.P. (2001). Effect of some medicinal plant leaves on the biology of *Heterodera cajani*. *Indian J. Nematol*. 31(2):143-147.

Singh, V.K. (2004). Management of *Heterodera cajani* on pigeonpea with nematicides and organic amendments. *Indian J. Nematol*. 34: 213-214.

Singh, V.K. (2008). Effect of organic amendments on the biology and management of root-knot nematode (*M. incognita*) in pea. IPS-MEZ Annual meeting and National Symposium on advances in microbial diversity and disease management for sustainable crop production. 13-15, Oct., 2008, (Abstract).

Singh, V.K. (2008). Management of root-knot nematode, *Meloidogyne incognita* infecting pigeonpea. *Indian J. Nematol*. 38 (1): 112-113.

Singh, V.K. (2008).Eco-friendly management of plant parasitic nematodes in vegetable crops. p. 440-447, *Insect pest and disease management* (ed.Dr.D.Prasad) Daya Publishing House, New Delhi.

Singh, V.K (2008). Effect of soil solarization for management of plant parasitic nematodes. *Ann. Plant Prote. Scie*. 16 (2) 541-542.

Singh, V.K. V.Koul, C.S Kalha, V.B.Singh, and V.S.Verma, (2008).New record of root-knot nematode *Meloidogyne incognita* infecting lentil in Jammu. *Journal of Research*, SKUAST-J., 7: (2) 267-268

Singh V.K. (2009). Pathogenic effect and management with organic amendments of root- knot nematode, *Meloidogyne incognita* infecting pea. 5th International Conference on Plant Pathology in the globalized era 659(S15) P.353 (Abstract).

Singh V.K (2010). Penetration and development of *Meloidogyne incognita* on pea with reference to its biology. *Journal of Research*, SKUAST-J., 9(2) 250-254.

Taylor, A. (1967). Introduction to research on plant nematology. FAO guide to the study and control of plant parasitic nematodes, pp. 133.

Upadhyay, K.D., Dwivedi, K. and Azad, C.S. (1987).Analysis of crop losses in pea and gram due to *Meloidogyne incognita. International Nematology Network Newsletter* 4(4): 6-7.

Hill Agriculture: Prospects, Constraints and Mitigations (2013) *Pages* **350–358**
Editor: **Dr. Shahid Ahamad**
Published by: **DAYA PUBLISHING HOUSE, NEW DELHI**

Chapter 24
Eco-friendly Management of Viral Diseases in some Vegetables

Ravindra Kumar and J.P. Tewari

Plant Pathology Laboratory,
Department of Postgraduate Studies and Research in Botany,
M.L.K. (P.G.) College, Balrampur, U.P.

Healthy plants look good, grow well and are productive. Plants remain healthy as long as conditions favour normal plant growth and development. Sometimes plants are unhealthy due to irritation caused by biotic and abiotic factors. The irritation may be somewhat continuous acting over extended period or it may occur nearly instantaneously. Continuous irritation cause disease; instantaneous irritation causes injury. Thus disease is an exception rather than rule in nature that frequently expressed by production of symptoms.

Common agents that cause plant disease include nonliving (abiotic) and living (biotic) factors. Non-living factors that cause disease include harmful temperature, moisture, light, nutrient, pH, air pollutants and pesticides. Living pathogens that cause disease include fungi, bacteria, viruses, viroids, mycoplasma and nematodes. Most pathogens are microscopic and derive their food by growing in or on the host plant. Viruses are unique among all pathogens being ultramicroscopic in size and obligate intracellular parasite in nature. Pathogen-host interaction in viral infections is more direct than in other diseases and could be characterized as genomic interaction.

Thus, disease is an abnormal condition in the plant produced by biotic or abiotic factors. The development of disease in plants commonly depends on the following factors:

1. Host genotype
2. Pathogen genotype
3. The environment

These factors are commonly called as disease triangle. Adding time to disease triangle as a fourth dimension gives a disease pyramid. If the four components of the disease pyramid could be quantified, its volume would be proportional to the amount of disease on a plant or in plant population. Reducing any component of disease pyramid through management (the host susceptibility, the virulence of pathogen, degree of favourableness of time or of the environment) will decrease the amount of disease in a plant. Disease development in cultivated plants is also greatly influenced by a fifth component; humans, human affect disease development in various ways. They affect the kind of plants, grown in a given area, their level of resistance, the number of plants planted, time of planting and density of plants. They also modify the environment of disease development in various ways.

In management of disease incidence one cannot exercise control on pathogen genotype because it changes continuously with time. We can modify environment and exercise some control on host genotype because in nature such adaptation in form of resistance is available, with certain limitations. It is very tedious to control or manage these limitations, which are not permanent (genotype). In such condition, chemicals that modify environment can be useful. These chemicals, pesticides and resistance genes played important role in success of green revolution.

When organic pesticides were discovered they were hailed for their positive contributions to increase food production and the reduction of serious human diseases such as malaria. These products today are considered by many to be environmental poison and health hazards. Thus on global level environmental issue is more important that cause increasing pressure for the reduction and eventual elimination of pesticides.

The complexity of many virus diseases has led to the development of a large number of approaches for their control. Among these use of leaf extracts of some higher plants have been reported to induce resistance against infection of few viruses in hypersensitive hosts. (Verma *et al.*, 1982, 1984 and 1998; Verma and Prasad, 1983) and non-hypersensitive hosts (Verma and Prasad, 1983). Many workers from time to time have investigated reduction of plant virus disease by application of several methods and found increased yield (Griffing, 1956). Simons (1960) have studied effects of some insecticides and physical barrier method on field spread of pepper vein banding mosaic virus. He also studied the effect of foliar spray of cytovirin on susceptibility and transmissibility of potato virus Y in pepper. Cook 1960 has studied use of resistant varieties against two virus diseases of chilli. In Israel, a significant reduction (40-90 per cent) of PVY and CMV incidence was obtained by oil spraying on bell peppers (Lobenstein *et al.*, 1970). Oil sprays have proved to be useful to prevent the spread of non-persistent viruses (Hein, 1971).

Black and Rolston (1972) have worked out that aphids were repelled and virus diseases were reduced in pepper plants by use of aluminum foil mulching. Trapping and yellow polythene sheet have been observed as an effective tool for the inhibition of aphid population and incidence of chilli mosaic viruses. Cohen and Marco (1973) has studied the transmission of pepper mottle virus from susceptible and resistant pepper cultivars. Soh *et al.* (1977) have also studied the inheritance of resistance to pepper veinal mottle virus in chilli. The use of a specially formulated mineral oil, JMS Stylet-oil, results in the control of virus diseases on such sensitive plants as crookneck squash and tomato without indication of phytotoxicity. This oil has been used in Florida on peppers, squash and tomato. The oil sprays were useful with the inhibitory effect of incidence of insect borne viruses and was also helpful to increase the yield (Asjes, 1979). A significant reduction of chilli mosaic viruses and PVY was also reported by oil sparaying on pepper (Walkey and Dance, 1979; Lobenstein and Raccah, 1980). Singh and Thakur (1980) have studied the reaction of some hot peppers (*Capsicum annum L.*) lines to cucumber mosaic viruses. Villalon (1981) has studied the breeding of peppers to resist virus diseases. The forms of resistance reported to control aphid borne viruses include immunity to PVY, pepper mottle and pepper vicinal mottle viruses (Rast, 1982; Ortega and Altega, 1988).

Murthy *et al.* (1981) reported the effect of prophylactic sprays of leaf extract on the infection of tobacco by tobacco mosaic virus.

Peshney and Moghem (1989) have reported the inhibition of tobacco mosaic viruses on chilli by different plant extracts. Cheema *et al.* (1991) have evaluated various plant products, insecticides and chemicals against cucumber mosaic virus on sweet pepper under green house conditions. Ravi (1991) has observed aluminum surfaced plastic film as an effective tool for the inhibition of vector population and reducing mosaic incidence of chilli. Green and Kim (1991) have reported the predominant role of non-additive gene action in controlling the expression of some traits. The genetic of resistance of CMV in *Capsicum annum* L. have been studied by Bansal *et al.* (1992). They have reported that perennial and Punjab gucchedar var. *Capsicum* were highly resistant to CMV in Punjab, India.

Duriat (1966) reported that CMV is one of the most prevalent and widespread virus infecting peppers. Paroda and Chadha (1996) have reported that some of the low yield of chilli in India (0.9t./ha.) in comparison to other varieties used were susceptible to various pests and diseases. Ariyaratne *et al.* (1996) worked out the resistance of *Capsicum* sp. genotypes to tobacco etch poty virus isolates from the Western hemisphere. Singh and Singh (1998) have reported that Punjab lal and perennial variety of *C. annum* L. were highly resistant to CMV. They have also reported that analysis of variance revealed significant difference for green fruit yield and CMV incidence of chilli indicating wide genetic variability among different genotypes. They have also studied the field reaction of some chilli genotypes for leaf curl virus in the Chhattisgarh of India.

Hobb *et al.* (2000) have reported many viruses among Solanaceous weeds and among these, potato virus Y (PVY), chilli veinal mottle virus (CVMV). tobacco etch virus (TEV), pepper vein banding virus (PVBV) and pepper veinal mottle virus (PVMV) belong to potyvirus group and cucumber mosaic virus belong to cucumo virus groups are economically most important virus having greater host range. Predominant role of non-additive gene action in controlling the expressions of these traits, were reported by Lohithaswa *et al.* (2000), and Grube *et al.* (2000).

Alegbejo (2002) have studied reaction of pepper cultivars to pepper veinal mottle potyvirus. Forty-two pepper (*Capsicum annum* L. and *C. frutescence* L.) cultivars were screened for resistance to pepper veinal mottle poty virus over a two-year period (1997 to 1998) during wet seasons at Samaru, Northern Nigeria. The cultivar 'Tea-14' was immune, the cultivars 'Kenba' and 'Hashin burgu' were moderately resistant. Thirty-three cultivars were moderately susceptible and six cultivars were highly susceptible. The immune and moderately resistance cultivars has characteristics that are highly valued as producers and consumers, high yield, bright yellow colour when ripe, high pungency and good aroma.

A wide variety of coloured plastic mulches, including red, yellow, silver, blue, grey, and orange have been used among various efforts to achieve specific production goals. Coloured plastic mulches have been used in many parts of the United States of America to enhance the yield. These mulches, however, have produced mixed results (Green and Dole, 2003).

Srivastava *et al.* (2004) studied on the management of mosaic disease of pumpkin by using leaf extracts and plant derivatives. Shukla *et al.* (2006) reported that effect of mulching on the spread of aphid transmitted watermelon mosaic virus in cucumber (*Cucumis sativus*). Ramana *et al.* (2005) have developed F_1 hybrids of chilli (*Capsicum annum*) with high yield and resistance to cucumber mosaic virus. He studied combining ability of resistance to CMV and green fruit yield in chilli, in full diallel crosses between eight parents involving six resistant and two susceptible genotypes.

Ekbote and Patil (2005) have studied the management of chilli virus complex disease in Karnataka and Dharward. It was controlled by using plant extracts of sorghum, coconut, neem and chilli leaves at 10 per cent concentration combined with insecticides monocrotophos (0.2 per cent) and dicofol (0.25 per cent). Totally four spray starting from nursery to fruiting stages were given sorghum leaf extracts (10 per cent) in combination with monocrotophos (0.2 per cent) and dicofol (25 per cent) and dicofol (0.25 per cent) treatment gave maximum disease control over other treatments and gave the highest dry chilli yield (7.57 q/ha). Shukla *et al.* (2006) studied the integrated management of viral disease of *Luffa acutangula* Roxb.

Mulches have been commonly, used for commercial vegetable production for more than 30 years. Excellent sources of information on specific characteristics and application of mulches are available on the word-wide-web. Most mulch materials are made of either high or low density polythene ranging from 0.3 to 0.76 inches (7.7 to 20.2 mm) in thickness, are 5 to 6 feet (1.5 m to 1.8m) wide and are available on rolls 555 to 1,338 yards (506 to 1233 m) long, depending on the thickness of the mulch (Lamor, 2001). The colour of mulch is an important determinant of the microclimate around a crop plant. A wide variety of other coloured plastic mulches, including red, yellow, silver, blue, grey and orange have been used in various efforts to achieve specific production goals. Each of these colours has distinct spectral reflectivity characteristics and thus modifies the radiation balance in and below a crop canopy. These coloured mulches affect not only the microclimate around a crop but have also been shown to influence insect behavior (Zitter, 1975, 1978).

The use of polythene sheet and yellow polythene mulches approach to decrease the spread of insect born virus diseases and to get virus free planting material are being practiced in vegetable crops (Vani, 1987; Vani *et al.*, 1989; Black and Rolston, 1972; Black, 1980). Plastic mulches have been reported to be helpful to reduce aphid population, delay of virus infection and reduction in virus incidence (Chen and Chen, 1983; Stapleton and Summers, 2002). Straw mulch plots also had less disease incidence and increased the yield than the un-mulched plots. The effect of straw mulching on the spread of TLCV and the influence of straw on the behavior of the vector whitefly was studied in Israel (Nitzany *et al.*, 1964; Cohen and Madjar, 1974; Vani *et al.*, 1989; Shukla *et al.*, 2005). Similar results have also been recorded by several workers; Lobenstein and Raccah, 1980; Shukla *et al.*, 2005). Some workers also used combination of the two treatments and obtained the decrease in the incidence of virus disease (Basky, 1984). The mulches not only decreased the incidence of virus disease in various crops but also increased their yield (Zatyko, 1979; Eulitz, 1977; Ravi, 1991; Whyman *et al.*, 1979; Hassan *et al.*, 1995; Tripathi and Varma, 2002).

Advances in modern biology, especially biotechnology, offers possible solution to previously intractable problems and difficult targets of disease resistance. Transgenic crops raised through genetic engineering appear to be more efficient and durable. Transformed plants with either plant derived genes or derived from pathogens usually exhibit good degree of resistance against pathogens. In India the first transgenic crop, *Bt* cotton developed for resistance to bollworm, was commercially permitted in 2002, from a small coverage of 30,000 hectare in 2002, the area under *Bt* cotton has increased to 3.5 million hectare in 2006. As a result, the average yield increased from 280 to 460 kg lint/ha. during 2002 to 2006.

Similarly, plants transformed with the virus coat protein gene or viral genes (other than coat protein) also show high level of resistance. Stark and Beachy, (1989), produced transgenic tobacco pants carrying a gene encoding coat protein of soybean mosaic virus (SBMV) a non-pathogen of tobacco. These transgenic plants showed resistance against two serologically unrelated poty viruses *viz.* potato virus-y and tobacco etch virus both are pathogens of tobacco. Thus, it could be utilized for induction of resistance in susceptible cultivars. Beachy *et al.* (1997) observed in some transgenic plants that accumulate the coat protein or a defective movement protein of TMV exhibit degree of resistance to infection by TMV that are dependent on accumulation of the protein.

Genetically resistant (naturally resistance) plants (cultivated or wild) are always a source of potential resistance genes for conventional plant breeding as well as for producing disease resistant transgenic plants. These resistant plants are more efficient, long lasting and reliable. Exploitation of non-host resistance/immunity is undoubtedly an excellent approach to plant disease management, but it is not an easy task to identify specific genes in non-host plant, which confer immunity to a large range of unrelated microbes.

Among all the measures available for disease management, no one is ultimate and permanent thus a new approach integrated disease management is more common utilizing all the possible ways of disease management.

Chemicals especially organic compounds that were popular in green revolution became hazardous now. Among all these circumstances certain botanicals or plant products or extracts of plants in different dilution alone or in combination with others. These plant or animal-based chemicals are biodegradable and offer no harm to ecosystem. Extracts of neem leaf together with cow urine is very popular in tea plantation due to no residual effect and effective in control of insects and microbes.

Table 24.1: Effect of Different Botanicals on Incidence of Watermelon Mosaic Viral Diseases and Yield of Ridge Gourd

Sl.No.	Treatment	Average Incidence of WMV MM+ WMV VB+ WMVC	Yield (q/ha)	Per cent Yield Increase Over Control
1.	*Clerodendruma culeatum*	8.31	6.45	49.30
2.	*Dillenia indica*	9.12	5.90	36.57
3.	*Grewia tenax*	9.42	5.40	25.00
4.	*Ocimum sanctum*	8.52	6.12	41.66
5.	*Phyllanthus emblica*	7.08	7.30	68.98
6.	*Solanum surratense*	9.18	5.80	34.25
7.	*Strychnos nox vomica*	8.71	6.00	38.88
8.	*Terminalia arjuna*	8.12	7.10	64.35
9.	*Vitex negundo*	9.35	5.70	31.94
10.	*Withania sominifera*	9.97	4.90	13.42
11.	Control	15.44	4.32	-

WMV MM: Water Melon mosaic virus-mild mosaic; WMV VB: Water Melon mosaic virus-vein banding; WMVC: Water Melon mosaic virus-chlorotic.

The data presented in Table 24.1 shows that the average of total incidence of the WMV MM, WMV VB and WMV C diseases were lower in *Phyllanthus* (7.08 per cent), *Terminalia* (8.12 per cent), and *Clerodendrum* (8.31 per cent) as compared to other leaf extract spraying treatments. Higher yield and least disease incidence were recorded in *Phyllanthus* and *Terminalia* (7.30q/ha and 7.10q/ha and 7.08 per cent and 8.12 per cent) followed by *Clerodendrum* (6.45q/ha and 8.31 per cent). *Ocimum* (6.12q/ha and 8.52 per cent), *Strychnos* (6.0 q/ha and 8.71 per cent), *Dillenia* (5.90 q/ha and 9.12 per cent) and *Solanum* leaf extract spray (5.80 q/ha and 9.81 per cent) were next best treatments followed by *Vitex* (5.70 q/ha and 9.35 per cent), *Grewia* (5.40 q/ha and 9.42 per cent) and *Withania* leaf extract spray (4.90 q/ha and 9.97 per cent) as compared to control. Table 27.1 indicated the highest yield increase over control was recorded in *Phyllanthus* (68.98 per cent), *Terminalia* 64.35 per cent) and *Clerodendrum* (49.30 per cent) leaf extract spray treatment as compared to control.

Conclusion

Plant diseases have been of great concern to the Indian agriculture of these; viral diseases are a major constraint in improving crop production, particularly in tropics and sub-tropics that provides ideal conditions for the perpetuation of the viruses and their vectors. Although plant viruses are simple in composition, the management of their diseases is much more difficult owing to complex disease cycles, efficient system of transmission and above all, non-availability of viricides. Efficient management of viruses require complete understanding of their relationship with the host the environment and the vector.

Indiscriminate use of the non-biodegradable compounds created certain problems, which are hazardous to environment and human health. In order to overcome these problems alternative remedies where sought from plants in form of biopesticides and non a day it is becoming a success story all over the world.

References

Asjes, C. J. (1978). Mineral olien op lelies on virus verspreideng tegen te gaan. *Bloembollencultior*, 88: 1046-1047.

Ariyaratne, I., Hobbs, H. A., Valverde, R. A., Black, L. L. and Dufresne, D. J. (1996) Resistance of *Capsicum* sp. Genotypes to tobacco etch poty virus isolates from the Western Hemisphere. *Plant Dis.*, 80: 1257-1261.

Alegbejo, M. D. (2002). Reaction of pepper cultivars to pepper veinal mottle potyvirus. *J. Vegetable Crop Production*, 8:19-26.

Black, L. L. and Rolston, L. H. (1972). Aphids repelled and virus diseases reduced in peppers planted on aluminum foil mulch. *Phytopathol*. 62:747(Abstr.).

Basky, Z. (1984). Effect of reflective mulches and oil sprays on mosaic virus incidence in cucumber, *Ecology* 7:243-248.

Bansal, R. D., Aulakh, R. K. and Hundal, J. S. (1992). Reaction of different genotypes of pepper (*Capsicum annum* L.) to cucumber mosaic virus. In 8th Meetting of Genetic and Breeding on *Capsicum* and Egg Plant. Sept. 7-10, 1992, Rome, Italy, pp. 131-137.

Beachy, R. N.; Bendahmane, M.; Heinlein, M.; Padgett, H.; Kahn, T.; Reichel, C. and Lim, C. O.(1997). Pathogen derived resistance to TMV in transgenic plants. Indian phytopathalogical society. Galden Jublee. IARI. N. Delhi 1997, Abstract.

Cook, A. A. (1960). Genetics of resistance in *Capsicum annum* to two virus diseases. *Phytopathol.*, 50: 364-367.

Cohen, S. and Marco. S. (1973). Reducing the spread of aphid-transmitted viruses in peppers by trapping the aphids on sticky yellow polythene sheets. *Phytopathol.*, 63:1207-1209.

Cohen, S. and Madjar, V. M. (1974). Prevention of the spread of tomato yellow leaf curl virus transmitted by *Bemisia tabacci* in Israel. *Bull. Ent. Res.* 64:193-197.

Chen, C. M. and Chen, C. C. (1983). Plastic cloth mulching for the control of virus in peppers, P. 106: In Abstracts of Experimental Report on Agriculture, Department of Agriculture and Forestry, Taiwan Provincial Government, R.O.C.

Cheema, S. S.; Chahal, A. S.; Bansal, R. S. and Kapoor, S. P. (1991). Efficacy of various by-products and chemical against cucumber mosaic virus on sweet pepper under green house conditions. *Indian J. Virol.* 7:169-175.

Duriat, A. S. (1996). Management of pepper viruses in Indonesia: problems and progress. *IARD Journal*, 18:45-50.

Eulitz, E. G. (1977). Aluminum foil for the control of water melon mosaic in vegetable marrow.

Ekbote, S. D. and Patil, M. S. (2005). Management of Chilli viral complex disease. *J. Mycol. P1 Pathol.*, 35: 177-178.

Griffing, B. (1956). Concept of general combining ability in relation to diallel crossing system. *Aust. J. Biol. Sci.*, 9: 463-493.

Green, S. K. and Kim, J. S. (1991). Characteristics and control of viruses infecting peppers a literature review. *AVRDC technical Bulletin* No. 18, 60 pp.

Grube, R. C., Zhang, Y., Murphy, J. E. Lowza Figuera, F., Lackney, V. K., Provvidenti, R. and John, M. K. (2000). New source of resistance to cucumber mosaic virus in *Capsicum frutescens*. *Plant Dis.* 84:885-891.

Green, L. and Dole, J. M. (2003). Aluminum foil, aluminum painted plastic, and degradable mulches increase yields decrease insect-vectored viral diseases of vegetables. *Hort. Tech.* 13: 176-184.

Hassan, S. A.; Zainal, A. R. and Ramlan, M. E. (1995). Growth and yield of chilli (*Capsicum annum* L.) in response to mulching and potassium fertilization. *Pertanika J. Trop. Agric. Sci.*, 1995: 113-117.

Hobb, H. A. D. M.; Eastburn. C. T.; Darey, J. D.; Kindhart, J. B.; Masiuan, D. J.; Voeglin, R. A.; Weingier and Coppin, N. K. Mc (2000). Solonaceous weeds as possible source of cucumber mosaic virus in Southern Illinois for aphid transmission to pepper. Plant Dis. 84(1): 1221-1224.

Lobenstiein, G.; Alper, M. and Levy, S.(1970). Field-tests with oil sprays for prevention of aphid-spread viruses in peppers. *Phytopathol.*, 6:212-215.

Lobenstein, G. and Raccah, B. (1980). Control of non-persistently transmitted aphid borne viruses. *Phytoparasitica*, 8: 221-235.

Lohithaswa, H. C.; Kulkurni, R. S. and Manyunath, A. (2000). Combining ability analysis for yield. Capsaisin and other quantitative traits in chilli (*Capsicum annum* L.) over environments. *Indian J. Genet. Plant Breed.* 60: 511-518.

Lamor, W. L. (2001). Vegetable production Using Plasticulture Food and Fertilizer Technology Center. http: II www agnet.org/library/article/eb 476html.

Murthy, P. N., Nagrajan, K. and Sastry, A. B. (1981). Effect of prophylactic sprays of leaf extract on the infection of tobacco-by-tobacco mosaic virus. *Indian J. Agri. Sci.* 51: 791-795.

Nitazany, F. E.; Geisenberg, H. and Kock, B. (1964). Test for the protection of cucumbers from a whitefly borne virus. *Phytopathol*. 54: 1059-1061.

Ortega, G. R. and Altega, L. M. (1988). Response of pepper to two Spanish isolates of CMV. *Capsicum News*. 7:65-66.

Peshney, N. and Maghem, D. G.(1989). Inhibition of tobacco mosaic virus on chilli by different plant extracts. *PKV. Res. J*. 13: 199-200.

Paroda, R. S. and Chadha, K. L. (1996). 50 years of crop science research in India, ICAR, New Delhi, pp. 574-592.

Ravi, K. S. (1991). Studies on pepper vein banding virus and other components of Murda Syndrome in chilli (*Capsicum annum* L.) Ph.D. Thesis, University of Agricultural Sciences, Hebbal Bangalore, Karnataka, India.

Rast, A. T. B. (1992). Resistant of *Capsicum* species to tobacco, tomato and pepper strains of tobacco mosaic virus. Netherlands. *J. Plant Pathol.* 88:163-170.

Ramana, C. V.; Reddy, K. M.; Reddy, M. K. and Sadashiva, A. T. (2005). Development of F_1 hybrid in chilli (*Capsicum annum*) with high yield and resistance to cucumber mosaic virus. *J. Mycol.Pl. Pathol.* 35: 179-182.

Simons, J. N. (1960). Effects of foliar sprays of Cytovirin on susceptibility and transmissibility of potato virus Y in pepper. *Phytopathol*. 50: 109-111.

Soh, A. C.; Yap, T. C.; Graham, K. M. (1977). Inheritance of resistance to pepper veinal mottle virus in chilli. *Phytopathol.* 67: 115-117.

Singh, S. J. and Thakur, M. R. (1980). Reaction of some hot pepper (*Capsicum annum* L.) lines to cucumber mosaic virus. *Indian J. Mycol. P1. Pathol.*, 9:276.

Singh, M. J. and Singh, J. (1998). Source of resistance to cucumber mosaic virus in chillies. *Plant Dis. Res.* 13(2): 184.

Stark, D. M. and Beachy, R. N. (1989). Protection against potyvirus infection in transgenic plants evidence for broad spectrum resistance *Bistechnology*, 7:1257-1262.

Stapleton, J. J. and Summers, C. G. (2002). Reflective mulches for management of aphids and aphid-borne virus diseases in late seasons cantaloupe (*Cucumis melo var. cantalupensis*). *Crop Protec.* 21: 891-898.

Srivastava, G.; Shukla, D.; Srivastava, S. and Tewari, J. P. (2004). Management of a mosaic disease of pumpkin by using plant extracts and plant derivatives. *Vegetos* 17:1-6.

Shukla, D.; Ansari, N. A.; Srivastava, G.; Shukla, S.; Chaturvedi, R. C., Srivastava, A. K.; Pandey, A.; Yadav, A. K. and Tewari, J. P. (2006). Integrated management of a viral disease of *Luffa acutangula* (Roxb.) caused by watermelon mosaic virus I. *J. Liv. World.* 13:44-47.

Tripathi, S. and Varma, A. (2002). Ecofriendly management of leaf curl disease of tomato. *Indian Phytopath.* 55: 473-478.

Villalon, B. (1981). Breeding peppers to resist virus diseases. *Plant Dis.* 65:557-561.

Verma, H. N., Awasthi, L. P. and Mukherjee, K. (1982). In "Advancing Frontiers of Mycology and Plant Pathology" (K.S. Bilgrami, P. S. Misra and R. S. Misra, eds.), *Today and Tomorrow Printers and Publishers*, New Delhi, pp. 255-264.

Verma, H. N. and Prasad, V. (1983). In "*Recent Advances in Plant Pathology*" (A. Husain, K. Singh, B.P. Singh and V. P. Agnihotri, eds.), 312, Print House (India), Lucknow.

Verma, H. N. Chowdhury, B. and Rastogi, P. (1984). Antiviral activity of leaf extracts of different Clerodendrum L. species. Z. Pflanzenks. Pflanzenschutz 91:34-41.

Vani, S. (1987). Studies on viral diseases of muskmelon and watermelon. Ph.D. Thesis, PG School, IARI, New Delhi.

Vani, S., Verma, A., More, T. A. and Srivastava, K. P. (1989). Use of mulches for the management of mosaic disease of muskmelon. *Indian Phytopathol*. 42:227-235.

Verma, H. N., Baranwal, V. K. and Srivastava, S. (1998). Antiviral substances of plant origin. In Plant virus diseases control (Eds. A. Hadidi, R. K. Khetarpal and H. Koganezawa), APS Press, The American Phytopathological Society, St. Paul, Minnesota.

Walky, D. G. A. and Donce, M. C. (1979). The effect of oil spray on aphid transmission of turnip mosic and yellow bean common mosaic virus. Plant Disease Reporter, 63:877-881.

Whyman, J. A.; Toscano, N. C.; Kido, K.; Johnson, H. and Mayberry, K. S. (1979). Effects of mulching on the spread of aphid transmitted watermelon mosaic virus to summer squash. *J. Econ. Entomol*. 72:139-143.

Zitter, T. A. (1975). Transmission of pepper virus from susceptible and resistant pepper cultivars. *Phytopathol*. 65:110-114.

Zatyko, L. (1979). Melen mulching with coloured plastic. Kertgaz da sag. 11:11-17.

Hill Agriculture: Prospects, Constraints and Mitigations (2013) *Pages* **359–378**
Editor: **Dr. Shahid Ahamad**
Published by: **DAYA PUBLISHING HOUSE, NEW DELHI**

Chapter 25

Major Virus Diseases of Vegetables Crops and their Management

Ranbir Singh and V.K. Razdan

Division of Plant Pathology, Faculty of Agriculture, Main Campus Chatha of Sher-e-Kashmir University of Agricultural Sciences and Technology, Chatha, Jammu – 180 009

Introduction

The important role of vegetables in human nutrition has long been recognized. They are especially important in developing countries with major problems of human nutrition since many vegetables are good sources of essential proteins, vitamins, minerals and complement the starchy staple foods well. In addition to such dietary benefits, vegetables are also high value crops and can often provide excellent income-generating opportunities to small farmers (Selleck and Opena, 1985). In quantitative terms, vegetables constitute the fourth largest group of commodities produced. Despite their relative importance and level of usage, the average yields of vegetables remain low due to several abiotic and biotic stresses that affect the productivity of vegetables. Among biotic factors, viral diseases of vegetables attract considerable attention because they impose significant production constraints, affecting both yield and quality, and are very difficult to control. Plant viruses are ultramicroscopic particles that multiply only within living cells and are potentially pathogenic while composed of nucleic acids surrounded by protein coat. More than half of the known viruses causes diseases of plants. All plant species are probably susceptible to one or more plant viruses. One virus may infect one or dozens of different plants, and one plant may be infected by one or several different viruses. Symptoms of virus infection may include stunting, yellowing, necrosis, mosaic and mottling, ring spots or abnormal structures. Economic damage consists of reduced yield and quality

Major Cucurbit Viruses

Squash Mosaic Virus (SqMV)

The disease can cause an important disease of melons and squash. The virus is seed-borne in muskmelon and is spread in nature principally by the spotted and striped cucumber beetles. The virus is carried within the seed and cannot be eliminated by hot water or chemical treatment.

Symptoms consist of pronounced chlorotic mottle, green veinbanding and distortion of leaves of young seedlings. On mature plants, leaves show intense dark green mosaic, blistering and hardening. Infected fruit coming from such plants show a strong mottled pattern with a lack of netting on melons.Control measures include selection of disease-free seed and cucumber beetle control.

Cucumber Mosaic Virus (CMV)

CMV is probably the most widely distributed and important virus disease of cucurbits. The virus overwinters in many perennial weed sources especially attractive to aphids Doolittle and Walker (1925) first observed CMV in chilli. Anjaneyulu and Appa Rao (1967) made detailed studies of chilli mosaic virus caused by Cucumo virus and noted that symptoms of this virus were similar to disease caused by TMV, PVY, PVX etc. The important vectors involved in transmission under field condition were different species of aphids like *A. gossypii* and *M. persicae*. Gahukar and Nariani (1982) studied and purified the three isolates of chilli mosaic disease collected from Delhi and suggested the isolates to be CMV strains belonging to cucumo virus group producing symptoms like vein-clearing, deformation, puckering and severe reduction of leaf lamina.

Geographical Distribution

CMV occurs worldwide. In Africa, CMV has been reported in eastern and southern African countries (AVRDC 1997; Nono-Womdim *et al.*, 1996), in Sudan (Magid, 1991), and in Tunisia (M'nari *et al.*, 1993).

The Virus

Cucumber mosaic virus (CMV) genus *Cucumovirus* belongs to the family *Bromoviridae*. CMV consists of 30 nm icosahedral particles with a tripartite genome that is encapsidated in three distinctive particles (Francki *et al.*, 1979). Numerous strains of CMV occur, often making the virus difficult to identify from symptoms alone. A number of strains have been assigned to two groups ToRS and DTL according to their antigenic properties (Devergne and Cardin 1973).

Symptoms

In cucumber, pumpkin and melon, CMV causes severe leaf mosaic and deformed, stunted, or mottled fruits. Colour breaking of squash fruit is usually seen, but is not unique for this virus; other viruses causing this symptom include watermelon mosaic viruses 1 and 2, squash mosaic virus and zucchini yellow mosaic virus. Early decline of muskmelon vines is usually attributed to CMV infection and should not be confused with collapse or "sudden wilt," which is a more complex disease and a plant-stress-related syndrome.

Host Range

CMV has one of the widest host ranges known among plant viruses. It attacks more than 750 plant species in 365 genera and 85 families (Douine *et al.*, 1979). Other vegetable hosts of CMV include beet, celery, hot and sweet pepper and tomato.

Transmission

CMV is transmitted by more than 60 aphid species in a nonpersistent manner (Francki *et al.*, 1979). CMV is not normally seed-transmitted in the vegetable species listed above. The virus is easily sap-transmissible.

Control

Cultivars resistant, partially resistant, or tolerant are available for cucumber and melon crops (Lecoq *et al.*, 1979; Pitrat and Lecoq, 1980).

Watermelon Mosaic Virus 2 (WMV-2)

It is the second most important cucurbit virus. This virus can infect and produce symptoms on all commercially grown cucurbits. This aphid-transmitted virus causes milder symptoms on the foliage of most infected plants like squash.

Geographical Distribution

WMV-2 has worldwide spread. The virus has been reported in eastern and southern African countries (Nono-Womdim, 1997) and in Morocco (Fisher and Lockhart, 1974a).

The Virus

Watermelon mosaic virus-2 (WMV-2) genus *Potyvirus* has flexuous rod-shaped particles 730–760 nm in length. WMV-2 contains a single-stranded positive-sense RNA (Purcifull *et al.*, 1984).

Symptoms

Fruit distortion and color breaking are still a problem on varieties like yellow straight-neck squash. Foliar symtoms in many cucurbit crops include chlorotic mottle and mosaic. Fruits are often stunted and distorted.

Host Range

The host range for WMV-2 is not limited to cucurbits, thus opening the possible overwintering of this virus in several leguminous species such as cloveWMV-2 has a narrow host range. It infects cucurbit crops. However, it can also infect some non cucurbitaceae such as pea (*Pisum sativum*) (Lecoq *et al.*, 1981).

Transmission

WMV-2 is transmitted in the nonpersistent manner by many aphid species. The virus is easily sap-transmissible. It is not seedborne.

Control

Resistance to WMV-2 has been reported in melon (Moyer *et al.*, 1985; Gilbert *et al.*, 1994) and in pumpkin (Gilbert-Albertini *et al.*, 1993). When resistant varieties are not available, overlapping of old and new cucurbit crops should be avoided.

Symptoms of PLRV on Foliage (Rolling of leaves, Stunting of plants) and Tubers (Net necrosis)

Potato Virus S

Smptoms

Potato virus S (PVS) is of increasing importance in potato. It remained unknown until the 1950's because its symptoms are very inconspicuous. PVS can cause yield loss up to 20 per cent. Seed potatoes are not yet certified for PVS, which contributes to its widespread distribution.

Most potato cultivars are symptomless. On some cultivars, if infected early in the season, will show a slight deepening of the veins, rough leaves, more open growth, mild mottling, bronzing, or tiny necrotic spots on the leaves.

PVS is a *Carlavirus* and is nonpersistently transmitted by aphids, including *Myzus persicae*, the green peach aphid. It is also mechanically transmissible and transmissible through tubers.

Management

PVS is very difficult to detect using visual cues. Insecticides are ineffective in controlling nonpersistently transmitted viruses. Crop oils may be used early in the season. Plants tend to be resistant to infection by PVS later in the season. Prevent mechanical spread within the field by sanitizing tools and minimizing movement through the field. Rogue (remove) any symptomatic plants.

Potato Virus X

Smptoms

Potato virus X is the type member of the *Potexvirus* family of plant viruses. Plants often do not exhibit symptoms, but the virus can cause symptoms of chlorosis, mosaic, decreased leaf size and necrotic lesions in tubers. PVX can interact with PVY and PVA to cause more severe symptoms and yield loss than either virus alone. The source of this virus is infected tubers. It is transmitted mechanically, not by an insect vector. Tobacco, pepper and tomato can also serve as hosts of PVX.

Management

The most important way to control this virus is to plant certified seed. Spread occurs through the use of tools and machinery. Sanitize all tools, rogue infected plants and limit within-field movement. Some varieties are more resistant to PVX than others.

Potato Mop Top Virus

Smptoms

Potato mop top virus (PMTV) is a *Pomovirus* and is transmitted by the pathogen (a protozoan with zoospores for infection) which causes powdery scab disease, *Spongospora subterranea*. PMTV is one of the causes of spraing disease, the other being *Tobacco rattle virus* (see below). "Spraing" means the

tuber has a rust-brown necrosis which occurs in rings or flecks. PMTV occurs more often in heavy, wet soils. Cool temperatures and wet conditions which are favorable for the germination of *S. subterranean* spores favour spread of the disease. The virus can survive with its vector in the soil for decades.

Management

Conditions that favor the spread of powdery scab, cool and wet conditions, also favour the spread of this virus. Viral infection may be reduced in some cases by improving drainage, reducing irrigation, or delaying planting until soils are warmer and drier. Spores infected with virus can survive in the soil for decades. Longer rotations may help reduce the incidence of powdery scab. In the absence of its vector, the virus is rapidly diluted.

Virus Diseases of Okra

Yellow Vein Mosaic of Okra

Disease Symptoms

Yellow vein mosaic virus causes severe damage in bhindi when the infection occurs in 30 days after sowing and is reported to occur all over the plants as well as in the lower hills of India (Armugam and Muthukrishnan, 1999). The viral nature of disease was established by Uppal *et al.*, 1940, who named is as yellow vein mosaic the Capoor and Varma 1950, described the symptomatology and host range. The characteristic symptoms of this disease are homogenous interwoven network of yellow vein enclosing islands of green tissues within. Initially the leaves exhibit only yellow coloured veins but in the later stages, the entire leaf turn completely yellow or at times fully cream coloured. The chlorophyll content of the loaf is partially or completely destroyed. Infected plants remains stunted and produce very few small leaves. Most of the affected plants develop thickening of veins on their lower sides. Hardly one or two fruits, often deformed, small, pale in colour and tough in texture are produced on infected plants.

Transmission and Host Range

Capoor and Varma (1950) had shown that, virus is not sap transmissible but under artificial condition grafting could transmit it. They reported that in nature the virus is transmitted by the vector white fly (*Bemisia tabaci*); even a single specimen of whitefly can transmit the virus. White fly requires 15 to 20 minutes to acquire the virus from the infected plants. White flies can inoculate the plants with in 15 to 20 minutes of inoculation feeding, incubation of virus in the vector has been found to be 6 to 8 hours. Whiteflies once fed for 12 hours on infection sources can remains viruliferous for the entire life.

Assessment of Losses

The plants that infected at 35 to 50 days following germination, the loss in the yield was estimated to be 83 and 49 per cent respectively. Sharma and Arora (1993) estimated the losses between 50 to 90 per cent because of yellow vein mosaic virus.

Management

According to Shivpuri *et al.* (2004) the seed treatment with Imidacloprid 70WS @ 5.0 g/kg seed was found economically, resulting in the highest C/B ratio of 5.90. Pun *et al.* (2000) studied the effect of virus inhibitory chemicals and neem products against YVMV and suggested that neem oil, neem seed kernel extract and barium chloride ($BaCl_2$) were found to be most effective and reduced the virus infection by 88.3 per cent, 86.7 per cent and 75.0 per cent respectively. Similarly among the plant species screened pre-inoculation spray with extracts of *Prosopsis chilensis* and *Bougainvillen spectabilis* were highly effective in reducing okra yellow vein mosaic virus infection (Pun *et al.* (1999), Debnath and Nath (2002) reported that four spraying of Metasystore, 15, 30, 45 and 60 days after sowing followed by Drumstic spray 60 days after sowing, resulted in lower degree of virus infestation. According to Singh *et al.* (1998) Monocrotophos and Cypermethrin at the concentration of 0.02 per cent spraying reduced the incidence of According to Singh *et al.* (1998) Monocrotophos and Cypermethrin at the concentration of 0.02 per cent spraying reduced the incidence of YVMV. Handa and Dattta (1993) given the application of Carbofuran and Phorate (3G-10G) at sowing 20 days later reduced disease incidence with consequent improvement in yield. Similarly According to Sastry and Singh,(1973) virus spread in *Hibiscus esculentus* was restricted by 4 sprays, each of parathion (0.02 per cent), Oxydemeton-Methyl (0.02 per cent) or Dimethioate (0.05 per cent) at 10-day intervals from germination; or one application of Phorate 10-G (15 kg/ha) at sowing. A higher incidence of *H. esculentus* YVMV occurred when sprays were delayed. Sprays approx. doubled yield.

Viral Diseases of Chilli

Chilli Mosaic Virus

Chilli mosaic virus (CMV) is an important disease of chilli and is responsible for the considerable loss if infection initiate at early stage of plant growth. The disease was first reported from India by Kulkarni (1924) and McRae (1924) and subsequently by Raychaudhuri and Jha (1954) recorded it from Indian Agricultural Research Institute, New Delhi.

Symptoms

As chilli mosaic disease is caused by number of virus, that contribute different symptoms. At the beginning Kulkarni (1924) reported mottling symptoms mostly pronounced in young leaves besides, blistering and distortion of leaves.

Causal Viruses

The causal viruses of mosaic symptoms are associated with at least eight viruses in India. Although in symptomatology, transmissibility, host range, these viruses are more or less similar but they differ in physical and genomic properties. The viruses responsible for causing chilli mosaic include, Potato virus X (Rishi and Dhawan, 1988), Potato Virus Y (Jeyarajan and Ramakrishnan, 1969), Pepper vein-banding virus (Prasadarao and Yaraguntaiah, 1979), Pepper veinal mottle virus; Tobacco mosaic virus (Ramakrishnan, 1961), Tobacco etch virus (Bidari and Reddy, 1991), Tobacco ring spot virus, Tomato spotted wilt virus (Bidari and Reddy, 1984) and Cucumber mosaic virus (Kandaswamy *et al.*, 1963).

Transmission

Nariani and Sastry (1958) reported two vectors namely *M. persicae* Sulz. and *Aphis evonymi* Fabr. of CMV and further they reported that a single viruliferous aphid (*A. gossypii*) could transmit CMV after feeding 30 seconds on a CMV infected plant and obtained maximum infection after 5 minutes of inoculation feeding (Nariani and Sastry, 1962).

Management

Varietal Screening

Khatri and Sekhon (1974) tested few chilli varieties by feeding the viruliferous aphids (*A. gossypii*) and observed that varietal reaction of chilli varieties NP-46, Anaheim and 129-12 exhibited severe mottling and leaf deformities, while AC-2207, Malagache, Agronomico No.8, Long red-I and II showed faint mottling, P-II was hypersensitive and K-1 was immune spraying with systemic insecticides will be effective in reducing mosaic incidence that have been tested in various aphid transmitted viruses (Devi and Reddy, 1995) including Pepper vein-banding virus (PVBV) and Cucumber mosaic virus (CMV) on chilli.

Chilli Leaf Curl Virus

Early infection resulted in 92.3 per cent yield reduction as compared with 74.1 to 28.9 per cent yield reduction in plant infected 35 to 50 days after transplanting. Singh *et al.* (1979) reported that the disease caused heavy loss in yield and quality of fruits. If the plants gets infected within 20-25 days after transplanting the loss in yield goes upto 80-90 per cent, but in case of later infection the loss is comparatively less.

Symptoms

The symptoms of CLCV consists of abaxial and adaxial curling of leaves accompanied by puckering and blistering of interveinal areas and thickening and swelling of the veins. In advance stages, auxiliary buds are stimulated to produce clusters of leaves, which are reduced in size having malformed or bushy appearance with stunted growth. Flowers and fruits are few being curled at the end.

Transmission

Chilli leaf curl virus is only transmitted by whitefly (*Bemisia tabaci*) in circulative non propagative manner (Shalaby *et al.*, 1997 and Kumar *et al.*, 1999).

Management

Dhanraj *et al.* (1968) reported that tobacco leaf curl virus severely attack all important crosses and exotic collection of Capsicum. On inoculation studies all the 23 mutants of *C. annuum* variety NP 46-A were infected by the enation strain of CLCV that developed typical leaf curl symptoms. Krishna *et al.* (1983) reported that Thimet (Phorate) at the rate of 0.1 per cent applied, after the seedling became established at 15 days interval gave the best control of insect vector followed by Furadon (Carbofuran) at the rate of 0.03 per cent.

Chilli Veinal Mottle Virus

The Virus

Chilli veinal mottle virus (ChiVMV) was first reported in Malaysia. It has flexuous particles 750 nm long and 12 nm wide (Ong *et al.*, 1979).

Geographical Distribution

ChiVMV has been a serious problem in most chilli-growing areas of Korea, Malaysia, Taiwan, and Thailand since 1979 (Brunt *et al.*, 1990). The virus has also been found in East Africa (Nono-Womdim *et al.*, 2001).

Symptoms

Symptoms caused by ChiVMV are variable depending upon the host, virus strain, time of infection, and environmental conditions. On tomato, typical disease symptoms include yellow foliar mosaic or chlorotic spots.

Host Range

The natural hosts of ChiVMV are sweet and hot pepper (*Capsicum* spp.), tomato (*Lycopersicon esculentum*) and African eggplant (*Solanum aethiopicum*).

Transmission

It is transmitted in a nonpersistent manner by several aphid species (Ong *et al.*, 1979) and also mechanically transmitted.

Control

It is a newly identified virus in Africa and control strategies have not been evaluated so far. However, in Asia, different measures that include the use of resistant or tolerant varieties (AVRDC, 1991), proper cultural practices that include intercropping with maize, or the use of reflective mulches to reduce the vector population (Ong, 1984) were found effective in controlling ChiVMV in pepper crops.

Viral Diseases of Tomato

Tomato Mosaic Virus

The Virus

Tomato mosaic virus (ToMV) genus *Tobamovirus* has rigid, rod-shaped particles 300 nm long and 18 nm wide. Each virus particle contains a single molecule of positive sense single strand RNA (van Regenmortel and Meshi, 1995). The virus exists in distinct strains. Pathotypes 0 and 1 have been characterized on tomato in Africa (Nono-Womdim *et al.*, 1996).

Geographical Distribution

ToMV has worldwide spread.

Symptoms

ToMV causes severe diseases in African nightshade, pepper and tomato crops. The symptoms vary considerably depending on variety and climatic conditions. Common strains will cause light and dark green mosaic on leaves. Some strains may cause chlorosis or necrosis of leaves. Fruits are stunted and may show circular, sunken, brown necrotic or chlorotic lesions.

Host Range

The primary hosts of ToMV are African nightshade (*Solanum americanum/villosum* complex, *S. scabrum*), pepper (*Capsicum* spp.) and tomato (*Lycopersicon esculentum*).

Transmission

ToMV is readily transmitted mechanically by machinery or workers from infected to healthy plants during various agronomic practices. The natural vector of ToMV is unknown. It is transmitted by seeds (up to 94 per cent in tomato). The virus is found outside the seed, sometimes in the endosperm, but it is not embryo-transmitted like normal, seed-transmitted viruses.

Control

Many commercial cultivars of pepper and tomato now available have resistance to ToMV. Other control measures include seed disinfection, avoiding planting in soil which contains ToMV from previous crops of tobacco, tomato or pepper. It is important to sterilize all equipment after each crop.

Tomato Spotted Wilt Virus

The Virus

Kendrick *et al.* (1951) first reported this virus. In India.TSWV on tomato was first reported by Toddy *et al.* (1975) and later by Bidari and Reddy (1984) and they described symptoms as bright yellow chlorotic rings on the leaves. This virus exists in number of strains and some strains could infect chilli most efficiently. The virus is transmitted by thrips vector like *Frankliniella fusca, F. occidentalis,* and *Thrips tabaci.*

Tomato spotted wilt virus (TSWV) genus *Tospovirus* belongs to the family Bunyaviridae. TSWV consists of roughly spherical, enveloped particles, ranging in diameter from 70 to 110 nm. The genome of TSWV consists of three single-stranded RNA segments (Francki *et al.*, 1991).

Geographical Distribution

TSWV is a cosmopolitan virus which was first found in South Africa in 1905 (Moore, 1933) and thereafter, reported in many other countries in subtropical as well as in temperate climate zones.

Symptoms

TSWV causes severe symptoms in pepper and tomato crops. Infected plants are stunted and old leaves turn yellow. Systemic necrotic patterns may occur. Infected fruits usually show characteristic green, yellow and red, slightly raised bulls eye rings.

Host Range

TSWV has a broad host range. At least 250 plant species are susceptible to the virus (Cho *et al.*, 1987).

Transmission

TSWV is transmitted in a persistent manner by several thrips species that include *Frankliniella fusca, F. occidentalis, F. schultzei, Scirtothrips dorsalis* and *Thrips tabaci*. It has been shown that the virus replicates within the thrips vector (Ullman *et al.*, 1993; Wijkamp *et al.*, 1993). Nonvector transmission of TSWV is by mechanical inoculation.

Control

Sources of resistance to TSWV have been identified in close relatives of pepper and tomato and have been used to develop resistant varieties (Black *et al.*, 1991; Kumar *et al.*, 1993; Nono-Womdim *et al.*, 1994; Thompson and van Zijl, 1996). Other efficient control methods include the elimination of thrips and host plants.

Tomato Leaf Curl Virus

Geographical Distribution

Leaf curl is a serious disease of tomato in many tropical countries (Czosnek *et al.*, 1990). It has been reported from south Africa (McClean,1940), Sudan (Yassin, 1983), Australia(behjatnia *et al.*, 1996) and Oman (Zouba, 1995). In India, this disease is major constraint in the tomato production and has been reported from Haryana (Varma and Poonam, 1978), Punjab) Butter and Rataul, 1981), Karnataka (Saikia and Muniyappa, 1989), Uttar Pradesh (Singh and Tripathi, 1991), New Delhi (Baranwal and Ahmad, 1997).

Incidence and Losses

The losses caused by leaf curl disease vary from place to place and season to season because of the complex relationship that exists between the virus, vector, host, virus source and environment.Yassin (1983) reported about 75 per cent reduction in tomato fruit yield. A yield loss of 90 per cent has been reported by Kalloo (1988). According to Rataul and Brar (1989), leaf curl disease causes 38-93 per cent loss in tomato yield, depending on the stage of the crop at the time of infection. Raghupathy *et al.* (1997) reported that when the plants are infected within 20 days of planting, the loss may go up to 92 per cent while infection of 35 and 50 days old crop results in 74 and 20 per cent loss respectively. In Akhnoor block of Jammu district Mahajan (2001) has reported 82.42 per cent loss due to leaf curl disease.

Symptoms

The symptoms of tomato leaf curl virus disease was first described in detail by Vasudeva and Samraj (1948). Who reported that infected tomato plants exhibits clearing of veinsn stunting, marked reduction in leaf size, shortening of internodes, curling of leaf margins inward or outwards, puckering of leaflets and bushy appearance. Similar symtoms were reported by Saikia and Muniyappa, 1989, and Zouba in 1995.TYLCV causes severe diseases in tomato crops. Infected tomato plants exhibit severe symptoms consisting of yellowing, puckering, and size reduction in the terminal leaves, curling of lower leaves, and stunting. The severity of foliar symptoms and reduction in yield depend on the age at which plants are infected.

The Virus

Tomato yellow leaf curl virus (TYLCV) genus *Bigeminivirus* has been classified in the family Geminiviridae. TYLCV consists of isometric, geminate particles measuring about 20 × 30 nm. The virus has a single undivided genome of circular *ss*DNA. Distinct viruses causing tomato yellow leaf curl disease are known to occur in Africa (Chiang *et al.*, 1997; Nono-Womdim *et al.*, 1996; Monci *et al.*, 2000; Pietersen *et al.*, 2000).

Host Range

TYLCV has been reported to have a host range including plant species belonging to Acanthacea, Asteraceae, Caricaceae, Euphorbiaceae, Fabaceae, Malvaceae, Pedaliaceae, Plantaginaceae and Solanaceae. In Tanzania, TYLCV was also detected in three newly identified hosts that are *Achyranthes aspera*, *Euphorbia heterophylla* and *Nicandra physalloides* (Nono-Womdim *et al.*, 1996). According to

Gameel (1980), *Bemisia tabaci* is known to reproduce on a wide range of cultivated plants as well as weeds while Yassin and Daffala (1980) reported that the host range of the virus includes upto 15 plant species. According to Conde and Connelly (1994) tobacco, meggplant, datura and zinnia are natural hosts of the virus.

Transmission

Varma (1963) reported that tomato leaf curl virus is transmitted by insect vector whitefly (*Bemisia tabaci*). The whiteflies are tiny sucking insects and are not always white. Approximately 28 plant viruses are known to be transmitted by whiteflies.The minimum latent period required by leaf curl virus in the vecto was found to be 6 hours (Reddy and Yaraguntaiah, 1981). TYLCV is transmitted in nature by the whitefly *Bemisia tabaci*. The virus is not sap and seed-transmissible.

Control

Certain strategies had been used to control tomato leaf curl virus which include cultural practices, control of the insect vector, and the use of resistant varieties Muniyappa and Saikia (1983) reported that covering the tomato nursery beds with nylon nets for 20-25 days and then transplanting in the field delayed the spread of disease up to 4-5 weeks. Moustafa *et al*. (1991) reported that covering the seed beds with insect proof muslin net cloth gave the best results regarding production of tomato seedlings free of tomato leaf curl virus in Egypt. Baranwal and Ahmad (1997) showed that *Clerodendrum aculeatum* leaf powder either as spray or soil application or both enhanced the growth of plants gave protection against tomato leaf curl virus by delaying the disease appearance.

References

Anjaneyulu, A. and Apparao, A. (1967). Natural occurrence of cucumber mosaic virus on chilli in India. *Indian Phytopath*. 20 : 380-381.

Arumugam, R. and Muthukrishnan, C. R. (1999). Screening of bhendi (okra) cultivars for resistance to yellow vein mosaic disease [India]. *Indian Journal of Horticulture*, 35:278-280.

AVRDC. 1997. AVRDC (1996). Progress Report. Asian Vegetable Research and Development Center, Shanhua, Taiwan.

Baranwal, V.K. and ahmad, N. (1997). Effect of *Clerodendrum aculeatum* leaf extracts on tomato leaf curl virus. *Indian Phytopathology* 50: 297-299

Behjantnia, S.a.A., Conde, B.D., Connely, M.I. and Rezaian, M.A. (1996). New potato spindle tuber viroid and tomato leaf curl Gemini viroid. *Phytopathology* 86: 880-886

Behl, M. K. (1990). Occurrence of unusual strain of PVY on bell pepper in Kangra Valley. *Indian J. of Mycol. and Pl. Pathol*. 20(2) : 127-129.

Bidari, V. B. and Reddy, H. R. (1984). Tomato spotted wilt virus (TSWV). A new record on chilli in India. *Current Sci*. 53(19) : 1045-1046.

Bidari, V. B. and Reddy, H. R. (1989). Etiological agent of chilli "murda" in Karnataka. *J. of Plantation Crops* 17(1) : 50-53.

Bidari, V. B. and Reddy, H. R. (1991). Incidence of chilli mosaic and its distribution in the commercial field of Karnataka. *J. of Plantation Crops* 19(1) : 21-25.

Black, L.L., H.A. Hobbs, and J.M. Gatti Jr. (1991). Tomato spotted wilt virus resistance in Capsicum Chinense PI 152225 and PI 159236. *Plant Disease* 75: 863.

Brunt, A., K. Crabtree, and A. Gibbs (1990). *Viruses of tropical plants*. CAB International, Wallingford, UK. 707 pp.

Butter, N.s. and Rataul, H.S. (1981). Nature and extent of losses in tomatoes due to tomato leaf curl virus transmitted by whitefly. *Indian Journal of Ecology* 8: 299-300.

Capoor, S. P. and Varma, P. M. (1950). Yellow vein mosaic of *Hibiscus esculentus* L. *Indian Journal of Agricultural Sciences*, 20: 217-230.

Chiang, B.T., M.K. Nakhla, D.P. Maxwell, M. Schoenfelder, and S.K. Green (1997). A new Geminivirus associated with a leaf curl disease of tomato in Tanzania. *Plant Disease* 81: 111.

Cho, J.J., R.F.L. Mau, W.C. Mitchell, D. Gonsalves, and L.S. Yudin (1987). Host list of plants susceptible to tomato spotted wilt virus. *Research Extension Series* 078: 10 pp.

Conde, B.D.and Connelly, M.I. (1994). Australian tomato leaf curl virus, a whitefly transmitted geminivirus causing a disease of tomatoes in Northern Australia. In: *Proceedings of the International Conference on Plant Protection in Tropics*, p. 235-237.

Czosnek, H., N. Navot, and H. Laterrot (1990). Geographical distribution of tomato yellow leaf curl virus. A fi rst survey using a specific DNA probe. *Phytopathologia Mediterranea* 29: 1–6.

Debnath, S. and Nath, P. S. (2002). Management of yellow vein mosaic disease of okra through insecticides, plant products and suitable varieties. *Annals of Plant Protection Sciences*, 10:340-342.

Devergne, J.C. and L. Cardin (1973). Contribution à l'étude du virus de la mosaïque du concombre (CMV). IV-Essai de classifi cation de plusieurs isolats sur la base de leur structure antigénique. *Annales de Phytopathologie* 5: 409–430.

Devi, P. H. S. and Reddy, H. R. (1995). Effect of antibiotics on aphid transmission of pepper vein banding virus and cucumber mosaic virus in chilli. *Indian J. of Hill Farming* 8(1) : 42-46.

Dhanraj, K. S., Seth, M. L. and Bansal, H. C. (1968). Reaction of certain chilli mutants and varieties to leaf curl virus. *Indian Phytopath*. 21(3) : 342-343.

Doolittle, S. P. and Walker, M. M. (1925). Further studies on over wintering and dissemination of cucurbit mosaic. *J. Agric. Res*. 31(1) : 58.

Douine, L., J.B. Quiot, G. Marchoux, and P. Archange (1979). Recensement des espèces végétales sensibles au virus de la mosaïque du concombre (CMV). Etude bibliographique. *Annales de Phytopathologie* 11: 439–475.

Fereres, A., Perez, P., Gemeno, C. and Pouz, F. (1993). Transmission of Spanish pepper and potato PVY isolates. *Environment Ento*. 22(6) : 1260-1265.

Fischer, H.U. and B.E.L. Lockhart (1974). Serious losses in cucurbits caused by watermelon mosaic virus in Morocco. *Plant Disease Reporter* 58: 143–146.

Francki, R.I.B., D.W. Mossop, and T. Hatta (1979). Cucumber mosaic virus. CMI/AAB Descriptions of Plant Viruses no. 213. Association of Applied Biology, Wellesbourne, UK.

Gahukar, K. B. and Nariani, T. K. (1982). Studies on an aphid borne mosaic disease of chilli. *Indian Phytopath*. 35(1) : 75-79.

Gameel, O.I.1980. Annual report of the Gezira Research Station, Sudan.

Gilbert, R.Z., M.M. Kyle, H.M. Munger, and S.M. Gray. 1994. Inheritance of resistance to watermelon mosaic virus in Cucumis melo L. HortScience 29: 107–110.

Gilbert-Albertini, F., H. Lecoq, M. Pitrat, and J.L. Nicolet (1993). Resistance of *Cucurbita moschata* to watermelon mosaic virus type 2 and its genetic relation to resistance to zucchini yellow mosaic virus. Euphytica 69: 231–237.

Gowda, K. T. P. and Reddy, H. R (1989). Aphid transmitted viruses infecting chilli. *Current Res.* Univ. of Agri. Sci. 18(5) : 71-72.

Greber, R.S., D.M. Persley, and M.E. Herrington (1989). Some characteristics of Australian isolates of zucchini yellow mosaic virus. *Australian Journal of Agricultural Research* 39: 1085–1094.

Henda, A. and Datta, M. G. (1993). Characterization of yellow vein mosaic virus of bhendi (*Abelmoschus esculentus* (L.) Moench. *International Journal of Tropical Plant Diseases*, 11: 117-123.

Jeyarajan, R. and Ramakrishnan, K. (1969). Potato virus Y on chilli (Capsicum annuum L.) in Tamil Nadu. *Madras Agric. J.* 56(12) : 761-766.

Joshi, R. D. and Bhargava, K. S. (1962). A vein banding mosaic virus disease of chilli. *Indian J. Microbiol.* 2 : 29-34.

Kalloo, G. (1988). *Vegetable breeding* Vol II, Chapter 2 CRC press INC. USA 1-94

Kandaswamy, T. K., Janaki, I. P., Ramakrishnan, K., Thangamani, G., Subba Raja, K. T. and Sellamal, S. (1963). A preliminary note on virus diseases of chilli in Madras State. Madras *Agric. J.* 50 : 110-111.

Kendrick, J. B., Anderson, L. D. and Dickson, R. C. (1951). Source and seasonal spread of certain viruses in pepper in Southern California. *Phytopathology* 41 : 20.

Khatri, H. L. and Sekhon, I. S. (1974). Studies on a virus causing mosaic disease of chilli. *Indian J. Mycol. and Pl. Pathol.* 4 : 121-125.

Krishna, A., Neema, A. G., Nadkarni, P. G. and Bhalla, P. C. (1983). Evaluation of chemical for control of leaf curl of chillies. *Proceedings of National Seminar on the Production Technology of Tomato and Chillies* (eds. Muthukrishnan, C. R., Muthuswamy, S. and Arumugam, R. J.) pp. 167-169, Tamil Nadu Agricultural University, Coimbatore, March 22-23, 1983.

Kulkarni, G. S. (1924). Mosaic and other related diseases of crops in Bombay Presidency. Agric. College Mag. 16 : 6-12.

Kumar R., Rai, N. and Lakpale, N. (1999). Field reaction of some chilli genotypes for leaf curl virus in Chattisgarh region of India. *Orissa Journal of Horticulture* 27(1) :100-102.

Kumar, N.K.K., D.E. Ullman, and J.J. Cho (1993). Evaluation of *Lycopersicon* germplasm for tomato spotted wilt tospovirus resistance by mechanical and thrips transmission. *Plant Disease* 77: 938–941.

Lana, A. O., Glimer, R. M., Wilson, G. F. and Shoyinka, S. A. (1975). An unusual new virus, possibly of the poty virus group from pepper in Nigeria. *Phytopathol.* 65(12) : 1329-1332.

Lecoq, H., J.M. Lemaire, and C. Wipf Scheibel (1991). Control of zucchini yellow mosaic virus in squash by cross protection. *Plant Disease* 75: 208–211.

Lecoq, H., M. Pitrat, and M. Clement (1981). Identifi cation et caractérisation d'un Potyvirus provoquant la maladie du rabougrissement jaune du melon. *Agronomie* 1: 827–834.

Lecoq, H., S. Cohen, M. Pitrat, and G. Labonne (1979). Resistance to cucumber mosaic virus transmission by aphids in *Cucumis melo*. *Phytopathology* 69: 1223–1225.

Lisa, V. and H. Lecoq (1984). Zucchini yellow mosaic virus. CMI/AAB Descriptions of Plant Viruses no. 282. Association of Applied Biologists, Wellesbourne, UK.

Lisa, V., G. Boccardo, G. D'Agostino, G. Dellavalle, and M. D'Aquilio (1981). Characterization of a Potyvirus that causes zucchini yellow mosaic. *Phytopathology* 71: 667–672.

M'nari, M., H. Jebari, and C. Cherif (1993). Study of strains of cucumber mosaic virus (CMV) on Tunisian melon and identification of resistant genotypes. *Cahiers Agricultures* 2(1): 60–62.

Magid, A. (1991). Host range and transmission of an isolate of cucumber mosaic virus from the Sudan. *East African Agricultural and Forestry Journal* 56: 63–67.

Mahajan, R. (2001). Investigation on leaf curl virus of tomato. *M.Sc. Thesis* in Plant Pathology. Sher-e-Kashmir University of Agricultural Sciences and Technology, Jammu.

Mc Clean, A.P.D. (1940). Some leafcurl diseases in South Africa. *Science Bulletin of South Africa* Department of Agriculture 225: 1-70

McRae, W. (1924). Economic Botany, Part-III. Mycology. Ann. Rept. Board. Sci. Advice, India 1922-23, 31-35.

Monci, F., J. Navas-Castillo, J.L. Cenis, A. Lacasa, and A. Benazoun (2000). Spread of tomato yellow leaf curl virus Sar from the Mediterranean basin: presence in the Canary Islands and Morocco. *Plant Disease* 84: 490.

Moore, E.S. (1933). The kromnek or Kat river disease of tobacco and tomato in the East province (South Africa). Department of Agriculture, *South Africa, Science Bulletin* 123: 5–28.

Moustafa, S.S., Nakhla, M.K., Fadl, F.A. and EL Safty, N. (1991). Effect of nursery treatment on the control of tomato yellow leaf curl virus disease. *Egyptian Journal of Agricultural Research* 69 : 807-820.

Moyer, J.W., G.G. Kennedy, and L.R. Romanow (1985). Resistance to watermelon mosaic virus II multiplication in Cucumis melo. *Phytopathology* 75:201–205.

Muniyappa, V. and Saikia, A.K. (1983). Prevention of the spread of tomato leaf curl disease. *Indian Phytopathology* 36: 183.

Nariani, T. K. and Sastri, K. S. M. (1958). Two additional vectors of chilli mosaic virus. *Indian Phytopath*. 11 : 193-194.

Nariani, T. K. and Sastry, K. S. M. (1962). Studies on the relationship of chilli mosaic virus and its vector, *Aphis gossypii* Glover. *Indian Phytopath*. 15 : 173-183.

Nono-Womdim, R. (1997). Vegetable pathology for African countries: A training manual. AVRDC Africa Regional Program. 65 pp.

Nono-Womdim, R., A. Palloix, K. Gebre-Selassie, T. Phaly, P. Gognalons, and G. Marchoux (1994). Sources of resistance to tomato spotted wilt virus in Capsicum species. Pages 237–239 in Proceedings of the 9th Congress of Mediterranean Phytopathological Union, 18–24 September, Kusadasi-Aydin, Turkey.

Nono-Womdim, R., I.S. Swai, M.L Chadha, K. Gebre-Selassie, and G. Marchoux (2001). Occurrence of chilli veinal mottle virus in *Solanum aethiopicum* in Tanzania. *Plant Disease* 85: 801.

Nono-Womdim, R., I.S. Swai, S.K. Green, K. Gebre-Selassie, H. Laterrot, G. Marchoux, and R.T. Opeña (1996). Tomato viruses in Tanzania: identification, characterization and disease incidence. *Journal of the South African Society for Horticultural Sciences* 6:41–44.

Ong, C.A. (1984). Reducing the spread of chilli veinal mottle virus using reflective surfaces. *MARDI Research Bulletin* 12: 200–204.

Ong, C.A., G. Varghese, and W.P. Ting (1979). Aetiological investigations on a veinal mottle virus of chilli (*Capsicum annuum* L.) newly recorded from Peninsular Malaysia. *MARDI Research Bulletin* 7: 78–88.

Pietersen, G., A.M. Idris, K. Krüger, and J.K. Brown (2000). Tomato curly stunt virus, a new Begomovirus of tomato within the tomato yellow leaf curl virus-IS cluster in South Africa. *Plant Disease* 84: 810.

Pitrat, M. and H. Lecoq (1980). Inheritance of resistance to cucumber mosaic virus transmission by *Aphis gossypii* in *Cucumis melo*. *Phytopathology* 70: 958–961.

Pitrat, M. and H. Lecoq (1984). Inheritance of zucchini yellow mosaic virus resistance in *Cucumis melo* L. *Euphytica* 33: 57–61.

Prasadrao, R. D. V. J. and Yaraguntaiah, R. C. (1979). The occurrence of pepper veinal mottle virus on chilli in India. *Mysore J. Agric. Sci.* 13(4) : 445-448.

Pun, K. B., Doraiswamy, S. and Jeyarajan, R. (2000). Screening of virus inhibitory chemicals and neem products against okra yellow vein mosaic virus. *Indian Phytopathology*, 53: 95-96.

Pun, K. B., Sabitha, D., Jeyaran, R. and Doraiswamy, S. (1999). Screening of plant species for presence of antiviral proteins against okra yellow vein mosaic virus. *Indian Phytopathology*, 52:221-223.

Purcifull, D.E., E. Hiebert, and J. Edwarson (1984). Watermelon mosaic virus 2. CMI/AAB Descriptions of Plant Viruses no. 293. Association of Applied Biology, Wellesbourne, UK.

Ragupathy, N., Thiruvudainambi, S. and Thamburaj, S. (1997). Reaction of tomato varieties to leafcurl. *South Indian Horticulture* 45: 277-278

Ramakrishnan, K. (1961). Potato virus X on chilli (*Capsicum* spp.). *South Indian Hort*. 7 : 41-42.

Rataul, H.S. and Brar, J.S. (1989). Status of tomato leaf curl virus research in India. *Tropical Science* 29: 111-118.

Raychaudhuri, S.P. and Jha, A. (1954). Chilli mosaic occurring in Delhi. *Proc. 41st Indian Sci. Congr. Assoc*. 256-259.

Reddy, K.S. and Yaraguntaiah, R.C. (1981). Virus – Vector relationship in leafcurl disease of tomato. *Indian Phytopathology* 34 : 310-313.

Rishi, N. and Dhawan, P. (1988). Field reaction of chilli germplasms to different viral diseases. *Pl. Dis. Res*. 3 : 69-70.

Saikia, A.K.and Muniyappa, V. (1989). Epidemiology and control of tomato leaf curl virus in Southern India. *Tropical Agriculture* 66: 350-354.

Sastry, K. S. M and Singh, S. J. (1973). Restriction of yellow vein mosaic virus spread in okra through the control of vector, whitefly (*Bemisia tabaci*). *Indian Journal of Mycology and Plant Pathology*, 3:76-80.

Selleck, G.W. and R.T. Opeña (1985). National programs: the need for increased emphasis on the development of vegetables and legumes. In *National Agricultural Programs and Future Needs*, edited by J.B. Petersen and P.W. Macgregor. FFTC Book Series no. 30. Food and Fetilizer Technology Center, Taipei, Taiwan. 184 pp.

Shalaby, A. A., Nakhla, M. K., Shafie, M. S., Mazyad, H. M. and Maxwell, D. P. (1997). Molecular characterization of tomato yellow leafcurl gemini virus isolated from pepper collected in Egypt. Annals of Agricultural Science Moshtohor 35 (2) : 819-831.

Sharma, B. R. and Arora, S. K. (1993). Improvement of okra. pp. 343-364. In: Chadha, K.L. and Kalloo, G. (eds.). Advances in Horticulture:Vegetable Crops Vol. 5. Malhotra Publication House, New Delhi, India.

Shivpuri, A., Bhargava, K. K., Chhipa, H.P. and Ghasolia, R.P. (2004). Management of yellow vein mosaic virus of okra. Journal of Mycology and Plant Pathology, 34:353-355.

Singh, B.R. and Tripathi., D.P. (1991). Loss due to leaf curl and spotted wilt disease of tomato. *Madras Agricultural Journal* 78: 34-36.

Singh, D., Verma, N. and Naqvi, Q.A. (1998). Control of yellow vein mosaic virus of okra by seed selection and insecticidal application. *Indian Journal of Virology*, 14:121-123.

Singh, S. J., Sastry, K. S. and Sastry, K.S.M. (1979). Combating leaf curl virus in Chilli. *Indian Horticulture* 24: 9-10.

Thakur, R. D., Chowfla, S. C., Khurana, S.M.P. (1988). Natural occurrence of a typical strain of potato virus Y on bell pepper in Himachal Pradesh. *Indian J. of Virology* 4(1-2) : 91-96.

Thompson, G.J. and J.J.B. van Zijl (1996). Control of tomato spotted wilt virus in South Africa. *Acta Horticulturae* 431: 379–384.

Toddy, J. M., Ponniahu, S. and Subramanyam, C.P. (1975). First record of Tomato spotted wilt virus from the Nilgris in India. *Madras Agric. J.* 62(3) : 162-163.

Ullman, D.E., T.L. German, J.L. Sherwood, D.M. Wescot, and F.A. Cantone (1993). Tospovirus replication in insect vector cells: immunocytochemical evidence that the nonstructural protein encoded by the sRNA of tomato spotted wilt Tospovirus is present in thrips vector cells. *Phytopathology* 83: 456–463.

Uppal, B. N., Varma, P. M. and Capoor, S.P. (1940). Yellow vein mosaic of bhindi. *Current Science*, 9:227-228.

Van Regenmortel, M.H.V. and T. Meshi (1995). Tobamovirus genus virus: taxonomy, classification, and nomenclature of viruses. *Sixth Report of the International Committee on Virus Taxonomy. Archives of Virology*, Supplement 10: 434–437.

Varma, J.P. and Poonam (1978). Protein from *Lycopersicon* species resistant and susceptible to tomato leaf curl virus. *Indian Journal of Experimental Biology* 16: 527-530.

Varma, P.M. (1963). Transmission of plant viruses by whiteflies. Bulletin of National Institute of Sciences. *Indian Bulletin* 24: 11-33

Vasudeva, R.S. and Samraj, J. (1948). A leafcurl disease of tomato. *Phytopathology* 18: 364-369

Verma, P. M. (1963). Transmission of plant viruses by whiteflies. *Bull. Nat. Inst. Sci.* 24 : 11-33.

Wang, H.L., D. Gonsalves, R. Provvidenti, and H. Lecoq. 1991. Effectiveness of cross protection by a mild strain of zucchini yellow mosaic virus in cucumber, melon, and squash. Plant Disease 75: 203–207. Weimer, J.L. 1931. Alfalfa mosaic. *Phytopathology* 21: 122–123.

Wijkamp, I., J. van Lent, R. Kormelink, R. Goldbach, and D. Peters (1993). Multiplication of tomato spotted wilt virus in its insect vector, *Frankliniella occidentalis*. *Journal of General Virology* 74: 341–349.

Yang, W.Z., C.H. Hsiao, and W.N. Chang (1986). Screening cucumbers for resistance to viruses and genetic studies on resistance to zucchini yellow mosaic virus. *Journal of Agricultural Research of China* 35: 192–201.

Yassin, A.M. (1983). A review of factors influencing control strategies against tomato leaf curl virus disease in Sudan. *Tropical Pest Management* 29: 253-256.

Yassin, A.m. and Daffalah, G.A. (1980). Leaf curl epidemic a threat to tomato cultivation in Sudan.In : Proceedings of the Second International Conference on the Impact of Viral Disease on the Development of African and Middle East Countries, Nairobi, 1-6 December.

Yassin, A.M. and M.A. Nour (1965). Tomato leaf curl disease in Sudan and their relation to tobacco leaf curl. *Applied Biology* 56: 207–217.

Yassin, A.M. (1983). A review factor influencing control strategies against tomato leafcurl virus disease in Sudan. *Tropical Pest Management* 29; 253–256.

Zouba, A. (1995). A geminivirus causing tomato leaf curl disease in Sultanate of Oman. *Indian Journal of Mycology and Plant Pathology* 25: 302-303.

Zouba, A., Azman, K.M., Razvi, S.A and al Wahaibi, A.K. (1993). Temporal increase of tomato leaf curl virus on staggered planting of tomato in the Sultanate of Oman. *South Indian Horticulture* 41: 28-32.

Hill Agriculture: Prospects, Constraints and Mitigations (2013) *Pages* **379–416**
Editor: **Dr. Shahid Ahamad**
Published by: **DAYA PUBLISHING HOUSE, NEW DELHI**

Chapter 26

Major Diseases of Rice and their Management in Hills

***Shahid Ahamad*[1] *and Upma Dutta*[2]**
[1]*Krishi Vigyan Kendra-Poonch, SKUAST of Jammu, J&K*
[2]*Horticulture Research Sub-Station, Bhadarwah, SKUAST of Jammu, J&K*

Rice (*Oryza sativa* L.) is one of the most important cereal crop of the world. About 90 per cent of rice produce in world is consumed only in Asia, while, China is the leading producer of rice (112.4 million tonnes per year) followed by India (87 million tonnes). India has largest area under rice (44.6 million ha.) cultivation in the world. It is the staple food crop and provides food security for its more than two-third population, and is a means of livelihood for millions of rural households (Mishra, 2005). The production of rice comprehended, 87 million tonnes during the year 2003-2004 from a cultivated area of 44.6 million hectares. It has ample cultivation in all the states under diversified agro-ecosystems, upto an altitude of 2200 m above mean sea level (a.m.s.l) in hilly tracks of north zone.

In the Jammu and Kashmir, rice is adopted and ascended as the most important cereal crop as staple food of the people, especially for the people of Kashmir valley. It occupies an area of about 0.259 million ha with a production of 0.495 tonnes, the annual contribution from Kashmir being 0.304 metric tonnes from an area of 0.143 million ha with an average production of 2.09 metric tones per hectare during 2003-2004 (Anonymous, 2004) which is far below the national productivity average.

Inspite of quite favourable adaphic and environmental conditions for rice cultivation in the valley, the yields have not been encouraging due to constraints in rice cultivation are essentially similar to those associated with conditions of temperate regions, such as low temperature, high altitude, short growing season, physiological disorder and biotic stresses. Among biotic stresses diseases are major abeyances in rice production. Many of them namely, blast (*Pyricularia grisea*), sheath blight (*Rhizoctonia solani*), brown spot (*Helminthosporium oryaze*) and glume discolouration are ascribed by Anwar (2000) however, blast becoming as serious problem in rice cultivation that causes heavy losses in the quality and quantity of grains. Several biotic and abiotic factors are responsible for dicolouration of rice seeds. Glume discolouration is term for the alteration in colour of mature seed from its original colour and seeds possess series of problems in seed certification programme. The symptoms of false smut and bunt are distinctly clear and the seeds affected by them could be easily identified. However, other abnormalities are caused by a number of fungi either singly or in association, invading seeds before or after harvest. All these maladies of rice seeds can be broadly termed as seed discolouration which are appeared externally on the glumes or internally in kernels or both.

Rice Blast Disease

The rice crop suffer from a number of diseases among them rice blast caused by one of the most devastating agricultural pathogens in the world, a fungus called *Magnaporthe grisea* (Hebert) Barr (anomorph: *Pyricularia grisea* (Cooke) Sacc.,), is one of the most important, causing significant losses in yield. Rice blast was probably first recorded as *rice fever disease* in China in 1637. It was later described as *imochi-byo* in Japan in 1704, and as *brusone* in Italy in 1828. The fungus is currently reported to be present in at least 85 countries. In 1996 rice blast was found in rice in California, and has since been found in grasses on golf courses in the mid western United States.

Biology and Pathology

Magnaporthe grisea is an *ascomycete* fungus. It is an extremely effective plant pathogen as it can reproduce both sexually and asexually to produce specialized infectious structures known as appressoria that infect aerial tissues and hyphae that can infect root tissues.

Asexual life cycle begins when the hyphae of the fungus undergo sporulation to produce fruiting structures called conidia which contain many spores. When these spores land on leaves and other aerial tissues of susceptible plants they germinate, developing the appressorium. The appressorium penetrates the plant cell by producing a penetration peg. Pressure in the appressorium increases and the structure explodes, forcing the penetration peg through the cell wall and into the cell. The fungus can then grow hyphae within the leaf and form lesions. Once established in the host plant the fungal hyphae can undergo asexual sporulation again.

Sexual reproduction occurs when two strains of opposite mating types meet and form a perithecium in which ascospores develop. Once released, ascospores can develop appressoria and infect host cells. Spores are transmitted between plants by the wind.

The fungus can kill or damage the plant in a number of ways. Aerial infection in seedlings is usually lethal because the young plants are unable to photosynthesize. Aerial infection in mature plants does not kill them, but reduced photosynthesis due to lesions on the leaf and use of photosynthate by the fungus greatly reduces yield. Infection of root and vascular tissues has the potential to kill the plant by cutting off the supply of water and nutrients to the root.

Binomial Name

Magnaporthe grisea (Hebert) Barr.

Economic Importance

Rice blast is an economically important disease and has received attention in all major rice growing countries because of its severe destruction. Several epiphytotics of the disease have been recorded in different parts of the world resulting in serious losses in yield. In Korea, it is the only rice disease that has ever caused serious losses in yield. The loss in yield during 1953, an epidemic year was estimated at about 800,000 tones in Japan. In India epiphytotics have been reported in the Tanjore area of Madras state in 1919. In the hills, blast may cause more than 65 per cent loss in yield. Its incidence on a large scale has been reported from the plains and also peninsular India.

Symptoms

Rice blast is caused by *Magnaporthe grisea* (Hebert) Barr. All the above ground parts of the plant can be attacked by the fungus at any growth stages. However, Seedling stage, rapid tillering stage after transplanting and flower emergence stage were identified as the most susceptible ones to blast. The disease can be described based on the part of the plant infected as follow:

Leaf Blast

On the leaves the lesion/spots first appear as minute brown specks, and then grow to become spindle-shaped, pointed at both ends. The center of the spots is usually gray or whitish with brown or reddish-brown margin. Fully developed lesions reach 1 – 1.5 cm long, 0.3-0.5 cm broad (Ou, 1985). Under favourable conditions, lesions enlarge and coalesce; eventually kill the leaves.

Collar Rot

Infection at the junction of the leaf blade and sheath in the typical brown "collar rot" symptom. A severe collar rot can cause the leaf to die completely. When collar rot kill the flag or penultimate leaf it may have a significant impact on yield.

Neck Blast

This occur when the pathogen infect the neck of the panicle to cause a typical "neck rot" or rotten neck blast" symptom. The infected neck is griddled by a grayish brown lesion and the panicle falls down if the infection is severe. If the neck blast occurs before the milk stage, the entire panicle may die prematurely, leaving it white and completely unfilled.

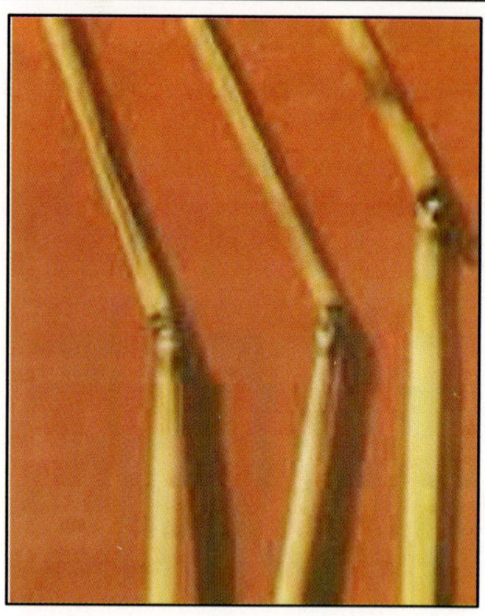

Source: Recent Trends in Plant diseases Management in India pp.468 Edited by Dr.Shahid Ahamad.

Panicle Blast

The pathogen also causes brown lesions on the branches on the panicles and on the spikelets pedicles, resulting in "panicle blast". Infection of the neck, panicle branches, and spikelets pedicles may occur together or may occur separately.

Node Blast

The fungus may also attack the stem at nodes, node blast in which the stem bend and break at the node causing spikelets sterility (Ou, 1985). Blast Symptoms of seeds themselves consist of brown spots, blotches.

Epidemology

- ✰ *Epidemiology* (*epiphytology*) deals with outbreak and spread of disease in a population.
- ✰ *Epidemiology* of plant diseases is essentially a study of the rate of multiplication of pathogen which determines its capacity to spread a disease in a plant population (Singh, 1998).
- ✰ *Epidemiology* studies population of the pathogens on population of host plants and the resulting disease under the interference of environment and human interferences.

Cause of Epidemic

The development of the blast in epidemic proportion is influenced by the presence of inoculums (*i.e.* Virulent pathogen), susceptible stage of the host, and period of favorable environmental conditions. It is obvious that there are other secondary causes that is determine by a net work of conditions, circumstances and factors which change over time. For example being, when a major part of the pathogen population become insensitive to fungicides, failure of disease control may occur which leads to an epidemic.

Factors Influencing/Favouring Severe Blast Epidemic

The disease is favored by long period of free moisture, high humidity, little or no wind at night and night temperature between 17.2-22.7 ºC. Leaf wetness from dew or other sources is required for infection. Spores are produced and released under high relative humidity (RH) conditions, with no spore production below 89 per cent RH. Sporulation increased with increasing RH above 93 per cent. The optimum temperature for spore germination, lesion formation and Sporulation is 25 - 27.7 ° C. Lesion produce spores for longer periods at 16.1 - 25 ° C than 27.7 ° C. Excessive nitrogen fertilizer, aerobic soils and drought stress favor blast. High nitrogen rate increase rice susceptibility to the disease (Ou, 1985).

Forecast for Rice Blast

The comparison of thirteen years (1984-1996 from 25th July to 7th September) meteorological date reveal that the number or rainy days in a week with >90 per cent RH were probably the critical factors in the development of rice blast epidemic during 1984 and 1992.Mean max., min. temperature (26-27 and 19-10 0C, respectively) and high humidity (RH87 per cent) or high rain fall (2766mm) during the past years favor the rapid blast development (Table 26.1). In general, most of the years, temperature, RH, Rainfall and wind velocity were more or less conducive/favourable for rice blast disease but not adequate for epidemic development

Overwintering

The primary source of inoculum to cause an epidemic is Overwintering of the fungus or from diseases seedlings. Overwintering can either be in rice straw, seeds, weeds, planting debris, soil, bamboo or bamboo grass or on alternate hosts in the form of mycelium which retains it variability up to the new growing season. The blast fungus survives as mycelia in plant residues, conidia or in living plant tissue (Ou, 1985); in tropical and subtropical areas, all three modes are considered important as sources of initial inoculum.

Host Phenology

In general, susceptibility to facultative saprophytes increases with the age, while susceptibility to obligate parasites decreases with age; with increasing age of the plant there was progressive decrease in susceptibility to infection show by many studies and the blast fungus prefers to attack the younger upper leaves of rice plants where as, the mature leaves become resistant.

Factors Influencing Disease Cycle

The climate as well as other factor influence each phases of the pathogen life-cycle and play a key role in determining the disease outbreak. The optimum weather parameter and the effect on the disease cycle are described as follows:

Infection

This is a composite term to describe sub processes, which include *conidia germination*, appresorium *formation* and *penetration*.

Conidial Germination

Leaf wetness is required for all of the infection sub processes; the relationship between infection and Leaf wetness is well known. If leaf wetness ends before completion of infection, the process is terminated. Increasing the leaf wetness period form 12 to 15 h is known to result in a 30 per cent increase in infection. The length of the leaf wetness period required for infection is influenced by temperature (Ou, 1985). Conidial germination and appressoria formation depend on the temperature of 25-28°C and RH near saturation or wetness duration.

Germination commonly begins within 3 hours of deposition if leaves are wet. Germination is delayed if conidia are exposed to a dry period, also pre infection processes are not affected by plant water stress.

Table 26.1: Weather Parameters during Rice Blast Epidemic and Non-Epidemic Years (25th July to 7th September) in Kangra District

Year	*Temp. Max (°C)*	*Temp. Min (°C)*	*Temp. Av. (°C)*	*RH (per cent)*	*Rainfall (mm)*	*Rainy Days*	*SSH*	*WV*	*Days RH> 90 per cent*
1984*	26.9	19.7	22.8	87	1517	6	2.8	4.4	47
1985	26.1	19.7	23.0	80	1692	6	3.0	5.0	24
1986	25.9	18.9	22.5	77	1827	6	3.8	4.4	12
1987	28.9	20.1	24.4	65	775	4	7.8	5.4	2
1988	26.0	19.5	22.9	81	2105	6	4.4	4.4	28
1989	26.7	19.2	23.0	73	1125	5	5.9	5.1	12
1990	25.7	19.7	22.4	80	1943	6	3.9	4.4	16
1991	28.0	20.1	24.0	71	1416	2	5.3	4.8	18
1992*	25.5	18.8	22.7	74	2766	6	3.7	4.4	27
1993	26.7	18.9	23.4	74	1125	4	5.8	4.6	14
1994	25.7	18.5	23.1	81	1529	6	3.2	3.9	25
1995	26.6	20.5	23.4	80	1253	6	3.8	3.9	27
1996	26.0	19.8	22.9	83	899	5	3.8	3.9	32

Source: Kapoor, Prasad and Sood, Indian Phytopathology, 57(4): 440-445(2004).

SSH: Sunshine hours; WV: Wind velocity.

* Blast Year.

Table 26.2: The Effect of Temperature on Spore Germination of *P. oryzae

Temperature (°C)	*Germination (per cent) after hr.*							
	2	*3*	*4*	*5*	*6*	*7*	*8*	*9*
13	0	0	0	0	0	0	0	0
15	0	13.37	37.30	44.46	47.35	50.53	52.54	55.54
18	0	15.40	41.30	50.55	54.28	60.42	62.71	62.32
20	19.21	40.87	55.50	65.76	81.32	90.12	92.65	93.24
23	19.82	43.04	56.06	67.78	80.95	89.50	92.16	93.65
25	21.74	39.90	56.67	77.31	85.60	87.90	90.26	90.74
28	25.56	40.53	58.62	76.52	83.70	90.52	92.32	93.89

* Average of two replications..

Table 26.3: Effect of Sunlight on Spore Germination and Appressorial Formation of *P. oryzae*

Exposure Time (min.)	Wet Spores				Dry spores			
	Germination (per cent)		Appressoria (per cent)		Germination (per cent)		Appressoria (per cent)	
	Light	Dark	Light	Dark	Light	Dark	Light	Dark
10	64.45	88.89	57.27	63.58	70.82	77.64	65.09	72.34
20	55.55	87.15	47.88	78.89	61.33	76.59	56.47	73.76
30	54.31	79.80	34.92	81.01	37.96	69.82	52.81	64.08
40	49.73	81.76	31.03	68.10	35.44	61.08	43.00	56.41
50	36.63	80.09	10.32	52.18	34.00	61.39	37.46	48.74
60	27.50	80.36	9.16	40.90	15.33	54.38	35.94	42.70

* Average of two replications.

Source: Kapoor and Singh (1977). Indian Phytopathology, 30: 370-373).

Effect of Sunlight on Spore Germination and Appessorial Formation of *P.oryzae*

- ☆ Sunshine strongly inhibits germination of both wet and dry spores.
- ☆ Inhibition increased with an increase in the duration of exposure to sunlight.
- ☆ Inhibition was slightly more in case of wet spores.
- ☆ The ambient temperature increase from 25 to 34°C.
- ☆ Formation of appressoria was also greatly reducer in wet spores as compared to dry spores exposed to sunlight.

Table 26.4: Effect of Temperature and Leaf Wetness on Infection of Rice Seedling by *P. oryzae*. (cv. Pusa 2-21)

Duration of Leaf Wetness (hrs)	No. of Infection Leaves/10 Seedling at °C				No. of Infection Lesion/10 Seedling at °C			
	18	20	23	25	18	20	23	25
8	0	0	0	0	0	0	0	0
12	0	8	10	13	0	12	23	13
16	10	20	18	17	12	14	23	16
20	10	12	17	18	12	26	25	16
22	13	10	15	20	14	16	26	16

* Average of two replications.

- ☆ The shortest period of leaf wetness needed for successful infection was 12 hrs. Within a temperature range of 20-25°C where as 16hrs at 18°C.
- ☆ Difference in the number of leaves showing infection and average number of lesion per leaf within temperature range of 20-25°C were negligible.
- ☆ No significant difference was found in the number of lesions produced on leaves exposed to varying duration of leaf wetness (12-22 hrs.) at a particular temperature between 20 and 25°C.

☆ However, number of leaves per seedling showing infection was greatest at 16hrs of leaf wetness at 20°C and 22hrs of leaf wetness at 25°C.

Aspersoria Formation and Host Penetration

Initiation of infection occurred from 5 to 7 hrs at 21 to 27°C and 8hrs at 18°C conidial deposition. Aspersoria formation occurred at an average of 11 h at 24°C but is rare at temperatures greater than 28°C. Various optimum temperatures for aspersorium formation have been reported to be between 15 and 25°C.

Temperature and dew period interact to affect infection. A minimum dew period of 6-8 h will initiate infection at the optimum temperature of 25° C. In the tropic, night temperature is almost always in the optimal range (Ou, 1985). In upland rice, longer dew periods were suggested as a factor in the higher incidence of blast compared with lowland condition (Ou, 1985).

Spore germination and aspersorium formation are stimulated in plants fertilized with high levels of nitrogen. Penetration rates are lower in plants with high levels of silica, possibly as a result of the mechanical barrier attributed to accumulation of silica in the epidermis (Ou, 1985). Application of N results in increased infection (Ou, 1985) what is unclear from the literature is whether N confers a direct physiological effect on the pathogen-host interaction or only has an indirect effect by changing plant morphology and consequently, micro-climate.

Dispersal and Spread

Blast conidia are commonly observed at high number over rice crops between 0000 h and 0600 h, and liberation is favored by darkness, high RH and wind speeds<3.5 ms^{-1}. Dispersal gradients for blast conidia are strongly related to dominant wind directions and follow the pattern found for other passively liberated spores. Spores may be released at temperatures above 19°C while rainfall on rice leaves is known to cause an increase in the conidial concentration over canopies. Rain, wind and air currents strongly affect the number of dispersing conidia.

The maximum distance that a blast conidium can travel remains controversial but is key to determining management strategies. Under average tropical conditions, blast conidia can be dispersed commonly within about 230 meters of the source. It is very possible that long-distance dispersal of conidia is closely related to weather patterns and low-level upper atmospheric disturbances.

This aspect of Epidemiology deserves more study in view of modern techniques for spatial analysis of weather and landscape patterns, and for monitoring of fungus and host.

Lesion Growth

Cultivars with different degrees of partial resistance to blast show differences in infection efficiency and lesion size. Plant age and leaf age, however, influence the expression of partial resistance effects; young leaves are commonly more susceptible. Lesion size is known to be affected by temperature with high alternating temperature of 32/25°C and 32/20°C resulting in initial rapid lesion growth and water stress increasing the size of lesions. Blast management through cultural practices will therefore be enhanced by minimizing water stress, especially at period of rapid leaf growth.

Sporulation

Under optimum conditions, conidiophores are formed in 4–6hrs with one conidium formed every 40 min. The number of spores produced by any one lesion has been found highest on the sixth day after lesion appearance.Mature lesions are capable of sporulating when relative humidity (RH) is higher than 89 per cent (with optimum RH>93 per cent). Maximum Sporulation occurs 7-12 d after inoculation, although Sporulation may continue at low levels for 60 days.

Sporulation is greater in lesions on leaves that were expanding when inoculated than on lesions from leaves that were fully expanded 3-4 d before inoculation. Lesions produced on plants inoculated at the tillering stage has a greater sporulation potential than on plants inoculated after the formation of the panicle primordium. In one study number of conidia per lesion varied from 4800 on IRAT13 to 33,000 on IR442-2-58.

Latent Period

Successful establishment after infection has occurred depends on the age and genotype characteristic of the host, crop nutrition, and environmental factors such as temperatures and soil moisture. Latent period varied from 13–18 days at 9-11ºC and 4–6 d at 26–28ºC.

Sheath Blight

Jammu region faces the wrath of sheath blight [*Rhizoctonia solani* Kuhn (Telomorph: *Thanatephorus cucumeris*)], paddy blast and Helminthosporiosis (*Helminthosporium oryzae* Brede de Hann) with sheath blight being the most important of the diseases.

The disease is characterized by the formation of lesions on leaf sheaths and culms at the water level, which become confluent giving characteristic banded appearance. The infection may spread up to the culms, killing all the leaves under favourable weather conditions. Losses up to 20 per cent in grain yield has been reported when disease invades at flag leaf stage (Singh, 1990) however, the resultant losses have been related with rice varieties cultivated.

Since 1910, Sheath blight of rice is known in East and South East Asia but was considered a minor disease (Ramakrishnan, 1971). However, built in inoculum and less attention paid has made this disease to be known world-wide which ultimately registered serious losses in temperate as well as tropical rice producing countries including (Singh, 1998).

In India, the disease was first reported by Paracer and Chahal (1963) from Gurdaspur (Punjab). Economic importance of the disease is tending to increase because of excessive use of fertilizers and introduction of high yielding varieties, which yield large number of tillers there by increasing the humidity under the plant canopy thus making the plant prone to the sheath blight. Jammu Division of Jammu and Kashmir is no exception where during the last few years the disease has been recorded in alarming proportions in most of the varieties sown in the region.

Distribution and Status of Disease

Sheath blight of rice was first described by Miyake (1910) who named the causal organism *Sclerotium irregulare*. Later, it was found widely distributed in rice growing countries from tropics to temperate regions and positioned next to blast in economic importance. Although, Butler (1918) mentioned this disease from India, Paracer and Chahal (1963) first described the disease in detail from Gurdaspur, Punjab. Rice sheath blight is now known to prevalent in all the rice growing states of India (Reddy and Reddy, 1986).

Number of workers have reported about the status and yield losses caused by sheath blight. Under condition of heavy severity, yield losses of more than 50 per cent due to the sheath blight disease have been reported in Japan, Vietnam, South Korea, Taiwan, China, US and India (Roy, 1993; Kim, 1994). Sharma and Teng (1990) reported that inoculation at flowering resulted in the lowest yield and highest yield loss. Peng and Tang (1990) studied that the disease index at wax ripeness was highly related to the percentage of tillers infected at jointing and yield loss was closely correlated with disease index at wax ripens. Saikia and Baruah (1990) studied that an increase of one per cent in sheath blight incidence resulted in a 0.38 per cent loss in grain yield. Li *et al*. (1996) studied the yield loss and dynamic economic threshold of sheath blight on hybrid early rice. Cu *et al*. (1996) reported that the estimated yield reduction from sheath blight in plots receiving the highest N rate ranged from 20 to 42 per cent. Sharma and Tang (1990) reported that yield loss at different stage of inoculation ranged from 23.01 to 32.15 per cent. Chahal *et al*. (2003) reported that when all the leaves of paddy were infected with sheath blight disease, the yield recorded was 18 g/plant compared to 35.9 g/plant in the healthy control. Araujo *et al*. (2006) studied the effect of sheath blight on some yield components and reported that the relationship between panicle length and sheath blight severity was linear and negative and further suggested that the estimated mean damage in grain weight/panicle was greater than on the empty spikelets and panicle length of the genotype tested.

Integrated Disease Management

Biological Control

During the past few years more attention than ever has been given to the development of biological methods to control plant diseases. It is indeed due to the concern for food of high quality without residues of pesticides and for a sustainable agriculture that will preserve the fertility of soil and eco-friendly environment. Various bio-control agents have been exploited by researchers for management of sheath blight of rice.

Bhuyan *et al*. (1994) reported the antagonist activity of *Trichoderma viride*, *Trichoderma harzianum* and *Aspergillus terreus* against *R.solani*. Baby and Manibhushanrao (1996) evaluated *Trichoderma* spp. and *Gliocladium virens* as well as some vesicular arbuscular fungi against sheath blight (*R.solani*) of rice and reported that *T.harzianum*, *T.longibrachiotum* and *Gliocladium virens* were equally effective in controlling the disease. Das *et al*. (1996) studied that *in vitro* sclerotial growth of *Rhizoctonia solani* Kuhn, causing sheath blight of rice gets significantly reduced by application of *Trichoderma viride*,

T.harzianum and *Aspergillus terreus* in 4 per cent maize-meal sand (MSM). *Trichoderma viride* was more effective than *T.harzianum* and *A.terreus*. Das and Hazarika (2000) studied the influence of seed treatment of rice with two antagonistic fungi *viz., Trichoderma viride* and *Trichoderma harzianum* alone and in association with 2 per cent (W/V) methyl cellulose and/or 0.1m $MgSO_4$ against sheath blight under pot condition. The seeds treated with *T.viride* and *T.harzianum* showed significant reduction in sheath infection. Both antagonists exhibited higher efficacy in reduction of sheath infection and subsequently increase in yield, when they were treated with either 2 per cent (W/V) methyl cellulose or 2 per cent (W/V) methyl cellulose and 0.1M $MgSO_4$. They further reported that *T.harzianum* was more effective than *T.viride* in reducing sheath infection and increase in yield. Laha and Venkataraman (2001) isolated bacterial antagonist from FYM, rice seeds, rice phyllosphere and rhizosphere and selected PF-9 (fluroscent *Pseudomonas*), B-44 (*Bacillus* sp.) and *Chitniolytic bacterium* for experimentation and reported that PF-9 was most effective in reducing disease severity either alone or in combination of PF-9 and B-44. Tang-Jia-Bin *et al.* (2001) evaluated efficacy of *Trichoderma* spp. against rice sheath blight pathogen, *R.solani* and out of 800 strains, 6 strains, were found to inhibit the growth of *R.solani*. T3 strain was reported to be the best antagonist who reduced the growth of the pathogen by 52.54 per cent. Jayaraj *et al.* (2004) reported that foliar application of *Bacillus subtilis* strain AUBS1 reduced sheath blight and triggered defence mechanism in rice. They concluded that field application of fluorescent *Pseudomonas* products reduced sheath blight disease severity just like fungicides. Whereas, propiconazole application increased the yield and *Pseudomonas* products failed to increase yield commensurate with observed reduction in disease incidence and severity of sheath blight. Khan and Sinha (2005) investigated the most promising isolates of *Trichoderma* sp. and suggested that soil treatment with *Trichoderma harzianum* was found more effecting in reducing sheath blight as compared to other isolates of *Trichoderma* spp. and commercial formulations of bio-agents. Singh and Sinha (2005) examined the efficacy of potential strains of *Pseudomonas fluorescens* against sheath blight of rice and reported that foliar spray with Pf -1 at seven days before pathogen inoculations resulted in maximum reduction in sheath blight severity and incidence while increase in grain yield. Saikia *et al.* (2006) studied the effect of *Pseudomonas aeruginosa* isolates in inducing rice resistance against *Rhizoctonia solani* and reported that the isolates showed plant growth promoting activity, induced systemic resistance and increased seed yield by production of salicylic acid and peroxidase.

Plant Extracts

Plant extracts have assumed special significance in the present day strategy of developing ecologically safe method of plant disease management. Plants and their constituents have shown value as less phytotoxic, more systemic, easily biodegradable and host metabolism stimulatory (Sharma *et al.*, 1999). An eco-friendly safe biopesticide will not only protect the crop from disease but also safeguard the environment from pollution.

Ashrafuzzaman *et al.* (1994) tested various plant extracts against *R.solani* and stated that extract from *Phalaris minor*, tomatoes affected germination, mycelial growth and sclerotial production of *R. solani*. Ansari (1995) tested the antifungal activity of ajwain (*Trachispermum ammi*), lemon grass (*Cymbopogen citrates*) tulsi (*Ocimum* sp.) mentha (*Mentha* sp.), *Rawolfia* sp., lawsome (*Lawsonia inermis*) and Samhalu (*Vitex trifolia*) against *Rhizoctonia solani* and reported that ajwain seeds and tulsi (leaves) extracts have shown fungicidal activity whereas other showed fungistatic activity. Leaf extracts of some other plants such as *Ramunculus asiaticus, Sourchus oleraceus* and *Mercurialis annua* were also effective against *R. solani* (Qasem and Abu-Blan, 1996) and provide possibilities of their use for management of rice sheath blight disease. Fungicidal activities of various plant extracts against sheath blight fungus have been reported by Kotasthane and Lokpale (1997). Sharma *et al.* (1999) reported that fifty per cent aqueous extracts of Bracken fern *(Diplazin esculentum)*, goat weed (*Agartum houstonianum*) and polygonum (*Polygonum plebeium*) were found to exhibit more than 90 per cent *in vitro* inhibition of the pathogen while *in vivo* results have also confirmed that Bracken fern extract followed by goat weed and sickle pod senna (*Cassia tora*) extract resulted in significant reductions in lesion length. Biswas and Roy choudhary (2003) studied the relative efficacy of some botanicals against sheath blight disease of rice. Gautam *et al.* (2003) *in vitro* evaluated twenty four botanicals belonged to the family Compositae *(Asteraceae)* for their fungitoxicity against *Rhizoctonia solani* Kuhn and found that maximum per cent inhibition of mycelial growth was observed in 1000µgml^{-1} concentration followed by 200 µgml^{-1} at 60h and 48h, respectively and none of the botanical affected the growth at 40 µgml^{-1}. Biswas (2006) evaluated the efficacy of Biotos at 0.25 per cent compared with other botanicals (Achook at 0.5 per cent and Tricure at 0.50 per cent) and suggested that Biotos at 0.25 per cent was superior in controlling sheath blight and increasing grain yield compared to Achook and Tricure.

Management through Chemical Fungitoxicants

Inspite of hazardous and pollutant effects, chemical fungitoxicants continue to dominate the management strategy for most of the plant disease including the rice sheath blight disease. Researchers have tried different groups of fungitoxicants, carbendazim, benomyl, thiophanate methyl, (Benzimedazoler), penconazole etc. (Ergosterol biosynthesis inhibitors) besides some fungitoxicants of other groups, alone or in combination to manage the disease in the field. Flutonil, a systemic fungicide which possessed both protective and curative attributes, was found effective against rice sheath blight by various workers. Das and Mishra (1990) reported that seed treatment with Topsin + 2 foliar sprays of the same fungicide at 0.1 per cent concentration resulted in lowest disease score and highest grain yield (50.9 q/ha). Eeckhout *et al.* (1991) reported significant decrease of disease incidence in plots treated with propiconazole at green ring stage while the disease severity was reduced most effectively by the application of pencycuron at booting and heading stage of crop growth. Mishra and Sinha (1991) evaluated several fungicides against *R.solani* and reported that carbendazim (5µml^{-1}) and propiconazole (10µgml^{-1}) resulted in 100 per cent inhibition of mycelial growth and were good fungicides for management of the sheath blight in field. Sarkar *et al.* (1991) tested seven fungicides under field conditions and achieved best results with Bavistin (carbendazim) and Rhizoex (tolilofosmethyl). Krishmaiah and Reddy (1992) evaluated three fungicides *in-vivo* and reported carbendazim as the most effective in reducing the disease followed by thiophanate methyl with disease index of 0 - 1.4 per cent as compared to 48.5 per cent in case of unsprayed check. Viji *et al.* (1997) studied the response of *Rhizoctonia solani* against fungal antagonists, *viz. Gliocladium virens, Trichoderma longibrachiatum* and *Trichoderma harzianum* and the systemic fungicides both *in-vivo* and *in-vitro*. The results indicated that the Benzimidazole group of fungicides (carbendazim and benomyl) were toxic to pathogen as well as antagonists, while the organophosphorus fungicides (edifenphos and IBP) were more toxic to pathogen alone. Akhter *et al.* (2001) tested six fungicides, namely Bavistin 50WP (carbendazim), Tilt (propiconazole) Contaf 5EC (hexaconazole), Anvil 5EC (hexaconazole), Forestin 50WP(carbendazim) and Thiovet 80WP (micronised sulphur) for sheath blight management and found that Contaf 5EC was best in reducing the per cent relative lesion length, per cent disease index and tiller infection. Anwar *et al.* (2002) investigated the management of rice sheath blight and rice blast and reported that best yield is obtained with spray + seed treatment of carbendazim and mancozeb. Ali and Archer (2003) evaluated fungicides of diverse group and their mode of action *in vitro* and *in vivo* for their efficacy against sheath blight disease of rice. Ten fungicides (azoxystrobin 250 SC, carbendazim 500 Flo2, cyproconazole 240 EC, fenpropimorph 750 EC, fludioxonil 2.5 SC, hexaconazole

250 SC, iprodione 50 WP, pencycuron 250 Flo, propiconazole 250 EC and tolclofos-methyl 500 Flo) were tested for their effect on radial colony growth of 6 isolates of *R. solani* and reported that except azoxystrobin, all fungicides were highly effective *in vitro* and among the fungicides tested *in vivo*, azoxystrobin, pencycuron and hexaconazole were highly effective against sheath blight disease.

Management through Different Dates of Sowing

Pathak *et al.* (2002) conducted study to validate an IPM module for pests and pathogens of rice and suggested that aside from pest management, cultural practices such as early sowing, higher plant density, regulated use of fertilizers and irrigation contributed to the improved yield of rice in IPM subjected fields. Yang *et al.* (2002) conducted field test to investigate the effects of early sowing (ES) plus early transplanting (ET) and conventional sowing (CS) plus conventional transplanting (CT) on the occurrence of diseases, insect pest and the grain yield and reported that occurrence of *Scirpophaga incertulas, Nilaparvata lugens* and rice sheath blight were significantly lower under ES plus ET than CS plus CT and ES plus ET increased yield and saved pesticides cost.

Integration of all the available practices of disease management discussed above can be helpful in devising an Integrated Management Strategy for the control of sheath blight of rice.

Seed Discolouration in Rice Grains

In Jammu and Kashmir, the glume discolouration of paddy (Plate 26.1) has perceived serious proportion both in grain (Plate 26.2) and seed production programme in temperate ecosystem area, where it appears to be surpassing the importance of even previously known major constraints. The appearance of glume discolouration by various organism before or after harvest the extent of which varies according to season and also within the season on short, medium and long duration rice cultivars grown under rice growing domain of Kashmir valley. They produce different types of symptoms, depending upon the organism involved in the degree of infection. Sometimes there are distinct black dots usually caused by fungus fruiting bodies are other structures on normal or bleached areas of the glume. In other cases there are brown or blackish blotches, which may be flecks, are large enough to cover the inter-glumes. Often the lesions are pale or greyish in the centre with a dark brown margin (Plate 26.1). About thirty three fungal species belonging to 2 genera are responsible to cause glume discolouration in India (Ahuja, 1987), which reduce yield and also affect the quality parameters. It reduces germination, causes coleoptiles or radicle decay, or both, resulting in chaffy grain and ultimately causing low yield. The infected plants produce partially filled discoloured grains with reduced viability and quality.

Damage

These microorganisms have various effects on the grains. When infection is by field fungi, the main effect are reduced viability and grain quality; seedling blights and other diseases may occur when the seeds are planted. When infection is by storage mode, besides a reduction in viability and grain quality, these may also be production of toxins. Uraguchi (1942) reported a toxin formed in yellowed rice due to *Penicillium* sp. Tran-Vy *et al.* (1958), however, reported that rice infected by *P. citrinum* was not toxic to rice. In view of present knowledge on the occurrence of mycotoxins in fungus infected cereal grains, the toxic aspect of fungus - infected rice grains should be further exposed.

***Curvularia lunata* on potato dextrose agar medium**

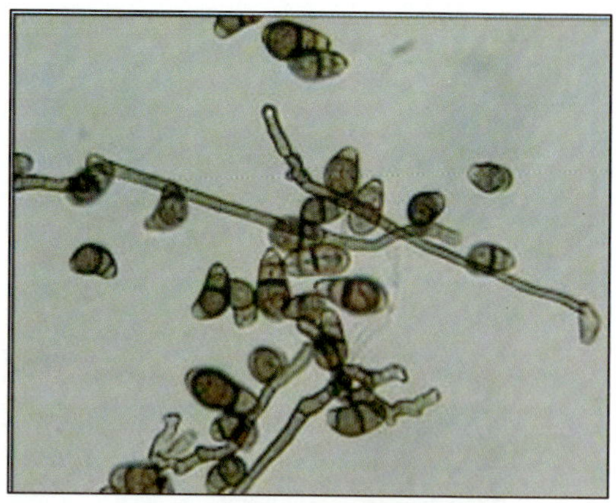

Mycelium with spores

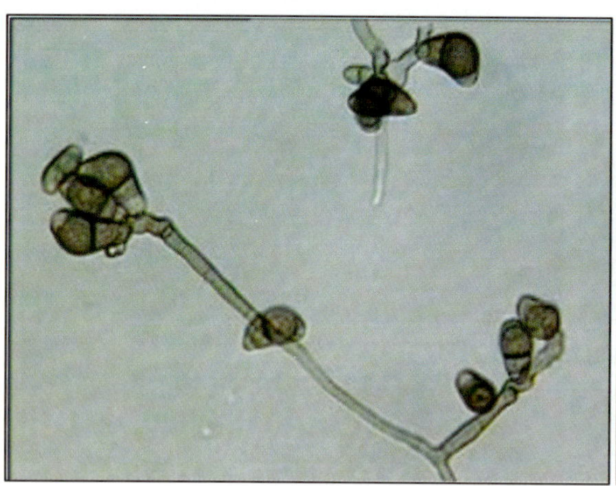

Conidiophore with bunch of conidia at 450 x

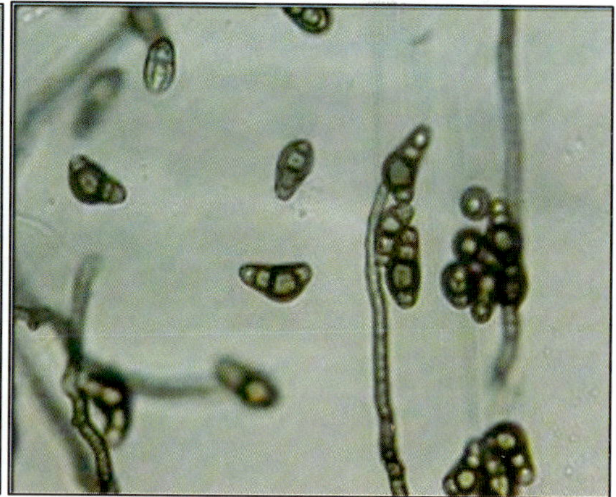

Conidia at 450 x

Plate 26.4: Morphological and Microscopic Features of *Curvularia lunata*

Occurrence and Distribution

The prevalence of micro-organisms causing glume discolouration of paddy varies widely. Tullis in (1936) has isolated different fungi associated with discoloured grain from USA. The most common were *Alternaria* spp, *Curvularia lunata, Alternaria padwickii* and *Phoma* spp. while *Drechslera oryzae* and *Fusarium* spp. were found rarely but *Cladosporium herbarum, Helicoceras oryzae* and *Penicillium* spp. were encountered occasionally. Baldacci and Picco (1948) from Italy reported that out of 50-55 per cent discoloured rice grains, 13.4 per cent exhibited *Alternaria* spp. 23.4 per cent *Cochliobolus miyabeanus,* 1.2 per cent *Fusarium* spp. and 4.9 per cent *Pencillium* spp. whereas, 30.8 per cent grains were not found

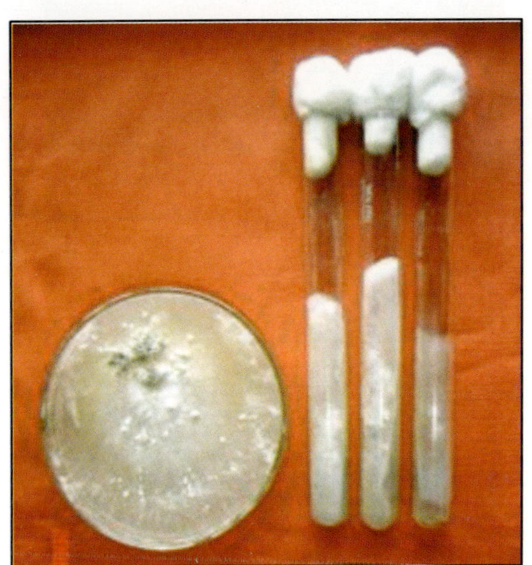

***Rhizoctonia solani* on potato dextrose agar medium**

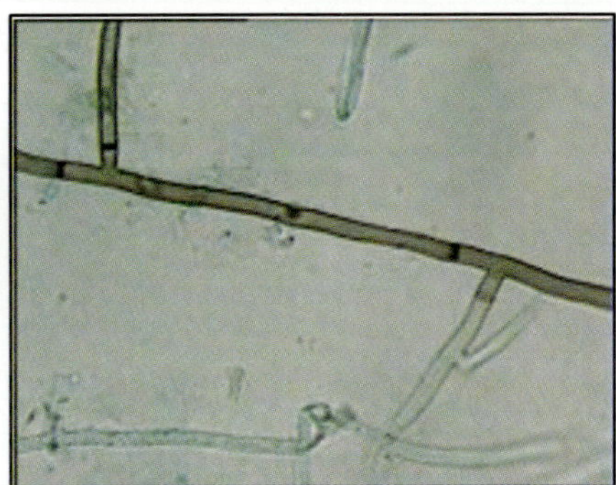

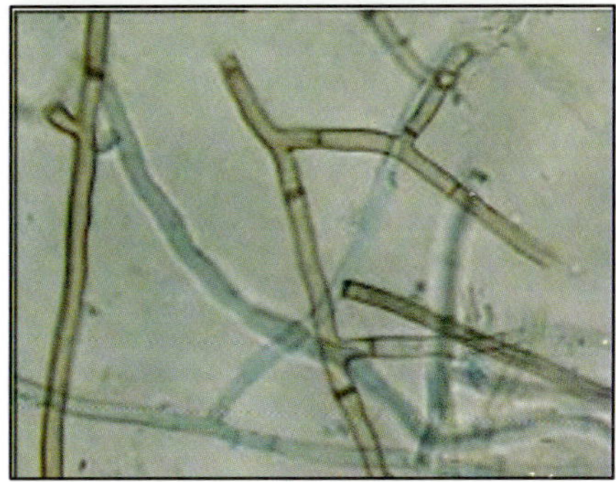

Mycelium at 450 x

Plate 26.5: Morphological and Microscopic Features of *Rhizoctonia solani*

affected with any fungus. Similarly, Padmanabhan (1949) reported *Alternaria padwickii* to be predominant over *Cochliobolus miyabeanus, Curvularia lunata* and *Nigrospora* sp. Subsequently, found that sixty per cent of discoloured grain in Hyderabad, India yielded five species of *Curvularia*. Besdies, *Helminthosporium oryzae, Curuvularia lunata, Alternaria oryzae, Rhizoctonia solani, Phyllosticta oryzae, Nigrospora* sp. *Fusarium moniliforme* from discoloured grain with variable occurrence pre-harvest stage of rice cultivar in Kashmir, India (Plates 26.3–26.11).

Sachan and Agarwal (1995) have been observed *Alternaria alternata* and *Helminthosporium oryzae* caused ash grey and black discolouration in seed respectively, whereas, due to *Curvularia lunata* and *Fusarium graminearum, Fusarium moniliforme,* eye shaped and light pink discolouration appeared in seed coat, endosperm and embryo of seed respectively.

Vaid *et al.* (1994) reported that glume discolouration is caused by complex of fungi with the species of *Alternaria, Curvularia, Fusarium, Helminthosporium, Nigrospora, Phyllostic* and *Sclerotium*. These fungal infection increased protein and decreased amylase content as well as decreased the hulling and milling quality of paddy.

Sharma *et al.* (1997) have been detected rice samples from various rice growth location in the districts of Himachal Pradesh *viz.*, Bilaspur, Chambal, Hamirpur, Kangra, Kullu, Mandi, Sirmaur and Una and observed that the extent of grain discolouration varied from 4.35 to 79.32 per cent. The fungi *viz., Alternaria alternata, Aspergillus niger, Curvularia lunata, Fusarium* spp. *Helminthosporium oryzae, Nigrospora oryzae, Phyllosticta glumarium, Penicillium* spp. and *Sclerotium oryzae* are attacked at booting,

Pyricularia grisea **on potato dextrose agar medium**

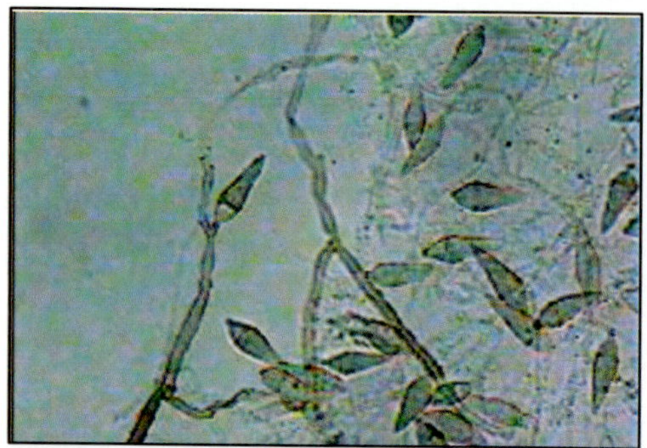

Mycelium and conidiophores alongwith conidia in culture

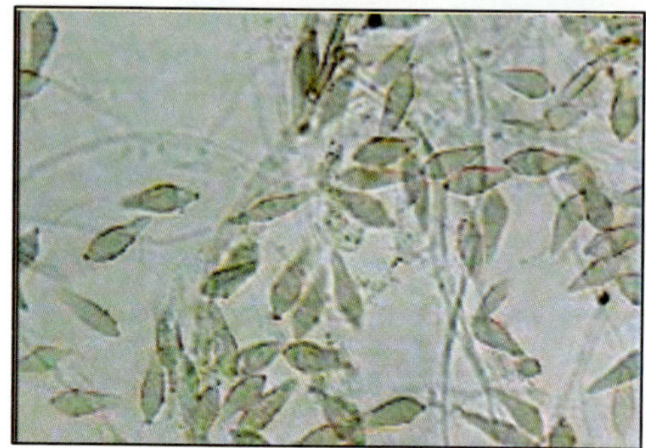

Conidia at 450 x

Plate 26.8: Morphological and Microscopic Features of *Pyricularia grisea*

Govindarajan and Kannaiyan (1982) observed that edifenphos and copper oxychloride are most effective fungicides against rice grain infection caused by seed borne pathogens *viz.*, *Helminthosporium oryzae*, *Nigrospora oryzae* and *Curvularia* sp.

Ray Pritima (1993) revealed that carbendazim @ 0.1 per cent was found effective in controlling the grain infection, even though there was significant variation in percentage of grain discolouration. Viswanathan and Narayanaswamy (1993) found that tricyclazole @ 2 g/kg and carbendazim @ 4 g/kg were most effective than other fungicides *viz.*, mancozeb, captan and thiram. When seeds were treated by shaking method with these fungicides or by wet application (10g seed + 1 ml water + fungicides), the seed germination increased by 64 per cent, while 68-90 per cent with wet application, moreover seed treated with these fungicides also improved growth and vigour of seedling.

Vaid *et al*. (1994b) have tested six fungicides @ 2 g/kg of seed to improve the germinability of discoloured grain. They found that carbendazim was most effective, hence, increased germinability of discoloured grain by 20-30 per cent followed by triadimefon. Deka *et al*. (1996) found that application of fungicides at the boot leaf stage followed by spray with common salt (10 per cent) 20 days after 50 per cent flowering is more effective in reducing grain discolouration that only fungicides at both stages. However, carbendazim (0.1 per cent) followed by common salt is found most effective.

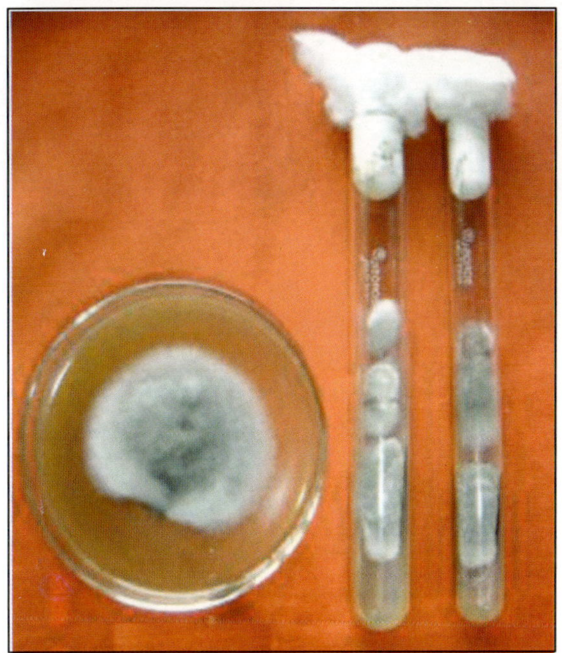

***Helminthosporium oryzae* on potato dextrose agar medium**

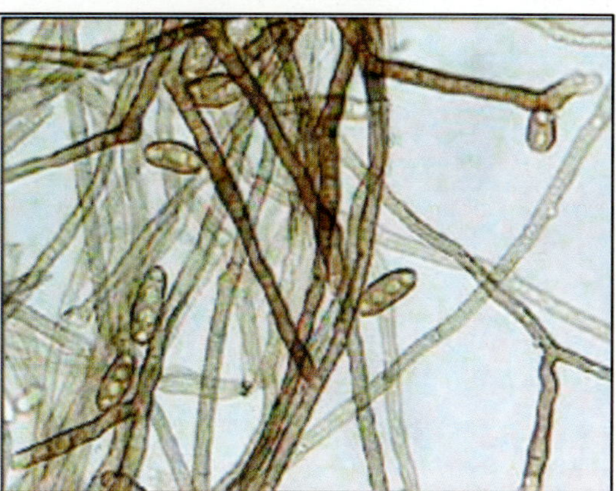

Mycelium and conidiophores alongwith conidia in culture

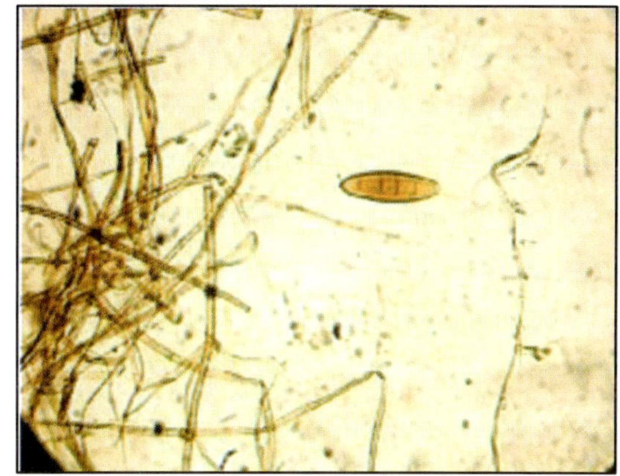

Conidium at 450 x

Plate 26.9: Morphological and Microscopic Features of *Helminthosporium oryzae*

Saifulla *et al*. (1996) revealed that mancozeb and tricyclazole were significantly superior over all treatments in highly discoloured seeds of paddy.

Malavolta and Takada (1997) have been reported tubeconazole (125 g/ha), chlorothalonil (1250 g/ha), tricyclazole (187.5 g/ha), propiconazole (125 g/ha) and kasugamycin + stannic triphenyle acetate (20 ml + 100 g/ha) were proved tobe arsenic against rice grain discolouration.

Santos *et al*. (2000) found that least incidence of grain spot of neck blast appeared by applying difenconazole and mancozeb followed by difenaconazole + tricyclazole. Tasleem-uz-Zaman *et al*. (2000) found the control of glume discolouration in paddy by using as foliar sprays as well as seed treatments of topsin M (thiophanate methyl), anvil (hexaconazole) and benlate (benomyl).

Malavolta and Bedendo (2004) reported that seed borne infection by *Bipolaris oryzae, Microdochium oryzae* and *Phoma sorghina* in rice, can be removed when procloraz is used for seed treatment. Subsequently chlorothalonil and imiricoctadine were also seen effective. Pizzatti *et al*. (2004), revealed the percentage of discoloured caused by *Cochliobolus miyabeanus, Magnaporthe grisea* and *Sarocladium*

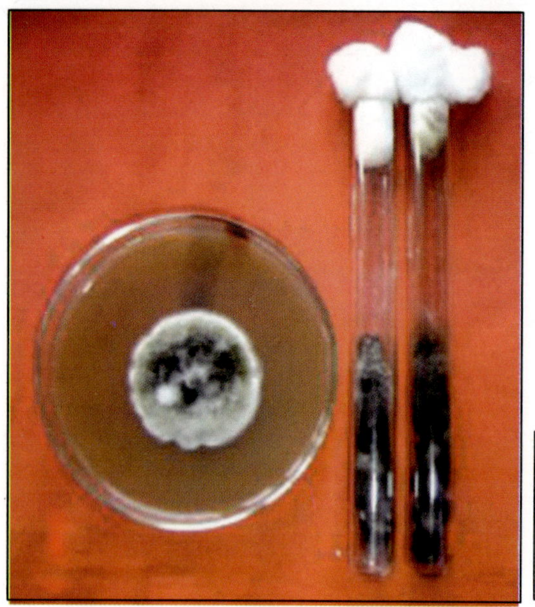

Alternaria sp. on potato dextrose agar medium

Mycelium in culture

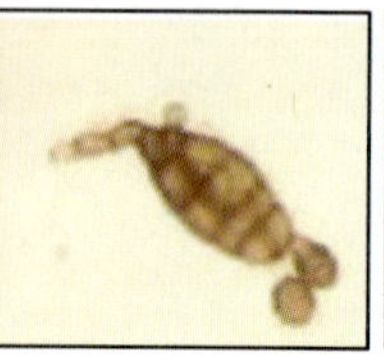
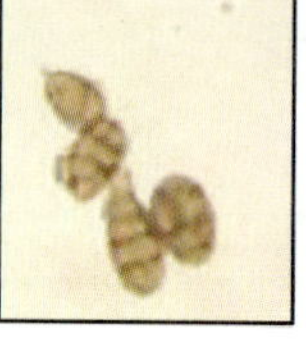
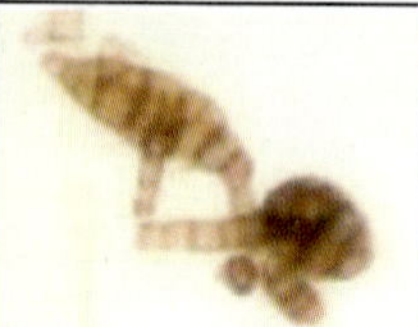

Conidia at 450 x

Plate 26.10: Morphological and Microscopic Features of _Alternaria_ sp.

spp. decreased by delaying the date of sowing and fungicidal spray with carbendazim. Shafait (2006) has been studied many fungi which cause glume discolouration of rice in Kashmir India (Plate-II-VIII) and recommended two sprays of of carbendazim 50 WP (0.1 per cent) bitertanol 25 WP (0.4 per cent) or tricyclazole 75 WP (0.1 per cent) to control the glume discolouration at pre harvesting stages of crop.

Sheath Rot of Rice

Sarocladium oryzae (Acrocylindrium oryzae) the incitant of sheath rot of rice was considered the main cause by several workers (Agnihothrudu, 1973; Amin *et al.*, 1974; Ahamed *et al.*, 1975; Nair and Rajan, 1978; Shahjahan *et al.*, 1977; Thrimurty *et al.*, 1980; Purkayastha and Ghosal, 1982; Kang and Rattan, 1983; Manibhushanrao *et al.*, 1986; Reddy *et al.*, 1986; Boa and Brady, 1987; Singh and Roy, 1987; Singh and Khan, 1988 and Singh *et al.*, 1996). Sheath rot caused by *Sarocladium oryzae* was also reported by several workers from different parts of the world (Zhuge *et al.*, 1985; Mbodj, 1989; Suparyono, 1990; Gill *et al.*, 1993; Jones *et al.*, 1993 and Mazzanti-di- Castanon *et al.*, 1993) from Pakistan, Argentina, Cameroon, Indonesia, Senegal and China, respectively. Severe incidence of this disease increases chaffy grain, reduction in yield as well as poor quality of the grain. Thrimurty *et al.* (1980) also reported that *Sarocladium oryzae* was found to be associated with sheath rot infected panicles of rice in Chhattisgarh region and also reported that sheath rot incidence on some popular rice varieties increased the number of chaffy grains in infected panicle than healthy.

Pathogen

The typical characters of *Sarocladium oryzae* (*Acrocylindrium oryzae*) (Sawada) Gams and Hawksworth (1975) the causal organism was described by several workers as: white, septate mycelium with sparsely branched conidiophore, slightly thicker than vegetative hyphae with terminal branches,

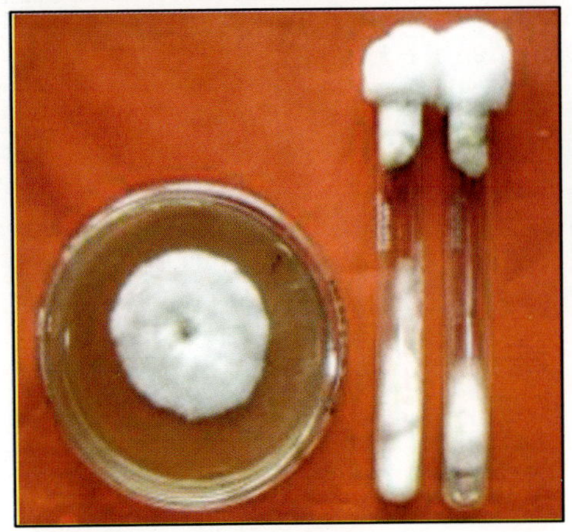

Fusarium graminearum **on potato dextrose agar medium**

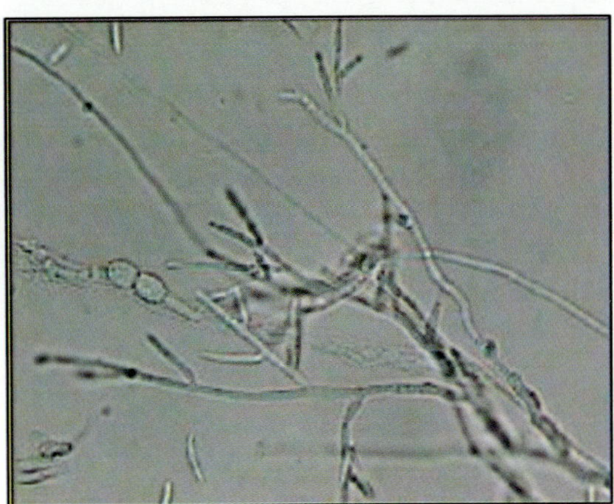

Conidiophores alongwith conidia in culture

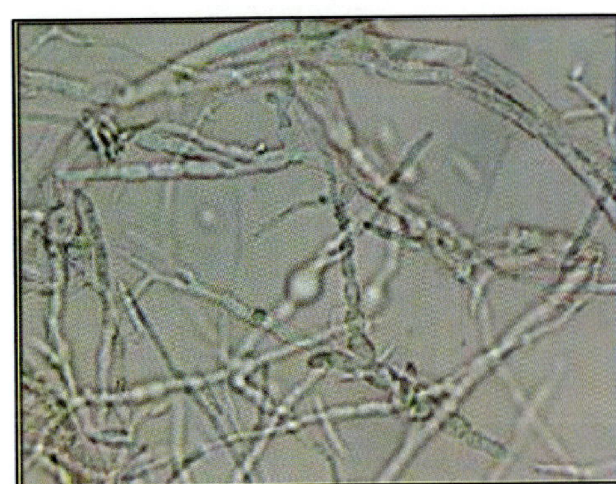

Intercalary chlamydospores in culture

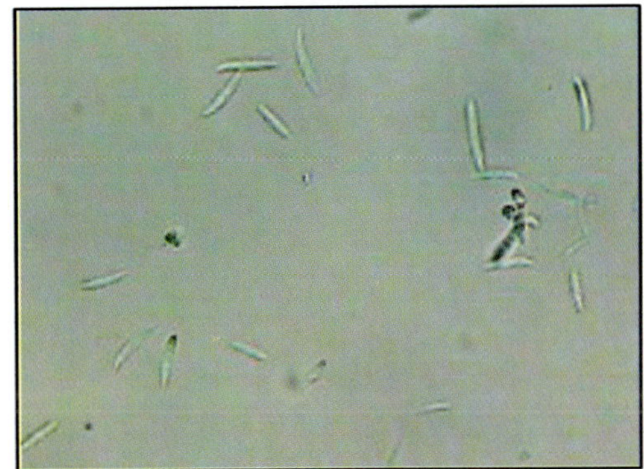

Macro conidia at 450 x

Plate 26.11: Morphological and Microscopic Features of *Fusarium graminearum*

phialides arising from conidiophores or directly from undifferentiated vegetative hyphae, 30-40 mm long and 1.5-2.0 mm wide at the base when formed at the appex of slender conidiophores, 6-20 mm long and 1.0-1.5 mm wide when arising in dense broom like fascicles, tips tapering to 0.6-1.0 mm in width, lacking any distinct collarette and conidia borne on tip, conidia formed in slimy masses, cylindrical with rounded ends, sometimes becoming slightly curved, hyaline, thin walled, smooth, one celled with a 3.5-7.0 x 0.8-1.5 mm. chlaymydospore absent, branches of conidiophores arises in whorls (Ou, 1985; Mew and Mishra, 1994 and Manibhushanrao, 1996).

Mode of Infection

Sheath rot pathogen mainly attacks the uppermost (flag) leaf sheath and infection was more facilitated in the presence of an injury (Kawamura, 1940 and Tasugi and Ikeda, 1956). Shahjahan *et al.* (1977) reported that the pathogen could enter the host through stomata or injuries and colonizes intercellular in the vascular and mesophyl region of the susceptible cultivars. Reddy *et al.* (1984) reported that pathogen produces cellulolytic and pectolytic enzyme *in vitro*. Such enzymes might have also a positive role in infection and colonization of the pathogen. *Sarocladium oryzae* could infect the rice plant at all stages of growth, but was most destructive at boot stage.

Zhuge *et al.* (1985) reported that the pathogen could infect the leaf sheath, leaf blade midrib and seeds. Reddy (1989) reported that an inoculum concentration of 8 x 10^4 spores/ml as the potential dose for effective infection and disease development. Further increase in spore concentration to the level of 10x10^4 spores/ml reduced the infection, which might be due to autolysis of the spores at high concentration. Manibhushanrao (1996) opined that histopathological studies need to be conducted in order to understand the exact mode of penetration and invasion of the host tissue, as little information is available.

Symptoms of Sheath Rot

According to Tasugi and Ikeda (1956) typical symptoms of sheath rot are characterized by the production of grayish brown blotches on the flag leaf sheath.

Agnihothrudu (1973) reported that grayish brown lesions with powdery masses of conidia were present in the flag leaf sheath. Dark brown to chocolate brown spots on flag leaf resulting in partial or incomplete emergence of panicles reported to be the most common symptom induced by *Sarocladium oryzae* (Subramanian and Ramakrishna, 1975). Under severe infection the panicle shows large number of chaffy and discolored grains (Amin *et al.*, 1974; Thrimurty *et al.*, 1980 and Arunyanart *et al., 1981*). Ou (1985) described as an oblong or sometimes-irregular spots of 0.5-1.5 cm in length with brown margins and gray centres or grayish brown throughout. The enlarged spots coaleses and coveres most of the leaf sheath. The young panicles remain within the sheath or partially emerge. Also an abundant whitish powdery growth inside affected sheaths and the rotting of the young panicles occurs.

Infected Debris

Singh and Raju (1981) found that *S. oryzae* survived for 10 months in the infected sheaths incubated under field conditions. Deka and Phookan (1998) reported that survival of *S. oryzae* varied with the type of soil investigated, survival was best (210 days) in sterilized soils at 30±1°C. Survival was also high (180 days) in naturally infected field soil and sterilized soil maintained outdoors.

Host Range of *S. oryzae*

The host range of *S. oryzae*, the rice sheath rot causal organism has not been widely studied. Few reports are available on the host range of this pathogen. Balakrishanan and Nair (1981) reported that field weeds *Cyperus diformis, Echinochloa crusgalli, Monochoria vaginalis* and *Cyperus teneriffe* were naturally infected hosts of *A.* (*Sarocladium*) *oryzae*. Subseqently among the 20 species of weed and wild rice tested, *Digitaria cilians, E. colona, E. crusgalli* var. *hispidula, Chloris barbata, Ischemum rugosum* and *Leptochloa chinensis* were reported to be infected by sheath rot pathogen (IRRI, 1981). *Echinochloa colona*, was also found to play a role as an alternate host for sheath rot pathogen (Rahman *et al.*, 1982).

Yield Loss

The pathogen mainly causes damage to the flag leaf sheath, which encloses the panicle and hence effecting the emergence of panicle, increasing the chaffy and half-filled grains depending upon the severity. Thus, the yield losses are caused by the infection. The damage caused by the fungus nearly 3-20 per cent losses were in general and sometimes as much as 85 per cent under severe infection (Chen,1957). Severe yield losses were also reported from abroad *viz.*, Taiwan (Chen and Chien, 1964), Philippines (Estrada *et al.*, 1984), Thailand (Surin and Disthaporn, 1977), United States (Shahjahan *et al.*, 1977). Mohan and Subramanian (1977) also reported nearly 6.2 per cent yield loss in

Jagannath, and 57.4 per cent yield reduction in Co39 due to sheath rot. Losses to an extent of 9.6 per cent to 26 per cent by way of reduction in grain yield were reported by Chakrabarty and Biswas (1978). Estrada *et al.* (1984) found 52.8 per cent yield loss due to sheath rot under severe infected conditions. Srinivasan (1980) reported that severe infection of sheath rot had reduced the yield up to 90 per cent under Indian conditions. Enormous reduction in yields were also reported from different states in India (Prabhakaran *et al.*, 1973; Chakrabarty and Biswas, 1978; Murlidharan and Venkatarao, 1980; Reddy and Ghosh, 1985; Manibhushanrao *et al.*, 1986; Lakshmanan *et al.*, 1991 and Reddy, 1991). Thrimurty *et al.* (1980) reported that sheath rot infection increased the chaffy grain percentage in some popular rice grown in Chhattisgarh region. Similar observation was also reported by Srinivasan (1980). In an elaborate study conducted at International Rice Research Institute, Philippines (1981), a positive relation was observed between the quantitative data on flag leaf sheath infection, panicle exscertion, grain filling and discolouration and with that of disease severity. Tikko (1985) and Surin *et al.* (1988) reported that sheath rot infection reduced the number of tillers/panicle, number of grains/panicle and also the 1000-grain weight and yield reduction were directly proportional to the severity. Singh *et al.* (1985) and Dhal *et al.* (1998) reported that yield losses varied from 1.7 to 54.7 per cent in 6 different cultivars under conditions of artificial inoculation in field. Similar reports of reduction in 1000 grain weight, poor grain filling was observed due to sheath rot (Vidhyasekaran *et al.*, 1984). Upadhyay and Diwakar (1984), Sachan and Agarwal (1995) and Jaykumar (1995), also reported grain discolouration due to sheath rot.

Disease Management

Different fungicides were tried by several researchers in order to know their effect on *S. oryzae* in *in-vitro*. Fungicides, Bavistin (Carbendazim), Benlate (Benomyl), Hinosan (Ediphenphos) and Dithane M-45 (Dithiocarbamate) inhibited the growth of *S. oryzae* (Mohan and Subramanian, 1977). Purkayastha and Roychaudhari (1977) tested the efficacy of 6 fungicides *i.e.* Benomyl, Vitavax, MBC, Kitazin, DDUP and Aureofungin against *S. oryzae* and found that among these fungicides MBC was most effective, in reducing the growth and sporulation. Chein and Huang (1979) reported that Bavistin (Carbendazim), Busan (TCMTB) and Benlate (Benomyl) were found effective *in vitro*. Raju and Singh (1981) tested the efficacy of 10 fungicides and found that 3 sprays of Bavistin, Benlate or Brestanol significantly reduced the disease severity. Corroborating with the above reports, Benomyl (Kannaiyan, 1979) and Carbendazim (Raina and Singh, 1980; Vishwanathan and Mariappan, 1980; Alagarsamy and Bhaskaran; 1986 and Singh and Roy, 1987) were also found to be highly effective in reducing the disease severity. Thrimurty (1986) studied on chemical control of sheath rot in Chhattisgarh region and found that Hinosan, Bavistin and Dithane M-45 significantly reduced *S. oryzae* infection as compared to other chemicals tested. Alagarsamy, *et al.*(1986) reported that Captafol at a concentration of 0.125 per cent reduce the disease severity and found to be best among the seven fungicides tested. In contradiction to the above reports, Lewin and Vidhyasekaran (1987) reported that fungicides could inhibit the test fungus in cultural conditions, but failed to reduce the disease severity in field conditions. Vidhyasekaran and Lewin (1987) concluded that sprays of Carbendazim every 3 or 5 days, Captafol at 3, 5 or 10 days intervals, controlled fungus completely. But fungicides sprayed at longer intervals were ineffective. On the contrary Ramabadran *et al.* (1990) stated that disease is difficult to control by chemical application. They also suggested that seed treatment with Benomyl and Panoctine (guazating) is useful to manage this disease. Karmakar *et al.* (1992) tested Carbendazim at 0.1 per cent, protected the plants against *S. oryzae* and reduced the disease severity. Viswanathan and Narayanasamy (1993) evaluated Mancozeb, Edifenphos and Tricyclazole for the control of sheath rot of rice, using 3 rice varieties (IR 20, Co 43 and IR 50). They found that Mancozeb + Tricyclazole gave the best control. Fungicide treatment also increased the number of productive tillers and grain yield. Das *et al.* (1997)

evaluated 6 fungicides (Edifenphos, Thiophonate methyl, Carbendazim, Kitazin, Validamycin and Mancozeb) for the control of sheath rot of rice. Two sprays of Thiophonate methyl or Carbendazim at 0.1 per cent were highly effective in controlling the disease.

Varietals Resistance

Use of resistant varieties in disease management practice is the least expensive, easiest, safest and one of the most effective means. In sheath rot management many cultivars were evaluated by several workers. Chen and Chein (1964) observed some degree of differences in susceptibility to the disease in field conditions in Taiwan. Ahamed *et al.* (1974) observed that Karikalan, Karuna, TKM6 and Sigadis showed high degree of resistance. Amin *et al.* (1974) and Raichoudhari and Purkayastha (1980) found that semi dwarf and dwarf varieties showed more susceptibility than tall cultivars. Generally, Japonica varieties were reported to be less susceptible to sheath rot (Chen and Chein, 1964; Chung, 1975; Chein and Thseng, 1982 and Manuel *et al.*, 1985). In contradiction to the above findings, Fang (1980) and Liang (1980) found Japonica *cvs.* to be more susceptible than Indica *cvs.* Lakshmanan (1991) also screened 87 breeding lines derived from *O. officinalis* and 15 lines showed high level of resistance under artificial inoculated conditions. Reddy and Ghosh (1993) also reported resistant rice to sheath rot. Naik (1976) reported IR-24, IR-26, CR-44-120-1, RP-825-71-4-1, RP-884-81-1 and RP-874-112-1-6 as resistant to sheath rot from Orissa. Amin (1976) screened 243 varieties and 1050 progeny lines. The *cvs.* Tetep, ARC 7117, Ramtulsi, Manoharsali, D-25-4, Sigadis, SR 26B, Intan, Zenith, Dissi Hatif, Raminad str 3 and Tadukan all tall varieties were highly resistant. Shahjahan *et al.* (1977) reported that Bluebella, Labella, Labonnet, Nova 66 and Starbonnet were found as highly susceptible to sheath rot. Kannaiyan *et al.* (1978) screened 130 rice varieties against sheath rot pathogen and found that AS5821, AS6975, AS2992, 1287, 3226 and 6403 were free from sheath rot infection, while 18 were resistant, showing disease incidence ranging from 2 to 20 per cent. Mohan and Subramanian (1972) tested reaction of 16 rices against *S. oyzae*, and reported that infection per cent ranged between 3.6 per cent in Jagannath to 43.6 per cent in Co39. It is reported that Taichung Sen Yu 229 was resistant. Chien and Thseng (1982) screened several cultivars against sheath rot from Taiwan. Estrada *et al.* (1979) tested 3 screening methods for sheath rot resistance evaluation. Raju and Singh (1978) reported that out of 46 varieties inoculated under field conditions Sigadis and Homthong were resistant and URR-171-12 and Ratna were moderately resistant. Mukerjee *et al.* (1980) tested 170 varieties against sheath rot, 5 varieties namely Masuri, Vishnubhog, Gajgor, Kala Namak and Usha, were found resistant to this disease. Among 17 varieties evaluated, Tainung 61 and Japonica varieties showed less per cent of infected panicles, brown grains and grain sterility than Indica varieties (Jian *et al.*, 1982). Purkayastha *et al.* (1983) reported that momilactone was present in higher concentration in tall cultivars resistant to sheath rot and in smaller amounts in susceptible semidwarf cultivars. The antifungal nature of this compound was also confirmed by Thin Layer Chromatography (TLC) and bioassay. Paromita *et al.* (1986) screened 83 germplasm lines against *S. attenuatum*, 7 cultivars were found resistant, 4 from early maturing group and 3 of the late maturing group. Phenylalanine ammonia lyase (PAL) activity decreased gradually upto 7 days in healthy and inoculated plants, but levels were higher in inoculated plants. All enzymes levels increased after gibberellic acid (GA_3) treatment, but decreased in plants treated with Aureofungin or Carbendazim. Shukla *et al.* (1995) evaluated 183 accessions against the major diseases of rice including sheath rot. Each accession was tested for a minimum of 2 seasons. Most accessions showed fairly high tolerance to sheath rot. Kumari *et al.* (1998) tested field tolerance levels of Indica rice varieties for major rice diseases. They reported that rich diversity in cultivated and the traditional rice varieties offered a pool of resistance genes against disease like blast, sheath blight, sheath rot, brown spot, false smut etc. that causes considerable damage to rice crop. Number of researchers (Subramanian and Ramakrishnan, 1975; Rajan and Nair, 1979; Raina *et al.*, 1980; Mukerjee *et al.*, 1981; Mishra and Mathur, 1983; Lewin and Vidhyasekaran, 1987; Singh, 1988; Shashidhar *et al.*, 1988; Nagraju *et al.*, 1991; Karmakar *et al.*, 1992 and Lakshmanan, 1993) also screened different cultivars for sheath rot resistance. Upadhyay and Diwakar (1984) from Chhattisgarh, reported that among several commercially grown rice varieties, Asha, Usha, Safri-17, Dubraj and Madhuri showed some resistance during 1981 and 1982 under field conditions. Sthapit (1991) observed that indigenous germplasm showed good tolerance for sheath rot in Nepal. Velazhahan and Ramabadran (1992) tested the effect of 5 levels of potassium (0,50,100,150 and 200 kg/ha) on the phenolic content of rice, at different stages of pathogenesis. Increasing the potassium level (0-200 kg/ ha.) reduced disease intensity and increased grain yield. The application of potash also increased the phenolic content. Photoperiod insensitive variety Rajavadlu was reported to be resistant by Kashikar *et al.* (1994). Hemlatha *et al.* (1999) tested a crude toxin preparation from *Sarocladium oryzae* as a molecular sieve, using *in vitro* techniques for selection of resistant somaclones of rice. Sakthivel and Gnanamanickam (1986) identified cerulenin as a toxic metabolite produced by *S. oryzae* in rice tissues showing sheath rot symptoms.

Efficacy of Plant Leaf Extracts on *S. oryzae* Growth

Bio-pesticides are becoming popular because of their eco-friendly nature. Botanical product extracts were tried by several researchers in order to know their bio-pesticidal efficacy against wide range of pathogens. On sheath rot causing pathogen also efficacy of botanicals formulations were tested and reported. Narsimhan *et al.* (1993) reported that neem seed kernal extract at a concentration of 5 per cent, when applied as foliar spray at booting stage followed with a repeated application after 10 days reduced the infection and increased yield. Its effect was comparable with that of Carbendazim treatment. In general, *Azadirachtin* formulations were found to be promising at most of the test locations in reducing sheath rot incidence (Dohroo and Gupta, 1995; Eswaramurty *et al.*, 1996 and Sivaprakasam and Jagannathan, 1996). Neem derivatives (neem oil, neem seed kernal extract) and leaf extract of *Vitex negundo, Accacia leucocephala* and *Polyalthia longifolia* reduced sheath rot significantly (Jagannathan and Sivaprakasam, 1996). Jeeva and Ramabadran (1993) screened 23 plants for inhibition of conidial germination of *S. oryzae*. They found that cold water extract of 4 plants (*Arachis hypogea, Caesalpinia pulcherrima, Euphorbia hirta* and *Ipomea crassicaulis*) were inhibitory to sheath rot pathogen and ethanolic extract from *I. crassicaulis* gave the maximum inhibition. Extracts of *Caesalpinia pulcherrima* and *Impomea crassicaulis* also reduced the abnormality incited by sheath rot pathogen in rice (Jeeva and Ramabadran, 1995). Rajappan *et al.* (1997) stated that leaf extracts of *Ipomea* spp. effectively controlled the growth of sheath rot pathogen. Narasimhan *et al.* (1998) stated that Neem oil and Pungam oil based EC formualtion [*Viz.* Neem oil 60 EC (acetic acid), Neem oil 60 EC (citric acid) and Neem oil + Pungam oil 60 EC (citric acid)] were evaluated for their efficacy against sheath rot pathogen. All the 3 formulations efficiently inhibited the mycelial growth of the pathogen *in vitro*. Rajappan *et al.* (1999) found that fresh and stored sample of neem oil based EC formulation (0.225, 0.4 and 0.9 per cent) showed a significant reduction in the mycelial growth of *S. oryzae* at all the 3 concentrations and storage duration (0, 3, 6 and 9 months) of the neem oil in comparison with control. Pramanick and Phookan (1998) reported that among 10 plant extracts tested against *S. oryzae in vitro* an aqueous extract of *Ocimum sanctum* was most effective followed by *Eucalyptus citriodora* and *Azadirachta indica*. Lesions on rice plant (*cv.* IR-36) were also significantly restricted by pre inoculation spraying with these 3 plant extracts. Rajappan *et al.* (2000) also reported that neem oil inhibited the growth of *S. oryzae.* Avdhesh and Satapathy (1977) reported that extract from different plant parts (*viz.* leaf, flower, stem and root) of *Vinca rosea* were found antifungal to *Helminthosporium nodorum, Sclerotium rolfsii, Pestalotia* sp., *Fusarium oxysporum,*

Colletotrichum sp. and *Aspergillus niger*. It inhibited the spore germination, sporulation and mycelial growth of fungi. Mishra and Tiwari (1992) tested the ethanol extract and essential oil of the leaf of *Polyalthia longifolia* against 5 rice pathogen *viz. Pyricularia oryzae, Rhizoctonia solani, Fusarium moniliforme, Aspergillus niger* and *Curvularia lunata*. They found that ethanolic extract was more effective, with broad-spectrum fungitoxicity than essential oil. Several reports were published by different workers on the efficacy of plant products in inhibiting the growth of different plant pathogens of important crops (Kishore and Dubey, 1986; Dubey and Dwivedi, 1991; Shah *et al.*, 1992; Yasmeen and Saxena, 1992; Singh and Tripathi, 1995; Gohil and Vala, 1996; Khan *et al.*, 1996; Shivpuri and Sharma, 1997; Grewal and Kang, 1988; Bansal and Gupta, 2000; Gomathi and Kannabiran, 2000 and Singh and Majumdar, 2001).

References

Agnihothrudu, V. (1973). *Acrocylindrium oryzae* Sawada – sheath rot on paddy. *Kavaka.* 1:69-71.

Ahamed, J.S., Kolandaisamy, S., Kannaiyan, S. and Venkta Rao, A. (1974). Reaction of certain rice cultivars in Tamil-Nadu to bacterial leaf blight and sheath rot. *The Farm Sci.* 1 : 4-6.

Ahamed, N.J., Kannaiyan, S. and Venkata Rao, A. (1975). First record of *Acrocylindrium* sheath rot disease of rice in India. *Science and Culture.* 41 : 544-545.

Ahuja, S.C. (1987). An inventory and bibliography of rice disease in India, HAU Rice Research Station, Karnal, Haryana. p. 150.

Akhter, S., Mian, M. S. and Mia, M. A. T. (2001). Chemical control of sheath blight disease (*Rhizoctonia solani*) of rice. *Bangladesh Journal of Plant Pathology*. **17**: 35-38.

Alagarsamy, G. and Bhaskaran, R. (1986). Effect of time of application of nitrogen and potassium in combination with nutrients and fungicides spray on sheath rot incidence and grain yield of rice. *Philippine Agriculturist.* 69 (1) : 99-105.

Ali Anwar (2000). Recent status of certain rice diseases under field conditions of temperate zone in Kashmir valley. National Symposium of IPS, IARI, New Delhi at Kernal, Haryana pp. 128.

Ali, M. A. and Archer, S. A. (2003). Evaluation of some new fungicides against sheath blight of rice caused by *Rhizoctonia solani*. *Bangladesh Journal of Plant Pathology*. **19**: 13-20.

Amin, K.S. (1976). Sources of resistance to *Acrocylindrium* sheath rot of rice. *Plant Dis. Reptr.* 60 : 72-73.

Amin, K.S., Sharma, B.D. and Das, C.R. (1974). Occurrence in India of sheath rot of rice caused by *Acrocylindrium oryzae*. *Plant Dis. Reptr*. 58 : 358-360.

Anonymous, (2004). Digest of Statistics 203-04. Directorate of Planning and Statistics, Planning and Development Department, Government of Jammu and Kashmir, Srinagar pp. 86-89.

Ansari, M. M. (1995). Control of sheath blight of rice by plant extracts. *Indian Phytopathology*. **48**: 268-270.

Anwar, A., Bhat, G. N. and Singhara, G. S. (2002). Management of sheath blight and blast in rice through seed treatment. *Annals of Plant Protection Sciences*. **10**: 285-287.

Araujo, L. G., Prabhu, A. S. and Silva, G. B. (2006). Single tiller method to estimate damage caused by sheath blight in yield components of rice. *Fitopatologia Brasileira*. **31**: 199-202.

Arunyanart, P., Sruin, A. and Disthaporn, S. (1981). Seed discolouration disease and its chemical control. *IRRN*. 6(3): 14-15

Arunyanart, P., Surin, A. and Disthaporn, S. (1981). Seed discolouration diseases and its chemical control. *International Rice Research Newsletter* 6 : 14-15.

Ashrafuzzaman, M.H., Khan, A.R. and Howlider, A.R. (1994). Effect of plant and seed extract on growth and sclerotial production of *R. solani* causing sheath blight of rice. *Advances in Plant Sciences*. **7**: 407-410.

Avdhesh, S. and Satapathy, J.N. (1977). Antifungal characteristics of *Vinca rosea* extract. *Indian Phytopath*. 30 (1) : 36-40.

Baby, U.I. and Manibhusharrao, K. (1996). Fungal antagonists and VA Mycorrhizal fungi for bio-control of *Rhizoctonia solani*, the rice sheath blight pathogen, pp. 1-9. **In** : *Recent Development in Biocontrol of Plant Pathogens* (Eds. K. Manibhushan Rao and A. Mahadevan). Today and Tomorrow printers and publishers, Karol Bagh, New Delhi.

Balakrishnan, B. and Nair, M.C. (1981). Weed hosts of *Acrocylindrium oryzae* Saw., a sheath rot pathogen of rice. *IRRN*. 6 (6) : 13.

Baldacci, E. and Picco, C.G. (1948). Ricerch sullo mycoflora delle coriosside Li riso dapo conversazione in magazine e in conidiozi experimentally. *Riso* **13** : 79-88.

Bansal, R.K. and Gupta, R.K. (2000). Evaluation of plant extract against *Fusarium oxysporum* wilt pathogen of fenugreek. *Indian Phytopath*. 53 (1) : 107-108.

Bhuyan, S.A., Das, B.C. and Bore, L.C. (1994). Antagonistic effect of *Trichoderma viride*, *Trichoderma harzianum* and *Aspergillus terrus* on *R. solani*. *Journal of Agri. Science Society of North East India*. **7**: 125-127.

Biswas, A. (2006). Efficacy of biotos: a new botanical against sheath blight disease of rice. *Environment and Ecology*. **24** : 484-485.

Biswas, A. and Roy choudhary, U.K. (2003). Relative efficacy of some botanicals against sheath blight of rice. *Journal of Mycopathological Research*. **41**: 163-165.

Boa, E.R. and Brady, B.L. (1987). *Sarocladium oryzae* associated with blight of *Bambusa* sp. in Bangladesh. *Trans. Br. Mycol. Soc*. 89 (2) : 161-166.

Chahal, K.S., Sokhi, S.S. and Rattan, G.S. (2003). Investigations on sheath blight of rice in Punjab. *Indian Phytopathology*. **56**: 22-26.

Chakrabarty, D.K. and Biswas, S. (1978). Estimation of yield loss in rice affected by sheath rot. *Plant Dis. Reptr*. 62 : 226-227.

Chen, C.C. and Chein, C.C. (1964). Some observations of the out break of rice sheath rot disease. *J. Taiwan Agric. Res*. 13 : 39-45.

Chen, M.J. (1957). Studies on sheath rot of rice. *Plant J. Agric. For. Taiwan*. 11 : 1-19.

Chien, C.C. (1980). Studies on sheath rot disease and its relation to sterility of rice. *Plant Prot. Bull. Taiwan*. 21 (1) : 7-16.

Chien, C.C. and Huang, C.H. (1979). The relation between sheath rot and the sterility of rice plant. *J. Agric. Res. China*. 28 : 7-16.

Chien, C.C. and Thseng, F.M. (1982). Rice variety and age in relation to the susceptibility to sheath rot. *J. Agric. Res. China.* 31 : 155-161.

Chung, H.S. (1975). Studies on sheath rot of rice caused by *Acrocylindrium oryzae* (Sawada) testing varietal reaction and culture filtrates of the causal fungus. *Korean J. Pl. Prot.* 14 : 23-27.

Cu, R.M., Mew, T.W., Cassman, K.G. and Teng, P.S. (1996). Effect of sheath blight on yield in tropical, intensive rice production system. *Plant Disease.* **80**: 1103-1108.

Dacthler, W.C. (1959). Research on conditioning and storage of rough and milled rice. Agricultural Research Service, US Department of Agriculture. Report No. ARS 20-27.

Das, B.C. and Hazarika, D.K. (2000). Biological management of sheath blight of rice. *Indian Phytopathology.* **53**:433-435.

Das, B.C., Bora, L.C., Phookan, A.K. and Bhagabati, K.N. (1996). Antagonistic effects of *Aspergillus terrus, Trichoderma harzianum* and *Trichoderma viride* on sheath blight of rice. *Oryza.* **33**: 62-65.

Das, S.R. and Mishra, B. (1990). Field evaluation of fungicides for control of sheath blight of rice. *Indian Phytopathology.* **43**: 93-96.

Das, S.R., Nivedita Nayak and Nayak, N. (1997). Efficacy of fungicides in controlling sheath rot of rice in Orissa. *Crop Res. Hisar.* 14 (3) : 497-500.

Deka, A.K. and Phookan, A.K. (1992). Some common weed hosts of *Sarocladium oryzae* in Assam, India. *IRRN.* 17 (60) : 25.

Deka, A.K. and Phookan, A.K. (1998). Survival of *Sarocladium oryzae* causing sheath rot of rice in acidic soils of Assam. *J. Mycol. Pl. Pathol.* 28 (1) : 55-56.

Deka, B., Ali, M.S. and Chandra, K.C. (1996). Management of grain discolouration of rice. *Indian J. Mycol. Plant Path.,* **26** : 105-106.

Dhal, A., Mohanti, A.K. and Lenka, S. (1998). Occurrence of sheath rot of rice in semi deep water situation in Orissa. *Env. and Ecol.* 16 (1) : 233-234.

Dohroo, N.P. and Gupta, S.K. (1995). Neem in plant disease control. *Agric. Rev.* 16 (3) : 133-140.

Dubey, R.C. and Dwivedi, R.S. (1991). Fungitoxic properties of some plant extract against vegetative growth and sclerotial viability of *Macrophomina phaseolina*. *Indian Phytopath.* 44 : 411-413.

Duraisamy, V.S. (1982). Fungicidal control of rice grain discolouration. *International Rice Research Newsletter* 7 : 14.

Eeckhout, E.V., Rush, M.C., Blackwell, M. and Van- Eeckhout, E. (1991). Effects of rate and timing of fungicide applications on incidence and severity of sheath blight and grain yield of rice. *Plant Disease.* **75**: 1254-1261.

Estrada, B.A., Sanchez, L.M. and Crill, P. (1979). Evaluation of screening methods for sheath rot resistance of rice. *Plant Dis. Reptr.* 63 (11) : 908-911.

Estrada, B.A., Torres, C.Q. and Bonman, J.M. (1984). Effect of sheath rot on some yield components. *IRRN.* 9 (2) : 14.

Eswaramurthy, S., Mariappan, V., Alagianagalingam, M.N., Subramaniam, K.S., Sankaralingam, A., Ebenezer, *e.g.* and Raja, P. (1996). Efficacy of *Ipomea cornea* in controlling rice sheath rot. *International Rice Research Notes.* 21 (1) : 50.

Eusuruoso, O.F., Komolafe, C.O. and Aluko, M.O. (1975). Seed borne fungi of rice (*Oryza sativa*) in Nigeria. *Seed Sci. Technol.* 3: 661-666.

Fang, H.C. (1980). Etiological studies on the sterility of rice plant and its control. *Pl. Prot. Bull.* 22 : 83-90.

Gams, W. and Hawksworth, D.L. (1975). The identity of *Acrocylindrium oryzae* Sawada and a similar fungus causing sheath-rot of rice. *Kavaka.* 3 : 57-61.

Gautam, K., Rao, P.B. and Chauhan, S.V.S. (2003). Efficacy of some botanicals of the family Compositae against *Rhizoctonia solani* Kuhn. *Journal of Mycology and Plant Pathology*. **33**: 230-235.

Ghosh, J.J. (1951). The effect of environmental factors on fungal deterioration of stored rice grains. *Science and Culture* **17**, 42-43.

Ghufran, S.M., Asghar, S.M.A. and Singh, A.P. (1980). Sheath rot spread in Bihar, India. *IRRN.* 5 (5) : 11.

Gill, M.A., Yasin, S.I., Khan, T.Z., Ullah, Q.A. and Butt, M.A. (1993). Sheath rot disease of rice in Punjab, Pakistan. *International Rice Research Notes*. 18 (2) : 23.

Gohil, C. and Vala, H. (1996). Study of effect of some medicinal plant on growth of *Fusarium moniliforme. J. Mycol. Pl. Pathol*. 26 : 110-111.

Gomathi, V. and Kannabiran, B. (2000). Inhibition effect of leaf extract of some plant on the anthracnose fungi infecting *Capsicum annum*. *Indian Phytopath*. 53 (3) : 305-308.

Govindarajan, K. and Kannaiyan, S. (1982). Fungicial control or rice grain discolouration. *International Rice Research Newsletter* **7** : 10.

Grewal, S.K. and Kang, M.S. (1988). Seasonal carryover of *Fusarium moniliforme* Sheld : causal organism of sheath rot of rice. *Phytopathologia Mediterranea*. 27 (1) : 36-37.

Gutierrez, S.A., Mazzanti-de-Castanon, M.A. and Cundom, M.A. (2002). Fungi isolated from rice seed in Argentina. *Fito Patologia* 37(2) : 156-156 [c.f. CAB Abstract, 2002-03].

Hemlatha, R.G., Jebaraj, S., Raja, J.A.J., Raguchander, T., Ramnathan, A., Samiyappan, R. and Balasubramanian, P. (1999). Employing a crude toxin preparation from *Sarocladium oryzae* as a molecular sieve to select sheath rot resistant somaclones of rice. *J. H. Biochem. BioTech*. 8 (2) 75-80.

Jagannathan, R. and Sivaprakasam, K. (1996). Effect of botanicals on managing sheath rot of rice. *IRRN*. 21 (1) : 49-50.

Jayakumar, G. (1995). Studies on rice (*Oryza sativa* L.) seed health evaluation for important pathogens in Chhattisgarh region. *M.Sc. (Ag.) Thesis*, Indira Gandhi Agricultural University, Raipur, C.G.

Jayraj, J., Yi, H., Liang, G.H., Muthukrishan, S. and Velazhahan, R. (2004). Foliar application of *Bacillus subtilis.* $AUBS_1$ reduces sheath blight and triggers defense mechanism in rice. *Zeitschrift-fur-pflanzenkrankheiten-und-pflanzenschutz*. **111**:115-125.

Jeeva, M.L. and Ramabadran, R. (1993). Effect of different extracts of non-host plants. *Madras Agric. J.* 80 (10) : 598-600.

Jeeva, M.L. and Ramabadran, R. (1995). Effect of plant extract on rice seed germination and seedling abnormalities caused by *Sarocladium oryzae*. *Madras Agric. J.* 82 (50) : 402-403.

Jian, J.Z. (Chien, C.C.), Zeng, F.M, (Thseng, F.M.), Chien, C.C. and Thseng, F.M. (1982). Rice variety and age in relation to susceptibility to sheath rot. *J. Agric. Res. China.* 31 (2) : 155-161.

Jones, M.P., Jeutong, F. and Tchatchoua, J. (1993). A survey of rice disease in Cameroon. *Plant dis.* 77 (2) : 133-136.

Kang, M.S. and Rattan, G.S. (1983). Sheath rot in the Punjab, India. *IRRN.* 8 (3) : 7-8.

Kannaiyan, S. (1979). Chemical control of sheath rot disease of rice. *IRRN.* 4 (3) : 14-15.

Kannaiyan, S., Jayaraman, V. and Palaniyandi, V.G. (1978). Reaction of certain rice cultivars to sheath rot disease. *IRRN.* 3 (1) : 7.

Karmakar, S.K., Mishra, S.K., Mohapatra, K.B. and Mishra, B. (1992). Response of rice cultivars against sheath rot disease caused by *Sarocladium oryzae* (Sawada) W. Gams and D. Hawksw, under protected and unprotected condition. *Orissa J. Agric. Res.* 5: 209-214.

Kashikar, M., Nanda, N.V., Hasan, V. and Kulkarni, N. (1994). Rajavadlu and sagar-samba released in the Andhra Pradesh, India. *IRRN.* 19 : 3-18.

Khan, A.A. and Sinha, A.P. (2005). Influence of different factors on the effectivity of fungal bio agents to manage rice sheath blight in nursery. *Indian Phytopathology*. **58**: 289-293.

Khan, M.I., Ashaq, M., Bhat, M.Y. and Fazal, M. (1996). A comparative study of the efficacy of some plant extract on mycoflora of chickpea seeds. *Ann. Pl. Prot. Sci.* 4 (1) : 85-87.

Kim, C.K. (1994). Blast management in high input, high yield potential, temperature rice ecosystems. **In** : *Rice Blast Disease*(Eds., Zeigler, R.S., Leong. S. and Teng, P.S). pp. 451 – 463. CAB International Wallingford, U.K.

Kishore, V. and Dubey, N.K. (1986). Volatile fungitoxicity in some higher plant as evaluated against *Rhizoctonia solani* and some other fungi. *Indian Phytopath*. 36 : 724-726.

Kotasthane, A.S. and Lokpale, N. (1997). Antifungal activity of some leaf extracts on the growth and sclerotial production of *R. Solani*. *Microbial Biotechnology*. 143-144.

Krishnaiah, N.V. and Reddy, C.S. (1992). Compatibility and cotoxicity of selected insecticides and fungicides against brown plant hopper and sheath blight in rice. *Indian Journal of Plant Protection*. **20**: 99-101.

Kumari, S.L., Joseph, T., Nayar, N.K. and Niza, T.J.R. (1998). Field tolerance level of indica rice varieties for major diseases of Kerala. *Crop Res. Hisar*. 15 (1) : 112-114.

Laha, G.S. and Venkataraman, S. (2001). Sheath blight management in rice with biocontrol agents. *Indian Phytopathology*. **54**: 461-464.

Lakshmanan, P. (1992). *Brovennia rehi* Lindinger, vector for the sheath rot pathogen. *IRRN.* 17 (2) : 23.

Lakshmanan, P. (1993). Efficient method of assessing resistance to sheath rot in rice. *Crop Prot*. 12 (3) : 189-192.

Lakshmanan, P., Kumar, S.M. and Indira, K. (1991). Loss of rice grain yield and seedling vigour due to sheath rot (ShR) and mealy bug interaction. *IRRN.* 16 (6) : 27.

Lewin, H.D. and Vidhyasekaran, P. (1987). Controlling sheath rot (ShR) in rice. *IRRN.* 12 (1) : 17.

Lewin, H.D. and Vidhyasekaran, P. (1987). Sources of resistance to sheath rot (ShR). *IRRN.* 12 (1) : 7.

Li, X.M., Hua, J.L., Song, A.Z. and Quan, M.F. (1996). Studies on yield loss and dynamic economic threshold of sheath blight on hybrid early rice. *Acta Phytophylacica Sinica*. **23**: 223-228.

Liang, W.J. (1980). Rearing method for tarsonemid mites (*Steneotarsonemus spinki*) free of rice sheath rot (*Acrocylindrium oryaze*). *Pl. Port. Bull. Taiwan*. 22 : 23-30.

Malavolta, V.M.A. and Bedendo, T.A. (2004). Efficacy of different fungitoxicants against rice grain discolouration. *J. Plant Path.*, **86** : 290-295.

Malavolta, V.M.A. and Takada, H.M. (1997). Chemical control of fungi causing rice grain discolouration. *Summa Phytopathologica* **23**(1) : 25-28 [c.f. CAB Abstract 1996-98).

Manibhushanrao, K. (1996). Rice Blast Disease. Daya Publishing House, Delhi. 180.

Manibhushanrao, K., Sreenivasa, S. and Chitralekha, R.S. (1986). Sheath rot disease of rice. *Zeitshriftfur Pflazenkrankheiten- und -Pflanzen Schutz*. 93 (3) : 319-329.

Manuel, W.W., Ganesan, K., Rajagopalan, C.K. and Mariappan, V. (1985). New source of resistance to major rice diseases. *IRRN*. 10 (1) :7.

Mazzanti-Di-Castanon, M.A. and Gutierrez-Di-Arriola, S.A. (1993). Sheath rot of rice *Sarocladium oryzae* in Argentina. *Fito Pathologia*. 28 (2) : 70-76.

Mbodj, Y. (1989). Research on the pathology of rice in Senegal. *I.R. commission Newsl.* 38 : 52-56.

Merca, S.D. and Mew, T.W. (1987). Movement of pests and control strategies. IRRI seed health units operation on rice. 241-251

Mew, T.W. and Mishra, J.K. (1994). *A Manual of Rice Seed Health Testing*. IRRI, Los Banos, Laguna, Philippines, p. 1-113.

Mishra, A.K. and Mathur, S.C. (1983). Sheath rot incidence in cultivars grown at different nitrogen levels. *IRRN*. 8 (2) : 5.

Mishra, D.S. and Sinha, A.P. (1991). Laboratory evaluation of fungicides against *Rhizoctonia solani* Kuhn, the cause of sheath blight of rice. *Agricultural Science Digest*, Karnal. **19**: 211-213.

Mishra, M. and Tiwari, S.N. (1992). Toxicity of *Polyalthia longifolia* against fungal pathogen of rice. *Indian Phytopath*. 46 : 321-323.

Misra, B. (2005). More crop per drop, pp. 41-46. **In** : The Hindu Survey of Indian Agriculture, Kasturi and Sons Ltd. Chennai.

Miyake, I. (1910). Studien aber die Pilze der Reispflanze in Japan, pp 1-101. *Sheath blight of rice*. **In**: *Sheath blight of rice* (Ed. K. Manibhushanrao). Trinagar, Delhi.

Mohan, R. and Subramanian, C.L. (1972). Reaction of rice varieties to and yield losses from sheath rot. *IRRN*. 2 (6) : 7.

Mohan, R. and Subramanian, C.L. (1977). The effect of certain fungicides on rice sheath rot. *IRRN*. 2 (6) : 14.

Mukerjee, P., Singh, B.P. and Paromita Mukerjee (1980). Testing of indigenous rice germplasms against sheath rot by artificial inoculation. *Indian Phytopath*. 33 (1) : 149.

Mukerjee, P., Singh, B.P. and Rahman, F. (1981). Testing of indigenous germplasms of rice against sheath rot by artificial inoculation. *Indian Phytopath*. 34 (3) : 287-290.

Murlidharan, K. and Venkata Rao, G. (1980). Out break of sheath rot on rice. *IRRN*. 5 (5) : 7.

Nagraju, R., Vasantha Kumar, H.L., Devaiah, B.M., Seshadri, V.S. and Naidu, B.S. (1991). Evaluation of rice genotypes for blast and sheath rot resistance in hilly regions of Karantaka. *Mysore J. Agric. Sci*. 25 (2) : 139-141.

Naik, R. (1976). Incidence of sheath rot in rice a potential problem for Sambalpur, Orissa. *IRRN*. 1 (1) : 19.

Naik, R. and Roy, J.K. (1975). Occurrence of sheath rot of rice, in Sambalpur. *Rice Path. Newsl*. 2:6.

Nair, M.C. and Rajan, P.K.S. (1978). Sheath rot of rice. *Agric. Res. J. Kerala*. 13 : 105-106.

Naito, H. (1953). Studies on mold microflora in stored rice grains, with special reference to osmotic pressure. Phytopath. Society of Japan **18**, 41-45.

Narasimhan, V., Rajappan, K., Ushamalini, C. and Kareem, A.A. (1998). Efficacy of new EC formulations of neem oil and pungam oil for the management of sheath rot disease of rice. *Phytoparasitica*. 26 (4) : 301-306.

Narasimhan, V., Sridhar, V.V. and Kareem, A.A. (1993). Efficacy of botanicals in managing sheath rot (ShR) of rice. *International Rice Research Notes*. 18 (3) : 33.

Ou, S.H. (1985). Rice diseases 2nd ed. Commonwealth Mycological Institute, Kew. Surrey, U.K. 368-380.

Padamanabhan, S.Y. (1949). Occurrence of fungi inside red kernels. *Current Science* **18** : 442-443.

Paracer, C.S. and Chahal, D.S. (1963). Sheath blight of rice caused by *Rhizoctonia solani* Kuhn, a new record in India. *Current Science*. **32**: 320-329.

Paromita Mukerjee, Singh, B.P., Singh, H.N. and Mukerjee, P. (1986). Response of indigenous rice cultivars against sheath rot under artificial inoculation. *J. Mycol. Pl. Pathol.* 16 (3) : 307-308.

Pathak, M., Garg, D.K., Jeswani, M.D. and Singh, R.N. (2002). Farmers participatory IPM in Basmati rice – a case study. *Annals of Plant Protection Sciences*. **10**: 386-388.

Peng, H.J. and Tang, D.Y. (1990). Studies on the suitable period and economic threshold for chemical controlling and epidemic dynamics of sheath blight in rice. *Acta Phytopathologica Sinica*. **20**: 153-158.

Pizzatti, C., Pedrali, D and Cortesi, P. (2004). Lack of correlation between incidence of rice grain discolouration and panicle brown spot. *J. Plant Path.*,**86** : 295-340.

Prabhakaran, J., Ragunathan, V. and Prasad, N.N. (1973). Occurrence of sheath rot of rice caused by *Acrocylindrium oryzae* Sawada. *Annamalai Univ. Agric. Res. Ann*. 4 : 182-183.

Pramanick, T.C. and Phookan, A.K. (1998). Effect of plant extract in the management of sheath rot of rice. *J. Agric. Sci. Soc. North East India*. 11 (1) : 85-87.

Purkayastha, R.P. and Ghosal, A. (1982). Sheath rot of rice. *Oryza*. 19 (2) : 78-83.

Purkayastha, R.P. and Roychoudhari, S. (1977). Studies on the effect of some systemic compounds on *Acrocylindrium oryzae* in relation to control of sheath rot disease of rice. *Indian J. Mycol. Res*. 15 : 27-32.

Purkayastha, R.P., Ghosal, A. and Biswas, S. (1983). Momilactone synthesis associated with differential resistance of tall and semidwarf rice cvs. to sheath rot disease. *Indian Phytopath*. 36 (2) : 322-327.

Qasem, J.R. and Abu-Blan, H.A. (1996). Fungicidal activity of some common weed against different pathogenic fungi. *Journal of Phytopathology*. **29**: 336-339.

Rahman, M.M., Shajahan, A.K.M. and Miah, S.A. (1982). *Echinochloa colona*, an alternate host of *Sarocladium oryzae* causing sheath rot of rice. *IRRN*. 7 (50) : 16-17.

Raichaudhari, S. and Purkayastha, R.P. (1980). Experimental observations on rice sheath rot disease. *IRRN*. 5 (4) : 13.

Raina, G.L. and Singh, G. (1980). Sheath rot outbreak in the Punjab. *IRRN*. 5 (1) : 16.

Raina, G.L., Gurjit Singh and Sidu, G.S. (1980). Sources of resistance to major rice disease in the Punjab, India. *IRRN*. 5 (3) : 6-7.

Rajan, K.M. and Nair, P.V. (1979). Reaction of certain rice varieties to sheath blight and sheath rot disease. *Agric. Res. J. Kerala*. 17 (2) : 259-260.

Rajappan, K., Mariappan, V. and Kareem, A.A. (1997). Effect of dried leaf extract of *Ipomea* on rice sheath rot pathogen and beneficial micro-organism. *Indian Phytopath*. 50 (3) : 329-331.

Rajappan, K., Mariappan, V. and Kareem, A.A. (1999). *In vitro* evaluation of a salvent free EC formulation of neem oil against the rice sheath rot *Sarocladium oryzae*. *Ann. Plant Prot. Sci.* 7 (1) : 101-103.

Rajappan, K., Mariappan, V. and Kareem, A.A. (2000). Effect of salvent free EC formulation of neem oil on beneficial micro-organism and fungal pathogen of rice. *Ann. Plant. Prot. Sci*. 8 (1) : 98-100.

Raju, C.A. and Singh, R.A. (1981). Studies on sheath rot of rice II : Chemical control. *Pesticides*. 15 (3) : 26-28.

Ramabadran, R., Velazhahan, R. and Vidhyasekaran, P. (1990). Management of sheath rot of rice. Basic Research for Crop Disease Management. 367-375.

Ramakrishnan, T.S. (1971). *Disease of rice*. Indian Council of Agriculture Research, New Delhi. p. 150.

Ray, P. (1993). Fungicidal control of rice grain discolouration. *Oryza* **30** : 367-369.

Reddy, A.P.K. and Reddy, C.S. (1986). *Present status of sheath blight disease and its control*, pp 118-127. **In:** Diamond Jubilee Souvenir 1925-85, ARS, Maruteree (APAU).

Reddy, A.P.K., Mich, S.A. and Mew, T.W. (1986). Disease resistance in rainfed low land rice. Progress in rainfed low land rice. Manila, Philippines, IRRI. 253-262.

Reddy, C.S. and Ghosh, A. (1985). Sheath rot incidence and yield losses in rice due to the joint infection of rice tungro virus and sheath rot fungus. *Indian Phytopath*. 38 (1) : 165.

Reddy, C.S. and Ghosh, A. (1993). Rice resistance to sheath rot identified. *Rice Biotech*. Q. 16.

Reddy, E.J.S., Reddy, P.C., Chary, M.P. and Reddy, S.M. (1984). *In-vitro* production of cellulolytic and pectolytic enzymes by *Acrocylindrium oryzae* an incitant of sheath rot of rice. *J. Mycol. Pl. Pathol*. 14 (3) : 213-217.

Reddy, K.S. (1991). Effect of sheath rot disease on rice panicle and other yield components. *Indian Agriculturist*. 35 (4) : 265-267.

Roy, A.K. (1993). Sheath blight of rice in India. *Indian Phytopathology*. **46** : 197 – 205.

Sachan, I.P. and Agrawal, V.K. (1995). Seed discoloration of rice : location of inoculum and influence on nutritional value. *Indian Phytopath*. 48 (1) : 14-20.

Saifulla, M. (1996). A new sheath rot disease of rice identified in Karnatka. *Agric. Sci. Digest Karnal.* 16 (3) : 149-150.

Saikia, L. and Baruah, B.P. (1990). Linear model for predicting yield loss in rice (*Oryza sativa*) due to sheath blight (*Rhizoctonia solani*) incidence. *Indian Journal of Agricultural Sciences*. **5**: 360-361.

Saikia, R., Kumar, R., Arora, D.K., Gogoi, D.K. and Azad, P. (2006). *Pseudomonas aeruginosa* inducing rice resistance against *Rhizoctonia solani*: Production of salicyclic acid and peroxidases. *Folia Microbiologica*. **51**: 375-380.

Sakthivel, N. and Gnanamanickam, S.S. (1986). Isolation of and assay for cerulenin produced by sheath rot pathogen *Sarocladium oryzae* (Saw.) Gams. *Curr. Sci.* 55 (19) : 988-989.

Santos, G.R., Dos, Lima Neto, A.F. and Didonet, J.C. (2000). Chemical control of up land rice diseases in the state of Tocantins. *Revesta cere* **47**(269) : 1-7 (c.f. CAB Abstract, 2000).

Sarkar, M.K., Sharma, B.D. and Gupta, P.K.S. (1991). Chemical control of sheath blight of rice caused by *Rhizoctonia solani*. *Journal of Mycopathological Research*. **29**: 161-163.

Schroeder, H.W. (1964). Sodium propionate and infrafed drying for control of fungi infecting rough rice (Oryza sativa). *Ibid* **54**, 858-862.

Shah, N.H., Khan, M.I. and Azam, M.F. (1992). Seed mycoflora of cowpea and its control with extract of *Argemone maxicana*. *Bioved*. 3 (2) : 167-168.

Shahjahan, A.K.M., Harhap, Z. and Rush, M.C. (1977). Sheath rot of rice caused by *Acrocylindrium oryzae* in Lauisiana. *Plant Dis. Reptr*. 61 : 307-310.

Sharma, N.R. and Teng, P.S. (1990). Effect of rice growth stages on sheath blight (Sh.B) development and yield loss. *International Rice Research Newsletter*. **15**: 19-20.

Sharma, O.P., Vaid, A. and Dubey, R.C. (1997). Mycoflora of discoloured rice grain and its pathogenic potential in Himachal Pradesh. pp. 273-282. In: *Himalayan Microbial Diversity Part-I* (Eds. S.C. Sati and J. Sexana). Today and Tomarrow's Printers and Publishers, New Delhi.

Sharma, R., Phookan, A.K., Bhagabati, K.N. and Sharma, R. (1999). Efficacy of some plant extracts in the management of sheath blight disease in rice. *Journal of Mycology and Plant Pathology*. **29**: 336-339

Shashidhar, H.E., Hittalmani, S., Kumar, H.L.V. and Shivashankar, G. (1988). Screening new rice selections for reaction to major diseases. *IRRN*. 13 (4) : 10.

Shukla, V.D., Chauhan, J.S., Variar, M., Maiti, D., Chauhan, V.S. and Tomar, J.B. (1995). Reaction of traditional rainfed rice accessions to brown spot, blast and sheath rot diseases. *Indian Phytopath.* 48 (4) : 433-435.

Sinclari, J.B. (1981). Fungicides sprays for the control of seed borne pathogens of rice, soybeans and wheat. *Seed Science and Technology* **9** : 697-705.

Singh, A.P., Asghar, S.M.A. and Ghufran, S.M. (1985). Assessment of yield loss due to sheath rot disease of rice. *Indian Phytopath*. 38 (4) : 753-755.

Singh, J. and Majumdar, V.L. (2001). Efficacy of plant extract against *Alternaria alternata*, the incitant of fruit rot of pomegranate. *J. Mycol. Pl. Pathol*. 31 (3) : 346-349.

Singh, J. and Tripathi, N.N. (1995). Fungitoxicity of some plants against *Fusarium oxysporum* causing wilt disease in lentil. *Rev. Pl. Pathol*. 75 (4) : 308.

Singh, K.D., Bhattacharya, H.C. and Pathak, A.K. (1996). Nature of perpetuation of sheath rot (*Sarocladium oryzae*) in rice seeds and its control. *J. Agric. Sci. Soc. North East India*. 9 : 76-78.

Singh, R. and Sinha, A.P. (2005). Influence of time of application of *Pseudomonas fluorescens* in suppressing sheath blight of rice. *Indian Phytopathology*. **58**: 30-34.

Singh, R. and Chand, H. (1995). Rice discolouration and its chemical control. *International Rice Research Newsletter* 10 : 16.

Singh, R.A. and Raju, C.A. (1981). Studies on sheath rot of rice. *IRRN*. 6 (2) : 11.

Singh, R.N. and Khan, A.T. (1988). Multiple resistance to bacterial blight (BB) and four fungal diseases. *IRRN*. 13 (2) : 9.

Singh, R.S. (1990). *Plant Diseases*. 6th Edition, Oxford and IBH Publishing Co. Pvt. Ltd., New Delhi. pp 444 - 447.

Singh, R.S. (1998). *Plant Diseases*. 7th Edition, Oxford and IBH Publishing Co. Pvt. Ltd., New Delhi. pp 489-492.

Singh, S.A. and Roy, L.G. (1987). Some observations on out break of a few disease of rice in Manipur. *Pl. Prot. Bull. India*. 39 (4) : 20-21.

Sivaprakasam, K. and Jagannathan, R. (1996). Effect of neem derivatives on sheath rot in rice. *International Rice Research Notes*. 21 : 76.

Srinivasan, S. (1980). Yield losses caused by sheath rot. *IRRN*. 5 (2) : 4.

Sthapit, B.R. (1991). Comparative performance of indigenous rice varieties for cold tolerance in the hills of Nepal. *IRRN*. 16 (6) : 11-12.

Subramanian, C.L. and Ramakrishnan, G. (1975). Sheath rot of rice. *FAO Plant Prot. Bull*. 23 : 49.

Suparyono, R. (1990). The potentials of biological control for fungal disease of rice in Indonesia. *Indonesia Agric. Res. and Dev. J.* 12 (1) : 1-5.

Surin, A. and Disthaporn, S. (1977). 'Rice abortion' or sheath rot a serious rice disease in Thailand. *IRRN*. 2 (5) 16-17.

Surin, A., Arunyanart, P., Dhitikiattipong, R., Rajanahusdin, W., Disthaporn, S. and Soontrajarn, K. (1988). Rice yield loss to sheath rot (ShR). *IRRN*. 13 (6). 34-35.

Tang-Jia-Bin., Ma, B.T., Wang, L.X., Li, P. and Chen, H. (2001). Biocontrol of *Rhizoctonia solani* with *Trichoderma* spp. *Chinese Rice Research Newsletter*. **9**: 4.

Tasleem-uz-Zaman Khan, Gill, M.A. and Khan, M.G. (2000). Seed borne fungi of rice from central Punjab and their control. *Pakistan J. Phytopath.*, **12**(1) : 12-14.

Tasugi, H. and Ikeda, Y. (1956). Studies on the sheath rot of rice plant caused by *Acrocylindrium oryzae. Rev. Appl. Mycol.* 37 : 162.

Thrimurty, V.S. (1986). Studies on chemical control of sheath rot of rice. *Pesticides*. 9 : 20-21.

Thrimurty, V.S., Veda, O.P., Satpute, R.G. and Kashyap, R. (1980). Sheath rot incidence and chaffy grain percentage of some popular rice. *IRRN*. 5 (5) : 7.

Tikoo, M.L. (1985). Effect of sheath rot (*Acrocylindrium oryzae*) on some yield component of rice. *Res. and Dev. Reptr*. 2 (2) : 24-25.

Tran-Vy, Truong-Van-Cham and Bio-Duy-Tam (1958). Experimental researcher on the toxicity of rice infected by *P.eitrinium* Thom. Proceeding of the symposium on phytochemistry, Kuala Lampur, 1957, 178-180.

Tullis, S.S. (1936). Fungi isolated from discoloured rice kernels. *Technical Bulletin*, USDA 540 : 12.

Upadhyay, R.K. and Diwakar, M.C. (1984). Sheath rot (ShR) in Chhattisgarh, Madhya Pradesh, India. *IRRN*. 9 (5) : 6.

Uruguchi, I. (1942). Pharmacological studies of fungi on a toxin formed in the yellowed rice (*Penicillium* sp.) report No. 1 (Sa).

Vaid, A., Sharma, O.P. and Kaushik, R.P. (1994a). Effect of grain discolouration

Velazhahan, R. and Ramabadran, R. (1992). Influence of potassium fertilization on sheath rot incidence and phenolic content of rice. *Madras Agric. J.* 79 (5) : 294-298.

Vidhyasekaran, P., Ranganathan, K., Rajamanicikam, B. and Radhakrishan, J. (1984). Quality of rice grains from sheath rot affected plants. *IRRN*. 9 (1) : 19.

Viji, G., Manibhushanrao, K. and Baby, U.I. (1997). Non – target effect of systemic fungicides on antagonistic microflora of *Rhizoctonia solani*. *Indian Phytopathology*. **50**: 324-328.

Viswanathan, R. and Narayanasamy, P. (1993). Chemical control of sheath rot of rice. *Madras Agric. J.* 80 (8) : 442-446.

Viswanathan, V. and Mariappan, V. (1980). Effect of seed treatment with fungicides in improving germination of sheath rot infected seeds. *IRRN*. 5 (6) : 12.

Yang, Y.G., Huang, M.J., Feng, J.S. and Yang, S.H. (2002). Effects of adjusting sowing and transplanting time of early rice (*Oryza sativa*) on decrease of diseases and insect pests. *Fujian Agricultural Science and Technology*. **5**: 23-24.

Yasmeen, S. and Saxena, S.K. (1992). Effect of fern extracts on seed mycoflora of chickpea. *Seed Res.* 20 (2) : 170-171

Zhuge, G.H., Lapis, D.B. and Chuke, K.C. (1985). Studies on rice sheath rot, pathogenicity of the pathogen and the factors affecting the disease. *Acta Phytopathologica Sinica*. 15 (1) : 1-8.

Zhuge, G.H., Lapis, D.B. and Chuke, K.C. (1985). Studies on rice sheath rot disease II. On the physiology of the pathogen *Sarocladium oryzae* (Sawada) Gams and Hawksworth. *Acta Phytopathologica Sinica*. 15 (2) : 81-86.

Hill Agriculture: Prospects, Constraints and Mitigations (2013) *Pages* **417–441**
Editor: **Dr. Shahid Ahamad**
Published by: **DAYA PUBLISHING HOUSE, NEW DELHI**

Chapter 27

Biotic Stresses in Maize and their Integrated Management in Hill Ecosystem

Shahid Ahamad[1], Naushad Alam[2] and M.H. Chsti[3]
[1]*Sher-e-Kashmir University of Agricultural Sciences and Technology of Jammu, K.V.K., SKUAST – J., Poonch – 185 131*
[2]*CSAUAT-Kanpur, KVK, Fatehpur*
[3]*RARS, SKUAST – J., Rajouri – 185 131*

In Jammu and Kashmir, maize being the staple food of intermediate zone is grown in 3 m ha with an acreage of approximately 2 m ha in Jammu division. The district wise area in Jammu division under this crop in Udhampur, Doda, Rajouri, Jammu, Poonch and Kathua is 55.58, 48.19, 44.89, 26.77, 23.67 and 15.89 thousand hectares, respectively (Directorate of Economics and Statistics J&K, 1999).

In recent years, the average yield of maize has declined and this is accounted by biotic, abiotic and nutritional stresses. Among the different stresses, biotic factors are the major constraints in the production and productivity of maize (Desai and Hegde, 1990). Due to the attack of several plant diseases caused by fungi, bacteria, viruses and possibly nematodes, yields are often reduced considerably (Rangarajan and Chakrabarty, 1969). Diseases are important constraints to maize production, wherever the crop is grown. The importance and magnitude of damage depend on interaction between the host plant and the pathogen under a given set of environmental conditions (Ahamad and Narain, 2007; Ahamad 2009 and Ahamad, 2012). Losses due to corn diseases are estimated at 9.4 per cent per year worldwide, 12 per cent in Asia and 14 per cent in Africa (Cramer, 1967). On a country basis, losses are estimated at 12 per cent in the USA and 13.2 per cent in India (Payak and Sharma, 1985). Several diseases, under favourable conditions, have the potential to completely destroy the crop or to render the produce unfit as human food or animal feed. The present findings are in accordance with Shurtleff (1980), Ullstrup (1977), De Leon (1984), Payak and Sharma (1982).

The incidence and economic impact of corn diseases in India vary from state to state (Table 27.1). A large number of pathogens infest maize in the tropical lowland, subtropical, intermediate transition zone and high land maize growing ecological zones covered by CIMMYT's mandate. In 1988, a study was conducted by CIMMYT where information was compiled on maize production along with perceived constraints for each of these defined zone (Tables 27.2 and 27.3). An attempt has been made to estimate losses caused by disease outbreaks and incidence in the crop in terms of shelled grain. Potential loss that can be caused by diseases has been worked out individually for 10 diseases (Table 27.4). Based on experimental work and other observations, estimates have been prepared for India as a whole (Table 27.5).

Table 27.1: Distribution of Maize Diseases in India

Diseases	*India*	*J.&K.*
Viruses		
SCMV/MDMV	+	
MStpV	+	
Downy Mildew		
Sorghum (*P. sorghi*)	+++	
Java (*P. maydis*)	+	
Philippines (*P. phillipinensis*)	+	
Sugarcane (*P. sacchari*)	++	
Brown stripe (*P. rayssiae*)	+++ +++	
Foliar Diseases		
Northern leaf blight	+++	+++
Southern leaf blight	+++	+++
Diplodia leaf blight	+	
Brown spot	+++	+++
Curvularia leaf spot	++	
Gray leaf spot	+	
Banded leaf and sheath blight	+++	+++
Diseases	*India*	*J.&K.*
Rust and Smuts		
Common rust	+++	+
Common smut	+	+++
Head smut	+	+++
Stalk/Root Rots		
Bacterial	+++	
Diplodia	+	
Fusarium	+++	+++
Charcoal	+++	
Pythium	+++	+++
Black bundle/late wilt	++	
Ear and Kernel Diseases		
Rhizoctonia	+++	
Fusarium	+++	+++
Aspergillus	++	++

Legend: +: Not important; ++: Moderately important; +++: Important.

Table 27.2: Percentage of Total Area in Tropical Lowland Maize with Reported Economic Losses to Diseases (Worldwide)

Diseases	*Per cent Losses*	*Diseases*	*Per cent Losses*
Maydis leaf blight (*B. maydis*)	25.5	Polysora rust (*P. polisora*)	33.3
Downy mildew (all genera and spp.)	30.5	Maize streak virus	17.6
Ear rots (*Fusarium* spp., *Stenocarpella.* spp, *M. phaseoli, C. maydis, A.strictus*)	26.7	Corn stunt complex (CSS, MBS, MRFV)	17.6

Source: CIMMYT Maize Program, 1988. Maize production region in developing countries, Mexico. D.F.

Table 27.3: Percentage of Total Area in Sub-tropical, Mid Altitude, Transition Zone and High Land Maize with Reported Economic Losses to Diseases (Worldwide)

Diseases	*Per cent Losses*
Turcicum leaf blight (*E. turcicum*)	58.6
Common rust (*P. sorghi*)	34.2
Downy mildew (all genera and spp.)	29.4
Maize streak virus	15.5
Corn stunt complex (CSS, MBS, MRFV)	1.8
Ear rots (*Fusarium* spp., *Stenocarpella* spp., *Aspergillus* spp.)	26.7
Stalk rots (*Fusarium* spp., *Stenocarpella* spp., *M. phaseolina, C. maydis, A. strictus, Pythium* spp. and *E. chrysanthemi*)	32.1

Source: CIMMYT Maize Program, 1988. Maize production region in developing countries, Mexico. D.F.

Table 27.4: Yield Losses by 10 Diseases Under Experimental Conditions in Susceptible Cultivars of Maize

Diseases	*Per cent Losses*	*Diseases*	*Per cent Losses*
Bacterial stalk rot	100	Pythium stalk rot	100
Late wilt	50.9	Charcoal rot	39.5
Maydis leaf blight	30.3	Turcicum leaf blight	66.0
Common rust	32.0	Brown spot	27.0
Brown stripe downy mildew	63.0	Banded leaf and sheath blight	40.5

Table 27.5: Estimation of Losses Due to Major Diseases of Maize in India

Diseases Group	*Per cent Losses*
Seed and seedling blight	0.1
Downy mildews	2.1
Foliar diseases (Turcicum leaf blight, Maydis leaf blight, Phaseospharia leaf spot, rust, Brown stripe downy mildew)	5.0
Stalk rots, root rots, ear rots	5.0
Sheath blight, smut	1.0
Viruses and other	13.2

Major Diseases and their Distribution

About 72 pathogens have so far been reported causing substantial damage in maize, 16 have been identified as major ones and are given here. These includes diseases which occur in both tropical and temperate environments.

1. Turcicum leaf blight (*Exherohilum turcicum*) occurs in Jammu and Kashmir (J&K), Himachal Pradesh (H.P.), Sikkim, West Bengal (W.B.), Meghalaya, Tripura, Assam, Rajasthan, Uttar

Pradesh (U.P.), Bihar, Madhya Pradesh (M.P.), Gujarat, Rajasthan, Andhra Pradesh (A.P.), Karnataka and Tamil Nadu (T.N.).

Table 27.6: Estimation of Losses Due to Major Diseases and Grain Loss of Maize in India

Diseases Group	*Per cent Losses*	*Annual Grain Loss (tones)*
Seed and seedling blights	0.1	7,923.5
Downy mildews	2.1	166,395.5
Foliar diseases(Turcicum leaf blight, Maydis leaf blight, Phaeosphoria leaf spot, Rust, Brown stripe downy mildew)	5.0	396.180
Stalk rots, root rots, ear rots	5.0	396,180
Sheath blights, smuts, Viruses and others	1.0	79.236
Total	13.2	1,045,915

2. Maydis leaf blight (*Drechslera maydis*), occurs in J&K, H.P., Sikkim, Meghalaya, Rajasthan, Punjab, Haryana, Delhi, U.P., Bihar, M.P., Gujarat, Rajasthan, A.P. and Maharashtra.
3. Common rust (*Puccinia sorghi* II, III stages only) occurs in J&K, H.P., Sikkim, W.B. Rajasthan, Punjab (winter), Haryana (winter), Delhi, U.P., Bihar, M.P., Gujarat, A.P., Maharashtra and T.N.
4. Phaseospheria leaf spot (*Phaseospheria maydis*) is found in Sikkim, W.B., U.P., Karnataka and Maharashtra.
5. Erwinia stalk rot (*Erwinia chrysanthemi* pv. *zeae*) occurs in H.P., Sikkim, WB. Rajasthan, Punjab, Haryana, Delhi, U.P., Bihar, M.P., Rajasthan and A.P.
6. Pythium stalk rot (*Pythium aphanidermatum = P. butleri*) is found in Sikkim, WB. Rajasthan, Punjab, Haryana, Delhi, U.P. and Bihar.
7. Fusarium wilt and stalk rot (*Fusarium moniliforme*) occurs in Rajasthan, U.P., Bihar, M.P., A.P., Karnataka and T.N.
8. Late wilt (*Cephalosporium maydis*) is reported from Rajasthan, U.P., Bihar and A.P.
9. Charcoal rot (*Macrophomina phaseolina*) occurs in J&K, Punjab, Haryana, Delhi, U.P., M.P., A.P., Karnataka and T.N.
10. Brown stripte downy mildew (*Sclerophthora rayssiae* var. *zeae*) occurs in J&K, H.P., Sikkim, W.B., Rajasthan, Punjab, Haryana, Delhi, U.P., Bihar, Meghalaya, M.P. and Gujarat,
11. Phillipine downy mildew (*Peronosclerospora phillipinensis*) occurs in Punjab, Haryana, Delhi, U.P., M.P. and Bihar.
12. Sorghum downy mildew (*Peronosclerospora sorghi*) occurs in Rajasthan, A.P., Maharashtra, Gujarat, Karnataka and T.N.
13. Sugarcane downy mildew (*Peronosclerospora sacchari*) is found in U.P. and Bihar.
14. Brown spot (*Physoderma maydis*) occurs in J&K, H.P., Sikkim, Rajasthan, Punjab, M.P., W.B., Rajasthan, Maharashtra and Karnataka.
15. Banded leaf and sheath blight (*Rhizoctona solani* f.sp. *sasakii*) occurs in J&K, Sikkim, Rajasthan, Punjab, M.P., Haryana, Rajasthan, U.P. and Delhi.
16. Cyst nematode (*Heterodeara zeae*) occurs in Rajasthan, M.P., Bihar and U.P.

Some Major Epidemics

Six major disease out breaks have been recorded in India since the sixties.

1. Bacterial stalk rot (*Erwinia chrysanthemi* pv. *zeae*) appeared in the foothills (*tarai*) area of U.P. in 1959 and in 1969 in H.P on Hybrid Ganga5.
2. Brown stripe downy mildew (*Sclerophthora rayssiae* var. *zeae*) appeared in 1962 and 1964 in WB hills, 1964 in western M.P. and south-eastern Gujarat, 1964 in north Bihar and in 1965 in the *tarai* region of UP and parts of Himachal Pradesh.
3. There was an out break of Sorghum downy mildew (*Peronosclerospora sorghi*) in Rajasthan, Karnataka and Tamil Nadu from 1968 to 1973.
4. Common rust appeared from 1973 onwards in the winter (*Rabi*) season on hybrids GS-2 and Hi-starch in Bihar.
5. Turcicum leaf blight (*Exherohilum turcicum*) virulent strain of *E. turcicum* appeared in 1974 on hybrid Ganga5, Deccan, Piox104, X114 and Kisan and in 1989–90 in Bihar.
6. Banded leaf and sheath blight (*Rhizoctona solani* f.sp. *sasakii*). In the late sixties there was an outbreak of this disease in HP, and other areas of western Himalaya.

Biotic Stresses are Responsible for Lowering Yield

Among the factors adversely affecting productivity, ubiquitous incidence of diseases in the pre-harvest stage is prominent. Since there is practically no possibility of increasing maize area, the productivity can only be raised by providing seed replacement with improved cultivars, better agronomic practices and protection against diseases and pests.

Changes in Diseases Spectrum Scenario

There is a major shift in disease pattern during the past years as the major diseases like TLB, MLB, BSDM and ESR are gradually becoming diseases of lesser economic importance now-a-days due to availability and use of sources of resistance in the newly developed hybrids and varieties. On the other hand the disease of minor importance became increasingly severe and assumed epidemic proportion in coming years. In early sixties the diseases like BLSB, Late wilt, Polysora Rust and Preharvest cob rots were considered as a disease of minor importance only, but due to changing the weather condition and pattern of disease spectrum these diseases are considered as major diseases not only in India but other countries of Asian region wherever maize is grown.

One of the cheapest and best method of disease control is the use of resistant varieties but the break down of resistance year after year resulting problem in gene pyramiding has restricted this approach. Chemical control is practically feasible but non judicious and untimely application has made this approach ecologically and economically unsuitable on the other hand biological control is ecologically and socially acceptable but difficulty in having an effective biocontrol agent as well as lack of awareness among the farmers has limited its inclusion in regular disease management schedule of crop. Considering all these constraints it is imperative formulate an IDM programme incorporating all the conventional disease management approaches in such a way so that the limitation of one approaches can be neutralized with other one.

Integrated Disease Management (IDM): Concepts and Principles

The Integrated Disease Management (IDM) involves the need based use of pesticide only when the disease incidence reach economic threshold levels and this will promote the build up of many biocontrol agent in the crop ecosystems. Thus IDM, a greener alternative to the conventional use of chemicals, is an attempt to promote natural, economic and sociological farming methods through the most effective combination of farming techniques and judicious and limited use of fungicide. In other words, IDM programme implies all the available disease management approaches including cultural, biological and chemical control with the main objective to keep the disease incidence below economic threshold level.

IDM: Components

Broadly IDM involves four components namely

1. Host resistance
2. Biological control
3. Cultural control and
4. Chemical control

Host Resistance

Resistance is the character of a plant which suppress pathogen and disease development. The use of resistant genotypes is a highly effective approach to suppress disease to tolerable levels. Resistance alone is sometimes sufficient to suppress disease to tolerable levels. In resistant genotypes, disease appears late, build up slowly and results in little damage to the crop.

Biological Control

In this method the pathogen actively is reduced through the use of other living organisms *e.g.* hyper-parasites, resulting in a reduction of disease incidence and severity.

Cultural Control

This involves deliberate manipulation of the crop environment to make it less favourable to harmful organism *e.g.* by disrupting their reproduction cycles, eliminating their food sources, or encouraging their natural enemies. Included in this method are such practices as intercropping, crop rotation, field sanitation, manipulation of sowing dates etc. some of these techniques provide only small benefits when integrated with other techniques, they significantly improve disease management.

Chemical Control

In IDM programme chemical control implies the judicious and need based use of pesticides (fungicides, insecticides and herbicides). Chemical control is essential in areas where diseases appear in the early stage of plant growth and environmental conditions are likely to spread them fast.

Foliar Diseases

Of the many foliar diseases occurring in India, three may be considered important based on geographical distribution and yield loss potential. Northern leaf blight *(Heiminthosporium turcicum = Exserohilum turcicum)* is a major problem in the region with cooler environment like J&K, H.P. etc under conditions favourable for disease development, susceptible materials suffer heavy blightening resulting in premature drying of leaves and lightweight kernels. It is a major disease in the mountainous areas and in the deep hillside areas of Rajouri and Poonch districts, especially in the cool and shady environment of intermediate zone of J&K. The highest disease intensity *i.e.*, 66 per cent was recorded in intermediate zone of Jammu region (Ahamad, 2005).

The foliar blight disease complex in cool conditions favoured for turcicum leaf blight. At the Regional Agricultural Research Station, Rajouri and MBRSS, Poonch, it was moderate to high. The natural disease pressure every winter cycle allowed for the selection of disease resistant germplasm. It is important that, where a disease outbreak is feared, spraying should be undertaken at an early stage of crop growth, *i.e.*, when it is not more than 30 days old (knee-high stage). For large scale high volume sprays 1.8–2.6 kg of fungicide in 900–1200 litre of water/ha is recommended, if the plant population is around 56,000/ha. Control of foliar blight disease can be achieved using host resistance which has been quite successful, chemical control is also feasible, it becoming profitable on high volume materials like popcorn, sweet corn, baby corn or seed crop.

Leaf Rust: *Puccina sorghi*

Rust pustules appear on the above ground plant parts especially on leaves. Round to elongated uredo-pustules are on both the surfaces of the leaf. They may coalesce and thereby produce scorching or drying of the leaves. Later on these pustules turn black due to formation of teleutospores. This disease is air born. The secondary infection take place by means of uredopustules which are wind born disseminated to the neighbouring plants. The major approach adopted has been the utilization of host resistance. Both hybrids and composites possessing resistance have been released.

Polysora Rust of Maize

Distribution

Among the rust diseases in maize Polysora rust or tropical rust or southern corn rust (*Puccinia polysora* Underw) is an important disease in tropical areas. It occurs in many parts of the world, and it is a recent introduction in the peninsular India in 1991 (Karnataka and Tamil Nadu during rainy season and Coastal districts of Andhra Pradesh during winter) on certain maize cultivars in Mysore distric (Payak, 1992). In recent years, the incidence of *P. polysora* has taken a heavy toll in majority cultivars grown in Karnataka namely Mysore, Mandya, Hassan, Kolar, part of Coorg, Shimoga and Chitradurga districts (Anon.2003) Polysora rust is a warm weather disease favoured by wet weather infection and disease development at 12- 27°C temperature (Hollier and King 1985, Melching, 1975). Rain drizzle or even heavy dews allow disease formation.

Symptoms and Causal Organism

Pustules are small circular and orange in colour. They are evenly spread over the surface Pustules on the midribs, ear husks and tassels may be elongated or irregular in shape. With heavy infection leaf may die prematurely and defoliation starts from the base towards upward. The disease resembles common rust. The pustule appear on leaf are light cinnamon brown, circular to oval 0.2-2.0 mm long densely scattered on the upper leaf. The uredospores are yellowish to golden, mostly ellipsoid or ovoid with minutely echinulate, 1-1.5 µm thick wall and sizes are 20-29, x 29-40µm. Pustule

Brown Stripe Downy Mildew (BSDM) (*Sclerophthora rayssiae* var. *zeae* Payak and Renfro)

It has been most severe in U.P., H.P., southern Rajasthan, Punjab, hilly part of W.B., J&K especially in areas that receive 100–200 cm of rains. A yield loss of 63 per cent was recorded in the *tarai* areas of U.P. As the leaves are infected by the pathogen, it drastically reduces the total photosynthetic area of the diseased plants, causing reduction in grain yield. The yield losses depend on susceptibility of the host, spread of the disease and environmental conditions prevailing in the growing season. This is more destructive disease in the intermediate zone of Jammu region. Disease symptoms have been observed only on leaves. In the initial stages, the lesions start developing on lower leaves as narrow chlorotic or yellowish stripes, 3–7 mm wide but variable in length. The stripes extend in parallel fashion, have well defined margins and are delimited by veins. The presence of downy whitish to creamy growth usually on the ventral surface of the infected leaves corresponding to- stripes is the most characteristic symptom. In cloudy weather, the growth is profuse and can be seen. Later, these chlorotic stripes turn brown and give a burnt appearance to the leaves. The Pathogen is an obligate parasite and cannot be grown in artificial media. It is mainly soil borne. Its seed borne nature is controversial. Some workers reported this pathogen as seed-borne but it was found later that *S. rayssiae* var *zeae* is not seed borne because no disease symptoms were observed in spite of providing ideal conditions to the pathogen grown from infected seeds. There seems to be no controversy about soil borne of the disease. Oospores remain viable for at least three years and this viable oospores material constitutes the main source of inoculums. Primary infection invariably occurs on lower proximate to the ground level. Once the primary infection becomes established, its spread is possible through the agency of sporangia. A definite relationship of disease development with rainfall pattern in different region ns has been mapped. In areas where rainfall varies from 40–60 cm, disease was observed from trace to low, in 60 to 100 cm rainfall areas it was from low to moderate, but in the areas having more than 100 cm rainfall, maximum disease ratings have been made. However, in areas of low rainfall, disease may be severe only when continuous spell of shower and cloudiness for a few days accompanied by about 25°C temperature are prevalent. The severity of the disease is influenced by temperature and moisture. Sporangia are produced at low temperature (20–22°C), while oospores are formed at high temperature. Soil temperature of 28–32°C favour disease development. Sporangial production and infection require a film of moisture for 12–96 hours. Zoospore germination occurs in the temperatures (15–30°C), with an optimum at 22–25°C. Sporangia and spore of the pathogen are disseminated by wind, rain and animals. The pathogen infects crab grass. Zinc deficiency predisposes plants to infection.

Integrated Disease Management

1. *Cultural Practices*: Several cultural practices reduce severity of diseases. These are designed to eliminate the causal pathogens from a particular area, to significantly reduce primary inoculum or to stimulate growth during the first month after planting.
2. Sow the seed before rainy season begins.
3. *Fungicidal management*: Metalaxyl (Ridomil) can be applied to seed, in furrows as granules or sprayed on foliage. Seed application (Apron 35 W.P., @ 0.25 per cent (2.5 g/kg seed) in a slurry is most economical and total control can be obtained. Systemic fungicides usually act on but one site in the fungus, whereas protective (non-systemic) fungicides act on several sites. Thus in the former case only one change in the fungus is required to develop tolerance.Treatment of seed with apron 35 w.p. @ 0.25 per cent is a good safeguard against

any possible seed-borne infection. Spraying of fungicides such as mancozeb @ 2.5–3.0 per cent as soon as disease symptoms appear to protect valuable and breeding materials. Sun drying of seeds lead to inactivation of mycelium present in seed and also reduce moisture levels. Seeds with less than 15 per cent moisture content produce healthy plants.

4. Use only disease resistant hybrids and composites in disease prone areas.

Sheath Blight

It affects the sheaths although they do infect leaves, husks and other parts. Two have been determined to cause significant damage, brown spot and banded leaf and sheath blight.

Brown Spot

This disease mainly occurs in sub-tropical and intermediate areas. It is distributed in the entire Himalayan area as also in states of Rajasthan and Karnataka. Resistance has been identified on the basis of natural incidence only and methods to induce artificial disease development in the field through resisting sporangia are to be worked out. A study conducted at Rajasthan, indicates that the yield loss in hybrid Ganga 5 was 27.0, while in the local open-pollinated variety it was Malan 24.5 per cent.

Banded Leaf and Sheath Blight (BLSB) *(Thanatophorus sasakii* Tu and Kimbrough =*Rhizoctonia solani* f.sp. *saakii* Exner)

It was first reported from Sri Lanka (Bertus, 1927). In India, it was first recorded by Ullstrup in the *Tarai* areas of U.P. (Payak and Renfro, 1966). Now it is present in J&K, H.P., U.P., M.P., Haryana, Punjab, Rajasthan, W.B., Meghalaya and Assam. This disease has been considered as one of the major diseases of maize (Payak and Sharma, 1981). Grain losses have been estimated in the range of 23.9–31.9 per cent in ten cultivars (Sangam Lal *et al.*, 1980). It is estimated 40.5 per cent loss in grain yield with 71 per cent disease index. The magnitude of grain loss may reach as high as 100 per cent if the ear rot phase of the disease predominates. The disease appears on plants at pre-flowering stage (40 to 50-day old plant) and within a period of 15 to 20 days spreads under favourable conditions from the lower most sheath to the ear shoot.*Rhizoctonia solani* Kunh is morphologically characterized by features like pale to brown colour of mycelium, branching near the basal septum in young growing hyphal, presence of a construction and formation of a septum in branch near the point of origin, absence of clamp connections, sclerotia of un-differentiated texture, young multinucleate hyphal cells with a prominent septal pore apparatus and rapid growth rate. The primary source of inoculums are sclerotia in the soil and grass hosts that grow in the vicinity of maize crop. Secondary spread of disease is by contact of infected leaves with parts of adjoining healthy plants. Secondary spread by basidiospores has not been observed in maize crop.

The optimum temperature for *in vitro* growth of the pathogens is 30°C and the highest level of disease is induced when RH is in the range of 90–100 per cent. At RH of 70 per cent lower, the disease development is negligible or absent.

It is caused by incited by *Thanatephorus sasakii* (Shirai) Tu and Kimbro; anamorph *Rhizoctonia solani f.sp. sasakii* Exner, has become increasingly severe and economically important disease of maize during last two decades or so. For the first time was reported from Srilanka under the name 'Sclerotial' disease. Though it was considered a disease of minor importance till it appeared in an epidemic form in the foot hills region of Himalayas especially in the district of Mandi in Himachal Pradesh. In India, in early sixties the disease was little more than a curiosity and a disease of minor importance in the western and central Himalayan foothill region. However, it became increasingly severe and assumed epidemic proportions in the next two decades. Presently, the disease is considered as a major disease not only in India but also in several countries of Tropical Asia wherever maize is grown.

Distribution

On maize the disease was first recorded from Sri Lanka (Bertus, 1927) under the name 'sclerotial disease'. In recent years the disease outbreaks have occurred in more countries and have assumed epidemic dimensions. In India, the disease was first recorded from Tarai region of Uttar Pradesh in 1960 (Payak and Renfro 1966). Now it has become increasingly severe and assumed epidemic proportions in the next two decades. Presently, the disease is considered as a major disease in India wherever maize is grown in warm and humid conditions. It is known to be present in severe intensity in the states of Himachal Pradesh, Uttar Pradesh, Uttarakhand, Haryana. Punjab, Madhya Pradesh, Rajasthan, Bihar, West Bengal, Sikkim, Meghalaya, Assam and Orissa.

Economic Importance

The disease was earlier reported as a minor disease on maize (Payak and Renfro 1966). The importance of the disease was only realized in early 1970s when an epidemic occurred in warm and humid foot hills area, particularly in the Mandi district of Himachal Pradesh (Thakur *et al.*, 1973). The disease causes direct losses, resulting in premature death, stalk breakage and ear rot indirect losses by reducing the grain yield. In India, losses in grain yield have been estimated in the range of 23.9 to 31.9 per cent (Sangam Lal *et al.*, 1980). It is estimated 40.5 per cent loss in grain yield with 71 per cent disease index. However, the magnitude of grain loss may reach as high as 100 per cent if the ear rot phase of the disease predominates. Payak and Sharma (1985) have reported that annually at least 1 per cent of the total grain yield is reduced by BLSB in India.

Losses to the extent of 11-40 per cent were reported while evaluating 10 different varieties of maize. Losses in grain yield showed a high positive correlation with premature death of plants and disease index. The disease caused drastic reduction in grain yield-to the tune of 97 per cent (Butchaiah, 1977) and exhibited a direct correlation with other yield parameters (Barua, 1979). Disease incidence varying from 2 to 80 per cent was observed affecting maize crop severely in Korea during previous two crop seasons (Kim *et at.*, 1992).

Symptoms and Signs

The disease generally appears at pre-flowering stage in 40-50 day old plants. The symptoms manifest on leaves, sheaths, stalk and ear. As the disease is soil borne it starts from the lowest leaf sheath or on leaves that are in contact with soil and travels up to the ear. The symptoms are more common on the leaf sheaths than on the laminate. The disease lesions are characterised by the presence of alternate bleached areas or zones that are initially water soaked and narrow, purple brown bands oriented perpendicular to long axis of leaves or leaf sheaths.

The hyphal masses on ears produce a caking effect and cement the husk leaves as well as styles (silk fibers) together. The mycelium develops above and between kernel rows. Sclerotia are found to be produced not only on sheaths, leaves and husk leaves but were also observed for the first time on glumes, in cupules, under the pericarp in caryopsis and silk fibres (Ahuja and Payak, 1982).

Causal Organism

The disease banded leaf and sheath blight incited by *Rhizoctonia solani* f. sp. *Sasakii* Exner (*Thanetophorus sasakii* (Shirai) Tu and Kimbro), is found in most parts of the world and is capable of attacking a wide range of host plants including maize, causing seed decay, damping off, stem canker, root rots, aerial blight and seed decay. Detailed work on maize pathogen (Ahuja and Payak, 1978) suggests need for critically determining the correct taxonomy of the pathogen. The teleomorphic stage of maize pathogen has not been recorded in nature or *in vitro* so far, whereas it has been observed on rice (Anonymous, 1974). Pathogenicity tests revealed that *R. solani* AG-1-1A was the most strongly pathogenic to leaf sheaths and stalks of maize, followed by *R. solani* AG-2-2. Differences were observed

Crop Affected by Banded Leaf and Sheath Blight in Maize

Water soaking | **Spreading upward** | **Symptoms on cob** | **Severely affected plant**

BLSB Symptoms

Ahuja and Payak (1978, 82, 86) found under laboratory tests that the growth of the pathogen was completely inhibited by carbendazirn, benodanil, thiobendazol and validamycin while field condition benodanil, validamycin and carbendazim were most effective in reducing disease severity. It observed that foliar spray with thiobendazole (60 W.P. @ 0.05 per cent followed by carboxin (75W.P. @0.1 per cent) and Duter (20 *w*–0.05 per cent) were most effective in reducing disease. Sharma and Hembram (1990) have worked out a simple and economical method to control this disease. The methodology comprised of mechanical stripping of third leaf sheath from the ground level which is effective in checking disease developing further up. This method was also found to be effective in minimizing the yield loss.

Stalk Rot

Stalk rots are the most serious and widespread group of the diseases in maize. The term "stalk rot" is often used to include stalk breakage, stalk lodging, premature death of plants and occasionally root lodging. In true sense, it is decay of the internal pith tissues of the stalk. Typically, the first sign of stalk rot is plant wilting. In initial stage, leaves become gray, ears drop and outer rind of lower stalk may turn brown. When outer stalk tissue is brown, pith tissue in the lowest internodes is rotted and pulled away from the rind, which may result in lodging. Plants are weak and those with rotted stalks always have rotted roots. There nature is often complex as a number of fungi, nematodes and sometimes bacteria are involved in causation of the disease(s). This group of disease is broadly divisible into two categories, *viz.*, pre-flowering and post-flowering types. Of the 12 diseases recorded so far, Pythium stalk rot (*P. aphanidermatum*) and bacterial stalk rot (*Erwinia chrysanthemi* pv. *zeae*) belong to pre-flowering type, while others such as late wilt, charcoal rot, Fusarium wilt and stalk rot, Acremonium stalk rot, Botryodiplodia stalk rot belong to the post-flowering type.Losses due to stalk rot may occur in the following three ways:

1. Yield loss may result from premature plant death, thus stopping normal grain fill. Total grain weight on stalk rot affected plants is less than the weight on healthy plant.
2. Another component of yield loss is that plants with stalk rot may lodge and not be harvested with mechanical equipment. Harvest is slowed if stalk rot is severe and losses also occur due to time loss during harvest.
3. Losses also occur with ear rot as a results of the ear on lodged plant coming in contact with soil. This results in reduced grain quality and potential dockage when the grain is marketed. Because losses due to stalk rot may occur in several ways, yield loss estimates are difficult to obtain.

Technique used to estimate losses are:

1. Yield of hybrids with practically nil natural stalk rot incidence have been compared to yields in years when stalk rot is severe (Koehler, 1960). These types of studies are confounded by year-to-year in average yield.
2. Another technique used is comparing inoculated vs. uninoculated plots. These studies, however, are confounded by natural infection in uninoculated plots, and by the fact that inoculation does not exactly duplicate the natural stalk rot condition.
3. In paired plant technique, grain yields of adjacent diseased and healthy plants are compared. In this, the problem in diseased plants have much less kernels than on healthy plants. The final problem with yield loss studies is that the losses are estimated on the basis of hand-harvested yield and do not take into account those losses due to lodging or ear rots.

Pre-flowering Stalk Rot

Erwinia Stalk Rot (*Erwinia chrysanthemi* pv. *zeae* Victoria, Arboleda and Munoj)

It is one of the most important disease of maize in tropical countries. At first, the upper leaves show signs of wilting. The symptom appear at the lower node and may remain confined at one or two internodes only. The inner tissues also get infected in severe cases. The infected tissues are at first soft but later on they turn into dry mass of shredded fibers. The plants may topple down at this stage. In some cases the infection (light to dark brown rotting) may spread rapidly throughout the leaf sheath and cause withering of leaf sheath and leaf under favourable environmental conditions. The stalks and ears may also be affected which emit an offensive odour. High disease incidence is linked with irrigation by sewage water, it is particularly favoured by high temperature (28°C and above) and the high ambient moisture which commonly prevails in the most of the maize growing areas 3 to 4 week after sowing. The disease can be minimized, using indigenously formulated compound bleaching powder containing 33 per cent chlorine @ 10 kg/ha as soil drench at pre-flowering in standing crop. Avoiding of water logging and proper drainage also helps in reducing disease incidence. Planting the crop on ridges rather than in flat soil is recommended. Spraying mixture solution containing 50 g Agrimycin and 2 kg of ceresan wet/ha will also control the disease.

Pythium Stalk Rot [*P. aphanidermatum* (Eds.) Fitzp.]

The disease is common during rainy season under hot and high soil moisture conditions. The most diagnostic feature of this disease is that the rot is contained mostly to a single basal internode leading to destruction of pith parenchyma and consequent weakening to the stalk. The affected plants topple over but do not die for upto 2 weeks after attack. The general symptoms of the disease are poor emergence or patchy growth of seedlings or the seedlings become pale yellow or dry up just after germination. On pulling out the affected plants, rotting of the seeds may be clearly seen. Stalk rot is first recognized when the plants first lodge but do not break over. Stalks often rot at the first, internode above the soil level and become soft and brown. This disease has been reported in *Kharif* season in areas having temperature ranging 30–35°C and RH (relative humidity) 80–100 per cent. When RH drop 50 per cent or below, no rot develop in the stalks. This explain why high disease incidence occurs when the field is water logged, low lying or poorly drained, plant age (pre-flowering stage) and high plant population (< 60,000/ha). The tooth pick method as described for post flowering stalk rot development can be used to inoculate plants at knee high stage (30–35 days) for screening genotypes (Payak, 1975), Payak *et al.* (1971), Payak and Sharma (1985).

Post-flowering Stalk Rot

Fusarium Stalk Rot [*Fusarium moniliforme* Sheld. = *Gibberella moniliforme* (Sheld.) Wineland]

The disease appears both on young and adult plants. In young seedlings the plants dry up in the early stage of the growth. In the adult plants the symptoms are not visible till maturity. At this stage drying of top leaves from tips begin showing water scarcity. Ultimately the entire plant dries up. The cob formation either does not take place or the grain formation is partial or wholly affected. The characteristic reddening of the internal tissues are discernible when affected plants are longitudinally split open. In some cases the red streak may also be observed. The pith becomes hollow and white to pale pink mycelial growth may also seen in the cavity thus formed in advance stage of disease development.

Disease Management

1. Sow the resistant varieties like GS-2 and Ganga-101 which are comparatively resistant.
2. Follow 2–3 years crop rotation.

Charcoal Rot [*Macrophomina phaseolina* (Mubl.) Ashby]

This is a common disease in warm and dry areas. A characteristic sign of disease is the presence of numerous, minute, black sclerotia, particularly on the vascular bundles and inside the rind of the stalk. This may cause the stalk to appear gray-black. The disease symptoms appear only near maturity. Water soaked, brown lesions appear on roots which turn brown later on. The disease is favoured by high temperature *i.e.* 30–42°C and low soil moisture. The pathogen over winters as sclerotia and may penetrate roots and lower stems during growing season.

Control

Avoid water stress after flowering of the plant.

Black Bundle Disease *(Cephalosporium acremonium* Corda)

The characteristic symptoms is the blackening of the vascular bundles as black dots on the cut ends of the stalk. Other important symptoms of this disease are a reddening or purpling of the leaves and stalks, lesions on basal portion of the stalk, multiple ear formation at node and excessive tillering. In severe cases, leaves dry and plants may wilt. Ears may rot or even the ears may not be formed.

Control

Seed treatment with carbendazim or benlate @ 2 g/kg seed.

Smuts

Two smut diseases are occurring in maize, the common smut *(Ustilago maydis*) and head smut (*Sphacelothica reilina*).

Common Smut (*Ustilago maydis* (DC) Cda.

The characteristic galls are formed on the infected tissues. The gall may appear on the stem, leaves, axillary's buds and parts of the male flower. When formed on the cob they cause extensive damage. The developing mycelium between the thin walled cells induces hypertrophy. The galls are light coloured in the beginning and later on become dark. As the galls are enlarging they appear light coloured or almost white. The individual flowers may be infected on tassel and even through individual flower parts may be transformed into galls. The light, shiny membrane ruptures and dry black spores masses are exposed. The black powdery masses are the chlamydospores of the fungus. Stem galls result in loss in yield and bending of the stalk.

Head Smut [*Sphacelothica reilina* (Kunh.) Clinton]

This attacks maize, sorghum and other grasses and is moderately destructive disease in the sub-temperate, intermediate zone of hilly areas of Rajouri and Poonch districts of J&K. It has occurred in severe form on Pioneer H-3054 and caused 25–35 per cent yield losses during *Kharif*-2005 in Manialdara region of Rajouri district (Ahamad, 2005, 2012). It appears with the formation of tassels and ears where partial or complete transformation of the inflorescence into sori may be observed. The tassel is partly or wholly converted into smut spores. Galls are first covered with a membrane that soon breaks open to expose a powdery mass and the vascular bundle of the host. If only the parts of the tassels and

Maize Corn Affected by Common Smut

ears are converted into galls, the floral bracts turn into leafy structures. Ears frequently abort and are replaced by leafy buds. It is externally seed borne and soil borne. Galls on the ear usually destroy it to a large extent while large galls above the ear cause much greater reduction in yield than do galls below the ear. Losses from corn smut are highly variable from one location to another and may range from a trace upto 35 per cent or more in localized areas (Ahamad, 2005). The spore retain viability for two years. Although, the disease may be externally seed-borne, the major source of infection is soil-borne inoculums. Only young plants are susceptible. Low temperature is favourable for infection.

Control Measures

1. Field should be cleaned off the trash after harvest. Since only few plants are affected in a field it is possible to locate and destroy the infected ears before they shed the ears.
2. Follow up crop rotation.
3. Seed treatment with captan or thiram @ 3 g/kg seed.
4. Grow resistant/moderately resistant var. like KH-517, KH-612, GS-2 etc.

Virus Diseases

Viruses are the most mysterious disease agents affecting both plants and animals, and hence the limited amount of studies and where there is the least information on their incidence and economic effects. The lack of information is more acute in the developing countries, where there is a lack of trained personnel and appropriate equipment, and mostly to the concept that when these viral diseases are present, nothing can be done to control them. The problem with such mentality, the incidence of these diseases continues to increase, especially under the present actively changing circumstances in agriculture when new germplasm is almost freely moving throughout all environments, agronomic practices are changing, and there is a conspicuous expansion of area where maize is being grown due to an increased demand. These changes in technology are bringing along an increase in problems, specially insects and diseases, that were not previously considered problems of any significance. Three viruses occurs on maize in India. These are maize mosaic virus I, maize mosaic (a strain of sugarcane mosaic virus) and vein enotion. The information is required to both pathologists and breeders to understand the complexity of the viral pathotypes, both in their genome and their evolution, to be considered when selecting for stable resistance to their pathogens.

Mosaic Virus (Mosaic Virus Transmitted by Leafhopper)

Small white fleck may occur on one side of the midrib near the base of the young leaf and is usually associated with whitening of veins. The specks elongated to form fine discontinuous stripes. Sometimes the stripes coalesce and form yellow bands of the leaves. Stripes may also be formed on sheaths, husks, and the stalks. The plants are pale green, stunted and show mosaic mottling on the leaves. Cobs are not filled well.

Control

Grow resistant hybrids.

Maize Dwarf Mosaic Virus (Sugarcane Mosaic Virus, Transmitted by Aphids, *Myzuspersicae*)

Early infested plants produce long chlorotic streaks along the veins at the base of the leaf lamina. With the growing age of plant, leaves in the whorl become chlorotic and in turn the chlorotic areas in leaves turn red or purple.

Control

Sow resistant varieties.

Maize Mosaic

This has been found to be prevalent in the several states and the incidence ranges from 2.2 to 10.6 per cent. Resistance of a high order even immunity (CM 103 and CM 104) has been identified and the improved released cultivars have not been reported to suffer much from virus diseases.

New Emerging Diseases In India

Cob Rots

The safety and integrity of the food supply are the paramount importance and are among the drivers of safe grain storage. In present scenario, the trade among the developing countries is expanding; hence quality of the produce is becoming a major concern. Demand for maize in both national and international market is high. Its uses are continuously increasing globally as important livestock, poultry feed and by industrial sectors. On the other hand the loss of grains is increasing with the increase of food production due to poor and improper storage as well as the pre-harvest fungi associated with maize. These fungi are responsible for microbial spoilage of the grins and production of mycotoxins in pre- and post-harvest stages.

Symptoms and Epidemiology

Fungi associated with the cob rots are – *Aspergillus flavus, Fusarium species, Gibberella zeae, Penicillium* etc. These field fungi infect the cobs either immediately or before harvest. The presence of mycotoxin traditionally regarded as indication of poor storage condition, but they may present in grains before coming to storage.

Fusarium Ear Rot

Fusarium ear rot caused by *F. moniliforme* occurs when harvest is delayed. The symptom of this disease is pink to reddish brown discoloration on the kernels and later on it spread on whole ear. As the disease progress, infected kernel becomes covered with a powdery/cottony pink mild growth. Kernels infected late in the season develop whitish streaks on the grains. Kernels become infected in several ways. The most common pathways is infection via silk channel. Air born spores present on residues can land on corn silks when it turns dark brown. Green silks are relatively resistant. Infection

follows some form of injury, bird damage, feeding of corn borers. Disease development and spread one favoured by dry warm weather. Under certain conditions that are stressful for maize plant as the fungus becomes pathogenic and causes disease. Stalk rot, ear rot, kernel rot can occur in infected tissue although many times infected tissue havě no symptoms.

Aspergillus Ear Rot

This rot caused by *Aspergillus* spp. and infection often follows drought stress and damage done by corn, earworms and corn borer and other insects. A tan sooty black, greenish or greenish yellow mold grows on and between kernels. The damage is most common at or near the tip of the ear; silk infection is favoured by high night and day temperature. Some times the susceptible maize genotypes got infected with *Aspergillus* sp. at the time of silking. The infection goes through silk to the maize seed/ grains but remained latent and cause decay at the time of

Hollier, C.A. and King, S.B. (1985).Effect of dew period and temperature on infection of seedling maize plants by *Puccinia polysora*. *Plant Disease*. 69:219-220.

Hulbert, S.H.and Bennetzen, J.C. (1990). Genetic fine Structure of a complex rust resistance locus in maize. *Phytopathology 80* : 1007.

Jines M. P., Balint-Kurti P., Robertson-Hoyt L. A., Molnar T., Holland J. B. and Goodman M. M. (2007).Mapping resistance to Southern rust in a tropical by temperate maize recombinant inbred topcross population.*TAG theoretical and Applied Genetics* 114:4

Koehler, B. (1960). Corn Stalk Rot in Illinois. *Illinois Agric. Experiment Stn. Bull.*, 658: 90.

Lal, S., Baruah, P. and Butchaiah, K. (1980). Assessment of yield losses in maize cultivars due to banded sclerotial diseases. *Indian Phytopath.*, 33: 440–443.

Kim, W.G., Cho, W.D. and Lee, Y.H. (1992). Anastomosis groups and pathogenicity of isolates:I of *Rhizoctonia solani* and *Rhizoctonia zeae* from corn. Role of fungi as frontiers of biosciences : *Proceedings of the Asian Mycological Symposium in Commemoration of the 20th Anniversary of Korean Society of Mycology*. October 1-4, 1992, Seoul, Korea, pp 147-149.

Lee, S.B., Kim, J.G., Han, M.S. and Han, H.J. (1989). Studies on the occurrence of corn sheath blight caused by *Rhizoctonia solani.* Influence of growth season and cultural environment on the occurrence of sheath blight disease. *J. Korean Soc. Grass/and* Sci. **9** : 174-178.

Lehman, P.S., Machado, C.C., Tarrago, M.T. and Correa, C.F. (1976). Evaluation of recommended soybean, maize and sorghum cultivars and soybean lines for resistance to dead patch. *Phytopathologica* **2** : 47-49.

Meena, R.L. Rathore, R.S. and Mathur, K. (2003). Efficacy of biocontrol agents against *Rhizoctonia solani* f. sp. *Sassakii* causing banded leaf and sheath blight of maize. *J. of mycol.pl. Pathol.* 33 (20) : 310-312

Melching, J.S. (1975). Corn Rust, Type, races and destructive potential. p. 90-115 In: *Proc.30th Annu. Corn and Sorghum. Res. Conf.*

Pandey, M.K. (1992). *Studies on the Banded Leaf and Sheath Blight of Maize.* M.Sc. Thesis. G.B. Pant Vniv. of Agric. and Tech., Pantnagar, 79 pp.

Payak, M.M. and Renfro, B.L. (1966). Diseases of maize new to India. *Indian Phytopath. Soc. Bull.* 3: 1 18.

Payak, M.M., Lal, S., Thind, B.S., Renfro, M.C., Diwaker, R.C., Sharma, R.C., Lilaramani, L. and Laxminarayan, C. (1971). Stalk rot of maize in India: Germplasm evaluation and assessment of losses. In: 7 th *Inter-Asian Corn Improvement Workshop*. Los Banos, Philippines.

Payak, M.M. and Sharma, R.C. (1979). Resistance to diseases in maize: The Indian experience *Crop Improv.*, 6: 75–85.

Payak, M.M. and Sharma, R.C. (1982). Disease resistance in maize in India. In: *Advancing Frontier of Mycology and Plant Pathology.* Today and Tomorrow's Printers and Publishers, New Delhi, pp. 229–237.

Payak, M.M. (1975). Epidemiology of maize downy mildews with special reference to those occurring in Asia. *Trop. Res.*, 8: 81–91.

Payak, M.M. and Sharma, R.C. (1985). Maize disease and approaches to their management in India. *Trop. Pest Management*, 31: 302–310.

Quazi, M.Z. and Iqbal, F.C. (1970). Effects of temperature on the growth of *Rhizoctonia solani. K. Agric. Pakistan* 21 : 447-451.

Raid, R.N., Pennypocker, S.P. and Stevenson, R.E. (1988). Characterization of *Puccinia polysora* epidemics in Pennsylvania and Maryland. *Phytopathology.* 78 *(5)* : 579-585.

Rangrajan, M. and Chakravarti, B.P. (1969). Bacterial stalk rot of maize in Rajasthan, effect on seed germination and varietal susceptibility. *Indian Phytopath.*, 23: 470–477.

Saha, L.R. (1990). *Handbook of Plant Protection*. Kalyani Publishers, 928 pp.

Sangam L., P. Barauh and Butchaiah, K. (1980). Assessment of yield losses in maize cultivars due to banded sclerotial diseases. *Indian Phytopathol*. 33 : 440-443.

Sangam L., P. Barauh and Butchaiah, K. (1980). Assessment of yield losses in maize cultivars due to banded sclerotial diseases. *Indian Phytopathol.* 33 : 440-443.

Sharma, G. (1996). *Integrated Management of Banded Leaf and Sheath Blight caused by Rhizoctonia solani Kuhn*. M.Sc. Thesis. G.B. Pant Univ. of Agric. and Tech., Pantnagar. 90 pp.

Sharma, R.C. and Hembram, D.(1990). Leaf stripping: A method to control banded leaf and sheath blight of maize. *Curr. Sci.* 59: 745-746.

Sharma, R.C. and Rai, S.N. (1998). Chemical control of banded leaf and sheath blight of maize. *Indian Phytopathol.*

Sherwood, R.T. (1970). Physiology of *Rhizoctonia solani*. In: *Rhizoctonia solani, Biology and Pathology* (Ed. J.Ro Parmeter, Jr.) pp. 69-92. Univ. Calif. Press, Berkeley.

Sinha, P.P. (1992). Chemical control of banded sclerotial disease of maize. *OrissaJ Agril. Res. 5: 54-57.*

Shurtleff, M.C. (1980). *Compendium of Corn Diseases*. The American Phytopathological Society St. Paul Menneseta, U.S.A., pp. 105.

Thakur, M.S., Sharrna, S.L. and Munjal, R.L. (1973). Correlation studies between incidence of banded sclerotial disease and ear yield in maize. *Indian J. Mycol. Plant Pathol.* 3 : 180-181.

Ullstrup, A.J. (1977). Diseases of corn. In: *Corn and Corn Improvement*, (Ed.) G.F. Sprague. Chapter 8.

Ullstrup, A.J. (1965). Inheritance and linkage of a gene determining resistance in maize to an American race of *Puccinia polysora. Phytopathology* 55; 425-428.

Zummo, N. (1988). Component contributing to partial resistance in maize to *Puccinia polysora. Plant Dis.* 27: 157-160.

Hill Agriculture: Prospects, Constraints and Mitigations (2013) *Pages* **442–449**
Editor: **Dr. Shahid Ahamad**
Published by: **DAYA PUBLISHING HOUSE, NEW DELHI**

Chapter 28

Major Disease of Rapeseed-Mustard and their Management in Hill Agriculture

Shahid Ahamad[1] and Nafees Ahmad[2]

[1]Krishi Vigyan Kendra, Poonch – 185 131
Sher-e-Kashmir University of Agricultural Sciences and Technology of Jammu
[2]Division of Extension Education, SKUAST-J., Chatha

India is a major oilseed growing country covering more than 7 million hectare. Oilseeds holds a key position in Indian economy and contributed 4 per cent of India's GDP and 10 per cent of value of all Agricultural commodities during last decade (Anonymous, 1997).

Among nine oilseed crops, rapeseed and mustard is the second most important group of oilseed crops producing one fifth of world's rapeseed and mustard. These crops account for 30 per cent of the total oilseeds production and 13 per cent of the country's gross cropped area. There has been phenomenal growth in rapeseed and mustard production and productivity during the last 12 years. But still there is a big gap to reach the achievable productivity levels. The lack of judicial disease management programme is considered to be one of the most important factor in limiting the productivity of oil seed crops. The total loss of crop in India is estimated about 18-25 per cent : out of which 26 per cent accounts due to plant diseases. In case of alternaria blight of rapeseed and mustard, it was found that the loss in seed yield ranges from 24.5-71.5 per cent depending upon the severity of pod infection. Yield loss of about 10-70 per cent have been reported in *Brassica juncea* due to Alternaria blight and white rust diseases (Kolte, 1985; Ahamad and Khan, 2007; Ahamad *et al.*, 2009; Kumar and Ahamad, 2009; Ahamad, 2012).

Diseases of Rapeseed-Mustard and their IDM Approaches

More than twenty disease are known to attack the rapeseed-mustard in India but only five diseases namely alternaria blight, white rust, downey mildew, powdery mildew and sclerotinia stem rot (Table 28.1) are considered to be of major consequences because of their wide distribution and heavy yield losses (Degenhardt *et al*,1974). The typical symptom, epidemiology, mode of perpetuation and their IDM strategies was discussed later in this chapter.

Table 28.1: List of Major Fungal Diseases of Rapeseed-Mustard

Name of Diseases	*Causal Organisms*
Alternaria blight	*Alternaria brassicae, A. brassicicola, A. parhani: A. alternata*
White rust/blister	*Albugo candida*
Downy mildew	*Perenospora parasitica*
Powdery mildew	*Erysiphe cruciferarum*
Slcerotinia stem rot	*Slcerotinia sclerotiorum*

Pathogen

The four *Alternaria* species *viz. Alternaria brassicae, A. brassicicola, A. raphani* and *A. alternata* can be easily distinguished by microscopic observation. The details are given below.

Alternaria Blight

It is the most destructive and wide spread disease of rapseed – mustard and is caused by four species of *Alternaria viz., A. brassicae* (berk) Sace., *A. brassicicola* (Schwein) Wiltsh., *A. raphani* Groves and Solko and *A.alternata* (Fr.) Keissl.

Symptoms

The initial infection on lower leaves starts as minute brown to blackish lesions which later spread rapidly to the upper leaves, stems and silique. Lesions on leaves due to *A. brassicicola* are sooty black, velvety, whereas those caused by *A. brassicae* are grey, dense and those caused by *A. raphani* are small with raised margins and are surrounded by yellow translucent halo (Verma and Saharan, 1994). The lesions on stem are of linear in shape. Siliquae may show sunken, dark brown to black circular lesions. The severely infected siliquae contain small and shriveled seeds.

Epidemiology

A temperature range of 15-20°C, relative humidity of 70-90 per cent intermitten winter rains and wind velocity around 2-5 km per h has been reported to be conducive to alternaria blight development (Verma and Saharan, 1994). A minimum period of 4 h leaf wetness and reduced light intensity favours the lesion development. Closed spacing (30x15), higher dose of nitrogen (80 kg N/ha) and frequent irrigations are factors favoring the severity of the blight incidence (Saharan, 1991).

Perpetuation

The pathogens over winters on diseased crop residues seed and weed hosts.

White Rust

The white rust (caused by Albugo candida) is a major problem of mustard (*Brassica juncea*). It is estimated that the floral infection can cause losses from 23-54.5 per cent (Saharan *et al.*, 1984). The mixed infection of white rust and downey mildew is becoming a serious problem causing 37-47 per cent less pods with 17-32 per cent reduction in grain yield (Bains and Jhooty, 1979).

Symptoms

The initial symptom appears as small, localized, isolated white raised pustules or blisters on leaves, stem, petiole and floral parts. Later the pustules coalesce to form larger patches. Matured pustules rupture to release a mass of white powder of spores. In some cases pustule are surrounded by yellow hallow.The mixed infection of white rust and downy mildew is conspicuous at floral stage of crop growth where white rust produces pustules and downy mildew produces a coating of fine white powdery mass of spores.

Epidemiology

The optimum temperature for germination is about 10°C. The oospore germinates at 10-20°C. The placement of oospores in soil is an imporant factor in determining the time of infection. Placement of oospore at depth of 7.5 cm causes early infection while deeper (15cm) and shallow (on the surface) placement delays the infection in the host (Kumar *et al.*, 1995). The mixed infection of white rust and downey mildew is favoured by temperature between 13.8-14.8°C and rainfall more than 91.9 mm (Bains and Jooty, 1979).

Perpetuation

The pathogen perpetuates through oospore formed on diseased plant debris lying in the soil. The oospore can survive more than 21 years (Verma and Petrie, 1975). Perennial weeds host may also act as source of primary inoculam.

Downy Mildew

Downy mildew caused by *Peronospora parasitica* (Pers. Ex. Fr.), occurs in almost all the brassica growing areas of the country causing upto 75 per cent seedling deaths at cotyledonary infection. In case of mixed infection with white rust the pod yields losses is estimated about 37-47 per cent (Bains and Jooty, 1979).

Symptoms

The symptom exhibits as yellow, irregular spots on the upper surface of the leaves and creamy white to light brown growth on the dorsal surface, opposite to spots. Severely diseased leaves dry up, shrivel and tear off easily. Stem swelling and malformation of floral parts are quite often found which is the result of mixed infection of downy mildew and white rust.

Epidemiology

The optimum temperature for formation and germination of conidia are 8-10°C and 8-12°C respectively. Severe infection and rapid disease development takes place when plants are exposed for 4-6 h leaf wet ness at 20°C and 6-8 h at 24°C.

Perpetuation

Pathogen perpetuates in the soil through oospores which are formed in abundance in the malformed tissues of the infected plants. Seed borne inoculums may also serve as source of primary infection. Wild host and self grown plants are also reported to act as source of primary inoculums.

Powdery Mildew

Symptoms

The disease appears in the form of dirty white circular patches on both sides of lower leaves. Severely diseased plants appears as dusted with white grancular, chalk like powdery mass. Later the infected leaves, stem and siliquae show minute, circular black or dark brown spherical, scattered or concentrated bodies of perithecia (Saharan and Kausik, 1981).

Epidemiology

The temperature favouring germination of conidia, germ tube growth and appressorium formation is 20-25°C. It is reported that mean temperature of 16-28°C, mean relative humidity of less than

Figure 28.1: Symptoms on *Brassica* spp.

Figure 28.2: Symptoms on Stems and Siliquae

60 per cent and low or no rain fall are most congenial weather factors favouring epidemic development (Sahara and Sheoran, 1988).

Perpetuation

The pathogen survives in the form of perithecia on the diseased plant parts. Off season plants and weed hosts may also survive as source of primary inoculums.

Sclerotinia Stem Rot

Symptoms

Rotting of stem near the soil region. In severe conditions the whole plant topple down and dry off.

Epidemiology

Disease is favoured by humid weather.

Perpetuation

The pathogen can survice in the soil for several years as sclerotia.

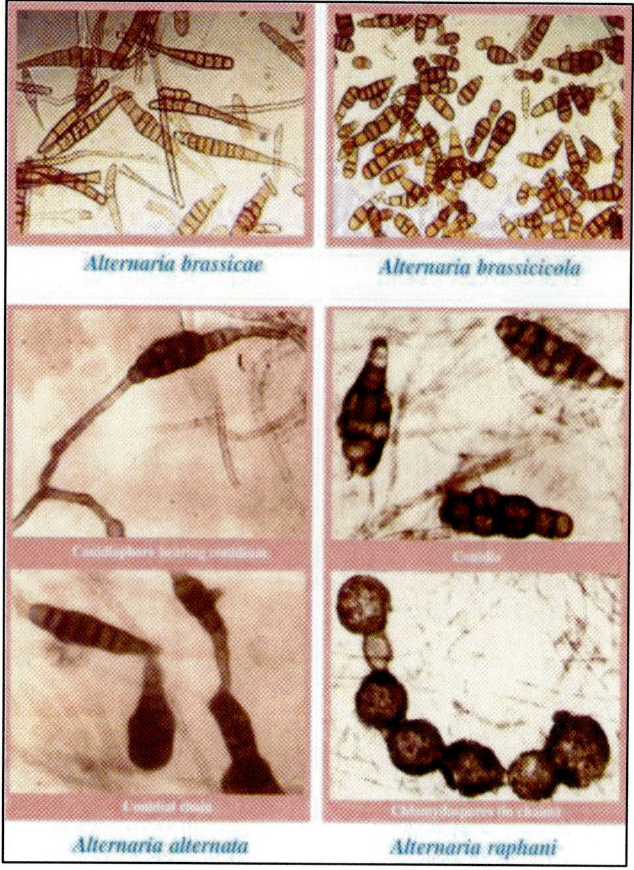

Figure 28.3: Morphological Characters of *Alternaria* spp.

Integrated Disease Management Approaches

Destructuion of Diseased Debris

1. Disease debris should be burnt after the harvesting of the crop.
2. Deep summer ploughing during the month of May – June to destroy the diseased debris.
3. Crop rotation with non cruciferous crops for atleast 3 years in case of severe infection.

Use of Resistant Varieties

1. Alternaria blight: The varieties PAB 93334, PAB 9511, RC 781, HC 9003, PBC 9221, PHR 1 and PHR 2 are found to be moderately resistant (Anonymous, 1997 and 1998).
2. White rust and downy mildew: The available resistant varieties are PWR 17, JYM 1, JYM 7 to 13, PWR 9538, PWR 9541, JMMWR 93-37, JMMWR 93-38, RH 8113, RH 9322, RH 8814, VSL 6, WDR 9601, PCR 10 and TERI (OE) (Anonymous, 1996, 1997, 1998).
3. Sclerotinia stem rot: The resistant varieties for sclerotinia stem rot includes PCR 10, PRO AGRO- HO-1, RGN 8006 and RW 8401 (Anonymous, 1997 and 1998).
4. Use of bold seeded true certified seed.

Adoption of Proper Cultural Practices

1. Depending upom the weather condition and diseased severity prevailing at particular area, sowing of the crop can be delayed or advanced.
2. Plant spacing should be appropriate to maintain proper plant population.
3. First irrigation should be given 40-45 days after sowing.
4. Field should be inspected at regular intervals. Off –typed plants, diseased plant, alternate host and weed host should be roughed out.
5. Judicious application of fertilizer at proper time of crop growth phase.
6. In case of root- rot diseases, the movement of soil from diseased field to healthy field should be restricted.

Need Based Use of Chemical

1. *Alternaria blight*: three sprays of Mancozeb (indofil M-45 0.25 per cent) at 60,80 and 100 days after sowing were found to be most effective in reducing the intensity of alternaria blight (anonymous, 1997).
2. *White rust*: Spray with Difolatan @ 0.2 per cent as soon as disease symptoms appears on the plant. Epidemic development can be checked by spraying Ridomil 0.25 per cent at 60 days followed by two sprays with Roveral 0.2 per cent at 80 and 100 days respectively.
3. *Downy mildew*: Seed treatment with Apron 35 @ 6g/kg seeds followed by Ist spray of Ridomil 0.25 per cent at 60 DAS and 2nd and 3rd spray of Roveral 0.2 per cent at 80 and 100 DAS (Anonymous, 1998). Spray with Zineb 0.2 per cent or Karathane 0.1 per cent for 2 to 3 times at 10 days intervals with the appearance of disease. Spray the crop with Ridomil MZ-72 @ 1.8-2.5 kg/ha in 750-1000 L of water against downy mildew in Toria as the disease starts appearing.
4. *Powdery mildew*: Spraying of Dinocap (Karathane 0.05 per cent) at 60,80 and 100 DAS was found most effective in minimizing the disease intensity (Anonymous, 1998).
5. *Sclerotinia stem rot*: Seed treatment with Bavistin @ 2.5g/kg when stem rot is a problem. Spraying of Roveral (0.2 per cent) or Bavistin (0.05 per cent) or Captan (0.02 per cent)were found quite effective (Anonymous, 1998).
6. Metasytox or Rogor can also be mixed with fungicidal solution @ 800-1000ml/ha for the control of aphids and other insects.

With the introduction of improved production technologies in the crop cultivation plant disease problem too have increased. Today it has become a challenge for farmers and Scientists to combat with the disease problem. If prior planning and suitable disease control strategies are adopted most of the disease can be managed so as to keep them below the economic threshold level. A crop may be affected by several diseases and it is not essential that a particular disease control method will be equally effective against all the diseases. Therefore, it is necessary to know the causal agent, mode of dispersal and disease cycle so as to develop such an strategy involving minimum activity and expenditure imparting complete or partial control of several disease simultaneously. Such a multi-disease management system can reduce the farmer expenditure and ease them in a application. It is essential for the success of IDM that it should be adopted in wide spread magnitude covering entire cropping areas. Since disease propagules spreads through wind, water, soil, insect- vectors etc. Therefore, if IDM is applied at community level it will be most effective with good economic returns.

Future Thrusts

Disease of rapeseed and mustard causes great destruction to the crop creating major economic loss. In spite of sufficient attention given to these diseases, no suitable method of its management has evolved so far. The only way to overcome the disease menace, to some extent, is the use of resistant varieties and crop rotation or intercropping. The discovery of the perfect state of the pathogen has provided substantial knowledge about the pathogens and has opened up new avenue for research on the disease and its casual organism. However, more efforts should be made to study:

1. Crop rotation, crop sequences and intercropping for natural suppression of the diseases.
2. Antagonism and biological control,
3. Survival of pathogen and factors adversely affecting the survival, and
4. Developing resistant varieties.

Referances

Ahamad, S. (2012). Recent Trends in Plant Disease Management in India. Kalyani Publishers, pp. 476.

Ahamad, S. and Khan, A. (2007). Integrated Management of white blister (rust) of rapeseed mustard edited by S.Ahamad and U.Narain, Published by Daya Publishing House, New Delhi, pp. 95-104.

Ahamad, S., Sharma, P. and Kumar, B. (2009). Integrated disease management on rapeseed mustard, In: Plant disease management for sustainable Agriculture, Edited by Shahid Ahamad, Published by Daya Publishing House New Delhi.

Anonymous (1996). Annual progress report 1996. All India Coordinated Research Project on Rapeseed – Mustard ICAR, Sewar, Bharatpur, Rajasthan, India.

Anonymous (1997). Annual progress report 1997. All India Coordinated Research Project on Rapeseed – Mustard ICAR, Sewar, Bharatpur, Rajasthan, India.

Anonymous (1998). Annual progress report 1998. All India Coordinated Research Project on Rapeseed – Mustard ICAR, Sewar, Bharatpur, Rajasthan, India.

Banis, S.S. and Jhooty, J.S. (1979). Mixed infection of *Albugo candida* and *Peronospora parasitica* on *Brassica juncea* inflorescence and their control. *Indian Phytopath*. 32: 268-271.

Degenhardt, K.J., Kondra, J.P. and Skoropad, W.P. (1974). Effect of Alternaria black spot on yield, oil content and protein content of rapeseed. *Canadian J. Plant Sci.*

Kadian, A.K. and Saharan, G.S. (1983). Symptomatology, host range and assessment of yield losses due to alternaria brassicae infection in rapeseed and mustard. *Indian J. Mycol. Pl. Pathol.* 13 (3): 319-323.

Kolte (1985). Diseases of annual edible oilseed crops. CRC, Boca Raton, Florida vol. 11, 135p.

Kumar, B. and Ahamad, S. (2009). Epidemiology and Management of Alternaria blight of mustard. In: Plant disease management for sustainable Agriculture, Edited by Shahid Ahamad, Published by Daya Publishing House New Delhi.

Kumar, V., Kaushik, C.D. and Gupta, P.P. (1995). Role of various factors in development of white rust disease of rapeseed- mustard. *Indian J. Mycol. Pl. pathol.* 25: 145-148.

Saharan, G.S and Kausik, J.C. (1981). Occurrence and epidemiology of powdery mildew of *Brassicas. Indian Phytopath*. 34: 54-57

Saharan, G.S. (1991). Assesment of losses, epidemiology and management of black spot disease of rapeseed- mustard. Proc. GCIRC 8th int. Rapeseed Cong. Saskatoon, Canada. 2: 465-470

Saharan, G.S. and Sheoran, B.S. (1988). Persistance and efficiency of some common fungicides against Alternaria brassicae (Berk). Sace., the causal agent of leaf blight of rapeseed and mustard. In 111rd Int. Symp. *Indian Phytopath.* Soc. Dec. 14-18, 1981.

Saharan, G.S., Kausik, C.D., Gupta, P.P. and Tripathi, N.N.(1984). Assesment of losses and control of white rust of mustard. *Indian Phytopath* (Abstr.) 14:11.

Verma, P.R. and Petire, G.A. (1975). Germination of oospores of *Albugo canadia*. *Canadian J. Bot*. 53: 836-842.

Verma, R.P. and Saharan, G.S. (1994). Monograph on Alternaria disease of crucifers research Branch Agri. Food. Canada, Saskatoon Research Centre, Tech. Bull. 1994-6E: 162p.

Hill Agriculture: Prospects, Constraints and Mitigations (2013)
Editor: Dr. Shahid Ahamad
Published by: DAYA PUBLISHING HOUSE, NEW DELHI

Pages 450–468

Chapter 29

Vegetable Diseases and their Management in Hill Agriculture

Shahid Ahamad[1], *Manoj Kumar*[2], *S.M. Wasi*[3] *and Naushad Alam*[4]

[1]*Krishi Vigyan Kendra-Poonch, SKUAST-Jammu*

[2]*Division of Vegetable and Floriculture, SKUAST-J*

[3]*Department of Agriculture, Bhagwant University, Ajmer – 305 001, Rajasthan*

[4]*KVK, Fatehpur, CSAUAT-Kanpur, U.P.*

Vegetables provide much needed nutritional security through balanced diet. India is the second largest producer of vegetables after China, but per capita consumption (140/day) is much below the minimum level (280/day) recommended by Indian Council of Medical Research. Hence, there is a need for further increase in the production to overcome the present inadequate availability and to keep pace with the ever-rising domestic demands. As the vegetables are more succulent and rich in nutrients, their successful cultivation is often affected due to numerous biotic and abiotic constraints, thereby, incurring high yield losses during post-production periods. The biotic constraints include the attack of various diseases caused by fungi, bacteria and viruses whereas nutritional and physiological disorders comprise abiotic constraints. Some of the previously unknown problems have become serious and are causing enormous losses to these crops. In the absence of suitable control measures, plant disease control has become heavily dependent on the fungicides to combat the wide variety of fungal diseases (Ahamad and Narain, 2007; Ahamad, 2009; Ahamad *et al.*, 2011; Ahamad, 2012).

Potato, tomato, chilli, cucurbits, pea, onion and crucifers are the major vegetables in hill agriculture. Various management strategies of some important vegetable diseases will be discussed hereunder.

Nursery Diseases of Vegetables

Vegetables such as tomato, chilli, cauliflower, cucurbits, onion etc. are raised through nursery. Damping off, a commonly occurring disease of nursery of vegetables appears usually in poorly managed nursery beds sown with infected seed.

Damping Off

One of the most common problems encountered in vegetable seedling production is "damping-off". Damping off is a disease that results in the rotting, collapse and finally death of seedlings just before or soon after they emerge. It often starts out in a few plants but can spread quickly throughout the entire flat and into neighbouring flats. One or more soil-borne pathogens may be responsible for this disease, depending on the environmental conditions in the greenhouse.

Soil-borne pathogens such as *Pythium* and *Phytophthora*, often called water molds, can be particularly destructive if the soil is kept too wet for a long period of time. Although temperature is important, different species of these so-called water molds can infect at different temperatures. Pythium is the most common water mold pathogen found on diseased greenhouse vegetable seedlings and is often associated with excessive nutrition or ammonium toxicity.

Rhizoctonia solani is another common soil-borne pathogen that causes seed rot, damping-off and root rot during the production of vegetable seedlings in the greenhouse. Most strains of *Rhizoctonia* infect in somewhat dryer soil than the water mold pathogens. Often, Rhizoctonia will girdle the stems above and below the soil line of susceptible seedlings resulting in the seedling toppling over. On some plants such as cole crop seedlings, the lower stems are completely girdled, and the stiff vascular tissue is all that is left holding up the seedling, and causing the symptom known as "wire stem".

The best control of damping-off is to avoid it altogether. Once damping-off has started in a seedling flat, it may be difficult to control. The following are a few tips that will help avoid or reduce the risk of damping off becoming established in the greenhouse production of vegetable seedlings.

Damping off is the cause for poor germination and guppy stand of the seedlings in nursery beds. The disease may occur as pre- or post-emergence damping off. There is failure of seedling emergence from the soil in the first phase resulting in patchy appearance. In post-emergence damping off, there is toppling down of the infected seedlings and infection occurs at ground level with water soaked tissue being apparent. Commonly occurring fungi reported to be responsible for damping off in nursery are *Pyhtium, Fusarium, Rhizoctonia* and *Phytophthora* spp. (Gupta and Thind, 2006).

Vegetable seedlings often do not grow well under humid conditions, particularly if the soil is cold and wet. The grower then has the choice of keeping weak, sickly plants or undertaking the costly operation of replanting. This kind of damage is usually caused by one or more soil-borne fungi that attack under conditions unfavorable for rapid seed germination and growth.

In pre-emergence damping-off, seeds may rot and seedlings may decay before they emerge. Young plants that do emerge may pale, curl, wilt and collapse from a rot at the soil line and below (post-emergence damping-off). The base of the stem is generally water-soaked at first, then turns gray to brown or black and then rots. Seedlings of plants such as cabbage, cauliflower, tomato and beans are girdled by brown or black sunken cankers. These plants may shrivel and become dark and woody, causing wirestem or collar rot, but they do not normally collapse. Transplants grow slowly or die.

Control

1. Buy the highest-quality seed of recommended varieties.
2. Treat seed before planting.
3. Plant in a light, well-drained, well-prepared, fertile seedbed at the time recommended for your area. If at all possible, pasteurize the soil – preferably with heat – before planting.

4. Apply a soil insecticide, such as diazinon, to the soil surface before planting, and immediately work it into the top 3 to 5 inches of soil. Follow the manufacturer's directions printed on the package label. This treatment controls root aphids, grubs, wireworms, ants, maggots, cutworms and webworms, all of which injure underground parts, allowing soil-borne organisms to enter and produce decay.
5. Be sure that flats, tools, plant containers, and benches are clean and free of plant pathogens. It is important that these materials be clean because they can introduce damping-off pathogens into clean soil, often with disastrous results. Tools, equipment, plant containers, flats, and other items can be disinfested by (1) steaming them at 180°F to 200°F for 30 minutes; (2) dipping and swabbing them with a formaldehyde solution (1 pint formaldehyde in 15 gallons of water); (3) dipping them in household liquid bleach solution (6 gallons household bleach in 100 gallons water) for 10 seconds – materials should first be brushed free of soil, sprayed with clean bleach solution, and allowed to dry; (4) fumigating with methyl bromide-chloropicrin mixtures; or (5) dipping or swabbing them with full strength alcohol (70 to 100 per cent) for 1 to 2 minutes – do not rinse.

Sanitation

Many outbreaks of damping-off can be attributed to poor sanitation practices. Disinfect equipment, benches, and particularly seedling trays with a 1-part bleach to 4-parts water solution or with another disinfecting products such as Virkon before planting.

Growing Medium

Use sterilized or heat-treated growing mediums. Soil-borne pathogens that cause damping off can naturally occur in peat moss and soil mixes. Ensure that growing medium is properly prepared and managed to provide good drainage, structure, aeration, and water-holding capacity.

Figure 29.1: Seed Treatment in Drum

Seed Treatment

Plant seeds that have been treated with a registered fungicide seed treatment to protect seeds and seedlings from infection by "damping-off" pathogens. Be sure to choose the proper seed treatment. Seed treatments such as Apron XL registered for protection against the water mold damping off pathogens will not necessarily control damping off caused by *Rhizoctonia*.

Seeding Depth

Do not plant seeds too deep and ensure that the right temperature is maintained for rapid seedling emergence and growth. Deep seeding cause stress on the seedlings as they grow through the soil and results in more tissue exposure below the soil line to the damping off pathogens.

Plant Density

Try not too over crowd or plant seedling too densely. Proper seeding rates to avoid thick plant stands, poor air movement and low light intensity, will help prevent disease from developing.

Fertilizer

Provide seedlings with the proper nutrients. Applying too much or too little fertilizer will affect the health of the seedlings and their ability to defend against disease.

Watering

Never over-water and always allow the seedling plugs to dry out just a little between watering. The longer the soil stays wet or saturated, the more likely water molds will infect and cause disease problems. Allowing soils to dry too much between watering may stress the seedlings and encourage damping-off caused by *Rhizoctonia*.

Many soil-borne diseases that start in the greenhouse can be easily introduced to the field on infected transplants. Once a soil-borne pathogen is introduced to a field it can quickly become established and may require more expensive management techniques in that field for years to come. By following a few simple tips in the greenhouse will result in healthy transplants ready to grow in a couple of months and avoid transferring diseases to the field.

Nursery should be grown on raised beds. Treating seeds before sowing with chemicals is the best method for controlling damping off. Fungicides *viz.* Captan, Thiram, Bavistin and organomercurials are recommended. The control is better if soil solarization of the nursery beds can be done for 2-3 weeks before sowing. One or two drenching with fungicides can be practiced for managing post-emergence damping off.

Disease Management in Field

Potato

Late Blight

It is the most destructive disease of potato throughout the world. Late blight has the tremendous potential to cause up to 70 per cent reduction in the yield in susceptible cultivar. The losses ranging from 20-25 per cent in Punjab, 40-50 per cent in Haryana, 15-20 per cent in Uttar Pradesh and 5-10 per cent in Bihar and West Bengal have been reported (CPRI, 1987). In epidemic years the losses very between 15-100 per cent depending on cultivar, crop stage and the fungicide sprayed.

Symptoms

Initially small water soaked lesions develop near the tips and margins of the leaves which rapidly grow into large brown to purplish black necrotic lesions. During morning hrs. whitish downy growth of the pathogen consisting of sporangiophores and sporangia can be seen on the edges of the lesions mostly on the underside of the leaves. Light brown to dark brown lesions appear on stem and petioles which may elongate later and girdle the affected parts. The affected stems become weak and collapse, causing death of the plant parts above the lesions. Since the disease is polycyclic in nature, the entire crop in a field may be killed in one or two weeks and the field gives blighted appearance. The infected tubers show irregular, small to large, slightly depressed areas of brown to purplish skin which extend deep into the internal tissue of the tubers. The disease is caused by *Phytophthora infestans.*

Management

Apart from host resistance, cultural practices should be adopted to reduce the risk of late blight. Since infected seed tubers stored in cold stores serve as the main source of primary inoculums, these should be culled and destroyed after taking out from stores before sowing. Cultivars such as Kufri Badshah, Kufri Jyoti, Kufri Chipsona provide moderate level of resistance to disease attack. Fungicides play an important role in managing late blight of potato. Several fungicides have been reported to provide good control of this disease. Copper oxychloride, mancozeb, zineb, propineb and chlorothalonil etc. (Malathi and Jeyarajan, 1995) have good efficacy as prophylactic applications. In the past two decades several systemic fungicides with significantly better efficacy have been found promising for late blight control. Metalaxyl, an acylalanine fungicide, has proved most effective against this disease. Oxadixyl, banalaxyl, ofurace and dimethomorph are other specific fungicides having high efficacy against it (Mohan and Thind, 1999). However, excessive and irrational use of these phenylamide fungicides may lead to the development of resistance in the pathogen. Metalaxyl resistant populations of *P. infestans* have been reported in many countries (Deahl *et al.*, 1993). To avoid the development of fungicide resistance, mixture formulations of systemic and contact fungicides are generally recommended. Ridomil MZ (metalaxyl + mancozeb) is one such mixture being commonly used against late blight (Thind and Mohan, 1997). Other mixtures which are in use for late blight control are ofurace + mancozeb, cymoxanil + mancozeb, dimethomorph + chlorothalonil, dimethomorph + mancozeb, oxadixyl + mancozeb, oxadixyl + copper oxychloride, banalaxyl + mancozeb and few others. Recently developed fungicides like azoxystrobin, fenamidone, famoxadone, iprovalicarb and mefenoxam (a new enantiomer of metalaxyl) and their combinations with contact fungicides like mancozeb have shown promising activity against potato late blight.

Black Scurf

It is a common disease of potato tubers and is of wide occurrence. It is prevalent where potatoes are cultivated. In India, the disease is observed both in hilly area and plains. Black scurf causes qualitative damage as it decreases the market value of the tubers both for table as well as seed purpose. Lakra (1992) reported that black scurf reduced seed germination by 11 per cent and affected yields by 68.2 per cent in the potato cultivar Kufri Chandramukhi. It is cause by *Rhizoctonia solani.*

Symptoms

Stem canker and scurf are the two distinct phases of the disease. In stem canker phase, the growing tips of the sprouts show browning. Sunken, circular or elongated brown necrotic spots may also be observed on the mature sprouts at the point of infection near mother tuber. Severely affected sprouts are killed. The most prominent symptom of black scurf is the presence of black crust on tubers due to the formation of sclerotia of the fungus. The pathogen produces a large number of sclerotia superficially on the surface of growing tubers. These sclerotia may be hard or spongy forming a black crust of scurf on the tuber surface. These are normally seated on the skin and do not cause any damage to the tuber inside. Black scurf phase is more common than stem canker in India.

Management

Treatment of seed tubers with chemicals before planting or before storing results in effective disease control. Dipping of seed tubers in 0.25-0.5 per cent solution of organomercurials like Emisan 6, Agallol or Aretan for 5-10 minutes is widely practised by the farmers. But these mercurial fungicides are being banned now because of toxicity and pollution problems. Ceratin oxathiin fungicides are also reported to be effective against stem canker phase of the disease. Tanii *et al.* (1982) reported significant control of the disease by dusting or spraying of tubers with validamycin. Khanna *et al.* (1991) reported good efficacy of carbendazim to control the disease. Dipping of seed tubers in 1 per cent acetic acid + 0.05 per cent zinc sulphate or in 3 per cent boric acid has also been reported effective. The treatment with boric acid is more effective, less toxic and safer than organomercurials which are not only hazardous but also affect emergence of sprouts especially when the temperature is > 300 C. Recently, Thind *et al.* (2002) have reported almost complete control of black scurf by dipping infected seed tubers in 0.25 per cent solution of Monceren, a phenyl urea based fungicide, for 10 minutes before sowing. Treating seed tubers in suspension of *Trichoderma harzianum* also provides good control of black scurf.

Common Scab

The disease causes tremendous economic losses as the affected tubers fetch low price in the market and are outrightly rejected for seed purposes due to bad appearance. The disease does not cause detectable yield loss but under severe root infection, yield may be reduced drastically. High disease incidence to the extent of 100 per cent has also been recorded (Shekhawat, 1990). It is caused by *Streptomyces scabies.*

Symptoms

The symptoms of common scab of potato appear on tubers as small brownish and slightly raised spots which later enlarge, coalesce and become corky. Multiple types of symptoms such as slight brownish roughening of tuber skin, proliferated lenticels with hard corky deposition, concentric series of wrinkled layers of cork around central black core, raised rough and corky pustules and deep pits (3-4 mm) surrounded by hard corky tissue (Paharia and Pushkarnath, 1963; Jeswani *et al.*, 1987) are produced.

Management

The tubers can be disinfected by dipping them in the suspension of organomercurials like MEMC @ 0.25 per cent for 10 minutes before sowing. Dipping of infected tubers in 3 per cent boric acid for 30 minutes is also quite effective. Green manuring with Dhaincha or Sunhemp and frequent light irrigations after tuberization also provides good control of common scab.

Tomato

Early Blight

The disease is caused by *Alternaria solani.* The pathogen can attack the crop at any stage but it is more destructive at maturity. The disease initiates in the month of April but is severe in the end of May. Severe epidemics can result in yield losses up to 78 per cent.

Symptoms

The symptoms appear as circular to angular, dark brown to black, concentric ring spots surrounded by yellow halo on lower leaves. Severe infection leads to premature leaf death, withering, dropping and eventually abscission. The whole leaf turns yellow and dries up when the lesions are more. The

Figure 29.3

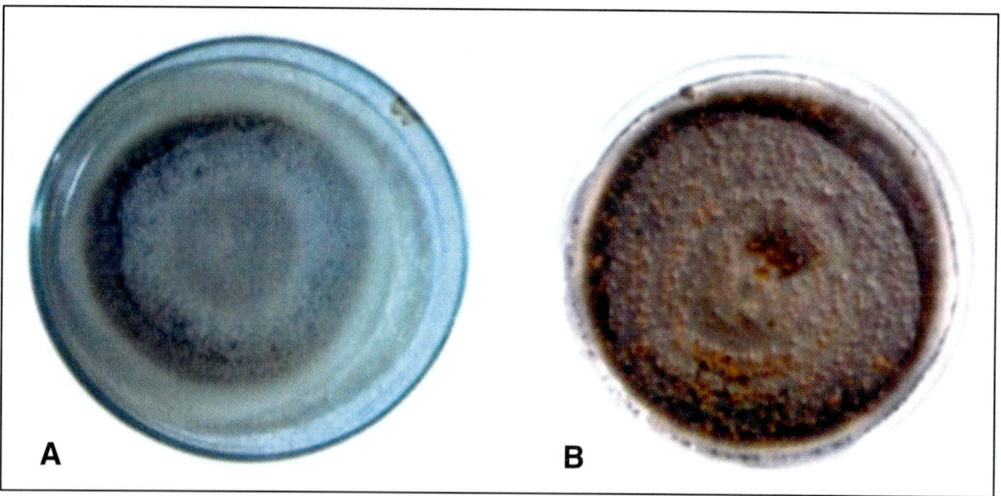

Figure 29.4: Pure Cultures of *C. capsici* Showing Pinkish Masses of Conidia
A: Pure culture; B: Pure culture showing masses of conidia

The pathogens are seed-borne and may also persist on alternate hosts such as other solanaceous crops (tomato, potato, and eggplant), cucumber, and many other cultivated crops and weeds. These pathogens will also readily persist in crop debris and in weeds, in some cases (*e.g. C. coccodes*) as resistant fungal structures called sclerotia. The pathogens can increase in number under continuous

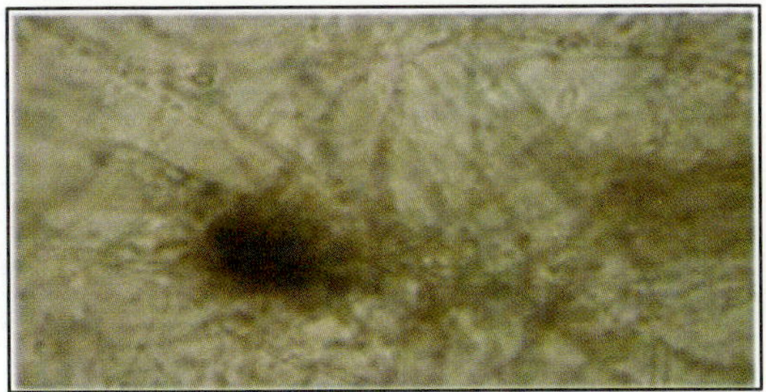

Figure 29.5: Mycelium Aggregation to Form Cushion

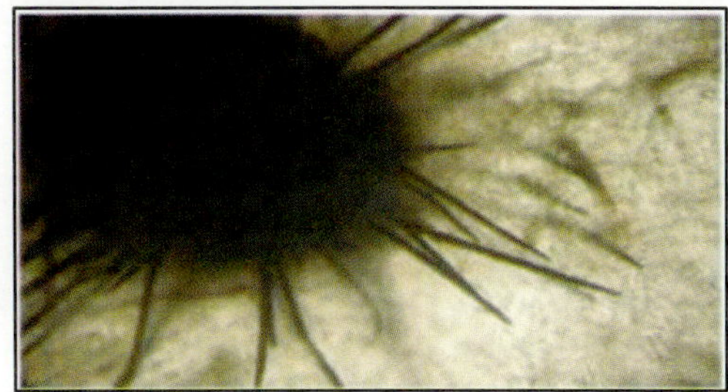

Figure 29.6: Acervulus Showing Setae and Conidia

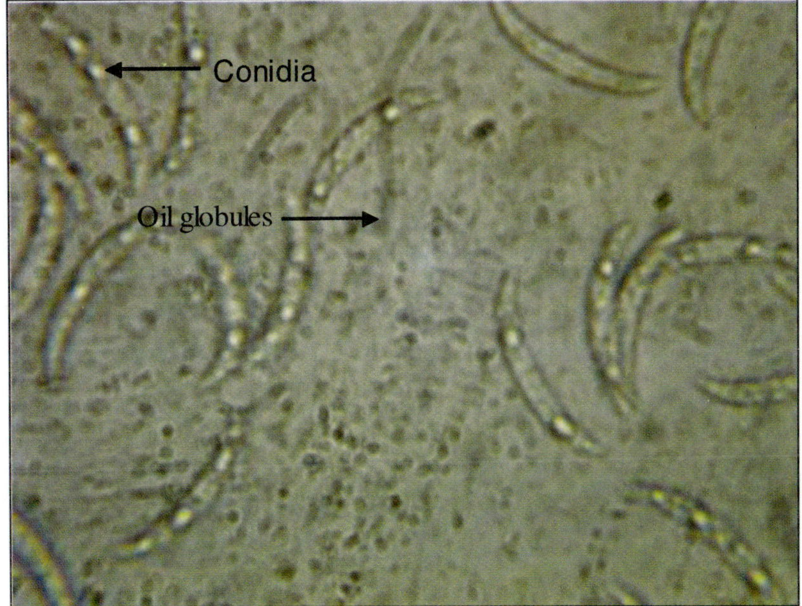

Figure 29.7: Sickle Shaped Conidia with Oil Globules

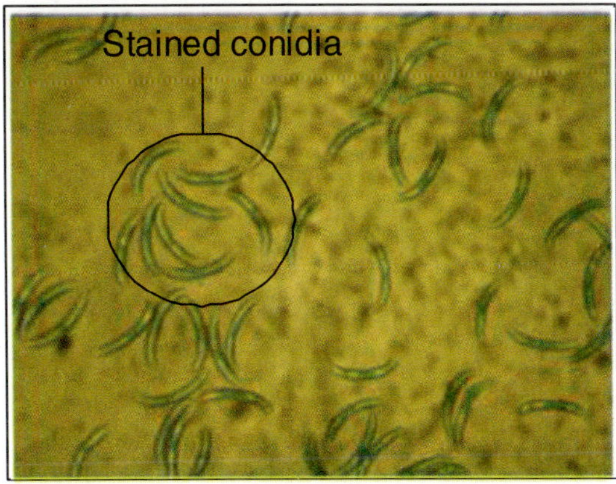

Figure 2.8: Stained Conidia with Oil Globules

cultivation of pepper, tomato or potato. Secondary cycles of anthracnose development during the growing season arise from spores produced on diseased fruit or leaves. Anthracnose fungi occur primarily on leaves and twigs (Figure 29.3).

On Deciduous Trees

These fungi over winter in infected twigs. Water-splash or wind-driven rain is required for dispersal of fungus spores or micro sclerotia on soil particles. Wounds in fruit are not required for infection but wetness is needed for spore germination and infection. The optimum temperature for fruit infection is 20–24°C with fruit surface wetness, although infection may occur from 10 to 30°C. However, the longer the period of fruit surface wetness the greater the anthracnose severity. Fruit that are at or near the soil surface are the most likely to become infected by rain-splash or direct soil contact. Overhead irrigation will favor development of anthracnose because of increased relative humidity and increased duration of dew periods.

Management

Control of the disease is through integrated management techniques. The disease should not be introduced on infected plants. Only seeds that are pathogen-free should be planted. Transplants should be kept clean by controlling weeds and solanaceous volunteers around the transplant houses. The field should have good drainage and be free from infected plant debris. If disease was previously present, crops Anthracnose caused by Colletotrichum sp. on Pepper should be rotated away from solanaceous plants for at least 2 years. Sanitation practices in the field include control of weeds and volunteer peppers plants. Resistance is available in some varieties of chili peppers but not in bell peppers. For bell pepper production, choose cultivars that bear fruit with as hotter ripening period which may allow the fruit to escape infection by the fungus. Wounds in fruit from insects or other means should be reduced to the extent possible because wounds provide entry points for *Colletotrichum* spp. and other pathogens like bacteria that cause soft rot. For late-maturing peppers, when disease is present, apply a labeled fungicide several weeks before harvest. Check your county Extension agent for a suitable, labeled product. The disease can be controlled under normal weather conditions with a reasonable spray program. At the end of the season, infected plant debris from the field must be removed or deep ploughed to completely cover crop diseases.

Select seed from anthracnose-free fruit or treat seeds with a fungicide. Hot water treatment is recommended to destroy seed-borne fungi. Soak seed at 52°C for 30 minutes. Following treatment, plunge the hot seeds into cold water, dry on paper and dust with thiram.

The most effective fungicides are chlorothalonil (1.5 per cent) and thiophanatemethyl (0.5 per cent) against the anthracnose and fruit disease. Freshly harvested seed withstands heat treatment better than one or two-year-old seed. Use healthy transplants. Sanitize flats if reusing them for transplant production. Broad-spectrum fumigants can be applied to soil in seedbeds to control the pathogens but annual re-application may be necessary if re-contamination of the soil by the fungus occurs.

Avoid potato, soybean, tomato, eggplant and cucurbits as rotation crops. Rotate with non solanaceous crops for three years. Mulch to reduce soil splash onto fruit and lower leaves. Minimize or avoid overhead irrigation to reduce periods of wetness on pepper fruit. Harvest fruit as soon as it ripens since anthracnose develops more readily as the fruit ages. Weed regularly and avoid injuring pepper fruit. Allow infested crop debris to decompose completely by deep ploughing of crop residues before planting again. If only a few plants are affected by the disease, these can be removed from the field and disposed of. Avoid planting over lapping pepper crops near by. Apply protectant fungicides to plants starting when the first fruit are set. This will prevent or minimize the occurrence of infections. Control by fungicide application will depends upon proper dosage and good coverage of fruit. Some sources of resistance have been identified; check with your local extension agent for possible resistant cultivars.

Biological control of fruit rot and dieback of chilli with plant products tested in many laboratories and field trials showed that the crude extract from rhizome, leaves and creeping branches of sweet flag (*Acorus calamus* L.), palmorosa (*Cymbopogon martinii*)oil, *Ocimum sanctum* leaf extract, and neem (*Azadirachia indica*) oil could restrict growth of the anthracnose fungus (Jeyalakshmi and Seetharaman, 1998; Korpraditskul *et al.*, 1999).

Anthracnose, Die-back or Ripe Fruit Rot:

This is one of the serious diseases of chilli in all the chilli-growing areas of the world. The disease is favoured by rainy season. The losses can reach up to 50 per cent. The disease is a major problem on mature fruits. However, the symptoms on immature fruits have also been found. The disease is caused by *Colletotrichum capsici*. The pathogen also causes dieback of chilli plants. The fruit rot phase of the disease can cause heavy damage in the field as well as during transit.

Symptoms

Dieback

There is necrosis of the tender twigs from the tip backwards. The entire branch or the entire top of the plant may wither away. The twigs are water soaked to brown, become grayish white or straw coloured in advanced stages of the disease. Large number of black dots called acervuli are formed on the affected twigs.

Ripe Fruit Rot

On mature fruits, circular to elliptical sunken spots appear with black margins and marked with concentric rings. Badly diseased fruits turn straw coloured from normal red. On this discoloured area, numerous acervuli are present. The diseased fruits may drop off prematurely.

Management

The seed should be obtained from healthy fruits. The seed should be treated with fungicides like Captan or Thiram @ 2 g/kg seed. The secondary spread can be checked by spray of fungicides like Dithane M-45, Blitox and Bavistin. Fenapanil and Triazole fungicides like Tebuconazole have also been reported to be effective against this disease (Datar *et al.*, 1990; Mohan *et al.*, 2004).

Wet Rot

The wet rot is one of the important fungal diseases of chillies, which is severe during the months of August and September. It is caused by *Choanephora cucurbitarum.*

Symptoms

Flower buds and tender young shoots are attacked by the pathogen under favourable humidity conditions. The infected portion shows the tendency of wet rot with luxuriant growth of the fungus on the infected portion, which appears as black pinheads. The infection spreads downward and the infected portion becomes limply, resulting in the death of the distal parts. The disease also appears on fruits causing soft rot with luxuriant growth of the pathogen with characteristic bead like black conidial heads.

Management

Control measures are the same as discussed under anthracnose.

Leaf Curl

This is a viral disease caused by tobacco leaf curl virus. The disease is very common wherever the chilli crop is grown. It is a major limitation in successful cultivation of both hot and bell pepper. The disease can cause heavy yield losses and reduce the quality of the fruit.

Symptoms

Initially the leaves become pale yellow in colour, followed by slight curling. The inter-veinal areas are puckered, there is shortening of internodes and reduction in size of the leaves giving the plants witches broom appearance. Severely infected plants remain stunted and assume bushy appearance. Fruit formation in susceptible cultivars is rudimentary and distorted (Mishra *et al.*, 1963). The disease is transmitted by whitefly, *Bemisia tabaci*. This vector spreads the maximum disease during the months of August to October.

Management

A combination of use of resistant varieties, barrier/trap crops and insecticides against the vectors has been found successful in reducing the disease intensity. Growing of marigold with capsicum has been reported to cause significant reduction in the incidence of viruses infecting pepper in Mexico (Chew-Madina-Veitia *et al.*, 1995). It is reported that the combined use of yellow traps followed by Cypermethrin spray @ 110 g a. i./ha reduces the population of *Bemisia tabaci*.

Cruciferous Vegetables

Black Rot

This disease is wide spread and occurs on both cultivated and wild crucifers where rainfall or heavy dew is plentiful. The pathogen can attack cabbage, cauliflower, turnip, radish etc. and if the favourable conditions prevail cabbage plants are unable to produce marketable heads. The disease is caused by the bacterium, *Xanthomonas campestris* pv. *campestris*.

Symptoms

The first disease symptoms appear as chlorotic lesions near the leaf margin which progress towards the centre in the form of "V" shaped yellow spots. In the affected portions, the veins and the veinlets turn brown and finally black which is characteristic of the disease. The vascular blackening of the petiole and stem causes systemic spread of the causal organism. Bacterial ooze may be seen on the surface of the diseased tissue. Curds also get infected and rot.

Management

Sprays of antibiotics are not much advocated because of resistance development in the pathogen and other hazards. However, some need-based recommendations are available. For the effective management of the disease, the seed treatment must be followed by three sprays of streptocycline (50 ppm) at transplanting, curd formation and pod formation stages. In the field, spraying of Streptocycline (100-200 ppm) has been reported effective against black rot infection (Bera, 1986). Sprays with copper fungicides at 10 days interval also help in reducing the spread of the diseased. Application of 10-12.5 kg/ha stable bleaching powder as soil drench is also effective (Saini and Parashar, 1981).

Alternaria Blight

The disease occurs on cabbage, cauliflower, radish, turnip etc. When too many spots are formed on the leaves, they die prematurely affecting the plant stand and in addition, the spots predispose the heads to various types of rots during transit and storage. The disease is caused by *Alternaria brassicae* and *A. brassicicola*.

Symptoms

Initial symptoms of the disease appear as small dark coloured areas, which spread rapidly and form circular spots with concentric rings. In humid weather, the fungus appears as bluish black growth in the centre of the spots. The cauliflower curds are also infected which show browning; beginning at the margins of the individual flower or flower clusters and later on the whole curd may become brown and rot. Brown to black spots appears on pods in the seed crop of cruciferous vegetables.

Management

The disease debris after harvest should be destroyed. The seed should be treated before sowing with Captan or Thiram @ 3 g/kg seed. Fungicides such as Dithane M-45, Blitox, Rovral and Kavach have been reported to give good control of this disease (Chand, *et al.*, 2005, 2006, 2007 and Ahamad and Narain, 2007). Recently Strobilurin fungicides have also shown promising activity against this disease.

Downy Mildew

The disease occurs on cole crops such as cabbage and cauliflower. In addition, it also occurs on radish. The disease is quite common in young plants and the destruction may reach up to 90 per cent. It causes heavy losses to the seed crop. The disease is caused by *Peronospora parasitica*. Under favourable conditions, the pathogen infects 50-60 per cent seed of cabbage. It is also responsible for the post-harvest spoilage of cauliflower heads.

Symptoms

The disease is characterized by the appearance of purplish brown spots on the under surface of the leaves. The upper surface of the leaf on the lesion is tan to yellow. Fungal growth appears on the under surface of these lesions. The stem shows dark brown and depressed lesions or streaks. In cabbage, the infection exposes the heads to soft rot during transit. The affected cauliflower curds turn brown at the top and the severely infected curds rot and fail to produce seeds.

Management

Use of chemicals is imperative once the disease has appeared on the growing crop. With the advent of systemic fungicides, a number of chemicals have been evaluated and recommended for the management of downy mildew. Heat treatment for seeds at 43-50°C for 20 minutes and dressing of seed with systemic fungicide Metalaxyl is helpful in reducing seed borne infection (Paulus *et al.*, 1978). Treatment of cabbage seedlings with Captafol @ 0.2 per cent or Polyram (Maneb) at weekly intervals significantly reduces the incidence of the disease in nursery (Apandi, 1980). Spraying phosphonic acid or a combination of Metalaxyl + Mancozeb or Cymoxanil + Mancozeb is sufficient to control downy mildew in cauliflower and cabbage respectively (McKay *et al.*, 1992).

Cucurbits

Powdery Mildew

The disease is caused by *Erysiphe cichoracearum*. The fungus attacks the leaves and stem of cucurbits *viz.* cucumber, pumpkin, bottle gourd, muskmelon and squashes.

Symptoms

The initial symptoms appear as white to dirty grey spots, which become powdery as they enlarge. This white powdery coating covers leaves, stem and other succulent parts of the plant. Severe infection may lead to premature defoliation and death of veins.

Management

Fungicides like Carbendazim, Dimethrimol, Pyrazophos, Triademefon and Tridemorph are quite effective in controlling powdery mildew. Strobilurin compounds developed recently also provide effective control of powdery mildew. Application of sulphur dust should be avoided as most of the varieties of cucurbits are highly sensitive to sulphur.

Downy Mildew

It is an important disease of cucurbits in different countries and affects its successful cultivation. Early infection of downy mildew can cause reduction in crop yield up to 60 per cent whereas late infection is less damaging. The loss of foliage as a result of early infection precludes normal flower set and fruit development. Maturing fruits fail to get proper colour, are tasteless and look sun burnt. The disease is caused by the fungus *Pseudoperonospora cubensis.*

Symptoms

At first, small water soaked lesions appear on the underside of the leaf. Such lesions when viewed from the upper side of the leaf appear as angular spots and grayish spore mass of the fungus is clearly visible on the underside of the spots. The lesions subsequently become necrotic. The severely infected leaves become dry and curl upwards giving blighted appearance to the veins. The disease appears in epiphytotics form during the months of April to June. Low night temperature (< 20°C) and long dew periods are the most important determining factors for the onset and development of downy mildew, The fungus perpetuates on off-season vines.

Management

Weekly prophylactic sprays of Chlorothalonil or Mancozeb give good control of this disease but these fail to check established infection. Spraying of Mancozeb and Metalaxyl mixture (Ridomil MZ) starting with the onset of conducive weather with repetition at 10 days interval are very effective, In recent years, Metalaxyl resistant strains of the pathogen have developed in many countries as a result of continuous use of this fungicide. Metalaxyl sensitive or tolerant strains of *P. cubensis* on cucumber and melons can be controlled by Dimethomorph. Cymoxanil + Mancozeb or Chlororthalonil as tank mixture also provide effective control (Robak, 1995). Recently developed Strobilurin fungicides have been found to be quite effective against this disease,

Cucumber Mosaic Virus (CMV)

Several cucurbits are attacked by viruses causing mosaic like symptoms. The nature of symptoms depends on the host and strains of virus associated with the disease. The losses are high, when the young plants are attacked. Maximum losses occur on squashes, bottle gourd, muskmelon, pumpkin and cucumber.

Symptoms

The characteristic symptom is yellowish of cotyledons and ultimately wilting of seedlings when the plant is attacked soon after emergence. In older plants, there are alternate green and yellow patches on the leaf. The spots are irregular in shape and enlarge rapidly. The diseased leaves are mottled, deformed, small and sometimes curl downwards. The veins and veinlets also turn yellow. Young fruits are rough, mottled, deformed and smaller in size. The virus is seed borne and is transmitted by aphids. The disease can also be transmitted through sap and mechanical methods. Some weed hosts serve as source of inoculum.

Management

Removal of alternate host plants, use of virus free seed and control of insect vectors through the use of insecticides are the possible measures. Neem derivatives could be very suitable for integrated virus disease management as they are primarily feeder poisons for nymphs/larvae of insect vectors of plant viruses (Singh and Krishna Reddy, 1995).

Pea

Powdery Mildew

It is the most common and serious disease-affecting pea in various countries. The disease becomes more serious during warm and dry days. Early sown varieties escape the disease. The disease is caused by *Erysiphe pisi.*

Symptoms

It first appears on leaves and then on older green parts of the plant including the pods. The characteristic symptom is the formation of white floury patches on both sides of the leaves, tendrils, stems and pods. These patches originate as minute discoloured specks from which a powdery mass radiates on all sides. In the advanced stages of the infection, all the aerial parts of the plant are found covered with white floury patches, which consist of mycelium and spores of the fungus. Seeds shrivel and loose taste.

Management

Various chemicals have been found effective in the management of pea powdery mildew both as seed treatment and spray. Seed treatment with Triademefon (Kapoor and Sugha, 1995), Bayleton (Singh, 1996) and Contaf (Paul and Rathour, 1998) was found effective in delaying the first appearance of the disease. Secondary spread is best controlled by spray of fungicides like Dinocap, Triademefon, Maneb + Tridemorph, Fenarimol, Mildex, Sulfex, Thiovit etc. The fungicides have to be sprayed periodically depending upon the prevailing disease intensity and weather conditions (Panja and Chaudhuri, 1994).

Rust

It is a serious disease of pea and occurs throughout the pea growing areas of the world. Two species of *Uromyces* are responsible for causing rust disease *i.e., Uromyces pisi*, a heteroceous type and *U. fabae*, an autoceous type. The later occurs in India and causes varying degrees of losses. Under artificial conditions, yield losses up to 30 per cent have been reported. Disease may become severe in K deficient soils.

Symptoms

The pathogen infects all the above ground parts of the plant. By the end of February, yellow to orange coloured round or elongated clusters appear on the leaves. These are aecial cups and bear aeciospores. Later, these clusters change to light brown in colour which is the uredial stage. Finally, the uredosori are replaced by the teliosori, which are dark brown to black in colour. Severely infected leaves may shrivel and shed off. Teliosori are also common on the stem and petiole besides leaves. Under favourable weather conditions, the entire plant may be killed.

Management

Chemical sprays are particularly useful to halt the early build up of the rust pathogen. In India, seed treatment with mercurial fungicides and Thiram has been recommended but mercurials are banned now. Fungicides like Dithane M-45 (0.2 per cent), Bayleton (0.05 per cent) and Calixin (0.2 per cent a.i.) have been found useful as spray. A total of 3 sprays given at 10 days interval are useful in reducing disease levels.

Onion

Purple Blotch

It has been reported to occur on onion, garlic, leek and other *Allium* spp. Losses due to the disease are heavy on seed crop. The disease is often aggravated by the thrip injuries (Mckenzie *et al.*, 1993). It is caused by *Alternaria porri*.

Symptoms

The disease is characterized as small (2-3 mm in diameter) water soaked lesions on the leaves or seed stalks that quickly develop white centres. These lesions enlarge, coalesce, become zonate and brown to purple under favourable conditions and extend both upward and downward. In moist weather, the surface of the lesion may be covered with black fruiting bodies of the fungus. Similar lesions may form on seed stalks; seeds either do not develop or are shrivelled. The bulbs can also be affected at harvest when the fungus enters through the necks or injuries, causing storage rot.

Management

The seed should be treated with fungicides like Captan or Thiram @ 3 g/kg seed before sowing. Dithiocarbamate fungicides like mancozeb and Zineb are reported effective in controlling purple blotch under field conditions (Gupta *et al.*, 1991). Addition of insecticides like Malathion and stickers like Triton in Mancozeb solution has been reported to be more effective against this disease by reducing the injuries due to thrips and providing better adherence of the fungicide to the waxy surface of the plant has good efficacy of triazole fungicides.

References

Ahamad, S. and Narain, U. (2007). *Eco-friendly Management of Plant Diseases*. Daya Publishing House, Delhi. pp.461-472.

Ahamad, S (2009). Plant Disease Management for Sustainable Agriculture, Daya Publishing House, New Delhi, pp373.

Ahamad, S., Ali Anwar and P.K.Sharma (2011). Plant Disease Management on Horticultural Crops, Daya Publishing House, New Delhi, pp405

Ahamad, S. (2012). Recent Trends in Plant Disease Management in India. Kalyani Publishers, pp. 476.

Apandi, E. S. (1980). *Peronospora parasitica* control on cabbage seedlings in the nursery. *Bull Penelitian Horticulture* **8**: 3-5.

Arvind A. Deshpande1 and Ram D. Rawal2 (2007). Resistant Sources of Chili (*Capsicum annuum* L.)Anthracnose Fruit Rot Disease [(*Colletotrichum capsici* (Syd.)Butler and Bisby] against Different Isolates Collected from Commercial Chili Growing Areas of India. First International Symposium on Chili Anthracnose, Hoam Faculty House, Seoul National University, Seoul, Korea, pp.46

Bera, S. C. (1986) Control of black rot disease of cauliflower by some chemicals. *Pesticides* **20**: 51-52.

Chand, G., Narain, U. Kumar, M. and Srivastav, M. (2005). Integrated management of *Alternaria* leaf Spot of broccoli. *Indian J. of Plant Pathol.*, **23:101**-103.

Chand, G., Narain, U. and Pandey, R. (2006). Efficacy of different fungicides against *Alternaria* leaf spot of broccoli. *Annals of Plant Prot. Sci.*, **14(2)**:*468-469*.

Chand, G., Narain, U., Kumar M. and Verma, S. (2007). Cultural management of *Alternaria* leaf spot of broccoli. *Annals of Plant Prot. Sci.*, **15(2)**:*510-511*.

Chew-Madinaveitia, Y. I., Zavaleia, M. E., Delgadillo, S. F., Velfivis, A. R., Pena, M. R. and Cardenas, S. E. (1995) Evaluation of control strategies for virus disease of pepper (*Capsicum annum* L.) *Phytopathologia* **30**: 74-84.

CPRI (1987). Annual Scientific Report. Pp. 91-97.

Datar, V.V. Sontakke, M.B., Purandave, N.D. and Shinde, N.N. (1990). Fungicidal control of anthracnose of chillies. *Indian J. Mycol. Pl. Pathol*. **20**: 156-58.

Deahl, K.L., Inglis, D.A. and Demuth, S.P. (1993). Testing for resistnace to metalaxyl in *Phytophthora infestans* isolates from North-Western Washington. *Am. Potato J.* **70** (11): 779-95.

Gupta R. P., Srivastava P. K. and Pandey U. B. (1991). Studies on the economical spray schedule of mancozeb for the control of purple blotch disease of kharif onion. *Indian Phytopath*. **44**: 537-38.

Gupta, S.K. and Thind, T.S. (2006). Disease Problems in Vegetable Production. Scientific Publishers, Jodhpur, pp. 576.

Jeyalakshmi, C., Seetharaman, K. (1998). Biological control of fruit rot and die-back of chilli with plant products and antagonistic microorganisms. Plant Disease Research, 13:46-48.

Jeswani M. D., Gadewar A. V. and Singh B. P. (1987). Studies on mixed infection of *Streptomyces griseus* and *Fusarium oxysporum* on potato tubers. *J Soil Ecol* **7:** 73-78.

Kapoor A. S. and Sugha S. K. (1995). Efficacy of different fungicides in controlling powdery mildew (Erysiphe pisi) of pea (*Pisum sativum*) in field. *Indian J Agric Sci* **65**: 771-73.

Khanna, R, Roy, S, Rai, R. P., Sharma, J, Jeswani, M. D. and Singh, P. H. (1991). Annual Scientific Report, CPRI, Shimla, pp. 74.

Korpraditskul, V., Rattanakreetakul, C., Korpraditskul, R., Pasabutra, T. (1999). Development of Plant Active Substances from Sweetflag to Control Fruit Rot of Mango for Export. In: Proceeding of Kasetsart University Annual Conference. Kasetsart University, Bangkok, p.34.

Lakra, B. S. (1992). Correlation of infection intensities of black scurf with yield components of potato. *Indian J. Mycol. Pl. Pathol*. **22**: 203-04.

Manandhar, J.B., Hartman, G.L., Wang, T.C. (1995). Anthracnose development on pepper fruits inoculated with *Colletotrichum gloeosporioides*. Plant Disease, 79.

Mckay A. G., Floyd R. M. and Boyd C. J. (1992). Phosphonic acid controls downy mildew in cauliflower curds. Aust. *J Expt Agric* **32**: 127-29.

Mishra M. D., Ray Chaudhuri S. P. and Jha A. (1963). Virus causing leaf curl of chilli (*Capsicum annum*). *Indian J Microbiol* **3**: 73-76.

Malathi, P. and Jeyarajan, R. (1995). Evaluation of chlorothalonil for potato late blight in Nilgiri. *Indian J. Mycol. Pl. pathol*. **25** (3): 211-13.

Mckenzie C L, Cartwright B, Miller M E and Edelson J V (1993). Injury to onions by *Thrips tabaci* Linderman (*Thysanoptera: thripidae*) and its role in the development of purple blotch. *Environ Entmol.* **22:** 1266-77.

Mohan, C. and Thind T.S. (1999). Persistence and relative performance of some new fungicides for effective management of potato late blight in Punjab. *Indian J Mycol Pl Pathol.* **29** (1): 32-37.

Mohan, C. and Thind, T.S. (2001). Relative performance of spray schedules of systemic and nonsystemic fungicides against late blight of tomato. *Pl. Dis. Res.* **16** (2): 247-50.

Mohan, C., Thind, T.S., Prem Raj and Arora, J.K.(2004). Promising activity of triazoles and other fungicides against fruit rot of chillies and Stemphylium blight of onion. *Pl Dis Res.* **19** (2): 200-03.

Paharia K. D. and Pushkarnath (1963). Nutritional requirements of potato-isolate of *Macrophomina phaseoli* (Maulb.) Ashby *Indian J Microbiol* **8:** 179-82.

Panja, B. N. and Chaudhuri S. (1994). Control of powdery mildew disease of pea economically with some systemic and non-systemic fungicides. *Pl Prot Bull Faridabad* **46:** 4-6.

Paul, Y. S. and Rathour, R. (1998). Integrated management of pea diseases in Himachal Pradesh. In: Abstracts of International Symposium on Sustainable Agriculture in Hill Areas. p. 132.

Paulus, A., Synder M. Gafney, J., Nelson, J. and Otto, H. (1978) New systemic fungicide controls downy mildew of broccoli. *California Agric* **32:** 12-13.

Ramachandran N, Madhavi Reddy K, and Rathnamma K (2007). Current status of Chili Anthracnose in India. First International Symposium on Chili Anthracnose, Hoam Faculty House, Seoul National University, Seoul, Korea, pp.25

Robak, J. (1995). Epidemiology and control of cucumber downy mildew *Psuedoperonospora cubensis. Biuletyn Warzywniczy* **43:** 5-18.

Saini, L. C. and Prashar, R. D. (1981). Comparative efficacy of stable bleaching powder with other antibacterial formulations in controlling black and soft rot of cauliflower. *Indian Phytopath* **34:** 465-69.

Shekhawat, G. S. (1990). Emerging trends in epidemiology and management of baceterial diseases of potato. In: Grewal J S, Shekhawat, G S, Bhattacharya S K and Singh R A (Eds.). Current Facets in.Potato Research. *Indian Potato Association*, CPRI, Shimla, pp 109-20.

Singh, A. (1996). Biology and management of Erysiphe pisi on pea. PhD Thesis submitted to the Department of Plant Pathology, HPKV, Palampur, p 73.

Singh, S. J. and Reddy, M. K. (1995). Neem based pesticides, potential agents for the management of viral diseases of vegetable crops. In: Neem for the Management of Crop Diseases. Ed. V. Mariappan. Pp 125-36. Associated Publishing Co, New Delhi.

Tanii, A., Akai, J. and Aota, T. (1982). Chemical control of several potato diseases transmitted through seed tubers. In: Potato in Developing Countries (eds. B. B. Nagaich *et al,*), CPRI, Shimla, pp. 271-78.

Hill Agriculture: Prospects, Constraints and Mitigations (2013) *Pages* **469–479**
Editor: **Dr. Shahid Ahamad**
Published by: **DAYA PUBLISHING HOUSE, NEW DELHI**

Chapter 30

Damping-off Disease in Vegetable Crops in Hill Ecosystem

Shahid Ahamad[1], S.M. Wasi[2], S.A. Khan[3], S.P. Gupta[4] and Manoj Kumar[5]

[1]*Krishi Vigyan Kendra-Poonch, SKUAST-Jammu*
[2]*Department of Agriculture, Bhagwant University, Ajmer – 305 001, Rajasthan*
[3]*N.D.U.A.T., Kumarganj, Faizabad*
[4]*Krishi Vigyan Kendra, R.S. Pura, SKUAST-Jammu*
[5]*Division of Vegetable Sciences, SKUAST-J*

Vegetables constitute a major portion of human diet in western countries. It is unfortunate that in India our food is so much cereal based that vegetables hardly find a place in the diet of the majority of the population. According to dietary standards, a man's diet must include 200-350 g of vegetables of which 200 g should be in the form of leafy material. But it has been stated that the production is approximately half of the requirement as per dietary standards. Vegetables being perishable commodity their shortage cannot bc mct by imports. Available cultivated area under grain production cannot be diverted for vegetable production for justifiable reasons. Since there are more input requirements for vegetable growing than for most of the food crops, there is hesitation on the part of the farmer to take these crops. The disease factor alone is deterrent, because the holdings of the farmers who grow vegetables intensively are very limited, forcing them to take one vegetable after another, often practicing mono-cropping, which leads epiphytotic development of some diseases (Ahamad, 1998; Ahamad and Yasmeen, 2000; Ahamad and Narain, 2007; Ahamad, 2009 and Ahamad, 2012). Diseases in the standing crop from the seedlings stage till harvest and spoilage caused by microorganisms during transit, storage and marketing are the major constraints in increased production and availability. This emphasizes the need for increasing the production and availability form the existing area under these crops and ensuring that the produce remains unspoiled before it reaches the consumer. Therefore, vegetables require more attention by the farmers and scientists' right from field to the market level and correct diagnosis of malady with suitable control measures need to be explored for better production of vegetables.

Damping off disease of seedling is widely distributed all over the world. It occurs in most soils, temperate and tropical climates and in greenhouses. The disease affects seed and seedlings of various crops. The amount of damage the disease causes to seedlings depends on the fungus, soil moisture and temperature. Cool wet soils favour the development of the disease. Seedlings in seed beds often are completely destroyed by damping off, or they die after transplanting. Frequently, germinating seeds are killed by damping off fungi before they emerge from the ground, which accounts for poor stands in many crops. Older plants are seldom killed by damping off fungi mainly because the development of secondary stem tissue. The greatest damage is done to the seed and seedling roots during germination either before or after emergence. Within days, a great number of seedlings are colonized by the fungi resulting in plant losses. Losses from the disease vary considerably with soil moisture, temperature, and other factors quite frequently. Plants selected from nursery beds with damping-off may develop root rot or stem canker several weeks later. This disease is common to most of the vegetables grown from transplanting or even when directly planted. The disease is responsible for the poor germination and stand of seedlings in nursery beds and often the infected seedlings carry the pathogens to the main field where transplanting is done. With some damping-off fungi, foliar blight may also occur.

Symptoms

Damping off in vegetables occurs in two phases.

Pre-Emergence Damping Off

In this phase of the disease, the young seedlings are killed before they emerge through the soil surface. If germination has occurred and the hypocotyl (emerging shoot) is water-soaked in appearance or visibly decayed, fungi are the most likely cause. In fact, the seeds may rot or the seedlings may be killed before the hypocotyl has broken through the seed coat. The radicle and the plumule, when they come out of the seed, undergo complete rotting. Since this happens under the soil surface, the disease is often not recognized by the farmer who attributes the failure of emergence to poor quality of the seed.

Post-Emergence Damping Off

Seedlings often develop a darkened or shriveled "wire stem" appearance at the base of the stem. This phase is characterized by the toppling over of infected seedlings after they emerge from the soil until the stem has hardened sufficiently to resist invasion by the pathogen. Infection usually occurs at the ground level or through roots. The infected tissues appear soft and water-soaked. As the disease advances the stem becomes constricted at the base and plants collapse. The top of the plant may appear healthy when it falls over but quickly wilts and dies. The roots may or may not be decayed. Generally, the cotyledons and leaves wilt slightly before the seedlings are prostrated, although sometimes they remain green and turgid until collapse occurs. In fields and nurseries the disease often occurs in a roughly circular pattern, radiates from initial infection points. This is because of the tendency of fungi to grow radially from the point of origin thereby causing large spots or areas in which nearly all the seedlings are killed (Wick, 1998). Other above ground symptoms of root rot include stunting, low vigor, or wilting on a warm day. Foliage of such plants may yellow and fall prematurely starting with the oldest leaves. The roots of a diseased plant will have some shade of brown or black. Healthy roots are fibrous appearing and are usually white or tan in colour. Besides this, damping off oftenly confused with plant injury caused by insect feeding (severe mite, aphid, scale infestation), excessive fertilization, high levels of soluble salts, excessive heat or cold injury, excessive or insufficient soil moisture, insufficient light or nitrogen, root feeding by nematodes or insect larvae or chemical toxicity in air or soil.

The Causal Organisms

More than 40 damping off organisms attack vegetables alone. The fungi responsible for damping off include species of *Pythium*, (*P. debaryanum* Hesse, *P. aphanidermatum* (Eds.) Fitz., *P. butleri* Subram., *P. ultimum* Trow and *P. arrhenomanes* Drechsler), *Phytopathora*, *Fusarium*, *Rhizoctonia*, *Sclerotium*. Other species of fungi that occasionally cause this disease include *Glomerella*, *Alternaria*, *Phoma* and *Botrytis*. These fungi are not host-specific.

Pythium spp.

The mycelium of *Pythium* is intercellular in the host. It consists of slender, coenocytic and much branched hyphae. Asexual reproduction is mostly by means of zoospores which are always differentiated in a vesicle extruded from the zoosporangium. The sporangium forms a beak, the tip of which softens and out of which flows the protoplasmic mass of the zoosporangium to form a spherical vesicle. Within this vesicle, the protoplasm cleaves to form zoospores. These zoospores are kidney-shaped with two flagella inserted in the groove. They rupture the wall of the vesicle and escape in the surrounding film of water. After swimming for a while the zoospores encyst and then germinate by a germ tube. Three distinct types of zoosporangia are seen in this genus:

1. Non-sphaerosporangiate type with filamentous, simple or branched sporangia which open at the tip to form vesicle,
2. Non-sphaerosporangiate type with irregularly thickened and lobed or toruloid hyphal sporangia, and
3. Sphaerosporangiate type with well defined, spherical, ovoid, lemon-shaped or pear-shaped terminal or intercalary sporangia which may be single or in short chains.

In sexual reproduction the oogonium arises terminally, less often intercalarly, on long or short unmodified hyphae. It is ovoid or spherical and smooth walled. The young oogonium is plurinucleate with homogenous protoplasm. Of the several nuclei all but one migrate to the periphery into the periplasms which surrounds the centrally placed egg (oosphere). This central uninucleate portion is also known as ooplasm. The nuclei passing into the periplasm gradually degenerate. The antheridium is also plurinucleate at first but only one nucleus functions in the fertilization of one egg. The branch bearing the terminal antheridium may arise at a distance from the same branch (monoclinous). Through a conjugation tube one male nucleus passes into the ooplasm which then forms a thick wall and becomes oospore. Germination of oospores may be delayed for a considerable time showing constitutive dormancy on germination the oospore produces a long germ tube or a short germ tube terminated by a zoosporangium. In some cases the oospore may directly convert into a zoosporangium emptying the zoospores through a short break or exit tube into a vesicle (Figure 30.1).

Disease Cycle and Environmental Relations

Pythium species are weak saprophytes and poor parasites. As saprophytes they cannot compete with other soil microbes for colonization of dead organic matter in soil. As plant parasites, they can infect only the juvenile or succulent organs of plants, tissues of which are devoid of secondary thickening. In soil these fungi are susceptible to antagonism (antibiosis and parasitism) of a number of bacteria such as *Bacillus*, *Pseudomonas* and *Streptomyces*, fungi such as *Trichoderma* and *Penicilllium*, and some amoaebae. In the host their activity is restricted to seeds and seedlings or fine feeder roots and root tips of perennial plants. Most plant parasitic species of *Pythium* enter the soil through the pre-colonized host residue carrying oospores and sporangia which are the survival structures. Vegetative growth of the fungus in the soil declines rapidly and is very limited.

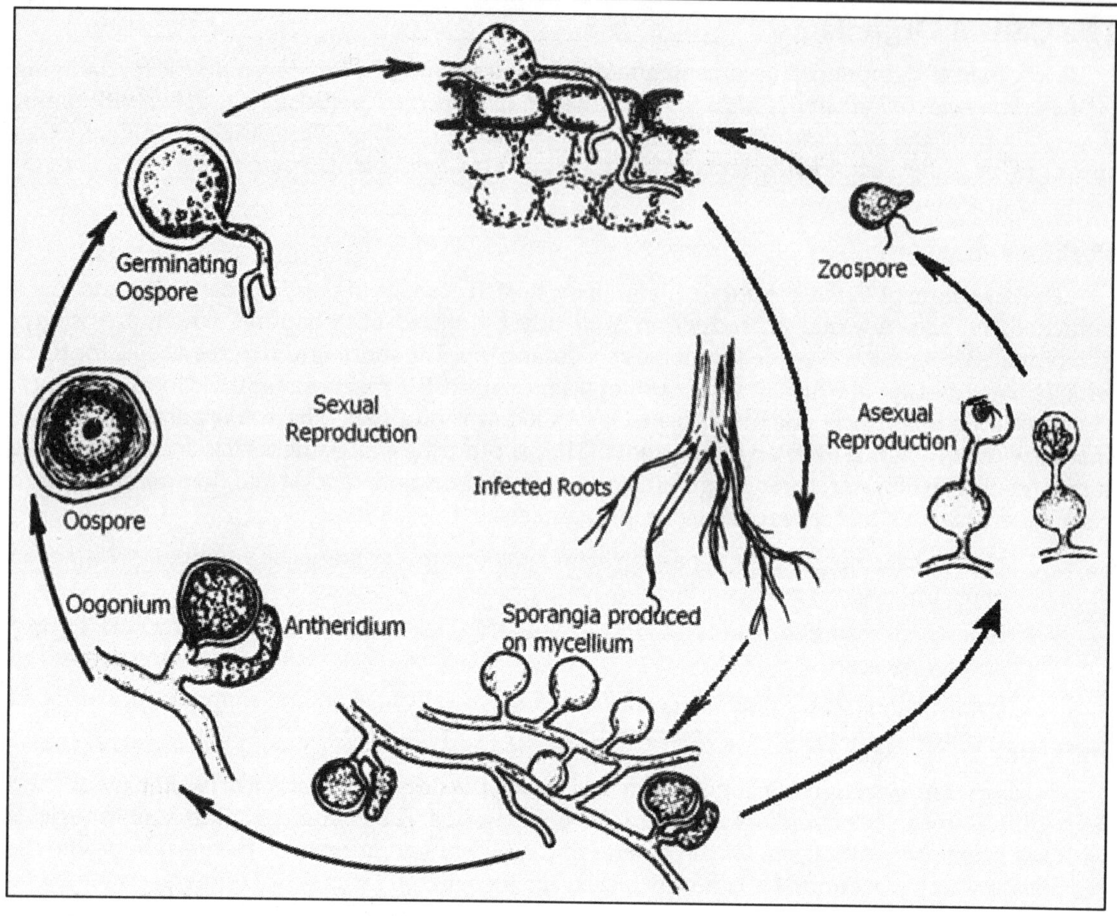

Figure 30.1: The Disease Cycle of *Pythium* Damping Off of Vegetable Crops

High soil moisture and relatively high soil temperatures favour rapid development of damping off. High soil moisture makes soil nutrients available to the oospores which germinate and produce zoospores. In the presence of high soil moisture there is rapid dissemination of the zoospores which attack the seeds and germinating seedlings. With the help of hydrolytic enzymes the fungus causes rapid breakdown of the host tissues prior to actual colonization of the tissues (killing in advance). After invading the host tissues the fungus rapidly forms oospores.

Factors Affecting Damping Off Disease Incidence

The best defense against damping off fungi is an effective and conscientious disease-prevention program. Soil moisture, timing and amount of irrigation, air and soil temperatures, method and timing of sowing, depth of soil over seed, soil pH, combinations of soil fungi and nematodes, timing and type of nutrients applied, type of organic matter, type of cover crop, history and pattern of pesticide use, and many other factors affect the incidence and severity of damping off. No single factor alone governs control of the disease, but good management will also take the following factors into account:

Soil Moisture and Drainage

The disease is more severe in ill-aerated, ill-drained soils. Such conditions are common in heavy and compact soils. Loose soils having a good proportion of sand exhibit less loss of damping off. Temperatures between 27°C and 31°C are ideal for this disease. High moisture, poor aeration and close planting lengthen the juvenile stage of the plants and pre-dispose them to damping off. An ideal nursery should be located on light, well-drained soil conditions. Wet soil generally favors damping off. Depth of irrigation is critical to young seedlings, especially during hot weather. It is important to irrigate deeply enough for water to reach seedling roots–a depth that increases steadily as the roots grow–but not so much that the soil is saturated. Too-shallow watering stresses tender roots in moisture deficient lower soil layers, while creating a warm, wet upper soil layer that favours the buildup of damping off fungi.

Soil Microflora

Nurseries are not alike in their makeup of soil organisms. Each has its own combination of soil microflora, consisting of bacteria, fungi, nematodes, and insects, and each combination influences the population of pathogenic fungi in the soil, the amount of infections, and the expression of disease symptoms.

Nutrition

Nitrogen applications made too early promote damping-off. Germinating seed and new seedlings do not need much supplemental nutrition; the endosperm contains sufficient food required by the seedlings to grow.

Management Strategies of the Disease

The best way to control damping off disease is to prevent it. There are many prevention techniques, and a combination of them is most effective. Certain cultural and sanitation practices can help in reducing inoculum load of the disease causing pathogens. It is essential to adopt timely control measures to avoid losses due to pathological problems during emergence. The following steps are necessary for successful integrated management to avoid, exclude or eradicate the pathogen from the place of activity.

Avoidance of the pathogen can be done by the adopting following practices:

- ☆ Selected area for planting a crop and raising a nursery should be free from phytopathogenic fungi responsible for pre and post emergence damping off disease.
- ☆ During summer season it is necessary to plough the soil deep in open sun to expose and check the growth of propagules of pathogens existing in the soil at high temperatures.
- ☆ One of the most important aspects in the control of damping-off is to ensure high quality seed. Avoid sowing of cracked or injured seeds.
- ☆ Planting should be done at right time to discourage the pathogen activity and enhance seed germination and emergence.
- ☆ Persistence of the pathogen can be checked by crop rotation involving non-host crops. Long crop rotations are helpful in managing the pathogens (Curl, 1963). Inter and mixed cropping check the severity of pathogens in soil which attack the crop even at early stages of emergence.

Cultural practices, such as thin planting, use of light soil in the nursery beds, light but frequent watering of the nursery, use of well-decomposed manure, recommended doses of nitrogenous fertilizers, proper drainage, wider row to row space and less population of seedlings per row may be employed in the nursery depending on the crop and pathogen associated with damping-off.

Exclusion of the pathogen can be done by seed treatment, soil treatment and rouging of diseased plants:

- Systematic fungicides like vitavax, carbendazim, benomyl may be used as seed treatment when the infection is within the seed. Contact/non-systemic fungicides like thiram, captan, mancozeb, organomercurials etc. are also used.
- Suitable soil treatment can be easily done in limited areas in nurseries which may control some specific pathogens infesting a particular crop.
- The diseased seed and seedlings should be removed and destroyed as soon as observed (Cook *et al.*, 1978).
- Proper cleaning of the seed is necessary before planting. Physical treatments like hot water treatment, solar energy treatments are also useful.

Eradication of the Pathogen

The pathogens can be eradicated by several methods.

Use of Seed Protectants

The incidence of damping-off can be reduced by treating the seed with a fungicide prior to planting. Fungicides will protect seed from the soil-borne pathogens, but have little effect against the seed-borne fungi. Pre-emergence damping-off of cabbage and cauliflower can be controlled by seed treatment with apron-70 (White *et al.*, 1984) or thiram (Sandu and Gill 1983). The seed protectants commonly recommended have been Agrosan GN (an organo mercurial), Captan, Botran, Thiram, etc. These fungicides are effective against species of *Pythium*. The dosase of Agrosan GN is 0.3 per cent by weight of seed (dry seed dressing). Approximately same dosase is used for Captan and Thiram (2.5 g/kg seed). Sometimes seed treatment with chemicals fails to check damping-off. This is because of multiplicity of the organisms involved.

- Periodical soil treatment of the nursery can be done by using formalin dust (15 parts of formalin and 85 parts of charcoal ash) @ 30 g/sq. foot (3-inch deep). Captan, Difolatan, Botran and Thiram can also be applied to the soil but these may prove costly. A 0.2-0.5 per cent suspension of these chemicals can be applied to the soil as drench.
- A more economical method is to burn 12 inch thick stack of farm trash over the nursery soil.

Biological Control

Effective fungicides are available to control damping off disease, but chemical control by seed treatment or soil drenching has several limitations besides environmentally hazardous (Richardson, 1991). Therefore, biological control as a mechanism to reduce soil borne plant pathogens is gaining importance in recent years due to chemicals that are used to control the diseases are expensive but also contribute to soil pollutants and adversely affect non-target species. Several microorganisms antagonistic to pathogens are available which can be used as seed or seed bed treatment. *Trichoderma viride*, *T. harzianum*, *T. hamatum*, *T. reesee*, *T. koningii*, have been reported to control species of *Pythium*, *Rhizoctonia*, *Fusarium*, *Sclerotium rolfsii*, (Bagyaraj and Govindan, 1996). Damping-off due to *Pythium indicum* in tomato was controlled by the application of *T. viride* (Krishnamoorthy and Bhaskaran,

1994). Chilli damping off due to *P. aphanidermatum* was controlled by seed treatment with conidia of *T. viride* and *T. hamatum* in nursery (Ramanathan and Sivaprakasan, 1994). Seed and soil treatment of chilli and nursery soil with *T. harzianum* and *T. reesee* effectively controlled damping-off and enhanced germination and emergence. Vescicular arbuscular mycorrhizae also control soil-borne pathogens and check seed rot and seedling damage problem (Jalali and Thaereja, 1981).

Soil Solarization

Soil borne pathogens are eradicated by solarization (Chen and Katan, 1980). Manomohan Das and Sivaprakasam (1994) have reduced the incidence of pre emergence damping off in chilli nursery after solarization of soil with transparent polythene sheets for 14 days. The percentage germination of seed was also enhanced. Soil solarization of nursery beds has been found very effective in reducing the damping off in tomato, chilli, brinjal and early cauliflower (Pandey and Pandey, 2004).

Integrated Disease Management

Integrated disease management provides a combination of cultural, biological and chemical tools to control and/or manage crop diseases effectively. Cultural controls keep *Pythium* from reaching the roots while biological and chemical controls inhibit or suppress *Pythium* in the root zone.

Integration of Biocontrol Agents and Fumigants

Fumigation with dazomet, methyl isothiocyanate/1,3-dichloropropene, and mixtures of methyl bromide and chloropicrin effectively decreases the populations of *Fusarium* and *Pythium*. Moreover, several biocontrol agents have also been used successfully for the control of damping off. Strashnow *et al.* (1985) reported that under green house condition the combined treatment of *T. harzianum* with lower dose of methyl bromide (equivalent to 200 kg/ha) completely controlled disease incidence of *R. solani* in bean seedlings. Under field condition the combination of 200 kg/ha methyl bromide and *T. harzianum* gave significant synergistic effect on damping off of carrot seedlings caused by *R. solani*. Besides this, *T. harzianum* also inhibited reinfestation of *R. solani* in fumigated soils.

Integration of Biocontrol Agents and Fungicides

It has been observed that in certain cases the biocontrol agents or the chemicals alone could not provide satisfactory result for the management of a particular soil borne disease. However, the integration of biocontrol agents with certain compatible chemicals may give synergistic effect and provide better disease control than either treatments alone (Sokhi and Thind, 1996). In radish, the conidial suspension of *T. harzianum* and benodanil was found effective to minimize pre-emergence damping off caused by *R. solani*. The control of damping-off by both seed treatment with *T. harzianum* and soil mix of benodanil was additive but not interactive (Lifshitz *et al.*, 1985). Bacterial species *viz. Pseudomonas cepacia*, *P. fluorescens* and *Corynebacterium* species as seed dressing in combination with captan provided effective control of the damping-off and root rot of peas caused by *Pythium* and Aphanomyces (Parke *et al.*, 1991).

AM Fungi and *Azospirillum* in Suppression of Damping-off

AM fungi are known to colonize a number of tropical plants including vegetables. AM association are known to help in the growth of various crops like carrot, tomato etc. (Sasal, 1991). Reddy *et al.*(2006) reported that *G. fasciculatum* proved as the benefactor and enhanced the plant growth, nutrition status, yield and reduction in disease severity. The dual inoculations (mycorrhiza with fungal pathogen) showed significant suppression in the progression of the pathogen and consequently reduction in the severity of the damping off. Mosse, 1973 indicated that mycorrhizal colonization induce chemical, physiological and morphological alterations in the host plant which may result in an increase in host resistance. The potential in biological suppression of soil-borne pathogens gives a wider vision of AM fungi, in that they act as an alternative strategy for the host plant in conditions that are deleterious to root growth. Therefore, the introduction and consequent management of such symbiotic colonization could be employed for the advantage of the crop.

Due to the imposition of competition with the pathogen for space, nutrition and host photosynthates (Harley and Smith, 1983) or the alterations of the physiology of the host which induces host defence mechanisms (Gianinazzi-Pearson *et al.*, 1996) the arbuscular mycorrhizal fungi (AMF) are able to suppress the damping off disease. Suppression of damping off by AMF has already been reported in cucumber (Rosendahl, C. N. and Rosendahl, S., 1990), ginger (Joseph, P. J., 1997) and pepper (Odebode *et al.*, 1997).

Damping-off of chilli (*Pythium aphanidermatum* (Edson) Fitz) was responsible for 90 per cent mortality of seedlings both in nursery and main field (Sowmini, R., 1961). Chilli seedlings pre-inoculated with native AMF recorded the least per cent disease incidence of 22.3 and this was significantly superior as against the control which recorded 65.1 per cent disease incidence (Kavitha *et al.*, 2003). Dual inoculation of AMF along with Azospirillum also reduced damping off by 72.7 per cent over control. This may be due to the effect of *Azospirillum* and AMF interaction which makes the plant healthier by way of enhanced uptake of nutrients and trigger the host defence mechanism.

Leaf Extracts against Damping-off

Plant extracts have also been successfully used to control emergence problems due to pathogens, such as *P. aphanidermatum* (Narayana Bhat *et al.*, 1994). Due to the presence of phenolic substances and resins, gummy and non-volatile substances the plant extracts are effective against *Pythium* spp. (Skinner, 1955). Narayana Bhat and Shukla (2001) evaluated the antifungal activity of 37 plants against *P. aphanidermatum* and reported that maximum inhibition (94.4 per cent) and least post-emergence damping off was recorded by *O. paniculata* extracts among all the plants. Tomato seeds soaked in 20 per cent leaf extract of *Bougainvillea glabra* or *Piper betle* for 6 hrs before sowing increased germination by 75 per cent and damping off due to *P. aphanidermatum* and *S. rolfsii* was also inhibited by the leaf extract of the two plants. Drenching of soil after sowing, with extracts of *Tamarindus indica* and *Leucaena leucocephala* also found effective against damping-off due to *P. indicum*.

Chemical Control

Although fungicides are slowly taking back seat in our fight against plant pathogens in view of some associated adverse effects on the environment, their role in managing several devastating plant diseases can not be overlooked (Thind, 2007). Prevent *Pythium* diseases by practicing integrated disease management strategies based on cultural and biological controls. Use fungicides as a last resort at the onset of disease. Fungicides have been commonly used for the control of plant diseases the world over since 19th century and in many cases have become an integral component of our crop production system. Among different methods of disease control, host resistance is still the most preferred choice although lack of its durability has been a persistent drawback. Fungicides, despite certain drawbacks, are considered to play a significant role in containing losses due to plant diseases in the coming years.

Effectiveness of fungicides to control damping-off is highly variable. Several fungicides are registered for use in vegetable nurseries to control soilborne diseases. Thiram and captan and organomercurial fungicides have proved to be highly effective in controlling seed-borne pathogens (seed rots, damping-off of seedlings due to *Pythium, Fusarium* spp. etc.) in different vegetables resulting in improved seed germination and crop stand. Phenylamide fungicides have a unique potential of curbing destructive plant pathogens belonging to the class Oomycetes such as *Pythium, Phytopathora* etc. Development of metalaxyl by the Ciba-Geigy Agrochemicals in 1977 came as a landmark discovery and a breakthrough was achieved in the effective control of diseases caused by Oomyceteous fungi (Urech *et al.*, 1977). The fungicide metalaxyl has systemic properties and may be used prior to sowing to reduce populations of *Pythium* and *Phytophthora* in the soil. Other fungicides having different chemistry but similar anti-oomycete activity spectrum such as prothiocarb and propamocarb (carbamates), hymexazol (isoxazoles) etc. can also be used but among these, metalaxyl is the most extensively used fungicide with protective and curative activity against majority of Peronosporales. Prothiocarb (superseded) and propamocarb, apart from controlling *Aphanomycetes, Pythium* and *Phytopathora* spp., also check downy mildew fungi. Hymexazole is used as seed dressing or soil drench to control soil-borne *Pythium* spp. and *Aphanomyces* spp. in various vegetables. The first post-plant fungicide application should be made when most seedlings have emerged and the seeds begin to drop from cotyledon leaves. A good all-purpose preventive treatment for damping off is a 50-50 mixture of captan and benomyl may be applied as a drench. Drenching of nursery soil with 0.3 per cent captan @ 5 litre/m^2 before sowing treated seeds provide excellent control of damping off. In order to provide protection from post emergence damping off nursery should be drenched with 0.2 per cent captan. If frequent applications of fungicides are planned, alternation of the captan-benomyl mix with other fungicides is advised to minimize the buildup of resistant pathogens. But in view of the resistance risk associated with most of the systemic, site-specific compounds, there is need to develop and employ an ideal IDM module ensuring integration of cultural, chemical and biological means of disease management options in a holistic manner.

References

Ahamad, S., Kushwaha, K.P.S. and Manzer, A. (1998). Economics in plant disease control. *Indian Farmers Times* 2: 35–39.

Ahamad, S., Narain, U. and Yasmeen (2000). Plant bios tress control: Approach for the century. *Pesticide World,* 12: 58–60.

Ahamad, S. and Narain, U. (2007). *Eco-friendly Management of Plant Diseases*. Daya Publishing House, Delhi. pp. 373.

Ahamad, S. (2009). Plant Disease Management for Sustainable Agriculture Edited by Dr. Shahid Ahamad, Published by: Daya Publishing House, New Delhi. pp. 373.

Ahamad, S. (2012). Recent Trends in Plant Disease Management in India. Kalyani Publishers, pp. 476.

Bagiraj, D. J. and M. Govindan (1996). Microbial control of fungal root pathogens. In Advances in Botany. Ed. K. G. Mukerji. APH Publishing Corporation, New Delhi. pp. 293-321.

Chen, Y. and J. Katan (1980). Effect of solar heating of soils by transparent polythene mulching on their chemical properties. *Soil Sci.* 130: 271-277.

Cook, R. J., M. G. Boosalis and B. Doupnik (1978). Influence of crop residues on plant diseases. In: *Crop Residue Management System*. Ann. Soc. Agron. 31: 147-163.

Curl, E.A. (1963). Control of plant diseases by crop rotation. *Bot. Rev.* 29: 413-479.

Gianinazzi-Pearson, V., Dumas-Gaudot, E., Gollotte, A., Jahiri, Alaouri, A. and Gianinazzi, S. (1996). *New Phytol*. 133: 45-57.

Harley, J. L. and Smith, S. E. (1983). *Mycorrhizal Symbiosis*. Academic Press, London.

Jalali, B. L. and M. L. Thaereja (1981). Management of soil borne plant pathogens by vesicular arbuscular mycorhizal model system. Indian Phytopathology. 34: 115-116.

Joseph, P. J. (1997). Ph.D. Thesis, Kerala Agricultural University, Thrissur, p. 192.

Kavitha, A. Meenakumari, K. S. and Sivaprasad, P. (2003). Effect of dual inoculation of native arbuscular mycorrhizal fungi and Azospirillum on suppression of damping off in chilli. *Indian Phytopathol*. 56: 112-113.

Krishnamurthy, A. S. and R. Bhaskaran (1994). *Trichoderma viride* in the control of damping off disease of tomato caused by *Pythium indicum*. In Crop Diseases- Innovative Techniques and Management. Ed. K. Sivaprakasam and K. Seetharaman. Kalyani Publishers, Ludhiana.pp. 199-203.

Lifshitz, R., Lifshitz, S. and Baker, R. (1985). Decrease in incidence of *Rhizoctonia* pre-emergence damping off by the use of integrated chemical and biological controls. *Plant Dis*. 69: 431-434.

Manomohan Das and K. Sivaprakasam (1994). Biological control of damping off disease in chilli nursery. In Crop Diseases. Innovative Techniques and Management. Ed. K. Sivaprakasam and K. seethraman. Kalyani Publishers, Ludhiana pp. 215-243.

Mosse, B. (1973). Plant growth responses of vesicular-arbuscular mycorriza. *New Phytol*. 72: 127-136.

Narayana Bhat, M., Sivaprakasam, K. and Jeyarajan, R. (1994). *Indian J. Forestry*. 17: 10-14.

Narayana Bhat, M. and Shukla, B.K. (2001). Evaluation of some leaf extracts against *Pythium aphanidermatum in vitro* and pot culture. *Indian Phytopath*. 54 (3): 395-397.

Odebode, A. C., Ladoye, A. O. and Osonuhi, O. (1997). *Int. J. Trop. Plant Dis*. 15: 85-92.

Pandey, K. K. and Pandey, P.K. (2004). Effect of soil solarization on disease management in vegetable nursery. *Journal of Mycology and Plant Pathology*. 34: 398-401.

Parke, J. L., Ramd, R. E., Joy, A. E. and King, E.B. (1991). Biological control of *Pythium* damping off and Aphanomyces root rot of peas by application of *Pseudomonas cepacia* or *P. fluorescens* to seed. *Plant Dis*. 75: 987-992.

Ramanathan, A. and K. Sivaprakasam (1994). Effect of seed treatment with plant extract, antagonist and fungicide against damping off disease od chilli. In: *Crop Diseases. Innovative Techniques and Management*. Edited by K. Sivaprakasam and K. Seetharaman. Kalyani Publishers, Ludhiana. pp. 221-226.

Reddy, B. N., Raghavender, C. R. and Sreevani, A. (2006). Approach for enhancing mycorriza-mediated disease resistance of tomato damping-off. *Indian Phytopath*. 59(3): 299-304.

Richardson, M.L. (1991). Chemistry Agriculture and Environmemt. Cambridge: *R Soc. Chem*. pp. 546.

Rosendahl, C. N. and Rosendahl, S. (1990). Symbiosis (Rehort), 9: 363-366.

Sandhu, K. S. and Gill, S.P.S. (1983). Effect of seed and soil treatments on damping off of cauliflower. *Pesticides*. 16 (11): 17-18.

Sasal, K. (1991). Effect of phosphate application on infection of vesicular-arbuscular mycorrizal fungi on some horticultural crops. *Scientific reports of the Miyagi Agricultural College* 39: 1-9.

Skinner, F.R. (1955). In: *Modern Methods of Plant Analysis*. Peach, K. and Tracey, M. V. (eds.) Vol. III. Springer Verlag Viertav Band. pp. 626-725.

Sokhi, S.S. and Thind, T.S. (1996). Application of fungicides in integrated disease management in vegetables. In: *Rational use of fungicides in plant disease control*. NATIC, Ludhiana. 133-142.

Strashnow, Y., Elad, Y., Sivan, A. and Chet, I. (1985). Integrated control of *Rhizoctonia solani* by methyl bromide and *Trichoderma harzianum*. *Plant Pathology*. 34: 146-151.

Thind, T.S. (2007). Changing cover of fungicide umbrella in crop protection. *Indian Phytopath*. 60 (4): 421-433.

Urech, P. A., Schwinn, F. and Staub, T. (1977). CGA-48988, a novel fungicide for the control of late blight, downy mildews and related soil-borne diseases. *Proc. Br. Crop Prot. Conf. – Pests and Diseases*, 42: 623-631.

White, J. G., Crute, I. R. and Wynn, E.C. (1984). A seed treatment for the control of *Pythium* damping-off diseases and *Peronospora parasitica* on brassicas. *Ann. Appl. Biol*. 104: 241-247.

Wick, R.L. (1998). Damping-off of bedding plants and vegetables. Department of Plant Pathology, Fernald Hall, University of Massachusetts, Amherst, MA. http://www.umass.edu/umext/floriculture/fact_sheets/pest_management/dampoff.html

Hill Agriculture: Prospects, Constraints and Mitigations (2013) *Pages* **480–498**
Editor: **Dr. Shahid Ahamad**
Published by: **DAYA PUBLISHING HOUSE, NEW DELHI**

Chapter 31
Role of Extension in Sustainable Agricultural Development

Vinod Gupta and P.K. Rai
Krishi Vigyan Kendra, R.S. Pura, SKUAST-J, Jammu

Sustainable development has caught the imagination and action all over the world for more than a decade. Sustainable agriculture is necessary to attain the goal of sustainable development. According to the Food and Agriculture Organization (FAO), sustainable agriculture "is the successful management of resources for agriculture to satisfy changing human needs while maintaining or enhancing the quality of environment and conserving natural resources". All definitions of sustainable agriculture lay great emphasis on maintaining an agriculture growth rate, which can meet the demand for food of all living things without draining the basic resources.

Organic farming is one of the several approaches found to meet the objectives of sustainable agriculture. Many techniques used in organic farming like inter-cropping, mulching and integration of crops and livestock are not alien to various agriculture systems including the traditional agriculture practiced in old countries like India. However, organic farming is based on various laws and certification programmes, which prohibit the use of almost all synthetic inputs, and health of the soil is recognised as the central theme of the method.

Adverse effects of modern agricultural practices not only on the farm but also on the health of all living things and thus on the environment have been well documented all over the world. Application of technology, particularly in terms of the use of chemical fertilizers and pesticides all around us has persuaded people to think aloud. Their negative effects on the environment are manifested through soil erosion, water shortages, salination, soil contamination, genetic erosion, etc.

The farming being practiced for the last three decades has increasingly been found non-sustainable. The system is oriented towards high production without much concern for ecology and the very existence of man himself.

Emerging Challenges

Agricultural development policies have been remarkably successful at emphasizing external inputs as the means to increase food production. This has led to growth in global consumption of pesticides, inorganic fertilizer, animal feed-stuffs, and tractors and other machinery.

These external inputs have, however, substituted for natural processes and resources, rendering them less powerful. Pesticides have replaced biological, cultural, and mechanical methods for controlling pests, weeds, and diseases; inorganic fertilizers have substituted for livestock manures, composts, and nitrogen-fixing crops; information for management decisions comes from input suppliers, researchers, and extensionists rather than from local sources; and fossil fuels have substituted for locally generated energy sources. The basic challenge for sustainable agriculture is to make better use of these internal resources. This can be done by minimizing the external inputs used, by regenerating internal resources more effectively, or by combinations of both.

Evidence is now emerging that regenerative and resource-conserving technologies and practices can bring both environmental and economic benefits for farmers, communities, and nations. The best evidence comes from countries of Africa, Asia, and Latin America, where the concern is to increase food production in the areas where fanning has been largely untouched by the modem packages of externally supplied technologies. In these complex and remote lands, some farmers and communities adopting regenerative technologies have substantially improved agricultural yields, often using only few or no external inputs (Bunch, 1991; GTZ, 1992; UNDP, 1992; Lobo and Kochendörfer-Lucius, 1992; Krishna, 1993; Shah, 1994; SWCB, 1994; Pretty, 1995).

All of these successes have three elements in common. They have made use of resource-conserving technologies such as integrated pest management, soil and water conservation, nutrient recycling, multiple cropping, water harvesting, and waste recycling. In all, there has been action by groups and communities at the local level, with farmers becoming experts at managing farms as ecosystems and at collectively managing the watersheds or other resource units of which their farms form a part. And there have also been supportive and enabling external government and non-government institutions, which have reoriented their activities to focus on local needs and capabilities.

Sustainability and Levels of Action

A necessary condition for sustainable agriculture is that large numbers of farming households must be motivated to use coordinated resource management. This could be for pest and predator management, nutrient management, controlling the contamination of aquifers and surface water courses, coordinated livestock management, conserving soil and water resources, and seed stock management. The problem is that, in most places, platforms for collective decision making have not been established to manage such resources (Röling, 1994a, 1994b). The success of sustainable agriculture therefore depends not just on the motivations, skills, and knowledge of individual farmers, but on action taken by groups or communities as a whole. This makes the task more challenging. Simple extension of the message that sustainable agriculture can match conventional agriculture for profits, as well as produce extra benefits for society as a whole, will not suffice.

Sustainability is commonly seen as a property of an ecosystem. But Sustainability can be seen from other perspectives, which are more relevant for extension. Environmental issues emerge from the human use of natural resources. Sustainability can therefore be defined in terms of human reasons, activities, and agreements. The definition of Sustainability then becomes part of the problem because people need to agree on how they define Sustainability and what priority they will give it (Pretty, 1994b).

In this approach, Sustainability is not a scientific, "hard" property which can be measured according to some objective scale, or a set of practices to be fixed in time and space. Rather, Sustainability is a quality that emerges when people individually or collectively apply their intelligence to maintain the long-term productivity of the natural resources on which they depend (Sriskandarajah, Bawden, and Packham, 1989). In other words, Sustainability emerges out of shared human experiences, objectives, knowledge, decisions, technology, and organization. Agriculture becomes sustainable only when people have reason to make it so. They can learn and negotiate their way towards Sustainability. In any discussions of Sustainability, it is important to clarify what is being sustained, for how long, for whose benefit and at whose cost, over what area, and measured by what criteria. Answering these questions is difficult, because it means assessing and trading off values and beliefs. Campbell (1994) has put it this way: "[Attempts to define Sustainability miss the point that, like beauty, sustainability is in the eye of the beholder.. It is inevitable that assessments of relative Sustainability are socially constructed, which is why there are so many definitions."

It is therefore crucial to focus on more than one system level (Fresco, Stroosnijder, Bouma, and van Keulen, 1994). At the farm level, there is the farm household. At the above-farm level, there are the collective stakeholders, who might or might not be organized for sustainable use of the whole resource unit. In an irrigation scheme, it is common for an irrigators' association collectively to manage water use at the scheme level. But when it comes to watersheds or other vulnerable resource units, it is usually impossible to identify an appropriate "platform" for decision making (Röling, 1994a, 1994b).

Resource-Conserving Technology Development and its Transfer

Although many resource-conserving technologies and practices have been widely proven on research stations to be both productive and sustainable, the total number of farmers using them is still small. This is because these technologies involve the substitution of management skills, knowledge, and labour for external inputs. The modern approach to agricultural research and extension, however, has been to emphasize comprehensive packages of technologies. Few farmers are able to adopt the whole modem packages of production or conservation technologies without considerable adjustments. Part of the problem is that most agricultural research still occurs on the research station, where scientists experience conditions quite different from those experienced by farmers.

This is true of many sustainability-enhancing innovations. Even though resource-conserving technologies are productive and sustainable, if they are imposed on farmers, then they will not be adopted widely. Alley cropping, an agro-forestry system comprising rows of nitrogen-fixing trees or bushes separated by rows of cereals, has long been the focus of research (Kang, Wilson, and Lawson, 1984; Attah-Krah and Francis, 1987; Young, 1989; Lal, 1989). Many productive and sustainable systems, needed few or no external inputs, have been developed. They stop erosion, produce food and wood, and can be cropped over long periods. But the problem is that very few, if any, farmers have adopted these alley cropping systems as designed. Despite millions of dollars of research expenditure over many years, systems that have been produced are suitable only for research stations.

Where these systems have had some success, however, farmers have taken one or two components of alley cropping and adapted them to their own farms. In Kenya, for example, farmers planted rows of leguminous trees next to field boundaries, or single rows through their fields; and in Rwanda, alleys planted by extension workers soon became dispersed through fields (Kerkhof, 1990). But the prevailing view tends to be that farmers should adapt to the technology. Of the Agro-forestry Outreach Project in Haiti, it was said:

Farmer management of hedgerows does not conform to the extension program. Some farmers prune the hedgerows too early, others too late. Some hedges are not yet pruned by two years of age, when they have already reached heights of 4-5 metres. Other hedges are pruned too early, mainly because animals are let in or the tops are cut and carried to animals.. Finally, it is very common for farmers to allow some of the trees in the hedgerow to grow to pole size. These trees are not pruned but are harvested when needed for house construction or other activities requiring poles. (Bannister and Nair, 1990).

Farmers were clearly making their own adaptations according to their own needs.

Incorporating Farmer Experimentation

The problem with agricultural science and extension is that it has poorly understood the nature of "indigenous" and rural people's knowledge. For many, what rural people know is assumed to be "primitive," "unscientific," or overtaken by development, and so formal research and extension must "transform" what they know so as to "develop" them. An alternative view is that local knowledge is a valuable and underused resource, which can be studied, collected, and incorporated into development activities. Neither of these views, though, is entirely satisfactory because of the static view of knowledge implied (Chambers, Pacey, and Thrupp, 1989; Röling and Engel, 1989; Warren, 1991; Long and Long, 1992; Scoones and Thompson, 1994). It is more important to recognize that local people are always involved in active learning, in (re)inventing technologies, in adapting their farming systems and livelihood strategies. Understanding and supporting these processes of agricultural innovation and experimentation have become an important focus in facilitating more sustainable agriculture with its strong locality-specific nature.

The problem with modem agricultural science is that technologies are finalized before farmers get to see them. If new technologies are appropriate and fit a particular farmer's conditions or needs, then they stand a good chance of being adopted. But if they do not fit and if farmers are unable to make changes, then they have only the one choice. They have to adapt to the technology, or reject it entirely.

The alternative is to seek and encourage the involvement of farmers in adapting technologies to their conditions. This constitutes a radical reversal of the normal modes of research and technology generation, because it requires interactive participation between professionals and farmers. Participatory technology development (PTD) is the process in which the knowledge and research capacities of farmers are joined with those of scientific institutions, whilst at the same time strengthening local capacities to experiment and innovate (Jiggins and De Zeeuw, 1992; Reijntjes, Haverkort, and Waters-Bayer, 1992; Haverkort, van der Kamp, and Waters-Bayer, 1991). Farmers are encouraged to generate and evaluate indigenous technologies and to choose and adapt external ones on the basis of their own knowledge and value systems.

But, of course, researchers and farmers participate in different ways, depending on the degree of control each actor has over the research process. The most common form of "participatory" research is researcher designed and implemented, even though it might be conducted on farmers' fields. Many on-farm trials and demonstration plots represent nothing better than passive participation (Pretty, 1994b). Less commonly, farmers may implement trials designed by researchers. But greater roles for farmers are even rarer. Fujisaka (1991) describes researcher-designed experiments on new cropping patterns in the Philippines. Even though farmers "participated" in implementing the trials, there was widespread uncertainty about what researchers were actually trying to achieve. Farmers misunderstood experiments and rejected the new technologies. The reason, as Fujisaka explains, was that "co-operation between farmers and researchers implies two groups continually listening carefully to one another.

Claveria farmers are avid listeners to.. researchers. The challenge is for all on-farm researchers to complete the circle."

Although this means that technology development must involve farmers, it does not mean that scientific research has no place. Research will have to contribute on many fronts, such as in the development of resistant cultivars, knowledge about the life cycles of pests, biological control methods, suitable crops for erosion control, and processes in nitrogen fixation. Such research also gives insight into complex processes such as the movement of nutrients in the soil and their accessibility for plants. But all these contributions must be seen as providing choices for farmers as they make farm-specific decisions and move the whole farm towards greater sustainability.

From Teaching to Learning and a Whole New Professionalism

The central principle of sustainable agriculture is that it must enshrine new ways of learning about the world. But learning should not be confused with teaching.

Teaching implies the transfer of knowledge from someone who knows to someone who does not know. Teaching is the normal mode of educational curricula and is also central to many organizational structures (Ison, 1990; Argyris, Putnam, and Smith, 1985; Russell and Ison, 1991; Bawden, 1992, 1994; Pretty and Chambers, 1993). Universities and other professional institutions reinforce the teaching paradigm by giving the impression that they are custodians of knowledge which can be dispensed or given (usually by lecture) to a recipient (a student). Where these institutions do not include a focus on self-development and on enhancing the ability to learn, they do not allow students to grasp an essential skill in the sustainable management of a complex agro-ecosystem. In that case, "teaching threatens sustainable agriculture" (Ison, 1990).

The problem for farmers is that they cannot rely on routine, calendar-based activities if they engage in sustainable farming. Their interventions must be based on observation and anticipation. They require instruments and indicators which make more visible the ecological relationships on and among farms. Technology for sustainable farming must emphasize measurement and observation equipment or services that help individual farmers assess their situations, such as soil analysis, manure analysis, and pest identification (Röling, 1993). It also has to focus on higher system levels. Predators and parasitoids to control pests often require a larger biotope than that of a small farm. Erosion control, water harvesting, biodiversity, access to biomass, recycling waste between town and countryside and between animal and crop production, all require local co-operation and co-ordination.

What becomes important is the social transition, or new learning path, that farmers and communities must take to support sustainable agriculture. This is much less obvious and often remains unrecognized by extensionists. Learning for sustainable agriculture involves a transformation in the fundamental objectives, strategies, theories, risk perceptions, skills, labour organization, and professionalism of farming. This learning path has four key elements:

1. *The information system.* Sustainable agriculture must be responsive to changing circumstances, so farmers need to invest in observation, observation equipment, record keeping, and monitoring procedures.
2. *Conceptual framework.* Sustainable agriculture is knowledge intensive, and so farmers must know about life cycles of pests and disease organisms and their recognition, biological controls, ecological principles, soil life processes, nutrient cycles.
3. *Skills.* Sustainable farming requires a whole set of new skills, including observation and monitoring, compost making, mechanical weed control, spot application of pesticides, and risk assessment.

4. *Higher system-level management*. Generally, sustainable management of the farm is not enough, and it is necessary to think at system levels higher than the farm and take part in the collective management of natural resources at those levels.

In educational systems, therefore, the fundamental requirement for sustainable agriculture is for universities to evolve into communities of participatory learners. Such changes are very rare. Farmer field schools can play a major role in transformation.

Box 1. The Key Principles of Farmer Field Schools

1. What is relevant and meaningful is decided by the learner and must be discovered by the learner. Learning flourishes in a situation where teaching is seen as a facilitating process that assists people to explore and discover the personal meaning of events for themselves.
2. Learning is a consequence of experience. People become responsible when they have assumed responsibility and experienced success.
3. Cooperative approaches are enabling. As people invest in collaborative group approaches, they develop a better sense of their own worth.
4. Learning is an evolutionary process, and is characterized by free and open communication, confrontation, acceptance, respect, and the right to make mistakes.
5. Each person's experience of reality is unique. As people become more aware of how they learn and solve problems, they can refine and modify their own styles of learning and action. *Sources*: Adapted from Kingsley and Musante, 1994; Van de Fliert, 1993; Kenmore, 1991; Stock, 1994.

A move from a teaching to a learning style has profound implications for agricultural development institutions. The focus is less on *what* we learn, and more on *how* we learn and *with whom* (see Box 1 for principles of farmer field schools used in the FAO IPM programme in Southeast Asia). This implies new roles for development professionals, leading to a whole new professionalism with new concepts, values, methods, and behaviour. Typically, normal professionals are single-disciplinary, work largely or only in agencies remote from people, are insensitive to diversity of context, and are concerned with themselves generating and transferring technologies. Their beliefs about people's conditions and priorities often differ from people's own views. The new professionals, by contrast, are either multidisciplinary or work in close connection with other disciplines, are not intimidated by the complexities of close dialogue with rural and urban people, and are continually aware of the context of interaction and development.

From Directive to Participatory Extension

Extension has long been grounded in the diffusion model of agricultural development, in which technologies are passed from research scientists via extensionists to farmers (Rogers, 1962, 1983). This approach is exemplified by the training and visit (T&V) system. It was first implemented in Turkey in 1967 and later widely adopted by governments (Benor, 1987; Roberts, 1989). It was designed to be a management system for energizing extension staff, turning desk-bound, poorly motivated field staff into effective extension agents. Extension agents receive regular training to enhance their technical skills, which they then hope will pass to all farmers through regular communication with small numbers of selected contact farmers.

But the contact farmers are usually selected on the basis of literacy, wealth, readiness to change, and "progressiveness," and so this sets them apart from the rest of the community. The secondary transfer of the technical messages, from contact farmers to community, has been much less successful than predicted, and adoption rates are commonly very low among noncontact farmers. Without a doubt, T&V is now widely considered as ineffective (Axinn, 1988; Howell, 1988; Moris, 1990; Antholt, 1992, 1994; Hussain, Byerlee, and Heisey, 1994).

Important lessons have been learned from the problems associated with T&V, and there is clearly a need to address the systemic issues facing extension (Zijp, 1993; Antholt, 1994). Extension will need to build on traditional communication systems and involve farmers themselves in the process of extension. Incentive systems will have to be developed to reward staff for being in the field and working closely with farmers. There must be a "well-defined link between the well-being of field officers and the extension system, based on the clients' view of the value of extension's and field workers' performance" (Antholt, 1992). Participation, if it is to become part of extension, must clearly be interactive and empowering. Any pretence to participation will result in little change. Allowing farmers just to come to meetings or letting a few representatives sit on committees will be insufficient.

There have been some recent innovations in introducing elements of farmer participation and group approaches into extension. Differences in impact between individual and group approaches have been well documented in both Nepal and Kenya. In western Nepal, Sen (1993) compared the rate of adoption of new technologies when extension worked with individuals or with groups. With groups, better communication between farmers and extensionists led to more adoption. When the individual approach was resumed after the experiment, adoption rates fell rapidly in succeeding years.

There are three major lessons for extension. First, it is important to make new things visible. An important role of extension is to make visible the state of the environment and the extent to which present farming practices are untenable. In addition, extension can demonstrate the feasibility of sustainable practices. Even more important is to give farmers the tools for observation and to train them to monitor the situation on their own farms.

The second lesson is the use of farmers' knowledge. The location-specific nature of sustainable agriculture implies that extension must make use of farmers' knowledge and work together with farmers. Often, indigenous practices which have been ignored under the impact of chemical farming can be fruitfully revived. Indigenous technology development practices and farmer experimentation can be an important "entry point" for introducing sustainable farming practices (Brouwers and Röling, in press).

PRA Tools and Techniques

PRA is mainly undertaken through interviews with key informants individually or in small groups; and by observing the community surroundings and its residents during group meetings and activities.

PRA users may select from a wide variety of tools and techniques in generating data about sustainable agriculture practices its users, stakeholders, and their community to help them plan appropriate communication strategies. These PRA tools and techniques include the following:

Geographical Data

1. Village resource map - a map of the village indicating locations and distribution of important resources, both natural and man-made, such as water pumps and irrigation canals, electricity, lakes and rivers, forests, farms, and so on.

2. Social map - a map showing social and political/power relationships among the families in a village (who is affiliated with whom).
3. Farm sketch and three-dimensional map - shows the layout of a farm indicating the crops and animals raised and their locations on the farm.
4. Transects - a sketch of the topography across the locality showing a cross-section of soil resources and crops, animals, and structures on the land.

Historical Data

5. Time lines - a line marked by important dates and events in the history of a village.
6. Trend lines - a line marked by important trends observed in the history of a village, *e.g.*, charting the volume of fish catch or farm yields and showing when marked changes in these occurred.
7. Seasonal calendar - a calendar showing the seasons (wet, dry; open, closed) and indicating the time (month) of the year when certain farm or fishing operations are done.

Community Communication Resources and Networks

8. Focus group discussions (FGDs) - an interview with selected seven to ten articulate and individuals with similar characteristics carried out by a skilled facilitator with the assistance of a documenter, and where exchanges (the discussion) among participants raises important insights and creative ideas from their perspective.
9. Key informant interviews - interview with selected members of the community who are knowledgeable about the topic of interview, such as the practices in a community or in an organization.
10. Observation - use of sight, hearing, and where appropriate, other senses in making sense of a process, event, or community and its residents' behaviour patterns.
11. Linkage diagram of communication and information networks, systems and channels of key interaction group/s - illustrates who talks or interacts with whom in a community on matters of concern, such as on farming practices, fishery regulations and their enforcement, and the like.

Community Perceptions and Problem Analysis

12. Listing and assessment of community needs, opportunities, problems, and solutions (NOPS).
13. Problem tree analysis of main problems - a diagram showing the key NRMA and related problems in a community connected by arrows and lines to show which problems are causes and which are symptoms or consequences.
14. Windows of perceptions (WOPS) - shown below indicating four window panes of perceptions

	We (in community) know	*We (in community) don't know*
Others (outside community or other sectors) know	a. FREE	c. BLIND
Others (outside community or other sectors) don't know	b. HIDDEN	d. DARK

(*a*) What the participants in the community know and others outside the community also know, or the "free" pane - for example, the fertile fishing areas around or near a floating village;

(*b*) What participants in the community know and others outside the community don't know, or the "hidden" pane, *e.g.*, illegal fishing practices of people in the village;

(*c*) What others outside the community might know but participants inside the community do not know, or the "blind" pane, *e.g.*, the fishing apparatuses used by outsider fishers within the area of jurisdiction of the village; and

(*d*) What both the community and others outside of it do not know, or the "dark" pane, *e.g.*, the exact extent of illegal fishing practices in the area of jurisdiction of a village.

15. Access and control profile -indicates who has access and control over resources in the locality, such as water resources, farming inputs, hybrid seeds, land resources, forest resources, and so on.
16. Livelihood mapping - indicates the types of livelihoods in a locality and their locations and distribution on the village or district map.
17. Gender analysis - shows a breakdown of tasks and responsibilities in farming and fishing and who performs these, whether men, women, or children.
18. Brainstorming - a listing of all possible solutions to problems whether feasible or not, realistic or not, in a search for creative solutions to community issues.

The Baseline Study

The baseline study is a research survey undertaken together data on users' awareness, knowledge, attitude, and practices (AKAP) in farming relevant to the use of their natural resources.

This survey is called a baseline study when undertaken before carrying out an information and communication strategy (ICS) because it establishes the initial AKAP levels of the farmers. The data that the baseline survey yields provides a basis for comparing AKAP levels after an ICS is carried out, showing if the ICS succeeds in bringing about desired changes in these behaviours.

The AKAP baseline survey is undertaken following the scientific method, applying the following stages:

1. *Formulation of the study purpose*: The objectives of the baseline AKAP survey usually includes determining the socio- demographic characteristics of Farmers; their perception, awareness and knowledge of their natural resources and selected sustainable agriculture (SA) ideas and practices; their attitude related to SA and community participation; their SA practices; access to communication and information resources and systems; and level of participation in ongoing SA and other community initiatives.
2. *Research justification/rationale*: This is part of the introduction of the study and explains the importance of carrying it out in the light of a participatory communication strategy in SA to be carried out.
3. *Literature review*: So that the survey researcher need not "re-invent the wheel" as the saying goes, related studies are reviewed and their research methods and findings are noted. This is to inform the researcher on what the baseline study can improve on or guard against.

4. *Designing a rigorous procedure for gathering and analyzing data*. This includes procedures in sampling and selection of respondents, determining data collection tools and procedures, questionnaire design, and data analysis.

   Questionnaire or interview schedule design requires selection of the right questions leading to attainment of the research objectives; proper wording and sequencing of questions; and pre-testing them to minimize errors in data gathering. Dummy tables, or tables without data entries, are then prepared for analysis.
5. *Data gathering*. This is the implementation of the research design drawn up in no. It includes training, supervision, and monitoring of interviewers and following a sampling plan and timetable for data gathering. Researchers should make sure that procedures are properly followed and all specified data are gathered.
6. *Data analysis and reporting.* When all the survey questionnaires or interview schedules are thoroughly accomplished, the researcher checks each one and edits them, eliminating those with erroneous and incomplete entries.

Responses are tallied and frequency tables are filled up. Statistical analyses are performed where relationships between variables are being investigated.

Pluralism and Extension

Common failures of extension approaches are analyzed which include issues such as the inadequacy of assessing extension effectiveness based on simple production and adoption targets; and extension institutions working under assumptions that do not take diverse activities under farming into account. This has led to rethink extension's role, based on an awareness of the ways in which farmers integrate technological information with other aspects of farm management, leading to the platform approach - a new way of conceptualizing extension. The complexity of rural development demands that farmers, input suppliers, local officials and other actors meet and negotiate to arrive at joint decisions for addressing complex problems of natural resource management. Extension can play a key role in this process by initiating, organizing, participating and following up on these platforms. This suggests that extension institutions should not strive merely to deliver research results but need to act as professional service organisations, being able to adapt priorities as ideas emerge from an ongoing dialogue between field staff and farmers.

This necessitates new attitudes toward structural reform. Extension planning should shift from its present emphasis on administrative directives, to identifying ways to strengthen links between the satisfaction of the extension agent and the satisfaction of the farmer.

Recommendations to support this reorientation of extension institutions are given which include the use of participatory methodologies to support platform approaches; flexible management to assist extension agents in meeting these challenges; finding ways to improve relationships and communication among farmers, researchers and extensionists; and the development of more qualitative indicators to evaluate the success of this new extension in meeting the needs of the poor.

Models

Participative Action Model

The Participative Action Management (PAM) model, developed by Prof. Shankariah Chamala, of the University of Queensland is a major paradigm shift in the way technology is developed and adopted by stakeholders. Adults learn collaboratively using adult-learning principles and action-learning processes.

In the traditional linear model, the development of scientific research and technology is seen as a top-down, centralised, mostly government run and technocratic approach to solving the agricultural production needs of clients. Technology transfer or extension is simply to tell or sell the technology. The client is viewed as passive. Extension agents use the "trickle down" approach to promote the spread of technology. Many technologies promoted under this model have not been readily adopted with reasons for lack of adoption being unclear. This approach suits research of a general nature, but may be of little use in specific practical situations.

PAM model is a management model where all relevant agencies, groups and individuals with a common interest in development or R and D come together. They form a platform to facilitate joint industry problem solving. The group works for the mutual benefit of the partners.

The PAM model is a working-together (or convergence) model where stakeholders' interests are focused on a specific issue, problem or opportunity. This convergence creates energy and the group plans and guides how this new energy is diverted. The group thus acts as a lens for the collection (convergence) of weak energy and distribution (divergence) of stronger energy.

The philosophy of the PAM model comprises of 10 principles:

1. PAM starts with a systems approach.
2. Principle of stakeholders' involvement.
3. Principle of convergence and directed divergence.
4. Empowerment is the cornerstone of the PAM model.
5. Individual rights and responsibilities.
6. Building empowering structures.
7. Networking with other agencies.
8. Encouraging action learning among groups.
9. Building group management capacities.
10. Sharing the credit/profits in a fair way between the project members.

Do Our Own Research (DOOR) Approach for Plant Nurseries

The test of a model is in utility. The PAM model has been used in the DOOR approach and has provided the underlying philosophy for the approach and shaped its management.

DOOR advocates a significant paradigm shift in technology transfer in horticultural research.

In Queensland dependence on the public sector for research solutions to Agriculture and Horticulture industry problems has proved useful for many of the single commodity industries, however less so for the nursery industry. In the nursery industry, generic research conducted by government institutions is often not specific enough to be adopted by the individual operator. Operators need practical solutions to particular problems. This uniqueness reflects the variation of options available in terms of species grown, media used, fertiliser, amendments and chemicals applied and the way water is supplied.

The Queensland Government is strongly advocating increasing industry self-reliance in many aspects of agriculture. The Do Our Own Research (DOOR)-approach which enhances the capacity of nursery operators to do their own research has been found to be the only cost-effective way to find solutions or to develop new opportunities.

DOOR promotes a major move away from the old way of providing research information, away from its dependency on externally generated research, to a new way of self-reliance with industry generating its own research information. This empowers nursery operators to conduct relevant and self-generated research. Operators own the research and so are more likely to adopt any solutions that are generated.

Relationship between the operator and the consultant is that of equal partners the common goal being resolution of an issue with emphasis on skill development in the operator without detracting from the supportive role of the consultant.

The DOOR implementation cycle starts with the recognition of a problem or opportunity. To find a solution or exploit that opportunity, the operator needs to undertake a number of activities.

Some of these activities can be the operator's responsibility: identifying the problem or opportunity, obtaining information, providing resources, implementing the trial, collecting data, validating recommendations. Designing experiments, analysing and interpreting data may be carried out by the DOOR consultant: Together, the operator and the consultant can carry out the remaining tasks: like clarifying the issue and selecting keywords, evaluating and making decisions, calculating cost-benefits, and formulating recommendations.

Certain paradigm shifts among DOOR participants are essential to the success of the DOOR project.

- from industry dependency on service agencies to inter-dependency.
- from passive learning to active learning.
- from competition amongst members to collective action and competition for the common good.
- from complete secrecy to sharing information for mutual benefit.
- from a one-to-one mode of action to collective/collaborative action.
- from win-lose approach to win-win approach.
- from individual problem solving to collaborative problem solving.
- from a personal efficiency perspective to industry bench marking.
- from competing with each other in Australia to complementing each other in overseas market development.
- from a limited market to an unlimited international market outlook.
- from closing the door on inhouse research to active participation in DOOR projects.
- from just taking actions to action learning mode.

Strategic Extension Campaign

The "strategic extension campaign" (SEC) approach has been developed and introduced by FAO in Africa, the Near East, Asia and Latin America. The approach is designed to improve the impact, with farmers, of an existing organised extension service.This methodology emphasises the importance of people's participation in identification of needs, strategic planning, systematic management and field implementation of agricultural extension and training programmes. Its extension strategies and messages are specifically tailored based on the results of a participatory problem identification process on the causes or reasons of farmers' non-adoption, or inappropriate practice, of a given recommended agricultural technology or innovation. The SEC technology transfer and application approach is need-based, demand driven, and has a problem-solving orientation. The programme follows a systems-approach which starts with a farmers' Knowledge, Attitude and Practice (KAP) survey. Using the KAP survey, the SEC functionally links extension with research and local media. It also demonstrates the impact of extension in terms of changes in knowledge level, attitude change, farming practices and productivity. One of the strengths of this approach is in orienting and training relevant extension personnel to apply a systematic, rational, and pragmatic approach to planning, implementing, managing, monitoring and evaluating the routine programmes of an agricultural extension service.

Studies of SEC methods applied to specific FAO-supported extension activities have reported positive changes in farmers' knowledge, attitudes and practices *vis-a-vis* is the recommended technologies as well as significant economic benefits.

Contributions of SEC to the improvement of the extension system, programme and its performance identified by Contado are:

1. Enhancing the agricultural extension planning process
2. Building cadres of extension programme planners and trainers;
3. Improving extension linkage with research;
4. SEC is needed most by small, resource-poor farmers;
5. Improving extension linkage with training;
6. SEC reduces extension system's workload and increases its coverage;
7. Encouraging partnership with and participation of local leaders and community based organisations,
8. Supports the revitalization of extension workers' professionalism;
9. SEC proves that extension programmes can be strategically planned, efficiently managed, and systematically monitored and evaluated;
10. SEC can contribute to improving and strengthening agricultural extension systems and programmes.

Challenges for Supportive Policy Processes

Policy making is commonly considered the prerogative of some central authority that formulates a policy, which is then decreed, imposed, and implemented regardless of conflicting knowledge and concerns. But policy is, in practice, often the net result of the actions of different interest groups pulling in complementary and opposing directions. This is particularly true with environmental problems because they are marked by uncertainty, complexity, and high stakes complexity, and high stakes (Funtowicz and Ravetz, 1993). There is therefore a growing tendency to see policy as a negotiated agreement resulting from interaction among citizens, in which central authorities play a facilitating role (Van der Poel and Van Woerkum, 1994). Policy is only effective if it is based on a widely shared consensus. From this perspective, it is easy to see why so many environmental policies which rely on coercion, control, and transfer have failed (Pretty and Shah, 1994; Pimbert and Pretty, 1994).

Table 31.1: Impact of IPM Programmes Involving New Participatory Approaches to Farmer Learning on Pesticide Use and Crop Yields

Country and Crop	*Average Changes in Pesticide Use (as per cent of conventional treatments)*	*Changes in Yields (as per cent of conventional treatments)*
Togo, cotton[1]	50 per cent	90-108 per cent
Burkina Faso, rice[1]	50 per cent	103 per cent
Thailand, rice[2]	50 per cent	no data
Philippines, rice[2]	62 per cent	110 per cent
Indonesia, rice[2]	34-42 per cent	105 per cent
Nicaragua, maize[3]	25 per cent	93 per cent[1]
USA, nine commodities[4]	No. of applications up volume applied down	110-130 per cent
Bangladesh, rice[5]	0-25 per cent	113-124 per cent
India, groundnuts[6]	0 per cent	100 per cent
China, rice[2]	46-80 per cent	110 per cent
Vietnam, rice[2]	57 per cent	107 per cent
India, rice[2]	33 per cent	108 per cent
Sri Lanka, rice[2]	26 per cent	135 per cent

[a] Even though yields are lower, net returns are much higher.

Sources: (1) Kiss and Meerman, 1991; (2) Kenmore, 1991: Winarto, 1993; van der Fliert, 1993; Matteson et at, 1992; FAO, 1994; (3) Hruska, 1993; (4) NRC, 1989; (5) Kamp et al., 1993; Kenmore, 1991; (6) ICRISAT, 1993

For sustainable agriculture to succeed, policy formulation must arise in a new way. Policy processes must be enabling and participatory, creating the conditions for sustainable development based more on locally available resources and on local skills and knowledge. Effective policy processes will have to bring together a range of actors and institutions for creative interaction and address multiple realities and unpredictability. What is required is the development of approaches that put participation, negotiation, and mediation at the centre of policy formulation so as to create a much wider common ownership in the practices. This is a central challenge for sustainable agriculture. The management of higher level systems, whether common grazing lands, coastal fisheries resources, communal forests, national parks, polders, or watersheds, requires social organization comprising the key stakeholders. All successful moves to more sustainable agriculture have in common coordinated action by groups or communities at the local level (Pretty, 1995). But the problem is that platforms for resource use negotiation generally do not exist, and so need to be created and facilitated (Brinkman, 1994; Röling, 1994a, 1994b).

Different methodologies are emerging to help stake-holders achieve collective resource management capacity. Well known are participatory rapid appraisal (PRA) and related methodologies (see chapter 6). In addition, the soft system methodology (SSM) developed for corporate environments is highly promising for resource use negotiation (Checkland, 1981; Checkland and Scholes, 1990). For stakeholders who have come to appreciate the fact that they share a problem, SSM takes them through a number of steps which allows them to create a "rich picture" on the basis of their multiple perspectives, reach some accommodation with respect to major causes of the problem, and hence decide on collective action. "Rapid appraisal of agricultural knowledge systems" (RAAKS) (Engel, 1995) is a related methodology for facilitating innovation as an emergent property of a knowledge network, comprising such actors as farmers, extension workers, researchers, NGO workers, and policy makers. This system provides stakeholders with different "windows" (such as mission, task differentiation, integration, articulation, coordination, performance) on their own collective practices which allow them to capture the potential synergy of their contributions to innovative performance.

A fundamental requirement if such approaches are to work is that stakeholders in a particular natural resource learn to appreciate that they have a common problem. Extension has an important role to play here by making visible the interdependence between stakeholders and the extent to which the resource unit on which they depend has been destroyed by their uncoordinated action and the collective impact of their individual activities. It is within policy contexts thus made conducive for sustainable agriculture that technology development and extension can be especially effective.

Policy Conclusions

- ✩ Extension programmes should highlight local development or adaptation of technologies: address the needs of specific categories of clients; support farmers' organisations and farmer to farmer extension; and influnce both collective and individual behaviour.
- ✩ Diversity of extension provision from agencies in the public, private, NGO and academic sectors gives clients greater choice of sources of information to support long term sustainability of their farming.
- ✩ Extension worker competence can be improved by training which addresses learning objectives specifically related to sustainability and by introducing management practices which support continued learning and professional development.
- ✩ Mass media coverage and relevance of content to rural audiences can be improved at low cost. Radio should be used specifically to promote farmer to farmer communication.
- ✩ Extension materials should be designed to offer options and problem solving strategies, facilitate decision making and technology adaptation and contain more explicit treatment of sustainability in relation to technical content.

References

Anyaegbunan, Chike, Paulo Mefalopulos, and Titus Moetsabi (1998). Participatory Rural Communication Appraisal: Starting with the People. Harrare, Zimbabwe and Rome, Italy: SADC Center of Communication for Development and FAO. 195 p.

Arboleda, Corazon (1981, 1991). Communication Research. Manila, Philippines: Chevalier Printing and Publishing Center and CFA Media Group.

Antholt, C. H. (1992). *Relevancy, responsiveness and cost-effectiveness: Issues/or agricultural extension in the 21st century*. Washington, DC: Asia Region, Technical Department, World Bank.

Antholt, C. H. (1994). *Getting ready for the twenty-first century: Technical change and institutional modernisation in agriculture.* World Bank Technical Paper 217. Asia Technical Department Series. Washington, DC: World Bank.

Argyris, C., Putoam, R., and Smith, D. M. (1985). *Action Science.* San Francisco and London: Jossey-Bass.

Attah-Krah, A. N., and Francis, P. A. (1987). The role of on-farm trials in the evaluation of composite technologies: The case of alley farming in southern Nigeria. *Agricultural Systems, 23*, 133-152.

Axinn, G. H. (1988). T and V (tragic and vain) extension. *INTERPAKS, INTERCHANGE, International Agriculture, 5* (3). Urbana: College of Agriculture, University of Illinois at Urbana-Champaign.

Bannister, M. E., and Nair, P. K. R. (1990). Alley cropping as a sustainable agricultural technology for the hillsides of Haiti: Experience of an agro-forestry outreach project. *American Alternative Agriculture, 5* (2), 51-59.

Benor, D. (1987). Training and visit extension: Back to basics. In Rivera and Schram, *Agricultural Extension Worldwide* (p. 137-149). New York: Croom Helm.

Brinkman, R. (1994). Recent developments in land use planning. Keynote address at the 75-year Anniversary Conference of the Wageningen Agricultural University. In L.O. Fresco *et al.* (Eds.), *Future of the land: Mobilising and integrating knowledge for land use options.* Chichester: John Wiley and Sons.

Brouwers, J., and Röling, N. (in press). Living local knowledge for sustainable development. Accepted for publication in D. M. Warren, S. Fujisaka, and G. Prain (Eds.), *Biological and cultural diversity: The role of indigenous experimentation in development.* London: Intermediate Technology Publications.

Bunch, R. (1991). *Low input soil restoration in Honduras: The Cantarranas farmer-to-farmer extension programme.* Sustainable Agriculture Programme Gatekeeper Series SA23. London: IIED.

Campbell, A. (1994). *Landcare: Communities shaping the land and the future.* Sydney: Alien and Unwin.

Chambers, R., Pacey, A., and Thrupp, L-A. (Eds.). (1989). *Farmer first: Farmer innovation and agricultural research.* London: Intermediate Technology Publications.

Checkland, P. B. (1981). *Systems thinking, systems practice.* Chichester: John Wiley.

Checkland, P., and Scholes, J. (1990). *Soft systems methodology in action.* Chichester: John Wiley and Sons.

Engel, P. (1995). *Facilitating innovation: An action-oriented approach and participatory methodology to improve innovative social practice in agriculture.* Published doctoral dissertation. Agricultural University, Wageningen.

Fresco, L. O., Stroosnijder, L., Bouma, J., and van Keulen, H. (Eds.) (1994). *The future of the land.* Chichester: John Wiley and Sons.

Fujisaka, S. (1991). Thirteen reasons why farmers do not adopt innovations intended to improve the sustainability of agriculture. In *Evaluation for sustainable land management in the developing world: Volume* 2. Technical Papers. IBSRAM Proceedings, 72(2), 509-522. Thailand: IBSRAM.

Funtowicz, S. O., and Ravetz, J. R. (1993). Science for the post-normal age. *Futures,* 25(7, September), 739-755.

GTZ (1992). *The spark has jumped the gap.* Eschborn: Deutsche Gessellschaft für Technische Zusammenarbeit (GTZ).

Haverkort, B., van der Kamp, and Waters-Bayer, A. (Eds.). (1991). *Joining farmers' experiments: Experiences in participatory development.* London: IT Publications.

Howell, J. (Ed.). (1988). *Training and visit extension in practice.* London: Overseas Development Institute.

Hruska, A. J. (1993). Care policy promotes IPM. *International Ag-Sieve,* 5(6), 4-5.

Hussain, S. S., Byerlee, D., and Heisey, P. W. (1994). Impacts of the training and visit extension system on farmers' knowledge and adoption of technology: Evidence from Pakistan. *Agricultural Economics, 10*, 39-47.

ICRISAT. (1993). Will the pod-borer become the farmer's pall bearer? *SAT News*, p. 7-10.

Ison, R. (1990). *Teaching threatens sustainable agriculture.* Gatekeeper Series SA21. London: IIED.

Jiggins, J. and de Zeeuw, H. (1992). Participatory technology development in practice: Process and methods. In C. Reijntjes, B. Haverkort, and A. Waters-Bayer (Eds.), *Farming for the Future.* Netherlands: Macmillan and ILEIA.

Kang, B. T., Wilson, G. F., and Lawson, T. L. (1984). *Alley cropping: A stable alternative to shifting agriculture.* Ibadan: IITA.

Kerkhof, P. (1990). *Agro-forestry in Africa: A survey of project experience.* London: Panos Institute.

Kingsley, M. A., and Musante, P. (1994). *Activities for developing linkages and co-operative exchange among farmers' organisations, NGOs, GOs and researchers: Case study of IPM in Indonesia.* Paper presented at IIED symposium, Brighton.

Kiss, A., and Meerman, F. (1991). *Integrated pest management in African agriculture.* World Bank Technical Paper 142. African Technical Department Series. Washington, DC: World Bank.

Krishna, A. (1993). *Watershed development in Rajasthan: The new approach.* Jaipur: Watershed Development, Government of Rajasthan.

Lal, R. (1989). Agro-forestry systems and soil surface management of a tropical alfisol. I: Soil moisture and crop yields. *Agro-forestry Systems, 8*, 7-29.

Lobo, C., and Kochendorfer-Lucius, G. (1992). *The rain decided to help us: An experience in participatory watershed development in Maharashtra State, India.* Ahmednagar: Social Centre.

Long, N., and Long, A. (Eds.) (1992). *Battlefields of knowledge: The interlocking of theory and practice in social research and development.* London: Routledge.

Moris, J. (1990). *Extension alternatives in Africa.* London: ODI.

Matteson, P. C., Gallagher, K. D., and Kenmore, P. E. (1992). Extension of integrated pest management for planthoppers in Asian irrigated rice: Empowering the user. In R. F. Denno and T. J. Perfect (Eds.), *Ecology and Management of Planthoppers.* London: Chapman and Hall.

Narayan, S. (2005), Organic farming in India: Relevance, problems and constraints. Occasional paper 38. Department of Economic analysis and research and National Bank for Agriculture and Rural Development (NABARD), Mumbai.

Niels Röling and Jules N. Pretty (1993) Improving agricultural extension. A reference manual.. Natural Resources Management and Environment Department. FAO corporate document repository.

Ornes, F. (1988). Community training centres for organic agriculture and appropriate technology. Dominion Republic. In C. Conroy and M. Litvinoff (Eds.), *The greening of aid.* London: Earthscan Publications.

Pimbert, M., and Pretty, J. N. (1994). *Parks, people, and professionals: Putting "participation" into protected areas management.* Geneva: UNRISD and WWF; London: IIED.

Pretty, J. N. (1994b). Alternative systems of inquiry for sustainable agriculture. *IDS Bulletin*, 25(2), 37-48. University of Sussex: IDS.

Pretty, J. N. (1995). *Regenerating agriculture: Policies and practice for sustainability and self-reliance.* London: Earthscan Publications.

Pretty, J. N. and Chambers, R. (1993). *Towards a learning paradigm: New professionalism and institutions for sustainable agriculture. IDS Discussion Paper* DP 334. Brighton: IDS.

Pretty, J. N., and Shah, P. (1994). *Soil and water conservation in the 20th century: A history of coercion and control.* Rural History Centre Research Series No. 1. Reading: University of Reading.

Reijntjes, C., Haverkort, B., and Waters-Bayer, A. (1992). *Farming for the future: An introduction to low-external-input and sustainable agriculture.* The Information Centre for Low-External-Input and Sustainable Agriculture (ILEIA). London: Macmillan.

Roberts, N. (Ed.). (1989). *Agricultural extension in Africa.* World Bank Symposium. Washington, DC: World Bank.

Roche, C. (1992). *Operationality in turbulence: The need for change.* London: ACORD.

Rogers, E. M. (1962 and 1983). *Diffusion of innovations.* New York: Free Press.

Röling, N. (1992, October). *Facilitating sustainable agriculture: Turning policy models upside down.* Paper presented at IIED/IDS Conference, Brighton.

Röling, N. (1993). Agricultural knowledge and environmental regulation: The crop protection plan and the Koekoekspolder. *Sociologia Ruralis,* 33 (2), 212-231.

Röling, N. (1994a). Platforms for decision making about ecosystems. In L. Fresco (Ed.), *The future of the land.* Chichester: John Wiley and Sons.

Röling, N. (1994b). Creating human platforms to manage natural resources: First results of a research programme. In *Proceedings of the International Symposium on Systems Oriented Research in Agriculture and Rural Development* (p. 391-395). Montpellier, France, November 21-25, 1994.

Röling, N., and Engel, P. (1989). IKS and knowledge management: Utilizing indigenous knowledge in institutional knowledge systems. In D. Warren *et al.* (Eds.), *Indigenous knowledge systems: Implications for agriculture and international development.* Studies in Technology and Social Change, 11. Ames: Technology and Social Change Program, Iowa State University.

Röling, N., and Jiggins, J. (1993). Policy paradigm for sustainable farming. *European Journal of Agricultural Education and Extension,* 7(1), 23-44.

Russell, D. B., and Ison, R. L. (1991, April). *The research-development relationship in rangelands: An opportunity for contextual science.* Plenary paper presented at the 4th International Rangelands Congress, Montpellier, France.

Scoones, I., and Thompson, J. (Eds.). (1994). *Beyond Farmer First.* London: IT Publications.

Sen, C. K. (1993). Nepal: Group extension. *Rural Extension Bulletin,* 3, 17-23. University of Reading, UK.

Shah, P. (1994). Participatory watershed management in India: The experience of the Aga Khan Rural Support Programme. In I. Scoones and J. Thompson (Eds.), *Beyond Farmer First.* London: IT Publications.

Sriskandarajah, N., Bawden, R. J., and Packham, R. G. (1991). Systems agriculture: A paradigm for sustainability. *Association for Farming Systems Research-Extension Newsletter,* 2(2), 1-5.

Stock, T. (1994). *The impact of integrated pest management farmer field schools in the Cordillera region of the Philippines.* Unpublished master's thesis. University of Reading.

Van der Poel, M. M., and Van Woerkum, C. J. M. (1994). *De positionering van communicatie in Beleid.* Deelrapport 1 (literatuuronderzoek) en Deel 2 (Weergave van bijeenkomsten, conclusies van het onderzoek en aan-bevelingen voor vervolgonderzoek). Wageningen: Agricultural University, Department of Communication and Innovation Studies, carried under a contract from the Government Information Office.

Warren, D. (1991, May). The role of indigenous knowledge in facilitating a participatory approach to agricultural extension. Paper presented at the International Workshop on Agricultural Knowledge Systems, Bad Boll, Germany.

Winarto, Y. (1993). Farmers' agro-ecological knowledge construction: The case of integrated pest management among rice farmers on the north-west coast of West Java. *Rural People's Knowledge, Agricultural Research and Extension Practice,* 1 (3), London: IIED.

Young, A. (1989). *Agro-forestry for soil conservation.* Wallingford: CAB International.

Zijp, W. (1993). Being a good communicator doesn't solve all of extension's problems. *Development Communication Report,* 80, 20.

Hill Agriculture: Prospects, Constraints and Mitigations (2013) *Pages* **499–513**
Editor: **Dr. Shahid Ahamad**
Published by: **DAYA PUBLISHING HOUSE, NEW DELHI**

Chapter 32

Opportunities and Constraint of Fisheries in Jammu and Kashmir

Prem Kumar[1], Sahar Masud[2] and Shiv Kumar Sharma[3]
[1]SMS (Fisheries Science); [2]Assistant Professor, F.V.Sc.,
Sher-e-Kashmir University of Agriculture Sciences and Technology of Jammu
[3]G.B. Pant University of Agriculture and Technology, Pantnagar

Introduction

Capture fisheries and aquaculture supplied the world with about 142 million tones of fish in 2008. Of this, 115 million tonnes was used as human food, providing an estimated apparent per capita supply of about 17 kg (live weight equivalent), which is an all-time high. Globally, fish provides more than 1.5 billion people with almost 20 per cent of their average per capita intake of animal protein, and 3.0 billion people with at least 15 per cent of such protein. Global capture fisheries production in 2008 was about 90 million tonnes, with an estimated first-sale value of US$93.9 billion, comprising about 80 million tones from marine waters and a record 10 million tonnes from inland waters. Aquaculture continues to be the fastest-growing animal-food-producing sector and to outpace population growth, with per capita supply from aquaculture increasing from 0.7 kg in 1970 to 7.8 kg in 2008, an average annual growth rate of 6.6 per cent. It is set to overtake capture fisheries as a source of food fish. While aquaculture production (excluding aquatic plants) was less than 1 million tonnes per year in the early 1950s, production in 2008 was 52.5 million tonnes, with a value of US$98.4 billion. It is estimated that aquaculture will meet more than 50 per cent of global food fish consumption by 2012. In the period 1970–2008, the production of food fish from aquaculture increased at an average annual rate of 8.3 per cent, while the world population grew at an average of 1.6 per cent per year.

Fish is one of the major sources of protein food for the Indian population also and the potential for culture production from various controllable small bodies of water in India is great. During 2008-09, fish production through aquaculture in India was 46.60MT while marine production was only 29.77MT.

Jammu and Kashmir is the north western state of India. The State occupies a strategic place on the country's map with its border touching Pakistan, Tibet and China. Agriculture is the predominant sector in the economy of Jammu and Kashmir. Directly and indirectly, it supports about 80 per cent of the population besides contributing nearly 60 per cent of the state revenue, which adequately explains the over-dependency of the population on agriculture.

Traditionally, fisheries in Jammu and Kashmir were developed and popularized as sport fisheries, primarily as a means of attraction for tourists. During the past few years, there has been a reorientation towards development of fisheries as a major food resource. Fisheries in Jammu and Kashmir has the potential to grow exponentially as the state is bestowed with a network of cold water streams, perennial rivers, lakes, reservoirs, sars and about 250 high altitude lakes spread over an area of 40 thousand hectares. The state has the scope to promote all types of fisheries in view of its unique agro-climatic conditions. Jammu and Kashmir is divided into three agro-climatic zones: Cold arid desert areas of Ladakh, temperate Kashmir Valley and the humid sub-tropical region of Jammu. Each has its own specific geo-climatic condition which determines the cropping pattern and productivity profile.

All the regions offer potential for promotion of different varieties of fisheries, while Kashmir valley the temperate zone offers potential for development of cold water fisheries, the tropical zone Jammu division offers potential for development of warm water fisheries. The areas of Kathua, Udhampur, Doda, Rajouri and Poonch in addition of warm fisheries have the potential for producing trout fisheries also. The Ladakh region is bestowed with the atmosphere suitable for cold water fisheries and for Brackish fisheries. It is, however, pertinent to mention that brackish fisheries are yet to be exploited.

At the present time, it appears that fish are an important 'natural asset' in the livelihoods of people in some parts of the region. This includes the role of aquatic resources food and subsistence fishing, eco-tourism and income from sport fishing and some small-scale aquaculture.

Aquatic Resources

The water resources in this large region include rivers and streams of various sizes, lakes, man-made reservoirs, floodplains and swamps, rice fields and man-made ponds and irrigation canals. These water resources have a diverse fish fauna that support the livelihoods of rural communities within the region. They also provide various opportunities for development of fisheries and aquaculture.

In the region, Inland fisheries exist in natural areas such as streams, rivers, swamps, lakes and in temporary waterbodies such as floodplains and seasonal ponds, and also in artificial and modified habitats such as irrigation systems, rice paddies, reservoirs and enclosed natural waterbodies (*e.g.* ox-bow lakes). The length of 27781 km of rivers/streams facilitates the farming of more than 40 million tonnes of fish. Out of total 27781 km of area under fisheries, the State has only 0.07 lakh hectares under reservoir area. There are 1248 lakes and water bodies and the water is spread into 39921.8 hectare of area. Some major lakes are shown in Table 32.1.

Fish and Fauna

The fish species distribution in the Himalayan streams depends on the flow rate, nature and substratum, water temperature, and the availability of food. In the torrential streams Sehgal (1988) identified several zones on the basis of the dominant fish species and the hydrographical features:

1. Headwater zone inhabited by rheophilic species of loaches and catfishes (*Nemacheilus gracilis, N. stoliczkae* and *Glyptosternum reticulatum*).

2. Large stream zone, formed by the joining of headwater streams, inhabited by *Diptychus maculatus* and*Nemacheilus* spp. In the upper reaches of the most torrential section of this zone, the rheophilic species of snow trouts, *Schizothoraichthys esocinus*, *S. progastus*, *Schizothorax richardsonii* and *Schizopygopsis stoliczkae*, occur. The intermediate reaches of the large stream zone are frequented by *Schizothorax longipinnis*, *S. planifrons* and. *S. micropogon*. The least rapid reaches of this zone are occupied by *Garra gotyla*, *Crossocheilus diplochilus*, *Labeo dero* and *L. dyocheilus*.
3. Slow moving meandering zone inhabited by a large number of cold or eurythermal species such as *Barilius* spp, *Tor* spp, catfishes, homalopterid fish (*Homaloptera* spp) and snakeheads (*Channa* spp).

Table 32.1: Major Lakes

Type	*Name*	*Altitude*
Flood plain lake	Wular lake	1537
	Dal Lake	1585
	Manasbal Lake	1587
Mountain Lakes	Gangabal	3570
	Konsernag	3670
	Vishansar	3817
	Kishansar	3677
	Alpather	3200
	Nundkul	3550
	Tarsar	3710
	Seshnag	3570
	Tulian	3750
Forest Lakes	Nilnag	2180
Brackish water Lakes	Tso Morari	4541
	Pangong Tso	4329
Siwalik Lakes	Surinsar	604
	Mansar	666

The need for shelter from the current has led to territoriality. Mahseers and Schizothoracines chase intruders to defend the limited food resource and available shelter. Water temperature is always an important limiting factor influencing geographical distribution and local occurrence within one water system. Cold stenothermic species such as the endemic schizothoracines (*Schizothoraichthys esocinus* and *Diptychus maculatus*) and exotic brown trout have an upper tolerance of about 20°C. Carps, mahseers and lesser barils have a wider tolerance and even survive water temperatures over 25°C. Schizothoracines and brown trout remain active in the near-zero temperatures which prevail in streams of the Lesser and Greater Himalaya during December and January. Hailstorms and drought conditions in the Lesser Himalaya may cause adverse conditions, leading to fish kill.

To cope with the steep fall in temperature in winter months schizothoracines migrate from headwaters to lower altitude where they represent a sizeable part in fish catches in large rivers and

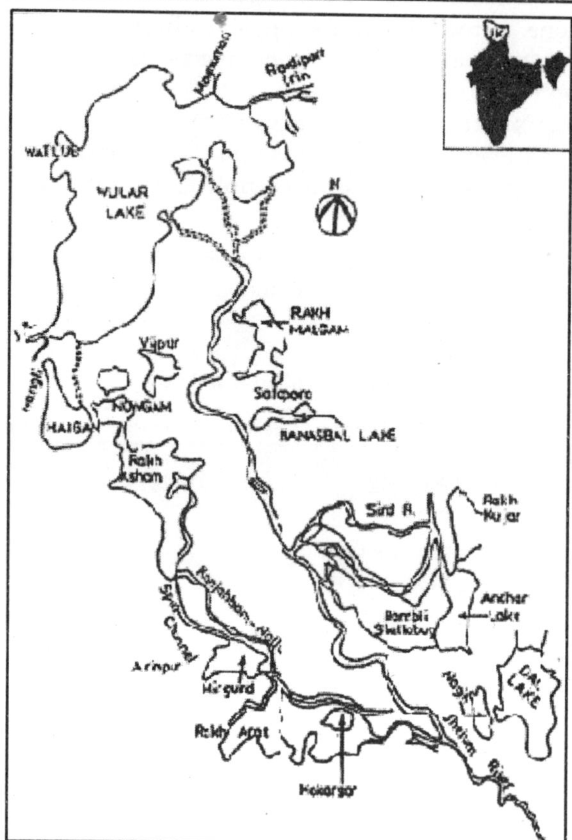

Figure 32.1: Floodplain Lakes and Major Wetlands in Kashmir Valley

their tributaries. The rise in temperature in Kashmir streams from near-freezing level to 10-17°C during May-June induces *S. richardsonii, S. longipinnis* and *S. curvifrons* to spawn. In the Sutlej River, *S. richardsonii* starts upstream migration with the rise in water temperature during March. During the upstream migration the fish still finds itself in waters of low temperatures, owing to the steady inflow of snow-melt water. To avoid it the species migrates into warmer side streams in which it spawns at temperatures of 17.5-21.5°C, such as in the Ravi River, where it spawns in May. In the upper Beas, the same species spawns only in July-August when the water temperature is warm enough, but some fish of the same species migrate downstream in the same river to spawn from October to December at 19.0 to 22.5°C (Sehgal, 1999).

The frequent occurrence of spates in some rivers hinders spawning of cold water fish. The low density populations of cold water fish in the upper reaches of the Sutlej and Chenab rivers may result from the passage of these rivers through deep and narrow gorges, and the presence of cold glacier- and snow-melt water.

As a result of a study of eleven rivers in the northwestern Himalayas, Sehgal (1999) noted the changes in the prevalent fish species. The eastern Himalaya drained by the Brahmaputra has a greater diversity of cold water fish than the western Himalayan drainage. 218 fish species are listed for the whole Himalayas (Menon, 1962).

The state has reported 125 fish species, of which only 11 are commercially important. The most common exotic species are rainbow trout, brown trout, common carp, while the indigenous fishes are: mahseers (*Tor putitora* and *Tor tor*), and schizothoracines (*Schizothoraichthys esocinus, S. progastus, Schizothorax richardsonii, S. niger and S. curvifrons*). Among these *Tor putitora, S. progastus* and *S. richardsonii* are preferred because of their wide range of distribution in the Himalayas.

In floodplain lakes of Kashmir Valley the major species are: *Schizothorax niger, S. micropogon, S. curvirostris, S. planifrons, Schizothoraichthys esocinus, Labeo dero, L. dyocheilus, Crossocheilus latius, Puntius conchonicus, Glyptothorax kashmiriensis*, and *Gambusia affinis*. Schizothoracines are highly valued fish, preferred to most other fish species. They feed on detritus, attached plant (including algal) coating of stones and rocks, and the associated invertebrate fauna. They grow slowly, and attain maturity at the age of 2 years (Sunder and Subla, 1984). All species of the group except *S. niger* exhibit spawning migration to the incoming streams and rivers and lay eggs in shallow pools amidst gravel and sand. *S. niger*, however, prefers clean and cold pockets of water in lakes for egg-laying, and it also lays eggs on the roots of willow trees (Vass and Raina, 1979). In 1959 common carp was introduced in Kashmir to augment the fish yield (Sehgal, 1989). Since then, this species has invaded all the meandering rivers, floodplain lakes and wetlands, and it has become a major commercial fish in Kashmir Valley.

In Kashmir, wetlands called *sars* cover more than 4000 ha and serve as a natural refuge for a wide variety of organisms. The wetlands are situated on floodplains of the River Jhelum. While some wetlands, such as Haigam and Hokarsar, are maintained by the State Government as bird sanctuaries, as they provide excellent habitats for water fowl, wild duck and geese migrating from China, Russia and other distant regions during winter, wetlands are also important habitats for fish. The principal species using wetlands are common carp, *Schizothorax niger, Crossocheilus latus, Puntius conchonius* and *Gambusia affinis*. The fish yield in wetlands varies from 15 to 30 kg ha^{-1} yr^{-1}. In wetlands minnows are caught with scoop nets in large numbers during summer months. The fish is sun-dried or smoked and it fetches a high price.

In the upper niche lakes (Gangabal, Nundkol, Kishansar and Vishansar) contain the exotic brown trout (Vass *et al.*, 1989). Lakes Gadsar and Zumsar have an endemic schizothoracine *Diptychus maculatus*. This fish is also present in the inflowing streams. Like other schizothoracines, it feeds on benthic plants, mainly on those attached to rocks and stones, and on the associated benthic invertebrates. It is fished for subsistence by herdsmen during summer. Brown trout is allowed to be fished with fly, but only by licensed anglers.

Twelve high-altitude lakes in Kashmir are situated at 3200 m to 3819 m a.s.l. They range from 1 ha to 157 ha in size. Only six of them contain fish: four (Gangabal, Nundkol, Kishangar and Vishansar) have the exotic brown trout. Lakes Zumsar and Gadsar have an endemic schizothoracine *Diptychus maculatus*. This fish is also present in the inflowing streams. It is fished for subsistence during summer by herdsmen. Brown trout is permitted to be fished with fly, but only by licenced anglers.

Fishing Methods and Fishermen

There are two basic types of fisheries in the Indian Himalayan rivers: subsistence fishery, and sport/recreational fishery. Fish production in mountain streams is low and therefore any commercial fishery is on a very limited scale. The low biological productivity results in the prevalence of small-sized fish, except in pools where fish have some shelter and resting place. Fish also reach a larger size in some cold water reservoirs and lakes.

The total fishermen population in the State as per livestock census 2003 was around 31000. Most fishing is done by cast nets. Other fishing methods use long-lines, scoop nets and traps, which account for 5-7 per cent of the total fishing gear. Fish production from floodplain lakes is not well documented. The number of fishing licence holders was 13026 ending March, 2009.

The fishing methods are simple, but well suited to the turbulent nature of the streams. Cast nets, drag nets, stake nets, bag nets and some other types are used, as well as different types of traps, nooses and harpoons. Poison is also used, such as sap of *Euphorbia rogleana*, powdered seed of *Xanthoxylum alatum*, and *Cascaria tormentosa*. The subsistence and commercial fisheries exploit carps (*Labeo* and *Tor* spp.), lesser barils (*Barilius* spp.), schizothoracines (*Schizothorax* and *Schizothoraichthys* spp.), garrids (*Garra* spp.) and sisorids (*Glyptothorax* and*Glyptosternum* spp.). The other genera are of small size and of low economic value. The exotic brown trout (*Salmo trutta*) has established itself in some areas of the Himalayas (Sehgal, 1999).

The Kashmir floodplain lakes Wular, Dal and Manasbal are situated in the Kashmir Valley at an altitude of 1537-1587 m a.s.l in which the major fish group is schizothoracines, represented by a number of species: *Schizothorax niger, S. micropogon, S. curvirostris, S. planifrons, Schizothoraichthys esocinus*. The following fish are also fished: *Labeo dero, L. dyocheilus, Crossocheilus latus, Puntius conchonicus, Glyptothorax kashmiriensis*. Schizothoracines are highly valued fish, preferred to most other fish species. In 1959 common carp was introduced in Kashmir to augment the fish yield. Since then this fish has become a major commercial species in Kashmir Valley. It is also the principal species in wetlands situated on floodplains of the River Jhelum. The other fish species utilizing wetlands are *Schizothorax niger, Crossocheilus latus, Puntius conchonicus* and *Gambusia affinis*.

Fish Production

Due to congenial conditions for development and sustenance of different variety of fisheries, the production is showing an increasing trend. The fish caught which was 184.67 thousand quintals in 2000-01 has reached to 192.70 thousand quintals in 2008-09. The production of famous Kashmiri Trout has increased by 888 quintals between 2000-01 and 2008-09. Of the 192.70 thousand quintals fish production for 2008-09, Kashmir Province accounts for 161.71 thousand quintals while as Jammu Province accounts for 30.99 thousand quintals. This indicates that 84 per cent fish is produced in Kashmir province (Table 32.2).

Table 32.2: Fish Caught (Species-wise) (Unit in Quintals)

Sl.No.	Year	Kashmir Province				Jammu Province	Total State
		Trout	Mirror Carps	Country Fish	Total		
1.	2000-01	792	114323	41295	156410	28257	184667
2.	2001-02	864	115405	42154	158423	29642	188065
3.	2002-03	921	116492	42737	160150	29950	190100
4.	2003-04	958	116547	42795	160300	30200	190500
5.	2004-05	1350	116587	42380	160317	30240	190557
6.	2005-06	1500	116675	42928	161103	30397	191500
7.	2006-07	1520	116835	43028	161383	30617	192000
8.	2007-08	1625	116942	43035	161602	30898	192500
9.	2008-09	1680	116987	43041	161708	30992	192700

Source: Directorate of Fisheries (J&K).

Table 32.3: State-wise Fish Production in 2008-09

		(Tonnes)	*(In 000' Tonnes)*
State/Uts	*Marine*	*Inland*	*Total*
Andhra Pradesh	291159	961618	1252777
Arunachal Pradesh	0	2880	2880
Assam	0	206150	206150
Bihar	0	300650	300650
Goa	82950	2284	85234
Gujarat	623055	142847	765902
Haryana	0	76285	76285
Himachal Pradesh	0	7793	7793
Jammu and Kashmir	**0**	**19270**	**19270**
Karnataka	218137	143005	361142
Kerala	583579	82574	666153
Madhya Pradesh	0	68466	68466
Maharashtra	395963	127138	523101
Manipur	0	18800	18800
Meghalaya	0	3959	3959
Mizoram	0	3760	3760
Nagaland	0	6175	6175
Orissa	135486	235973	371459
Punjab	0	104770	104770
Rajasthan	0	25700	25700
Sikkim	0	168	168
Tamil Nadu	365280	168885	534165
Tripura	0	36000	36000
Uttar Pradesh	0	349274	349274
West Bengal	186789	1323123	1509912
A and N Islands	32785	167	32952
Chandigarh	0	244	244
D and N Haveli	0	48	48
Daman and Diu	14060	81	14141
Delhi	0	511	511
Lakshadweep	11586	0	11586
Pondicherry	36550	3306	39856
Chattisgarh	0	158699	158699
Uttarakhand	0	3163	3163
Jharkhand	0	75816	75816
Total	2977379	4659582	7636961

P: Provisional

Source: Department of Animal Husbandry, Dairying and Fisheries, Ministry of Agriculture, Govt. of India.

The fish output is mostly confined to the valley - Baramulla (47 per cent) and Srinagar (36 per cent) districts, accounted for the major share of the state's fish output. Out of the total fish produce, about 80 per cent of the fishing activity takes place in the lakes and the rest from rivers. There are four types of fish available in Jammu and Kashmir, *viz.*, trout, mirror carp, country fish and Jammu fish. The predominant variety is mirror carp forming 61 per cent of the total output. The other three contribute 40 per cent of the total fish production.

Trout fishing holds great potential in terms of increased income rather than output through proper harvesting.

Available data show that prior to the introduction of common carp the fish yield was low. With its introduction in Kashmir, yields have increased spectacularly. During 1974-77 Pandit and Qadri (1986) recorded a mean annual catch of 1640 T. Common carp constituted 1380 T, while schizothoracines represented 387 T of the total catch. The average fish yield in Lake Dal was estimated to be 16.5 kg ha^{-1}, and that in Lake Wular 17-25 kg ha^{-1} (Vass and Zutshi, 1983).

The fish catch during the year 2008-09 has reached to 1.927 lakh quintals recording an increase of 4.35 per cent over the figures for 2000-01. Fish catch over the years has been encouraging but in overall fish production of India, J&K occupies 22nd position among the different states. Production of famous Kashmiri trout has recorded increase of 112.12 per cent between 2000-01 and 2008-09.

Status of Aquaculture

In the Indian Himalayas the cultivation of fish contributes little to the overall freshwater fish production. Earlier every facility was created for fish cultivation in the Indian Himalayas produced fish for stocking streams and lakes primarily to meet the requirements of sport fishing and commercial fishery was also dependent to some extent on the stocking of lakes and reservoirs with fry and fingerlings. While form last few years fish hatcheries in the J&K have been raising eyed-eggs, fry and fingerlings of brown and rainbow trout, and fry and fingerlings of common carp, grass carp, rohu, mrigal and silver carp for stocking, only recently have some hatcheries started producing seed for stocking the indigenous mahseers and schizothoracines.

Trout

At the beginning of the 1990s farming of trout was being carried out on several fish farms in which Kashmir was having four fish farms. The establishment of these aquaculture farms and hatcheries were completely based on the availability of water in required quantity and quality, *i.e.* from rheocrene springs and snow-melt/glacier-fed streams. Most farms receive water supply from snow-melt/glacier-fed streams. The old facilities at Laribal, Harwan and Achabal (the last receiving seepage water of the Bringhi stream and hence not a true spring-fed facility) in Kashmir were established when the stream water did not have the present level of silt load. Deterioration of catchment soils by inappropriate agricultural practices and deforestation, and pollution are reducing the water quality, harming the cold water fish stocks and hatcheries and adversely affecting the performance of fish farms. At present a heavy load of silt reaches these farms from February to September and only from October to January the silt load is low. Even some of the more recently established facilities are also facing the same difficulty.

Natural spawning and the time for egg stripping of brown and rainbow trout is related to the changing of the photoperiod. After receiving snow/glacier-melt water, brown trout attains full maturity in mid-November to mid-December in trout farms of Kashmir. In spring-fed farms Brown trout matures in December-January. Rainbow trout matures in mid-February to March. The average number of eggs per kilogram of female body weight varies between 1234 and 1342 in brown trout and 1649 to 1850 in rainbow trout. The wide application of the improved technology has resulted in achieving 80-90 per cent success in several hatcheries in Kashmir. In Kashmir farms the conventional diet of trout consists of raw fish which results in nutritional deficiency.

The principal factors responsible for fry losses are silt load, hailstorm-cooled water, diseases such as myxosporidiasis, octomiasis, ichthyophthiriasis, and nutritional deficiencies. Yearlings and 1+ age group are raised in growing ponds and raceways. Trout is fed there on a conventional diet consisting of local raw fish, mutton, slaughter house waste, dry silk-worm pupae, etc. The most serious loss of growing trout is due to nutritional imbalance and non-observance of prophylactic measures.

Schizothoracines

The culture of schizothoracines is still in its experimental stage. Like the mahseers this group of cyprinids has shown a sharp decline in catches all along the Himalayas due to indiscriminate fishing and environmental degradation. While Das (1978) believes that the introduction of exotic common carp has adversely affected the schizothoracine fishery in the lacustrine environment of the Kashmir Valley lakes, Sunder and Subla (1984) are of the opinion that in the Jhelum River the presence of the common carp has had no impact on schizothoracines. But observations of Sehgal (1977) and Natrajan (1988) support the Das (1978) opinion. Sehgal (1974) started experiments with collection and artificial fertilisation of eggs from Schizothorax planifrons, S. curvifrons and S. plagiostomus. These fish make spawning runs from Lake Wular into two inflowing streams. Subsequently, egg-taking of

Schizothorax esocinus **(Snow trout)**

S. plagiostomus and raising of post-larvae and fry was successfully carried out using mature specimens caught from several streams of Himachal Pradesh (Jhingran and Sehgal, 1978). Eggs of *S. niger* from Lake Dal were artificially fertilised and achieved a hatching rate of 10-40 per cent (Vass *et al.*, 1978). The result of induced breeding and artificial fertilization and incubation of *Schizothoraichthys esocinus* in a running water system gave cumulative hatching of 30-55 per cent. Further rearing of *S. esocinus* to fry and fingerling on artificial feeds met with some success (Raina *et al.*, 1986). There is still work to be done on schizothoracine seed production to achieve a fully viable fingerling stage ready for release into rivers and lakes.

Exotic Carps

In the Himalayan region two German phenotypes of common carp (mirror carp and scale carp) are commonly produced in aquaculture. These species are cultivated primarily to produce seed for extensive pond culture and for stocking of lakes and reservoirs. Common carp is produced in fish farms of the State Fisheries Departments. The common carp is successfully bred in cement tanks, in rectangular cloth containers fixed in ponds (hapas), and in earthen ponds. Since common carp eggs are adhesive and need a spawning bed, aquatic plants such as Hydrilla and Najas, palm leaves, dried pine and cedar needles and banana leaves, are used as a substrate in different parts of the Himalayas.

Grass carp (*Ctenophyrangodon idella*) is doing well in aquaculture at lower niche and areas of Jammu and showing highest growth. Seed is produced in different hatcheries of Fisheries Department and demand is increasing rapidly.

Silver Carp (*Hypothalmicthys molitrix*) is also used in composite fish culture at some portion and its growth is showing potential to use it for major species for aquaculture. The main constraint is preference of silver carp is less in comparison to other fishes.

Indigenous Carps

Many new ponds are coming under different schemes like RKVY and PM package in different District of Jammu and Kashmir in which Department of Fisheries is helping the farmers to adopt composite fish farming. The seed is available in plenty of all the Indian major carps like Catla, Rohu and Mrigal. They are showing good potential for development and needed to be given proper attention so that hurdles of slow growth rate due to inbreeding, inproper feeding, poor management could be removed.

Ornamental Fisheries

The 1st underground Aquarium cum Awareness Centre at Baghi Bahu Jammu was established at an estimated cost of Rs. 10.00 crore. During 2008-09 an amount of Rs. 47.70 lakh was realized on account of entry fee. Other such unit at Gagribal is being upgraded. Another centre under the Department of Fisheries is made at Akhnoor in Jammu with the facility of breeding of ornamental fishes. But not much progress has been seen at the farmer's level in adopting ornamental fish breeding and its marketing.

Infrastructure Facilities for Fisheries Development

There are 18 state-owned trout hatcheries located in the districts of Srinagar, Anantnag, Pulwama, Baramulla, Budgam, Doda, Rajouri and Kargil and these have played an important role in boosting trout culture. The hatcheries have a dual role: one, as an agency for commercial production and two, as an agency for the stocking of the rivers and lakes and distribution of fingerlings to private farms. Besides, the Department also runs a mahaseer hatchery at Anji and fish farms at Srinagar, Budgam,

Anantnag, Pulwama, Baramulla, Kupwara, Jammu, Kathua, Udhampur, Doda, Poonch, Leh, Kargil and Rajouri. This infrastructure gives considerable boost to the fisheries activities in the state.

The State National Fish Seed Farms are located at Kathua, Anji, Muradpore, Year Kashmir Province Jammu Province Ghomansion, Nowabad (Jammu), Narbal, Trigam, Pandach, Kohru (Kashmir), Shey and Damsana (Ladakh). Some fish farms have the latest technical trout fish farming, *viz.*, Laribal, Kokernag, Beerwah, Mammer (Kashmir), Phalni, Gatha, Shalimar (Jammu) Wakha (Kargil) etc. The state government has also taken steps to develop private fish farms. 811 fish farms among rural masses have been identified, two thirds of which are in Jammu. Ponds which had been neglected were badly silted.

Trout Fisheries in the State of J&K received a major thrust during the last two decades with the establishment of Mother Trout Fish Farming Project at Kokernag under European Union Assistance. Besides this, the Department has another Trout Hatchery at Laribal where quality Trout seed of Rainbow Trout and Brown Trout is being produced. Trout culture is under taken under modern technology of breeding and rearing to ensure better survival at different stages of the fish. Moreover, with the facilities of highly equipped Trout feed mill imported from Holland, the Department has been able to produce quality Trout feed to achieve better conversion ratio and healthy salable stock.

The Department proposes to increase the trout hatching capacity by way of establishing necessary infrastructure in the shape of Trout hatcheries at suitable units like Aliyalpora (Shopian), Beja (Bhaderwah), Tchansar (Kulgam) and Panzath (Anantnag), Khag (Budgam), Mammer (Ganderbal) and with the establishment of these new Trout hatcheries, the Trout egg production is expected to increase to 5 million per year.

Issues, Constraints and Remedial Measures

The major constraints in fishery sectors are listed below:

1. The natural waters of the State are under great stress due to pollution, encroachment, siltation as a result of which the area of these waters has shrunk considerably, thereby affecting the biotic life especially the Fish fauna. In most of the water bodies, fast growth of water hyacinth and other weeds and heavy siltation year after year has destroyed the ecology and affected fish production. Remedial measures by way of removing the encroachments and rid these water resources of the menace of pollution, need to be taken to safeguard not only these waters but also the biotic life from its extinction. The breeding ground of various fish species especially schozothorax have been damaged. Kashmir Valley lakes Dal and Wular in Jammu and Kashmir are being reduced in size through the process of eutrophication, explosive growth of aquatic plants and encroachment of agriculture into their margins.
2. The river system in the region and other water bodies is critically short of stocking of fish. The feeding channels of most of the beels are lost due to construction of dam and embankments. The leaseholders try to exploit these water bodies as much as possible but are least bothered about making any improvement. Most of the river and other water bodies are deficient of nutrients, which in turn adversely affected the growth of fish.
3. The existing infrastructure is not sufficient to hold the exponential growth in fish farming. Emphasis will be laid to increase the hatching and rearing capacity of fish farms and Trout hatcheries. The supply of quality fish seed is a key factor to the expansion of fish farming. New ponds are coming every day. Fish seed demand at present is strong, unsatisfied and expected to expand.

4. In the hatchery, Production - care of broodstock, induced spawning, egg hatching, seed nursing, seed grading and quality control must be carried carefully.
5. Seasonality and inconsistency of seed supply is prevailed in the area. The fish seed unavailability of different species during the whole year is hampering to harvest the total potential of fish farming. Farmers have to look out to the different states for seed supply.
6. One of the important challenges faced by this sector is the lack of training to fish farmers, only 1183 in J&K being trained annually as against over five lakh in the country. At present, the state has only two Fish Farmers Development Agencies as against a total of 414 in 26 States of India.
7. There is inadequate research support for development of paddy-cum-fish culture and other Integrated Farming Systems.
8. Inadequate support for seed Production- Private local farmers need to be supported to open private fish hatcheries for continues and proper supply of fish seed.
9. Impacts of releases of cultured seed stocks need to be studied.
10. The degradation of hatcheries took place especially where water quality deteriorated and the silt load in streams increased.
11. Deterioration in quality of seed stocks. Due to the very high fecundity of these species, often relatively small numbers of broodstock are maintained, which contribute to the next generation. Furthermore, the emphasis of hatchery managers is on the production of seed quantity, with little if any attention given to issues related to the genetic basis of seed quality. As a result, effective population sizes are commonly low, resulting in inbreeding and unconscious (usually negative) selection and genetic drift, which are likely to have brought about a steady decline in the culture performance of these stocks.

Extension

Extensive extension work is needed in fisheries field. One to one approach is needed to understand the problem of fish farmers of different region as there is big difference at each place topographically, culturally and availability of inputs. This is one of the major activity of the fisheries department but still its demanding more to encourage the farmers for getting good fish production in State. Progressive fish farmers must be provided not only technical knowhow regarding fish farming but also live demonstrations at regular basis. Tours of outside state, ahead in fish farming must be arranged for Fish farmers group at demand and at regular basis.

Future Potential of Fisheries Development

1. Ladakh Region has vast water resources with a high potential for fish culture. Very few of these water bodies have been tapped. The steps taken by department of the state need to be accelerated.
2. The panchayats in the state have nearly 500 ponds under their control which, due to lack of cooperation between the Department and the panchayats have not been properly utilized for fish production. Improving these ponds can augment fish production.
3. The water of Jammu and Kashmir is suited for the fishing of carps-catla, rohu, silver carp and mirror carp. Properly harvesting of the potential in the area can fulfill the local needs.

4. Fishing in manmade reservoirs is an important activity in many states and can become one of the important sources of revenue for the state. J&K however, has only one manmade reservoir in Salal. The state needs to give proper attention to this area to increase fish output.
5. Strengthening of Extension system is needed.
6. The state has extensive inland water bodies, particularly in the valley, which provide excellent habitat for almost any kind of temperate fish. The lakes cover an estimated 0.3 lakh hectare, predominantly in the valley (about 98 per cent of total). River Jhelum flowing, over 162 km in the valley, with its extensive tributaries has enough potential to sustain fish production.
7. The Indus river system has carps, catfishes, the exotic rainbow and brown trouts. The trouts of Kashmir are very rich and attract sport fish enthusiasts.
8. The state offers a favourable habitat for sport-fish like trout in its cold-water streams, particularly in the Lidder and Sindh valleys.
9. Paddy-cum-fish culture is gaining rapid ground in the tropics and subtropics but not in temperate climate. Fish farming as an adjunct to paddy cultivation has a lot of potential either as integrated simultaneous crops or as different crops in the same lands in alternate seasons. Such possibilities with particular reference to compatible fish species should be studied scientifically and a package of technical and management practices evolved for propagation among farmers.
10. To improve marketing of fish, particularly the fish harvested in Jammu region, which is closer to the Punjab, the possibility of selling fish in the neighbouring Punjab districts of Gurudaspur, Amritsar, Jalandhar and Ludhiana should be explored and private enterprise encouraged for the marketing of fish.
11. Infrastructure support by way of purchase of refrigerated containers or vehicles and working capital, should be extended to private sector.
12. To attract fishing enthusiasts from within and outside the country, catch and release sport should be launched and fishing festivals or tournaments organized. In short, fishing should be made part of tourism promotion in the state.
12. The state is having a chain of canals in which fish culture must be tried as it has been found successful in many other countries.
13. Extensive survey of areas where fish farms have already been constructed should be conducted to know the actual factors affecting the fish farmers to close their farms.
14. The average holding size in the state which was 0.76 hectare during 1995-96 had declined to 0.67 hectare during 2000-01 indicating fall of 0.09 hectare in the size of land holdings. So, semi-intensive or intensive fish culture should be introduced to complete utilization of land invested for making fish pond.
15. Fish marketing faces several problems. Fish is a commodity of highly perishable nature requiring the provision of ice plants, cold storage and refrigerated vehicles. There is only one fish-marketing project in progress and more are required.
16. Ornamental fish culture have tremendous scope which is needed to be harvested. Very few farmers have given priority for ornamental fish culture and their breeding due to which mostly ornamental fishes available at the shops are imported from different states like West Bengal, Delhi and Punjab etc.

17. Recreational and Sport Fisheries need to be further strengthened and farmers are needed to be encouraged for establishing new trout ponds to attract and earn from the tourists.
18. Development of endemic fish fauna/hill stream fisheries.
19. To increase quality seed production for augmenting natural water resources of the state for sustainable fisheries and achievement of self sufficiency in fish production.

The J&K state in view of its diverse agro-climatic features, plethora of streams, lakes and hilly torrents, reservoirs, sars has the potential to grow all types of fisheries. The main objectives are to popularize modern fish farming, creating employment opportunities and diversifying aquaculture practices and providing assistance to fish farmers engaged in aquaculture. It appears to be a neglected and diminishing 'natural asset'. To give boost to this sector of the economy, more efforts are needed from both at central and at state level to strengthen the existing infrastructure to provide necessary input to farmers for successful cultural practices. The aquatic resources are part of the larger ecological/ agricultural, social, economic and institutional systems. Therefore management of these resources necessarily has to consider. This is an important activity allied to agriculture and can strengthen the productive base of the agricultural economy and generate self-employment, an aspect which has not been explored properly and fully till now.

References

Anon. (1989). Status of fish seed production and its perspectives in Arunachal Pradesh. *Fishing Chimes*, December 1989: 45-46.

Anon. (1993). Fisheries development of Himachal Pradesh. *Fishing Chimes*, October 1993: 35.

Anon. (1997). Successful induced breeding of golden mahseer. Another breakthrough at Lonavla Farm of Tata Electric Companies. *Fishing Chimes*, March 1997: 9.

Das, S.M. (1978). Ecology and management of freshwater in the highlands of India. In: Proceedings of the National Seminar on Resources, Development and Environment in the Himalayan Region, New Delhi, 1978: 255-61.

Graham, J.M. (1949). Some effects of temperature and oxygen on metabolism of the speckled trout (*Salvelinus fontinalis*). *Canadian J. Res.* 27(D): 270-88.

Howell, G.C.L. (1916). The making of the Himalayan trout waters. *J. Bombay Nat. Hist. Soc*. 17: 61-68.

Jhingran, V.G. and K.L. Sehgal. (1978). Coldwater Fisheries of India. Inland Fisheries Society of India, Barrackpore, West Bengal. 278pp.

Joshi, C.B. (1982). Artificial breeding of golden mahseer, Tor putitora at Bhimtal (U.P.) *J. Inland Fish. Soc. India* 13(1):73-74.

Joshi, L.M. and B.N. Saxena (1989). Artificial propagation of Tor putitora (Hamilton). In: National Workshop on Research and Development Needs of Coldwater Fisheries. Abstract No. 16.

Joshi, C.B., K.L. Sehgal and K.C. Malkani (1989). Experimental trials on feeding of *Tor putitora* with formulated diets at Bhimtal in Kumaon Himalayas. *Indian J. Anim. Sci*. 59(1): 206-9.

Kumar, K. (1992). Indo-Norwegian trout farming project, Kullu Valley. *Fishing Chimes*, June 1992: 27-29.

Mitchell, F.J. (1918). How trout were introduced in Kashmir? *J. Bombay Nat. Hist. Soc*. 26(1): 295-99.

Mohan, M. (1996). Coldwater farming among the Himalayas. Fish Farming International, September 1996: 56-57.

Natrajan, A.V. (1988). International approaches to production and management of fishery resources of large rivers and their significance to India. ICAR Spl. Publ. 59pp.

Nayak, P.K. and B.K. Mandal (1989). Feasibility of paddy-cum-fish culture in hills. *Indian J. Hill Farming* 2(1): 69-78.

Pandey, K.D. (1996). Propagation and conservation of mahseer fish with special reference to coldwater of U.P. hills. *Fishing Chimes*, October 1996: 7.

Raina, H.S., Shyam Sunder, K.K. Vass and R.K. Langer (1986). Artificial breeding and nursery rearing of S. esocinus (Heckel) in Kashmir. *Proc. Nat. Acad. Sci. India* 56(B): 335-38.

Sehgal, K.L. (1974). Report on the researches carried out at the Coldwater Fisheries Research Centre in Himachal Pradesh and Kashmir from 1966-73. CIFRI Spl. Rep. 13pp.

Sehgal, K.L. (1977). Report on Dal lake, Srinagar, Kashmir with suggestions for development of its fishery. *CIFRI Bull.* 24. 13pp.

Sehgal, K.L. (1989). Present status of exotic coldwater fish species in India. Asian Fisheries Society, Indian Branch Spl. Publ. 1. 132pp.

Sehgal, K.L. and K. Kumar (1977). Final project report on induced breeding and rearing of mahseer (Tor putitora) seed in running water ponds. CIFRI (mimeographed). 12pp.

Sehgal, K.L., A.R. Wani, G.M. Bhatt and K. Kumar (1976). Experiments on the efficiency and cost of dry compound pelletized feed in relation to conventional feed in Kashmir trout farms. *J. Inland Fish. Soc. India* 8: 1-12.

Sunder, S. and B.A. Subla (1984). Fish and fisheries of R. Jhelum, Kashmir. *Zoologica Orientalis* 1(2): 34-39.

Tripathi, Y.R. (1977). Artificial breeding of Tor putitora (Ham.). *J. Inland Fish. Soc. India* 9: 161.

Vass, K.K., H.S. Raina and S. Sunder (1978). On the breeding behaviour of S. niger in Dal lake, Srinagar. *J. Bombay Nat. Hist. Soc.* 76: 179-84.

Hill Agriculture: Prospects, Constraints and Mitigations (2013) *Pages* **514–525**
Editor: **Dr. Shahid Ahamad**
Published by: **DAYA PUBLISHING HOUSE, NEW DELHI**

Chapter 33

Parasitic Diseases of Sheep and Goat in North-western Himalayan Region of India

Kamal Hashan Bulbul
Sher-E-Kashmir Agricultural Science and Technology-Kashmir
Krishi Vigyan Kendra, Budgam, Srinagar – 190 001

Introduction

While sheep and goat, like other animals, suffer from infectious and noninfectious diseases, the most serious losses, especially in farm flocks, are due to internal parasites. Theses losses result not so much from deaths from gross parasitism, although fatalities are not infrequent, as from loss condition, unthriftiness, anaemia, and other effects. However parasitism and gastro-intestinal nematode infection in particular, is arguably the most serious constraint affecting sheep and goat production worldwide. Economic losses are caused by decreased production, cost of prevention, cost of treatment, and the death of infected animals in North-western Himalayan region in India. Losses due to parasites are much less sensational, but they are constant and especially in farm flocks and they are exceed those due to bacterial diseases. It is difficult by any form of major survey or other estimation to establish precise figures on losses incurred in production from infection and disease. Even minimal accuracy of loss estimates is difficult because production diseases or disorders may result from interaction with nutritional and environmental stresses, management methods, concurrent diseases, genetic predispositions, or other factors.

Problems with nematode parasitism are often classified as production disease (*i.e.*, chronic subclinical condition affecting productivity such as weight loss, reduced weight gain, reproductive inefficiency, etc.). The nutritional status of the animals is influenced by gastro-intestinal (GI) nematodes which cause anorexia, anaemia, hypoproteinaemia, diarrhea and intestinal obstruction. Most of the gastro-intestinal (GI) parasites browse on mucosa feed on ingesta and digest food or blood suckers.

What ever the case may be, the intruders rob the host nutrients indirectly causes the anoestrus in dairy animals. Gastro-intestinal parasitism due to nematode is a major problem in live stock production as heavy losses are inflict in the form of decrease production, morbidity, mortality, less feed conversion ratio, high cost of treatment, drug resistance and possibility of predisposition of animals to fatal diseases as lower body resistance. The control of nematode parasites traditionally relies on grazing management, anthelmentic treatment, or both. However, grazing management schemes are often impractical due to expense or to the hardiness of infective larvae on pasture.

Fasciolosis is a widespread parasitosis responsible for immense economic losses in goat industry in terms of condemnation of livers, decreased milk and meat production, loss of weight and poor carcass quality. This disease entity causes losses in terms of morbidity and mortality in fluky areas. Among the parasitic diseases fascioliosis, caused by *Fasciola gigantica* and *F. hepatica* is one of the major problems where ever farm livestock are maintained (Kendall, 1954 and Sarrwar, 1956). In India, the disease has been reported from various part of the country (Thapar, 1956; Bali, 1973; Dar *et al.*, 1988; and Pandit *et al.*, 1989). A perusal of relevant literature reveals that only scanty information exists on aspects pertaining to fascioliosis in North-eastern region of the country (Endrejat, 1964; Roy and Tendon 1992 and Talukdar1996.) in goat, which represents a variety of climatic conditions ranging from tropical and humid to subtropical and super humid.

Lapage (1959) have mentioned that vitamin A, D and B complex are important vitamins to develop resistance to internal parasites. Quiquanodon (1978) have quoted that Cobalt deficiency leads to parasitism, since it is used by animals to synthesize Vitamin B-12. Literature further reveals that Co deficiency has direct relationship with pathogenic effect of *Haemonchus contortus* infection in sheep.

Parasitic Diseases of Sheep and Goat

The principal internal principal internal parasites of sheep and goats are round worms, tape worms, flukes and protozoal parasite. Besides this, the external parasites also cause ill effect in productivity in terms of morbidity and mortality. The external parasites are tick, lice, mite etc. Lice are widespread and most important external parasites of goats and goat lice are probably more easily eliminated, usually at less expense in money and labour, than any other groups of insect parasite of livestock. Some important parasites occur in North- Western Himalayan region in India, their mode of infection, symptoms and treatment and control are given bellow.

In temperate regions, lice arc most abundant during the colder months and often are very difficult to find in the summer. Lice are largely host-specific, living on one species or several closely related species. Sheep may become infested with the sheep biting louse, *Damalinia ovis*, and 3 species of sucking louse: the sheep foot louse, *Linognathus pedalis*; the face and body louse, *L. ovillus*. The foot louse of sheep is so named because, except in very heavy infestations, it is confined to the hairy parts of the foot. The face louse is usually found on hairy parts of the sheep's skin; as populations increase, they spread to other parts of the body. *Linognathus* spp forms clusters, often on the flanks of coarse-wooled sheep. Slippage of wool is common. *Linognathus stenopsis*, the goat sucking louse, is found on both shorthaired and Angora breeds of goats. It has been reported from sheep in various parts of the world. *Damalinia caprae*, the goat biting louse, is most frequently found on short-haired goats, whereas chewing lice on Angora breeds are more likely to be *D. limbatus* (the Angora goat biting louse) or *D. crassipes.*

Table 33.1: Common Parasites of Sheep and Goats and their Transmission, Symptoms and Treatment Control

Disease	*Parasite*	*Mode of Infection*	*Symptoms*	*Treatment and Control*
Fasciolosis	*Fasciola gigantica* and *F. hepatica*	Ingestion of metacercariae encysted on herbage especially in low lying and marshy areas.	**Acute form:** Acute cases resulting from massive invasion of the liver by large numbers of young fluke and animal die without showing any clinical symptoms within few days of infection (Clunies Ross and Gordon, 1936). In this form sheep die suddenly, blood stained froth appears at nostrils and blood is discharged from anus and may occur in the form of outbreak.	**Treatment** Triclabendazole @ 10mg/kg b.w Oxycloxanide @ 15mg/kg b.w. Closantal @7.5mg/kg b.w. Rafoxanide @ 15mg/kg b.w. Nitroxynil @7.5mg/kg b.w. Benzimidazoles like, Luxabendazole @7.5-10mg/kg b.w. in sheep Albendazole @7.5mg/kg b.w. **Control** i) Preventing sheep and goats from grazing in marshy areas ii) Preparation of silage and hay from grasses cut from marshy and low lying areas. iii) Prophylactic treatment of sheep and goats in endemic areas.
Immature paramphistomosis	*Paramphistomum cervi, Cotylophoron cotylophorum, Gastrothylax crumenifer, Fischodirus elongatus*	Ingestion of metacercarae encysted on herbage especially in low lying and marshy areas.	Profuse foetid diarrhoea, weakness, depression, dehydration, anorexia and bottle jaw	**Treatment** Oxyclozanide 20mg/kg b.w (two doses three days apart) Niclosamide 100-150mg/kg b.w (As a single dose or two doses three days apart) Niclofolan 4-6mg/kg b.w. Resorantel 65mg/kg b.w. Brotianide 15mg/kg b.w **Control** i) Preventing sheep and goat from grazing in marshy areas ii) Preparation of silage and hay from grasses cut from marshy and low lying areas
Tapeworm infection	*Moniezia expansa, M. benedeni*	Ingestion of infected oribatid mites along with herbage	Diarrhoea, loss of weight, anaemia, weakness Unthriftiness and digestive disturbances. Tapeworm segments are often found singly or in string of various length in the droppings of sheep and goat running on pasture and showing clinical evidence of parasitic infestation. The occurrence of monieziosis is more in kids and lambs than adult sheep and goat.	**Treatment** Niclosomide Lambs and kids @75-150mg/kg b.w. Praziquantel Lambs & Kids @15mg/kg b.w. Fenbendazole Lambs & Kids - @5mg/kg b.w. Bithionol Lambs & Kids-@200mg/kg b.w.

Contd...

Table 33.1–*Contd...*

Disease	*Parasite*	*Mode of Infection*	*Symptoms*	*Treatment and Control*
				Control i) Regular dosing with broad spectrum anthel-mintics at proper times keeping in view epidemiology of parasitic infections in the area.
Roundworm infection (gastro-intestinal worm)	*Haemonchus contortus, Trichos-trongylus axei T. capricola, T. vitrinus., Ostertagia* spp., *Cooperia curticei, C. oncophora, C. punctata and C. pectinata Bunostomum trigono-cephalum, Marshallagia* spp. *Nematodirus spathiger, N. fillicollis, N.abnormalis Oesopha-gostomum columbianum, Chabertia ovina, Trichuris ovis*	Ingestion of eggs containing infected larvae or 3rd stage larvae with herbage	Unthriftness, lack of vitality, bloom, reduced feed intake, loss of weight and animals may pass dark green or black, soft and foul smelling faeces which eventually become very thin. In severe cases there is anaemia, submandibular edema and in terminal cases body coat becomes dry, dehydrated with sinking of eyes. In haemonchosis, the animals become anaemic. *Chabertia ovina* causes ulceration and haemorrhages in iarge intestine leading to anemia. *Ostertagia* species said to show progressive loss of condition with intermittent diarrhoea and stunted appearance. The unthriftiness, pronounced diarrhoea, progressive weakness and anorexia are the main symptoms of trichostrongylosis. However, bone weakness, manifested by lameness and spontaneous fracture of the leg bone, especially femur and hummers has been reported occasionally in sheep heavily infested with small trichostrongyles. Mainly anaemia, edema and unthriftiness are found in hookworm infestations. Anorexia, severe blackish to greenish diarrhoea, prostration and mortality in lamb is the main symptoms of *Nematodirus* infestation. In lamb the first sign is marked and persistent diarrhoea which results in exhaustion	**Treatment** Fenbedazole - @5mg/kg b.w. Oxfendazole - @15mg/kg b.w. Levamisole - @7.5mg/kg b.w. Morantel tartarate -@10mg/kg b.w. Ivermectin, Doramectin & Moxidectin -@200µg/kg b.w. **Control** i) Proper disposal of animal excreta by allowing it to ferment in heaps. ii) Alternate grazing of pasture with hosts of different species (goat / sheep). iii) Periodic deworming

Contd...

Table 33.1–*Contd...*

Disease	*Parasite*	*Mode of Infection*	*Symptoms*	*Treatment and Control*
			and death unless the animals are removed from the infected pasture in oesophagostomosis. The faeces usually have a dark green colour and contain much mucus and some times blood.	
Lungworm infection	*Dictyocaulus filarial* (lung worm), *Muellerius capilaris, Protostrongylus* spp.	Ingestion of 3rd stage larvae or infected snails and slugs with herbage	Clinically animals show cough, dyspnoea, loss of condition, fever and evidence of toxemia if, secondary bacterial infection occurs	**Treatment** Ivermectin 0.2mg/kg b.w. Levamisole 7.5mg/kg b.w Tetramisole 15mg/kg b.w Albendazole 5mg/kg b.w Fenbendazole 5mg/kg b.w Fabental 5mg/kg b.w **Control** i) By vaccination with irradiated larvae of *Dictyocaulus filaria* in sheep and goat. ii) Other control measures similar to those prescribed for gastro-intestinal nematode infection.
Lice	*Damalinia ovis, D.caprae, D. limbata* (angora goat) *Bovicola painei, Linognathus ovillus, L. pedalis, L. stenopsis*	Transmitted primarily by contact between hosts	They cause irritation on the affected part of the body. The sheep and goat become restless and do not fed or sleep well and may injures or damage their hairs and wools by biting and scratching the parts of their body irritated by the lice. In sheep, wool is damaged and soiled by the faeces of lice. The coat become rough and shaggy and if the irritation is severe the hair may become matted. The foot louse of sheep is found most frequently around the dew-claws and severe infection may produce lameness. Anaemia is occurred in heavy infection.	**Treatment** Ivermectin, Doramectin & Moxidectin, - @200μg/kg b.w. -Dipping with 0.25% Comphous (Co-ral). Ausuntol- @500mg/L of water Taktic- @4ml/L of water Cyprol- @1ml/ L of water (body spray) and 20ml/L of water (House spray) -Application by garden watering can may be used in small groups of sheep crowded together (Ronnel 24%). - Synthetic pyrethroids may be value in the control of louse. **Control** i) Prophylactic treatment with above mentioned drugs in late spring (1st to 31st May) and

Contd...

Table 33.1–*Contd...*

Disease	*Parasite*	*Mode of Infection*	*Symptoms*	*Treatment and Control*
				late autumn (1st to 30th Nov.) ii) Spraying of animal houses and pastures with insecticides. iii) Jetting, spraying or showering techniques are often applied
Tick	*Ixodes ricinus, Boophilus microplus, Rhipicephalus* sp.		Anaemia, dermatosis, tick worry, tick toxicosis, sweating toxicosis and tick paralysis are the main significant symptoms in tick infestation. Tick pyaemia is caused by *Staphylococcus areus* in sheep. Economically slaughter value of the hide or fleece may be reduced.	**Treatment** Ivermectin, Doramectin & Moxidectin, - @200μg/kg b.w. s/c - Ausuntol- @1g/L of water -Taktic- @2ml/L of water -Cyprol- @1ml/ L of water (body spray) and 20ml/L of water (House spray) -Butox -@2-3ml/l of water **Control** i) Prophylactic treatment with above mentioned acaricides and insecticides drugs in late spring (1st to 31st May) and late autumn (1st to 30th Nov.). ii) Spraying of animal houses and pastures with acaricides. iii) Burning of pasture iv) Ploughing of cultivated land. v) Rotational grazing
Mange	*Psoroptes ovis, Sarcoptes scabiei* var *ovis, Sarcoptes scabiei* var *caprrae, Chorioptes bovis* var. *ovis, Sarcoptes, Demodex canis* var. *ovis*	Highly contagious disease and apart from direct contact with infected animals inert materials like bedding and grooming tools may act as carriers	In sheep scab, presence of small papules, general erythema of skin, scratching, biting, loss of hair or wool, presence of thick brown scabs, thickening and wrinkling of skin. Infected sheep become restless, scratch themselves, and rub against fences and other members of the flock. The itching becomes more intense when the sheep are warm. The wool is disturbed; first slender tag come loose, then the fleece becomes matted, and sheep pulls out pieces with its mouth. In neglected cases animals show emaciation, anorexia and weakness and in severe	**Treatment** Ivermectin, Doramectin & Moxidectin, Sheep and goat @200μg/kg b.w. Or Spraying or dipping with insecticides Like Diazinon, 0.01%; Propetamphos, 0.0125%; Coumaphos, 0.1%; Phoxin, 0.05%; Amitraz, 0.05%, Flumethrin, 0.05% etc. **Control** i) Prophylactic treatment with above mentioned drugs in late spring (1st to 31st May) and late autumn (1st to 30th Nov.) ii) Spraying of animal houses and pastures with acaricides

Contd...

Table 33.1–*Contd...*

Disease	*Parasite*	*Mode of Infection*	*Symptoms*	*Treatment and Control*
			cases death may occur. Sarcoptic mange is occurred only on the non-wooly portion of the body of sheep. Scab lesions may occur on all parts of the body that are covered with wool or hair, but occur most frequently around the shoulder and along side of the body in wooled sheep and along the back, sternum and dorsal aspect of tail in hairy sheep.	
Babesiosis	*Babesia motasi, B.ovis, B. taylori and B. foliata*	Transmitted from infected to susceptible animals by bites of ticks	In acute cases there is rise in temperature to 41-42°C, anorexia, depression, weakness, cessation of rumination, fall in milk yield, followed by severe anaemia, icterus and haemoglobinuria and prostration. Acute ovine babesiosis may lead to death.In sub-acute & chronic disease runs milder course, in which fever is mild and haemoglobinuria is absent but anaemia may be marked. In chronic form, about 1 percent red blood cells are infected. Recovered animals become immune.	**Treatment** Diminazene aceturate 2-3.5mg/kg b.w by deep i/m injection Pirevan - @ 2ml of 0.5% solution per 10 Kg B.W. Trypan blue (an azo dyes)- @10-25 ml of 1% solution through i/v route only. All other drugs used in cattle are useful.

In general anorexia, loss of body weight gain, reduce productivity *viz.* milk yield, body weight, wool growth etc. loss of blood (haematophagus nematodes) protein losing gastro-enteropathy, intestinal lymphangiectasia, hypoproteinaemia, hypoalbominaemia, anemia are the significant clinical observation in parasitism.

Economic Effect

The anorexia and reduce feed intake losses of blood and plasma proteins into the gastro-intestinal tract, alteration in protein metabolism, depressed levels of minerals, depressed activity of some intestinal enzymes and diarrhoea all contribute low weight gain, wool growth and milk production. Wool production is suppressed as a result of the reduced in corporation of amino acids into by wool follicles. The quality of wool is reduced and a break in wool growth may be seen heavily parasitized sheep and goats.

General Control Measure of both Endo and Ecto-parasites

The following measure can be taken for control of parasitism in sheep and goat in northern Himalayan region as well as other parts of our country.

Table 33.2.: Insecticides Used in Sheep and Goat to Control the Lice and Ticks

Trade Name	*Active Ingredient*	*Product Type*	*Approved Species*	*Label Use*	*Labeled Dosage*	*Withdrawal Time*
BackSide™	1 per cent Permethrin Butoxide	Pour-on Spray	Sheep	Sheep ked and licehouse flies mosquitoes, black flies	1/2 oz/100 lbs.and 3 oz for sheep Do not treat more than 2 weeks.	0 days slaughter
BackSide™Plus	1 per cent Piperonyl Butoxide	Ready-to-use	Sheep	Sheep ked and lice	3 oz for sheep Do not treat more than every 2 weeks.	0 days milk discard
Boss®	5 per cent permethrin	Pour-on	Sheep and Goats	Lice, and keds	1.5 ml/100 lbs. 18 ml maximum for sheep	0 days slaughter 0 days milk discard
Catron IV	Permethrin	Aerosol	Sheep and Goats	Maggots screwworms ear ticksface flies	One can will treat 1-10 animals depending upon severity of fly-strike.	5 days slaughter swine
Co-Ral®	Coumaphos	Spray or dip	Sheep	Maggots, ked, lice		15 days slaughter
Diazinon	Diazinon	Concentrate spray	Sheep **Do not use on goats**	Ticks ked, lice	4 oz/25 gal water for low pressure spray; 2 oz/25 gal water for high pressure spray	14 days slaughter
Ectiban ®-EC	5.7 per cent permethrin	Concentrate Pour-on or spray	Sheep and Goats	Mange, lice, ticks	Use spray method For mange, lice, ticks, etc., thoroughly wet the animal	0 days slaughter
Ectrin®	permethrin	Pour-on	**Sheep** Do not treat lactating goats	Lice keds ticks	For sheep:1/4 oz./50 lbs (up to 3 oz) pour along back line over neck, shoulders and rump Residual activity for 28 days	2 days
Expar®	Permethrin	Concentrate pour-on spray	**Sheep**	Lice, flies fleas, mites	For sheep:1/4 oz./50 lbs (up to 3 oz)pour along back line over neck, shoulders and rump Residual activity for 28 days	0 days

Contd...

Table 33.2–*Contd...*

Trade Name	*Active Ingredient*	*Product Type*	*Approved Species*	*Label Use*	*Labeled Dosage*	*Withdrawal Time*
Ultra Boss®	5 per cent permethrin and 5 per cent piperonyl butoxide	Pour-on	Sheep	Horn flies, face flieslice, and keds	3 ml per 100 lbs.	0 days slaughter withdrawal 0 days milk discard
Cydectin® Oral Drench for sheep	Moxidectin	oral	Sheep	Control of external parasites in sheep	1 ml/11 lbs.	7 days sheep slaughter
Synergized DeLice	Permethrin-synergized	Pour-on or spray Ready-to-use	Sheep	Sheep ked and lice	Pour along back 8 cc per 50 lbs. After shearing	0 days withdrawal
Ivomec® Sheep Drench 0.8 per cent solution	Ivermectin	oral	Sheep	Larval stages of nasal bot	3 ml/26 lbs.	11 days slaughter sheep

Table 33.3: Deworming Schedule for Sheep and Goat in Northern Himalayan Region of India

Sl.No.	*Parasitic Diseases*	*Timing of Deworming*	*Area of Application*
1	Fascioliosis	Late winter/Early spring (15th February to 14th March) Early summer/Mid summer (15th June to 14th July) Mid autumn/Late autumn (15th October to 14th November)	Marshy and low laying areas where sheep and goat are fed mainly on paddy hay during winter is the very important area to deworm the small ruminants but can be followed in all parts of Himachal Pradesh and Kashmir Valley of J&K.
2	Immature paramphistomosis	Late winter/Early spring (15th February to 14th March) Early summer/Mid summer (15th June to 14th July) Mid autumn/Late autumn (15th October to 14th November)	Marshy and low laying areas where sheep and goat are fed mainly on paddy hay during winter is the very important area to deworm the small ruminants but can be followed in all parts of Himachal Pradesh and Kashmir Valley of J&K.
3	Round worm infestation	Late spring (1st tc 31st May) Late autumn (1st to 30th November)	In all areas of J&K and Himachal Pradesh
4	Tape worm infestation	Late autumn (1st 30th November)	In all areas of Himachal Pradesh and Kashmir valley of J&K
5	Ecto- parasites like ticks, lice, mite infestation	Late spring (1st to 31st May) Late autumn (1st 30th November)	In all areas of Himachal Pradesh and Kashmir valley of J&K

1. *Stocking rate*: Decrease stocking of small ruminants in pasture also reduce the parasitic infestation.
2. *Pasture resting*: Animals may be with held from pasture for at least for 4 months during winter and 3 months under hot and humid condition. Gaddi tribe of HP and Gujjars and Bakerwal of Jammu and Kashmir are dominant tribes used to follow this practice as during summers they migrate along with their sheep and goat to alpine pasture and during winters they used to migrate to plains of northern India. This method helps in reduction of larvae to acceptable level as it is well known that high as well low temperature is detrimental to infective larvae.
3. *Alternative grazing with host of same species*: Susceptibility varies with age thus it is logical to graze the kids and lambs first in pasture there after adult sheep and goats are allowed to graze on these pastures.
4. *Alternative grazing with different species*: Susceptibility varies with different species thus it is logical to graze the horse first in pasture there after cattle or sheep and goat are allowed to graze on these pastures. Alternative grazing between ruminants/equines can be practice. Similarly cattle and sheep grazing can be practiced.
5. *Grazing time*: Graze animals after sun has dried the dew on pasture and do not let your animals on pasture immediately after rains. It is well proven that infective larvae is abundant in pastures during monsoon and post-monsoon (Sharma *et al.*, 2004; Yadav *et al.*, 2004) thus during this period animals should be fed with leaves of fodder trees as well as straw feeding should be encouraged.
6. *Grazing height*: It is well established that infective larvae need moisture and thus it remains on the lowest part during summer; therefore, do not allow sheep and goat to graze very close to ground.

References

Bali, H.S. (1973). Incidence of helminth parasites in sheep in Bihar. J. Anim. Hlth. Prod. 1:35-39.

Dhar, D. N., Sharma, R. L. and Rain, O.K. (1988). Fascioliasis in animals in the Kashmir Valley. J. Vet. Parasitol. 2: 31-35.

Endrejat, E. (1964). Helminths and helmithiasis in Assam. Indian Vet. J. 41: 538-542.

Kendall, S.B. (1954). Fascioliasis in Pakistan. Ann. Trop. Hed. Parasitol. 48:307-313.

Lapage, G. (1959). Monning's Veterinary Helmithology and Entomology. 4th edn. Balliere, Tindall and Cox, London. Pp 511.

Pandit, B.A. Mir, A.S., Nasreen, S. and Khan, A.A. (1989). The epidemiology of ovine fascioliosis in Jammu and Kashmir State, India. J. Vet. Parasitol. 3: 13-15.

Quiquandon, H. (1978). 12 balles pourun veto. Tome II. Lere partie. Editios Agriculture et Vie. Pp 239.

Roy, B. and Tandon, V. (1992). Trematodiosis in north-east India: A study on spectrum of digenetic trematodes among pigs, buffaloes, cattle, goats and sheep. Indian J. Anim Hlth. 31:1-14.

Sarwar, M.M. (1956). Epidemiology and chemotherapy of fascioliosis in goats. Healer. 3: 4-10.

Sharma, D., Katoch, R., Agnihotri, R.K. and Pathak, V. (2004). Epidemiological studies on gastro-intestinal helmints of sheep in North-Western Himalayas. Presented at 15th National Congress of Indian Association for Advancement of Vety. Parasitol. 25-27th October, 2004, Patnagar.

Thapar, G. S. (1956). Systemic survey of helminth parasites of domesticated animals in India. Indian Vet. J. 25: 211-271.

Talukdar, S. K. (1996). Prevalence of Helminthic infections of goats in Assam. J. Vet. Parasitol. 10:83-86.

Yadav, A., Khajuria, J. K. and Raina, A. K. (2004). Prevalence of gastro-intestinal parasites of sheep and goats of Jammu. Presented at 15th National Congress of Indian Association for Advancement of Vety. Parasitol. 25-27th October, 2004, Patnagar.

Index